T0315080

Ecohydrological Interfaces

Ecohydrological Interfaces

Edited by

Stefan Krause
University of Birmingham, Birmingham, UK

David M. Hannah
University of Birmingham, Birmingham, UK

Nancy B. Grimm
Arizona State University, Tempe, AZ, US

This edition first published 2024

Registered Office(s)
John Wiley & Sons, Inc., 111 River Street, Hoboken, NJ 07030, USA
John Wiley & Sons Ltd, The Atrium, Southern Gate, Chichester, West Sussex, PO19 8SQ, UK

For details of our global editorial offices, customer services, and more information about Wiley products visit us at www.wiley.com.

Wiley also publishes its books in a variety of electronic formats and by print-on-demand. Some content that appears in standard print versions of this book may not be available in other formats.

A catalogue record for this book is available from the Library of Congress

Hardback ISBN: 9781119489672; ePub ISBN: 9781119489665; ePDF ISBN: 9781119489689;
oBook ISBN: 9781119489702

Cover Images: Courtesy of Stefan Krause; Courtesy of David M. Hannah; Caio Pederneiras/Shutterstock; Nattapon Ponbumrungwong/Shutterstock
Cover Design: Wiley

Set in 9.5/12.5pt STIXTwoText by Integra Software Services Pvt. Ltd, Pondicherry, India
Printed and bound by CPI Group (UK) Ltd, Croydon, CR0 4YY

C9781119489672_091023

Contents

Preface

Ecohydrological interfaces represent the dynamic transition zones at ecosystem boundaries that control the fate, transport, and transformation of water, matter, energy, and organisms between adjacent systems. In contrast to more stationary concepts of *boundaries* (as separators of different ecosystems or subsystems) or *ecotones* (boundaries that have a defined thickness and share characteristics with each of the systems they separate), *ecohydrological interfaces* dynamically evolve at ecosystem boundaries in space and time where they can emerge, disappear, expand, or contract. *Ecohydrological interfaces*, such as groundwater–surface water interfaces including hyporheic and riparian zones, soil–atmosphere interfaces, or the dynamic interfaces between phreatic and vadose zones, are hotspots of ecohydrological and biogeochemical processes. As such, they harbour key ecosystem functions of biogeochemical cycling, water purification, and buffering of thermal extremes, and they support biodiversity and ecosystem resilience.

This state-of-the-art edited volume provides novel insights into the diverse nature of different types of ecohydrological interfaces, experimental and modelling tools to study the mechanisms controlling their functioning, and integrated approaches to restore and protect their functions. The 16 chapters of this book are therefore organized into 4 sections. Section 1 provides an introduction to the conceptual background and the role and spatial-temporal organization of ecohydrological interfaces in the environment. Section 2 discusses the principles of ecohydrological interface controls on energy, water, and solute fluxes across neighbouring ecosystems for a selection of different ecohydrological interfaces, and Section 3 provides an overview of quantitative experimental and model-based methods for characterizing the functioning of ecohydrological interfaces. Section 4 analyses the functioning of ecohydrological interfaces in a globally changing environment and discusses how the functioning of ecohydrological interfaces can be restored and protected in practice.

Section 1 introduces the concept of *ecohydrological interfaces* as a new way to investigate the often non-linear system behaviour stimulated by complex interactions between hydrological, biogeochemical, and ecological processes across system boundaries. Chapter 1 provides an overview of the principles of *ecohydrological interfaces*, their functioning and role in a landscape context. It discusses how ecohydrological interfaces are defined and delineated in space and time by their specific functioning rather than system properties, which distinguishes them from ecotones or ecosystem boundaries. Based on this introduction, Chapter 2 provides novel insights into the role of biological activity for enhancing hydrological and biogeochemical activity at *ecohydrological interfaces*. Reflecting on the growing body of functional analyses performed by microbial and invertebrate ecologists, this chapter specifically discusses mechanisms of how biological activity (such as of bioturbating species) is affected by hydrological conditions but can also trigger and control *ecohydrological interface* functioning.

Section 2 provides a detailed overview of the mechanisms by which ecohydrological interfaces control fluxes of energy, water, and solutes. These are discussed for different types of groundwater–surface water interfaces including in riparian corridors (Chapter 3), stream hyporheic zones (Chapter 4), groundwater–lake (Chapter 5), and coastal and marine interfaces with groundwater (Chapter 6).

Section 3 discusses new developments in quantitative methods for characterizing the functioning of the ecohydrological interface across a wide range of spatial and temporal scales for processes covering (i) *ecohydrological interface* fluxes of water, (ii) characterization of concentrations and fluxes of reactive substances and their biogeochemical cycling at ecohydrological interfaces, and (iii) biological process dynamics at *ecohydrological interfaces*. In addition to laboratory and field experimental methods ranging from micro-plate-reader scale to remote sensing, a special focus is given to advances in integrated approaches to quantify and predict *ecohydrological interface* behaviour.

Chapter 7 therefore provides an overview of point- to reach-scale methods for identifying water and energy fluxes across *ecohydrological interfaces* in the field, followed by Chapter 8, which introduces novel methods of using heat as a tracer for tracking and quantifying hydrological fluxes across *ecohydrological interfaces*. Chapter 9 provides a comprehensive overview of capabilities and limitations of active and passive in situ sampling approaches at groundwater–surface water interfaces, with Chapter 10 discussing advances in automated in situ sensing technologies for monitoring nutrient and organic matter cycling at freshwater interfaces.

Chapter 11 introduces the use of terrestrial diatoms as tracers at ecohydrological interfaces with Chapter 12 discussing the use of experimental ecosystems for quantifying metabolism at sediment–water interfaces. Using advanced data analysis techniques, Chapter 13 provides new insights into the use of diel solute signals for assessing ecohydrological and biogeochemical processing at *ecohydrological interfaces*, followed by Chapter 14, which explores the use of molecular methodologies for monitoring pathogenic viruses at *ecohydrological interfaces*.

Section 4 analyses the role of *ecohydrological interfaces* in globally changing river basins, providing a systematic overview of interface impacts and responses to global environmental pressures, as well as increasing evidence of facilitation of *ecohydrological interface* functions in environmental restoration and engineering schemes. By providing an overview of general global environmental pressures, Chapter 15 identifies threats and risks to the functioning and health of *ecohydrological interfaces*, whereas Chapter 16 introduces new, integrated concepts for restoring ecosystem functioning of *ecohydrological interfaces*.

This section concludes the discussion of the functioning of *ecohydrological interfaces* by developing a new paradigm of how research at *ecohydrological interfaces* can be used not only to improve mechanistic process understanding but also to facilitate novel approaches in interdisciplinary research.

Stefan Krause, David M. Hannah, Nancy B. Grimm
Birmingham, UK; Tempe, AZ, USA; May 2023

List of Contributors

Benjamin W. Abbott
Brigham Young University
Provo
Utah
USA

Ion Gutierrez Aguirre
Department of Biotechnology and Systems Biology
National Institute of Biology
Ljubljana
Slovenia

M. Antonelli
Wageningen University
Department of Environmental Sciences, Hydrology and Quantitative Water Manag-ement Group
The Netherlands
Luxembourg Institute of Science and Technology
Department Environmental Research and Innovation
Catchment and Eco-hydrology research group
41 rue du Brill
L-Belvaux
Luxembourg
Wageningen University
Department of Environmental Sciences
Hydrology and Quantitative Water Manag-ement Group
Droevendaalsesteeg 3a
Building 100
PB Wageningen
The Netherlands

Alba Argerich
Oregon State University
Department of Forest Engineering, Resources & Management
Corvallis
USA
School of Natural Resources
University of Missouri
Columbia
Missouri
USA

Ian Baker
Small Woods Association
Station Road
Coalbrookdale
Telford
TF8 7DR
UK

Mukundh N. Balasubramanian
BioSistemika LLC
Ljubljana
Slovenia

Viktor Baranov
LMU Munich Biocenter
Großhaderner Str. 2
Planegg-Martinsried
Germany
Ludwig Maximillian University Munich

Tom Battin
Ecole Polytechnique Fédérale de Lausanne
School of Architecture, Civil and Environmental Engineering
Lausanne
Switzerland

S. Bernal
CSIC
Blanes
Spain

Susana Bernal
Center for Advanced Studies of Blanes

Mike Blackmore
Wessex Rivers Trust
Phillips Lane
Salisbury
SP1 3YR
UK

Phillip J Blaen
School of Geography, Earth and Environmental Sciences
University of Birmingham
Edgbaston
Birmingham
B15 2TT
UK
Birmingham Institute of Forest Research (BIFoR)
University of Birmingham
Edgbaston
Birmingham
B15 2TT
UK
Yorkshire Water
Halifax Road
Bradford
BD6 2SZ
UK

Michael E. Böttcher
Leibniz-Institute for Baltic Sea Research
Seestrasse 15
D-Warnemünde
Germany

Chris Bradley
School of Geography, Earth and Environmental Sciences
University of Birmingham
Edgbaston
Birmingham
UK

Athena Chalari
Silixa
Watford
England
UK

Francesco Ciocca
Silixa
Watford
England
UK

Thibault Datry
IRSTEA

Jake Diamond
RiverLy
INRAe
Villeurbanne
France

Jennifer Drummond
Center for Advanced Studies of Blanes

Jan H. Fleckenstein
Helmholtz-Center for Environmental Re-search (UFZ)
Department of Hydrogeology
Leipzig
Germany

Jan Fleckenstein
Department of Hydrogeology
Helmholtz-Center for Environmental Re-search – UFZ

Leipzig
Germany

J.N. Galloway
Department Ecohydrology
Leibniz-Institute of Freshwater Ecology and Inland Fisheries
Berlin
Germany

Nancy B. Grimm
Arizona State University
School of Life Sciences
Tempe
AZ
USA

David M. Hannah
School of Geography, Earth and Environmental Sciences
University of Birmingham
Edgbaston
Birmingham
B15 2TT
UK

Ben Christopher Howard
School of Geography, Earth & Environmental Sciences
University of Birmingham
Birmingham
B15 2TT
UK and Birmingham Institute of Forest Research

Matjaž Hren
BioSistemika LLC
Ljubljana
Slovenia

Nicholas Kettridge
School of Geography, Earth & Environmental Sciences
University of Birmingham
Birmingham
B15 2TT
UK and Birmingham Institute of Forest Research

Kieran Khamis
School of Geography, Earth and Environmental Sciences
University of Birmingham
Edgbaston
Birmingham
B15 2TT
UK

Julian Klaus
Luxembourg Institute of Science and Technology
Department in Environmental Research and Innovation
Belvaux
Luxembourg
Department of Geography
University of Bonn
Bonn
Germany
Luxembourg Institute of Science and Technology
Department Environmental Research and Innovation
Catchment and Eco-hydrology research group
41 rue du Brill
L-4422 Belvaux
Luxembourg

Julia L.A. Knapp
Department of Earth Sciences
Durham University
Durham
UK

Katarina Kovač
BioSistemika LLC
Ljubljana
Slovenia

Stefan Krause
School of Geography, Earth and Environmental Sciences
University of Birmingham
Edgbaston
Birmingham
B15 2TT
UK

Birmingham Institute of Forest Research (BIFoR)
University of Birmingham
Edgbaston
Birmingham
UK

Marie Kurz
Drexel University

Marie J. Kurz
Environmental Sciences Division
Oak Ridge National Laboratory
Oak Ridge
Tennessee
USA

Scott T. Larned
National Institute of Water and Atmospheric Research
Christchurch
New Zealand

Jörg Lewandowski
Leibniz-Institute of Freshwater Ecology and Inland Fisheries (IGB)
Ecohydrology Department
Berlin
Germany

Charlotte E.M. Lloyd
Organic Geochemistry Unit
Bristol Biogeochemistry Research Centre
School of Chemistry
University of Bristol
Cantocks Close
Bristol
BS8 1TS
UK

Ulf Mallast
Department Monitoring- and Exploration Technology
Helmholtz Centre for Environmental Research- UFZ
Permoserstr. 15
Leipzig
The Netherlands

Eugènia Martí
Centre d'Estudis Avançats de Blanes
Biogeodynamics and Biodiversity Group
Blanes
Spain

N. Martínez-Carreras
Luxembourg Institute of Science and Technology
Department Environmental Research and Innovation
Catchment and Eco-hydrology research group
41 rue du Brill
L-Belvaux
Luxembourg

Gudrun Massmann
Institute for Biology and Environmental Sciences
Carl von Ossietzky University of Oldenburg
Ammerländer Heerstraße 114–118
Oldenburg
Germany

Karlie McDonald
University of Birmingham
School of Geography, Earth and Environmental Sciences
Birmingham
UK

Karin Meinikmann
Julius Kühn Institute
Institute for Ecological Chemistry
Plant Analysis and Stored Product Protection
Berlin
Germany

Clara Mendoza-Lera
University of Koblenz-Landau

Alexander Milner
University of Birmingham

Florentina Moatar
RiverLy
INRAe
Villeurbanne
France

Nils Moosdorf
Leibniz Centre for Tropical Marine Research
Fahrenheitstr. 6
Bremen
Germany

Mike Müller-Petke
Leibniz
Institute for Applied Geophysics
Hannover
Niedersachsen
Germany

Aaron Packman
Northwestern University

Laurent Pfister
Luxembourg Institute of Science and Technology
Department in Environmental Research and Innovation
Belvaux
Luxembourg

L. Pfister
Luxembourg Institute of Science and Technology
Department Environmental Research and Innovation
Catchment and Eco-hydrology research group
41 rue du Brill
L-4422 Belvaux
Luxembourg

Gilles Pinay
Centre national de la recherche scientifique
Observatoire des Sciences de l'Univers de Rennes
Université de Rennes 1
Rennes
France
Environnement
Ville & Sociétés - UMR CNRS
Lyon
France

Michael O. Rivett
University of Birmingham
School of Geography, Earth and Environmental Sciences
Birmingham
UK
University of Strathclyde
Department of Civil and Environmental Engineering
Glasgow G1 1XJ
Center for Applied Geoscience
Eberhard Karls University of Tübingen
Tübingen
Germany

Donald O. Rosenberry
U.S. Geological Survey
Sunrise Valley Drive Reston
VA
USA

Janine Rüegg
Stream Biofilm and Ecosystem Research Laboratory
École Polytechnique Fédéral de Lausanne
1015 Lausanne
Switzerland
Current address: Interdisciplinary Center for Mountain Research
University of Lausanne
Lausanne
Switzerland

Francesc Sabater
University of Barcelona
Department of Ecology
Barcelona
Spain

Hanieh Sayedhashemi
RiverLy
INRAe
Villeurbanne
France

Jonas Schaper
Geography Department
Humbold University Berlin
Germany

Jacob Schelker
University of Vienna
Department of Limnology & Bio-Oceanography
Vienna
Austria

Christian Schmidt
Helmholtz-Center for Environmental Research (UFZ)
Department of Hydrogeology
Leipzig
Germany

Albert Sorolla
Naturalea
Castellar del Vallès
Spain

Valentina Turk
National Institute of Biology
Marine Biology Station
Piran
Slovenia

Sami Ullah
School of Geography, Earth & Environmental Sciences
University of Birmingham
Birmingham
UK and Birmingham Institute of Forest Research

Loes van Schaik
Department of Environmental Sciences
Wageningen University and Research
The Netherlands

Jesus Gomez Velez
Vanderbilt University

Adam S. Ward
Indiana University Bloomington

Hannelore Waska
Institute of Chemistry and Biology of the Sea
Carl von Ossietzky University of Oldenburg
Ammerländer Heerstraße
Oldenburg
Germany

Glenn Watts
Environment Agency
Scientific and Evidence Service
Bristol
UK

C.E Wetzel
Luxembourg Institute of Science and Technology
Department Environmental Research and Innovation
Catchment and Eco-hydrology research group
41 rue du Brill
L-Belvaux
Luxembourg

Anne Zangerlé
Ministry of Agriculture, Viticulture and Rural Development
Luxembourg
Luxembourg

Jay P. Zarnetzke
Michigan State University

Section 1

1

Ecohydrological Interfaces as Hotspots of Ecosystem Processes

Stefan Krause[1], Jörg Lewandowski[2], Nancy B. Grimm[3], David M. Hannah[1], Gilles Pinay[4], Karlie McDonald[1], Eugènia Martí[5], Alba Argerich[6], Laurent Pfister[7], Julian Klaus[7], Tom Battin[8], Scott T. Larned[9], Jacob Schelker[10], Jan Fleckenstein[11], Christian Schmidt[11], Michael O Rivett[1,16], Glenn Watts[12], Francesc Sabater[13], Albert Sorolla[14], and Valentina Turk[15]

[1] *University of Birmingham, School of Geography, Earth and Environmental Sciences, Birmingham, UK*
[2] *Leibniz-Institute of Freshwater Ecology and Inland Fisheries (IGB), Ecohydrology Department, Berlin, Germany*
[3] *Arizona State University, School of Life Sciences, Tempe, AZ, USA*
[4] *Centre national de la recherche scientifique, Observatoire des Sciences de l'Univers de Rennes, Université de Rennes 1, Rennes, France*
[5] *Centre d'Estudis Avançats de Blanes, Biogeodynamics and Biodiversity Group, Blanes, Spain*
[6] *Oregon State University, Department of Forest Engineering, Resources & Management, Corvallis, USA*
[7] *Luxembourg Institute of Science and Technology, Department in Environmental Research and Innovation, Belvaux, Luxembourg*
[8] *Ecole Polytechnique Fédérale de Lausanne, School of Architecture, Civil and Environmental Engineering Lausanne, Switzerland*
[9] *National Institute of Water and Atmospheric Research, Christchurch, New Zealand*
[10] *University of Vienna, Department of Limnology & Bio-Oceanography, Vienna, Austria*
[11] *Helmholtz-Center for Environmental Research (UFZ), Department of Hydrogeology, Leipzig, Germany*
[12] *Environment Agency, Scientific and Evidence Services, Bristol, United Kingdom*
[13] *University of Barcelona, Department of Ecology, Barcelona, Spain*
[14] *Naturalea, Castellar del Vallès, Spain*
[15] *National Institute of Biology, Marine Biology Station, Piran, Slovenia*
[16] *University of Strathclyde, Department of Civil and Environmental Engineering, Glasgow G1 1XJ, UK*

1.1 Introduction

The study of system boundaries has been a mainstay in ecological and hydrological research (Cadenasso et al. 2003; Strayer et al. 2003; Yarrow and Marin 2007). Interdisciplinary research has highlighted the importance of ecosystem boundaries, many of which are "hotspots" of ecological, biogeochemical, or hydrological processes (Caraco et al. 2006; McClain et al. 2003; Peipoch et al. 2016; Pinay et al. 2015).

We introduce *ecohydrological interfaces* as a new concept to support the quantitative analysis of non-linear system behaviour stimulated by the complex and multi-faceted interactions of hydrological, biogeochemical, and ecological processes across system boundaries. Ecohydrological interfaces represent the dynamic transition zones that may develop at ecosystem (or subsystem) boundaries and control the movement and transformation of

Ecohydrological Interfaces, First Edition. Edited by Stefan Krause, David M. Hannah, and Nancy B. Grimm.

organisms, water, matter, and energy between adjacent systems (referred to by Hedin et al. (1998) as "control points"). In contrast to stationary *boundaries* (separators of different ecosystems or subsystems) or *ecotones* (boundaries that have a defined thickness and share characteristics with each of the systems they separate), ecohydrological interfaces are non-stationary, emerging for a limited time and then disappearing, expanding, and contracting, or moving around within a boundary or ecotone. Differing from boundaries and ecotones, which are delineated primarily based on system properties (Cadenasso et al. 2003; Strayer et al. 2003; Yarrow and Marin 2007), ecohydrological interfaces are defined by their specific functioning (for example, the dynamic extent of surface water mixing in streambed environments forming hyporheic zones as ecohydrological interfaces with distinct redox environments and ecological niche functions and behaviour).

Ecohydrological interfaces are manifold, including (1) soil–atmosphere interfaces, (2) capillary fringes as interfaces between phreatic and vadose zones, (3) interfaces between terrestrial upland and lowland aquatic ecosystems, (4) groundwater–surface water interfaces, including those associated with riparian or hyporheic zones, biofilms, and surface water–benthic zone interfaces (Figure 1.1). Ecohydrological interfaces provide key ecosystem functions and services (Belnap et al. 2003), including water purification, thermal regulation, and maintenance of biodiversity (Freitas et al. 2015; Krause et al. 2011a; Perelo 2010). They increase ecological resilience by providing refuge for organisms during extreme events or source areas for recolonization after disturbances (Clinton et al. 1996; Crump et al. 2012; Kumar et al. 2011; Stubbington 2012).

In this chapter we aim to uncover the organizational principles – the main drivers and controls, and their interactions and feedbacks – that determine the development and capacity of ecohydrological interfaces to transform the flow of energy, water, and matter between adjacent ecosystems. We compare the characteristics of transformation processes at different ecohydrological interfaces in freshwater ecosystems, including groundwater–surface water, groundwater–vadose zone, and benthic–pelagic interfaces, to determine common or unique features of their non-linear process dynamics. Based on a comparison of the organizational principles of different ecohydrological interfaces, we propose a roadmap for the development of multi-scale conceptual models of ecohydrological interface processes and their interactions that can be expanded to other types of ecohydrological interfaces not covered here.

1.2 Transformation of Energy, Water, and Matter Fluxes Across Ecohydrological Interfaces

Ecohydrological interfaces developing in aquatic ecosystems (e.g. between groundwater and surface water or groundwater and the vadose zone) extend from the micro-scale (e.g. interfaces at microbial biofilms) to kilometre scale (e.g. aquifer–river interfaces). Despite their varied dimensions, these interfaces share common properties: (1) abrupt changes in aggregate state (e.g. solid, liquid, or gas phase), and (2) steep gradients in physical and biogeochemical conditions (Naiman 1988; Naiman and Decamps 1997). The steep physical, chemical, and biological gradients in ecohydrological interfaces often correspond to distinct types and enhanced rates of biogeochemical processes (McClain et al. 2003; Yarrow and Marin 2007), and have significant impacts on ecosystem responses and resilience to

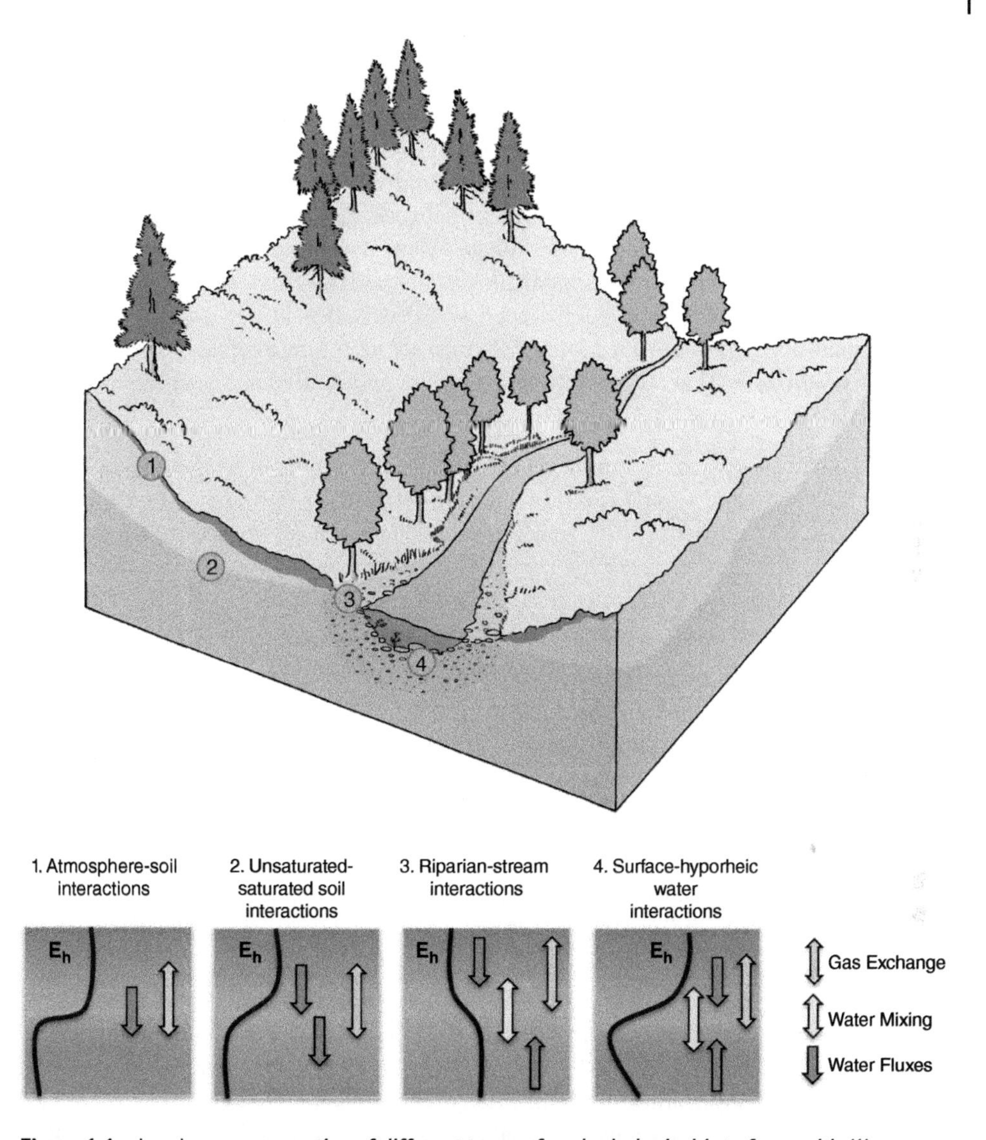

Figure 1.1 Landscape perspective of different types of ecohydrological interfaces with (1) atmosphere–soil interfaces, (2) unsaturated–saturated soil interfaces, (3) riparian–stream interfaces, and (4) hyporheic zone interfaces and characteristic profiles of water fluxes, mixing, gas exchange, and redox conditions (E_h).

environmental change (Brunke and Gonser 1997). Examples of specific conditions at ecohydrological interfaces that facilitate transformative processes include:

- ***Steep redox gradients*** across groundwater–surface water interfaces as a result of enhanced biogeochemical activity (Krause et al. 2013; Lautz and Fanelli 2008; Lewand owski et al. 2007; Trauth et al. 2015).
- ***Step changes in microbial metabolic activity*** (Argerich et al. 2011; Haggerty et al. 2009) and high concentrations of bioavailable organic carbon, nitrogen, or phosphorus at

riparian–wetland interfaces (Schelker et al. 2013), at groundwater–surface water interfaces (Zarnetske et al. 2011a, 2011b), and in biofilms (Battin et al. 2003, 2007, 2016), resulting in micro-zonation of denitrification (Briggs et al. 2015) and enhanced interface denitrification rates where microbial denitrifiers are concentrated (Harvey et al. 2013).
- ***Co-existence of multiple aggregate states (solid, liquid, gas phase)***, across which energy and matter are transferred, such as between the atmosphere and porous soil matrix (Shahraeeni et al. 2012), atmosphere–water interfaces (Assouline et al. 2010), unsaturated and saturated soil compartments (Li and Jiao 2005), and between the soil matrix and soil water or air in soil macropores (van Schaik et al. 2014).
- ***Shifts between physical and biological controls of solute transport*** across water–organism interfaces (Larned et al. 2004; Nishihara and Ackerman 2009; Nishizaki and Carrington 2014).

While there have been recent improvements in understanding how ecohydrological interfaces control energy and water fluxes (in particular, between groundwater and surface water (Boano et al. 2014; Cardenas 2015; Krause et al. 2011a)), critical knowledge gaps remain with respect to how they affect reactive transport, solute mixing, and biogeochemical cycling across system boundaries (Krause et al. 2011a; Puth and Wilson 2001). Our understanding of the spatial and temporal organization of driving forces (e.g. hydrostatic pressure distribution, concentration gradients, turbulence intensity) and controls (e.g. interface transmissivity, roughness) of ecohydrological interface fluxes and reactivity are at an early stage (Gomez-Velez et al. 2012, 2014; González-Pinzón et al. 2015; Zhang et al. 2015).

Many ecohydrological interfaces are spatially heterogeneous and temporally dynamic (Kennedy et al. 2009; Roskosch et al. 2012). While the physical (structural) boundaries between adjacent and interacting systems (e.g. between groundwater and surface water) are usually clearly defined and stationary, dynamically developing ecohydrological interfaces (e.g. hyporheic zones) are defined by their functioning and may change in time with regard to their spatial extent and activity (Boano et al. 2010, 2014; Cardenas and Wilson 2006, 2007; Gomez et al. 2014; Stubbington 2012; Trauth et al. 2015). However, some structural boundaries around which ecohydrological interfaces evolve can themselves be dynamic, such as migrating bedforms and flexible and compressible benthic organisms (Harvey et al. 2012; Huang et al. 2011; Larned et al. 2011; Ren and Packman 2004), further complicating the identification and delineation of ecohydrological interfaces.

Patterns and dynamics of ecohydrological interface activity include the development of hotspots (zones of enhanced activity: Frei et al. 2012; Krause et al. 2013; Lautz and Fanelli 2008; McClain et al. 2003) and hot moments (periods of increased activity: Battin et al. 2003; Harms and Grimm 2008; McClain et al. 2003) that disproportionately alter the fluxes of water, energy, and matter. Hotspots or "control points" (Bernhardt et al. 2017) have captured the attention of many researchers, who study how they affect nutrient turnover (Lewandowski et al. 2007; Moslemi et al. 2012), ecosystem productivity (Poungparn et al. 2012), pesticide degradation (Klaus et al. 2014), and the bioavailability of metals, such as mercury, to organisms at higher trophic levels (Sizmur et al. 2013). Yet, when and under what conditions ecohydrological interfaces represent hotspots or control points, or what makes them behave as such, has not always been clearly determined.

We have, for instance, only just begun to understand how biological activity (e.g. earthworm and chironomid burrowing, stream periphyton growth, or riparian plant root growth) can create small-scale ecohydrological interfaces that are hotspots of microbial and biogeochemical activity (Baranov et al. 2016; Hölker et al. 2015). Furthermore, the concept of hot moments entails long periods of relatively low activity punctuated by pulses of rapid activity. These temporal dynamics suggest that some ecohydrological interfaces can be ephemeral. We now turn to these and other gaps in our understanding of ecohydrological interfaces.

1.3 Critical Gaps in Understanding Ecohydrological Interfaces

We currently lack an overarching framework that integrates the factors that drive and control transformation processes at ecohydrological interfaces. Perceptions and conceptualizations of boundaries, and with that ecohydrological interfaces, are often scale-dependent (Cadenasso et al. 2003; Strayer et al. 2003). At large scales, some ecohydrological interfaces (e.g. between aquifers and rivers) may be conceptualized as discrete boundaries, causing abrupt transitions with step changes in processes across the boundary (Figure 1.2A). However, down-scaling reveals three-dimensional gradients within interfaces (e.g. in hyporheic zones), and transient or gradual changes of physical or biogeochemical properties (Figure 1.2B). Acknowledgement of the context and scale-dependent view of ecohydrological interfaces is important because the scale in which ecohydrological interfaces are investigated can preclude the detection and quantification of physical, chemical, and biological activity at other scales (Atkinson and Vaughn 2015). Further, temporal variation in the shape or spatial extent of interfaces and the steepness of gradients within them suggests that our conceptualizations of interfaces vary over temporal as well as spatial scales – as, for instance, shown for transient behaviour of hyporheic zones in response to hydrological forcing (Malzone et al. 2016).

Clear delineations of the spatial and temporal extent of ecohydrological interfaces are further complicated by discipline-specific perspectives on interface properties, processes,

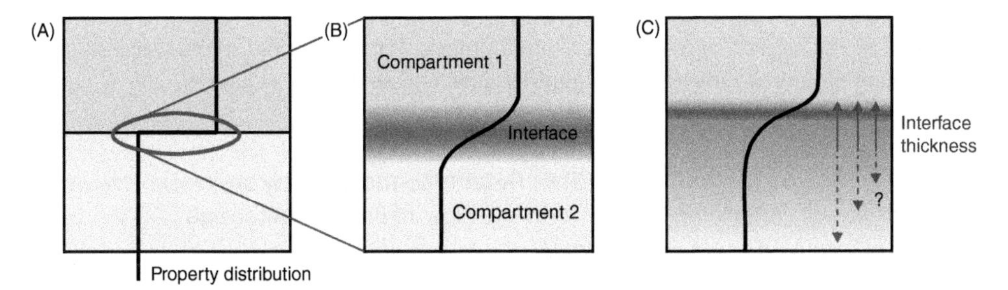

Figure 1.2 Conceptual model of ecohydrological interfaces connecting two adjacent contrasting environments (Component 1 and Component 2) with scale-dependent representation of gradients of chemical, physical, and biological properties (solid black line). (A) Large scales exhibit step functions in interface properties, where interfaces appear as two-dimensional layers of zero depth; (B) zoomed into smaller scales with steep gradient of chemical, physical, and biological properties and a three-dimensional interface zone with some depth dimension; (C) difficulties are frequently encountered in determining the upper and lower boundary and depth of the interface zone, especially where property distributions blend into background properties due to their non-linearity.

and functions (Figure 1.2C; Harvey et al. 2013; Yarrow and Marin 2007). Based on discipline-specific perceptions, *hyporheic zones*, for instance, are defined by the spatial extent of groundwater and surface water mixing (hydrology), the extent of steep chemical gradients (biogeochemistry), or the abundance of benthic and hypogean taxa (ecology), resulting in significantly different perceptions of their extent (Krause et al. 2011b, 2014b). Recent studies of benthic systems have focused on the dynamics and ecological effects of *multi-layered interfaces* (e.g. small-scale diffusive boundary layers nested within larger-scale roughness layers, within larger benthic boundary layers; Larned et al. 2004; Nikora 2010) and on micro-zonation of biogeochemical processes, e.g. redox micro-zones (Briggs et al. 2015). Views of the capillary boundary at the *groundwater–vadose zone interface* differ between ecologists focusing on matric potential effects on plant available water and water uptake, (bio)geochemists interested in redox chemistry differences between pore water and adsorption to mineral surfaces (Alexander and Scow 1989; Baham and Sposito 1994), and groundwater hydrologists and hydrogeological engineers concerned with water table depths. Such discipline-specific perceptions of ecohydrological interfaces can limit the transferability of process understanding and the exchange of data and knowledge across disciplinary boundaries.

Detailed understanding of the drivers and controls of enhanced interface activity is critical for evaluating the functional significance of ecohydrological interfaces. Examples include the shift from aerobic to anaerobic respiration in hyporheic zones, which is controlled by residence time of hyporheic water and nutrients in the streambed (Briggs et al. 2014; Zarnetske et al. 2011a), or temperature thresholds triggering bacterial activity (Bourg and Bertin 1994). Here we pose four critical questions (spanning scales and crossing disciplinary boundaries) that must be answered to understand the role of ecohydrological interfaces in ecosystem functioning:

1) What environmental conditions determine the capacity of ecohydrological interfaces to transform the flow of energy, water, and matter between adjacent ecosystems?
2) How are ecohydrological interfaces organized and how do they evolve in space and time?
3) What mechanisms (drivers and controls) determine the spatio-temporal organization of ecohydrological interfaces?
4) How do the impacts of hotspots and hot moments at ecohydrological interfaces upscale to ecosystem ecohydrological, biogeochemical, and ecological processes?

1.3.1 What Environmental Conditions Determine the Capacity of Ecohydrological Interfaces to Transform the Flow of Energy, Water, and Matter between Adjacent Ecosystems?

Ecohydrological interfaces have been described as intensive modifiers of energy, water, and solute fluxes and biogeochemical cycling (Harvey and Fuller 1998), that exhibit hotspot characteristics (Krause et al. 2013; Lautz and Fanelli 2008; McClain et al. 2003) and non-linear behaviour (Briggs et al. 2014; Zarnetske et al. 2011a, 2011b). To understand why ecohydrological, biogeochemical, and ecological transformation processes in ecohydrological interfaces often differ from their neighbouring ecosystems, it is necessary to review the physical, chemical, and ecological interactions that characterize them.

1.3.1.1 Physical Properties

Contrasts in interface material properties from the adjacent environmental systems (sometimes coinciding with aggregate state boundaries such as between liquid and gas phase, or with changes in transmissivity) affect velocity and direction of exchange fluxes (Figure 1.3). Impacts of ecohydrological interfaces on exchange fluxes can vary from complete cessation, if the interface is impermeable (Figure 1.3A), to unaffected (Figure 1.3B) or even accelerated exchange. The geometry of property distributions at ecohydrological interfaces (such as hydraulic conductivities at groundwater–surface water interfaces) may cause hysteretic behaviour that is dependent on exchange-flow direction (Figure 1.3C). For example, surface water flow velocities decrease when infiltrating into the streambed,

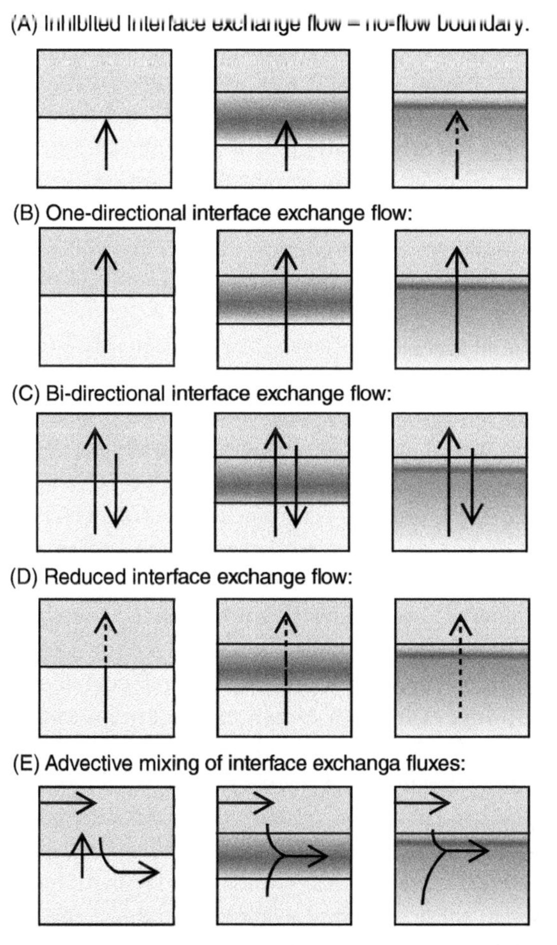

Figure 1.3 Conceptual model of the scale-dependent complexity of ecohydrological interface exchange-fluxes in systems with interfaces representing thresholds with infinitesimal thickness (left), steep gradients with abrupt property changes (centre), or variable (transient and abrupt) property changes between interface and adjacent environments (right), with one-directional flow ceasing at (in) the interface zone (A) or crossing the interface (B), bi-directional exchange fluxes across the interface (C), flow reduction across the interface pathway (D), and the advective mixing of interface exchange fluxes with intra-compartmental fluxes (E).

while groundwater up-welling through the streambed may accelerate towards the interface with surface water. Reduced flow velocities and increased residence times that have been observed at many ecohydrological interfaces (Figure 1.3D) can substantially enhance biogeochemical processing (Briggs et al. 2014; Zarnetske et al. 2011b). Quantifying the spatio-temporal variability of biogeochemical processing in heterogeneous interface zones of variable activity will require a shift from the current focus on mean residence times to residence-time distributions that are dynamic (Botter et al. 2011; Pinay et al. 2015).

In many cases, exchange fluxes at ecohydrological interfaces interact with larger flow systems in the adjacent ecosystem (Figure 1.3E). At aquifer–river interfaces, for instance, exchange fluxes interact across multiple scales. Hyporheic exchange here can be affected by regional groundwater flow, causing complex and nested patterns of exchange fluxes (Gomez-Velez et al. 2014; Trauth et al. 2015) and thus, spatially complex and temporally dynamic ecohydrological interfaces. In this context, we have only begun to understand the impacts of interacting drivers and controls of interfaces exchange. Following the previous example, this includes how streambed transmissivity (Krause et al. 2013) and pressure variations caused by interface topography, such as riparian micro-topography (Frei et al. 2012), bedforms (Cardenas et al. 2004), or meanders (Boano et al. 2010), overlap in their impacts on hyporheic exchange fluxes (Boano et al. 2010, 2014; Gomez-Velez et al. 2014) and dynamically evolve in time due to variability in atmospheric and hydrodynamic forcing (Malzone et al. 2016).

1.3.1.2 Ecological (Including Microbiological) Processes

Ecohydrological interfaces between groundwater, surface water, and vadose zones can have large effects on ecological conditions in the adjacent systems (Cadenasso et al. 2003; Pinay et al. 2015). Thermodynamically controlled microbial processes drive biogeochemical transformations in these subsurface systems, and in turn, biota respond to the chemical gradients that result from their activity. A classic example is aerobic respiration, which in subsurface zones is largely carried out by microorganisms. As they consume oxygen and organic carbon, microbes create conditions that favour transition to anaerobic metabolism. Although some microorganisms are facultative anaerobes, others are excluded once oxygen concentration drops below a threshold. In fact, a sequence of terminal electron accepting processes, each with their suite of microbial specialists, ensues along redox gradients that characterize anoxic environments (Morrice et al. 2000). Aquatic macrophytes, benthic biofilms, and riparian vegetation may exude or release organic matter during metabolism or upon death or decomposition, which provides an energy source for microbial metabolism. Community structure and elemental composition of primary producers may influence biogeochemical turnover and location of biogeochemical hotspots at ecohydrological interfaces, as they are likely to release organic matter at different rates and with different chemical composition. Hence, in addition to altering nutrient availability and stoichiometry, aquatic macrophytes, benthic algae, and pelagic phytoplankton colonies may induce hotspots of microbial metabolism (de Moraes et al. 2014).

Aquatic and wetland plants influence the saturated substrate where fine-scale microenvironments develop around their root systems, altering the oxygen concentrations, nutrient uptake, sediment structure, and microbial activity of riparian and hyporheic zones. For example, exudates from the roots of a wetland shrub, *Baccharis* sp., fuel microbial respiration, including denitrification, in streamside sediments and riparian zones (Harms and Grimm 2008; Schade et al. 2001). The size of the ecohydrological interface zone in which these root exudates drive microbial metabolism tends to be restricted to a few centimetres around the root zone (Schade et al. 2001). Vascular plants influence not only the interstitial water of the sediment but also the water column, through mutualistic interactions with phytoplankton and bacterial communities (Brodersen et al. 2014), and the atmosphere, by respiration and gas exchange (Xing et al. 2006). Ecological impacts on ecohydrological interface functioning are not restricted to living organisms. Large woody debris alters streambed topography and enhances groundwater–surface water interactions and supply of organic carbon, thus supporting habitat complexity and biotic activity (Krause et al. 2014a; Warren et al. 2013). The nutrients and pollutants that had previously been absorbed by biota are now released during decomposition and can stimulate localized hotspots of increased resource availability (Krause et al. 2014a), or invertebrates can induce the development of biogeochemical hotspots through the regeneration of nutrients (Grimm 1988a).

The morphology, physiology, and productivity of benthic autotrophs (e.g. algal and cyanobacterial mats, seagrasses, corals growing on the bottom of streams, lakes, and coastal marine ecosystems) are strongly influenced by the hydrodynamic and chemical conditions in surface water–benthic interfaces. Mass transport across these interfaces is often the rate-limiting step for nutrient acquisition and gas exchange by the organisms (Jumars et al. 2001; Larned et al. 2004), and hydrodynamic forces imposed by these interfaces affect the organism stature and biomechanical properties (Albayrak et al. 2014; Statzner et al. 2006). While interface conditions clearly affect benthic autotrophs, the opposite is also true. Benthic autotrophs function as roughness elements that modify flow structure and as biogeochemical reactors that alter water chemistry (Folkard 2005; Larned et al. 2011; Reidenbach et al. 2006). The picture that is emerging from recent studies of surface water–benthic interfaces is a flow–organism feedback system consisting of responses by organisms to flow conditions, flow modifications induced by the organisms, subsequent responses by the organisms to the modified flow, and so forth (Dijkstra and Uittenbogaard 2010; Larned et al. 2011; Nikora 2010). Similar feedback systems should apply to the heterotrophic organisms in sedimentary systems, as described later. Such feedbacks are an important source of non-linearity in process rates at ecohydrological interfaces.

Recently, ecohydrological research has considered biota at higher trophic levels, such as macroinvertebrates and aquatic vertebrates, and their capacity to alter the physical-chemical characteristics that regulate the rate of activity and ecosystem functioning at ecohydrological interfaces (Coco et al. 2006; Layman et al. 2013; Patrick 2014). Lewandowski et al. (2007), Roskosch et al. (2012), and Baranov et al. (2016), for instance, describe a system of interactions and feedbacks between chironomids and aquifer–lake ecohydrological interfaces. In these studies, chironomid activity had direct impacts on hydrodynamics and biogeochemistry, while physical-chemical conditions, such as temperature, affected chironomid pumping behaviour (Roskosch et al. 2012) and hence, subsurface flow pattern and biogeochemical processing rates. Similarly, vertebrates may alter streambed topography, for example, through nest-building activities, which lead to changes in connectivity and fluxes across the surface water–sediment interface (Collins et al. 2014), through their movement (Hippopotamus) or beaver dam construction (Naiman et al. 1994). Additionally, fish induce biogeochemical hotspots by excretion (Grimm 1988b; Vanni 2002) and nutrient release following their death and decomposition (Levi and Tank 2013).

1.3.1.3 Thermodynamics and Biogeochemistry

At stationary boundaries, matter and energy fluxes may be absorbed, transmitted, reflected, transformed, amplified, or unaffected. Boundaries can be highly permeable to some substances, and represent reactive filters for others (Belnap et al. 2003; Cadenasso et al. 2003; Strayer et al. 2003). We propose that these concepts of flux behaviour at boundaries can be extended to non-stationary ecohydrological interfaces, which develop dynamically in space and time. Processing rates at ecohydrological interfaces are controlled by both mass transport and reaction kinetics, with *transport-limited* conditions arising when reaction rates are faster than mass-transport rates (Cornelisen and Thomas 2009; Larned et al. 2004; Sanford and Crawford 2000). Conversely, process rates tend to be kinetically controlled (*reaction limited*) when mass-transport rates are faster than reaction rates (Argerich et al. 2011; Nishihara and Ackerman 2009; Sanford and Crawford 2000). Increased biogeochemical activity is often attributed to the spatial and temporal coincidence of reactants in a mixing zone (Figure 1.4A) (McClain et al. 2003); however, enhanced turnover may also be controlled by high reactivity in interfaces (Figure 1.4B), resulting directly from the chemical gradients at the interface (Krause et al. 2013; Trauth et al. 2015). It has yet to be established how the mixing of reactants at ecohydrological interfaces influences interface redox conditions and controls residence-time distributions of different reactants, and hence, biogeochemical processing rates. Possible approaches to achieve this involve combinations of residence-time distributions and dimensionless numbers used to describe the transport vs. reaction relationships of flow systems, such as the Damköhler number or Péclet number describing diffusion/advection ratios (Pinay et al. 2015). Furthermore, the reaction significance factor (RSF) approach has been applied for quantifying reaction vs. transport limitation in single hyporheic flow paths within basin-scale assessments of the number of river excursions through the hyporheic zone (Gomez-Velez et al. 2015; Harvey et al. 2013).

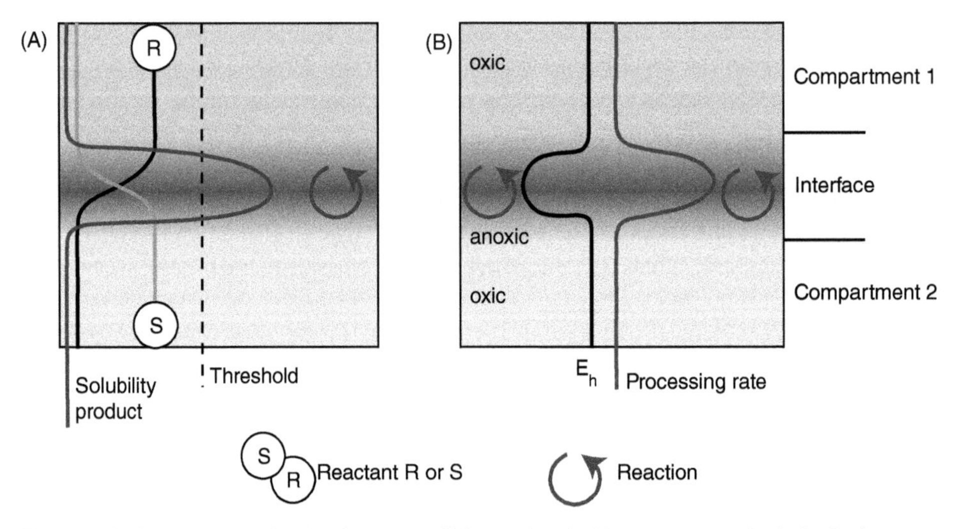

Figure 1.4 Examples for the development of biogeochemical hotspots at ecohydrological interfaces, hosting distinctly different reaction properties and hence biogeochemical processes than its adjacent environments: (A) Enhanced reactivity directly resulting from the interaction of interface exchange fluxes such as the precipitation of a reactant at the ecohydrological interface due to exceeding its solubility product. (B) Enhanced reactivity as an intrinsic property of the interface environment, such as anoxic areas in hyporheic or riparian zones.

Despite these advances, predictions of biogeochemical processing at ecohydrological interfaces remain challenging, since not only can biogeochemical turnover be enhanced, but also the type of processes and chemical reactions may differ distinctively from adjacent ecosystems (Naiman et al. 1988).

1.3.2 How are Ecohydrological Interfaces Organized in Space and Time?

Complex microhabitat structure and biological activity create ecohydrological interface heterogeneity (Hanzel et al. 2013; Lewandowski et al. 2007), with interface processes often varying over a wide range of spatial and temporal scales (Belnap et al. 2003). Hyporheic exchange flows, for instance, include sinuosity-driven flows in meandering streams (Boano et al. 2010; Gomez-Velez et al. 2012) and bedform-driven flows caused by streambed features such as riffles and pools (Käser et al. 2013; Thibodeaux and Boyle 1987; Tonina and Buffington 2007), small-scale ripples and dunes (Cardenas and Wilson 2007), and flow obstacles such as dams or wood (Briggs et al. 2012; Krause et al. 2014a; Sawyer et al. 2011).

Scale-dependent drivers of the spatial and temporal organization of ecohydrological interface properties are complex. Mixing of chemical reactants in ecohydrological interfaces may involve the transport of multiple reactants from source areas to the interface (Figure 1.5A; e.g. Zarnetske et al. 2011a, 2011b), or the mixing of a reactant already present at the interface with another reactant that is transported into it (Figure 1.5C; e.g. Krause et al. 2013; Lewandowski et al. 2007). In many cases, just a fraction of the mass flux crosses the ecohydrological interface. Often mass fluxes return to their original compartments (Figure 1.5B, D); e.g. surface water infiltrates into the hyporheic zone and exfiltrates back into the stream after passage through the bed.

Ecohydrological interfaces are frequently characterized by non-linear temporal dynamics, including tipping points, caused by rapid changes in thermodynamic or biogeochemical characteristics at the interface, such as the shift from aerobic to anaerobic metabolism (Briggs et al. 2014, 2015; Harvey et al. 2012; McClain et al. 2003; Zarnetske et al. 2011a, 2011b) or biogeochemical responses to fast changes in soil water content (Fromin et al. 2010). Also, rainfall pulses in dryland environments can result in rapid and non-linear increases in microbial respiration at the soil–air interface (Collins et al. 2014) or in the vadose zone–groundwater interface of riparian zones during dry seasons (Baker et al. 2000; Harms and Grimm 2008). In both of these examples, ecohydrological interfaces come into existence when water is added (i.e. rainfall impinges on the soil surface, or the groundwater table rises into previously dry riparian soil), such that biogeochemical processes are stimulated rapidly. However, the cumulative long-term effects of such hot moments on ecohydrological interfaces, as well as their subsequent contribution to system behaviour at a global scale (Kreyling 2014) still need to be investigated in detail.

1.3.3 What Mechanisms (Drivers and Controls) Determine the Spatio-temporal Organization of Ecohydrological Interfaces?

Spatial patterning in the properties of an ecohydrological interface can result directly from interface processes and thus, may partly be explained by the functioning of the ecohydrological interface. Examples include redox patterns in hyporheic zones resulting from oxygen depletion by hyporheic biogeochemical processing (Krause et al. 2013; Zarnetske

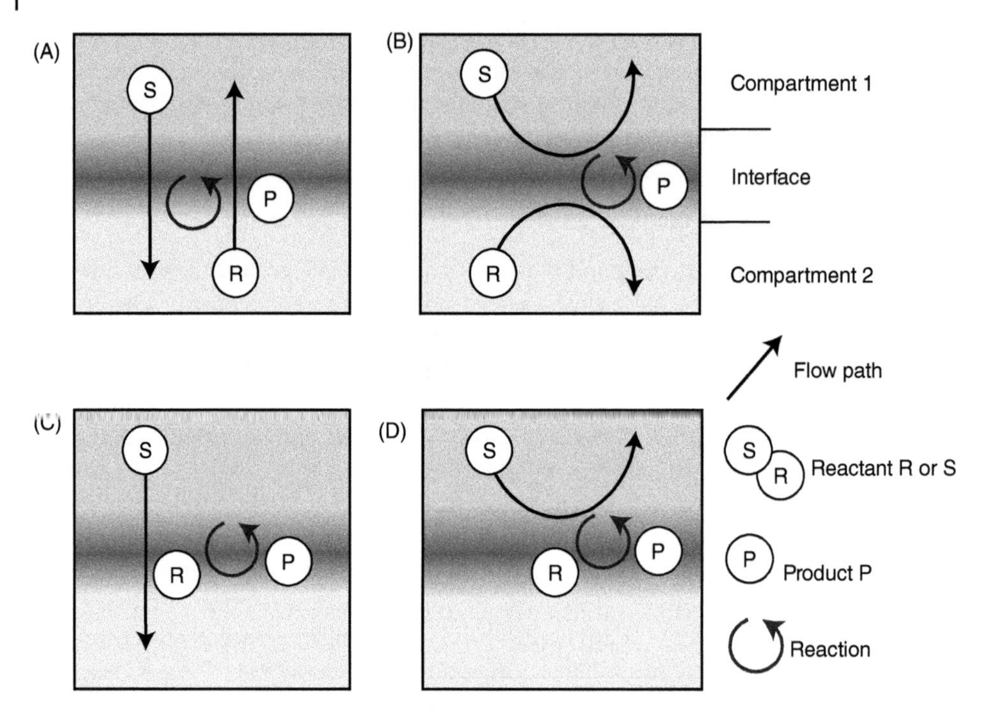

Figure 1.5 Enhanced ecohydrological interface reactivity as a function of exchange flow patterns at/in/across the interface with fluxes carrying reactant R and S to meet and mix at the interface (A) with not all but just a fraction of the reactants mixing at the interface due to tangential fluxes (B) and transport of reactant S into the ecohydrological interface already containing autochthonous reactant R, results in the processing of S and R to product P (C) with some of the external reactant (S) returning to the compartment it originated from (D).

et al. 2011a, 2011b). In other cases, the origin of spatial variability is independent of actual interface processes (e.g. spatial variability in hydraulic conductivity can control patterns of exchange fluxes in ecohydrological interfaces). Spatial patterns of solute concentration in ecohydrological interfaces may be controlled partially by the spatial organization of properties in the adjacent ecosystems (Figure 1.6). For example, spatially homogeneous physical properties in hyporheic zones (Figure 1.6A, B, C) or around chironomid burrows (Figure 1.6E) will facilitate ecohydrological interface activity that is controlled primarily by interface exchange fluxes and mean residence times (e.g. Zarnetske et al. 2011a, 2011b). In contrast, a heterogeneous matrix in surrounding ecosystems (Figure 1.6D, G) will add further complexity, making it important to quantify not only exchange fluxes and residence times but also their distributions (Gomez-Velez et al. 2014). In addition to spatial heterogeneity, patterns may evolve with time as interface processes progress. For instance, chironomid pumping can affect property distributions at the sediment/burrow wall interface (Figure 1.6F, H), where they have been shown to induce gradients of decreasing oxygen concentration with increasing distance from the tube (Figure 1.6F) or increasing concentration of soluble reactive phosphorus with increasing distance from the tube walls into the adjacent sediment (Figure 1.6H) (Baranov et al. 2016; Lewandowski et al. 2007).

Disentangling the impacts of different drivers and controls on processes in ecohydrological interfaces remains a challenge, partly due to combined effects and feedbacks between

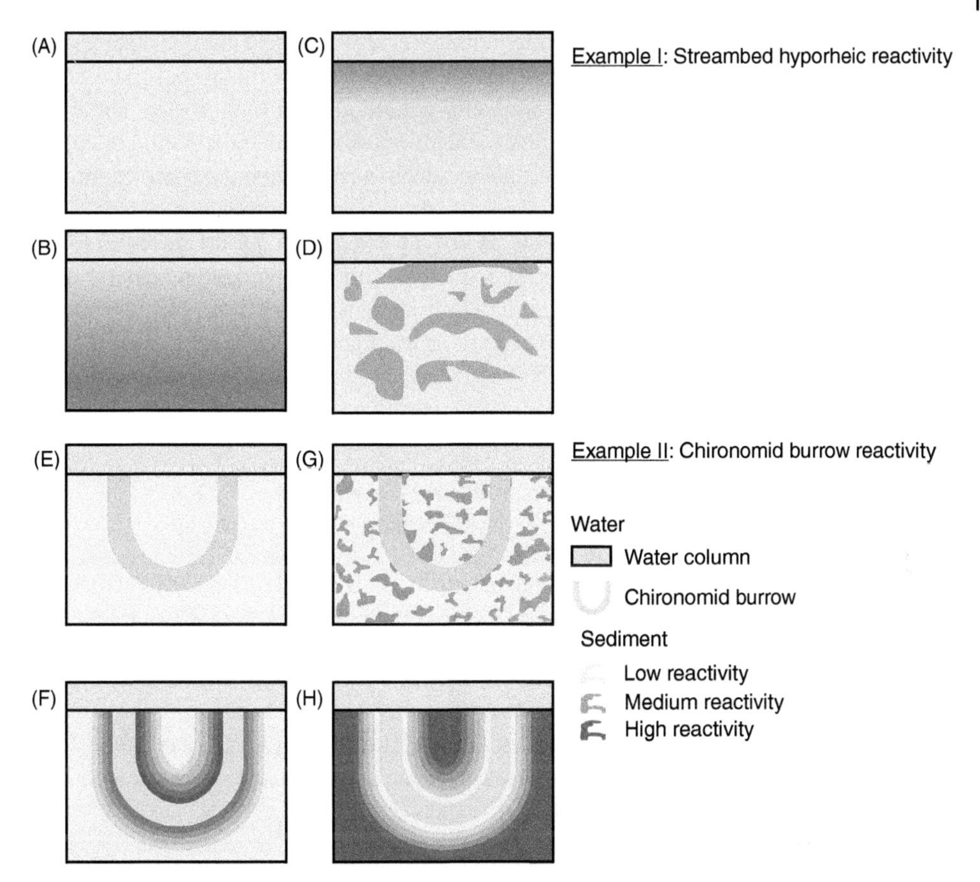

Figure 1.6 Variable characteristics and heterogeneity of ecohydrological interfaces as result of differences in passive or active organizational mechanisms structuring interface properties, with: Example I, passive controls – streambed properties controlling hyporheic zone reactivity: Homogeneously low or high ecohydrological interface reactivity (concentrations) at hyporheic zones resulting from continuously low streambed reactivity (A), or depth decreasing (B), or increasing (C) streambed reactivities, in contrast to spatially heterogeneous streambed properties, subsequently causing spatial variability at the interface (D). Example II, active controls – chironomids as engineers of interface complexity: No effect of chironomids and homogenous (E) and heterogeneous (G) distribution of biological, chemical, and physical properties within the sediment matrix and at the burrow wall interface; chironomid pumping induced gradients of decreasing oxygen concentration from the tube into the adjacent sediment (F), and increasing soluble reactive phosphorus concentration from the tube into the adjacent sediment (H).

hydrological, biogeochemical, and biological processes that may be additive, synergistic, antagonistic, or undetectable. To use freshwater microbial biofilms as an example, biogeochemical turnover in biofilms is related to their biomass (Haggerty et al. 2014; Singer et al. 2010). Hence, biofilm growth causes biogeochemical turnover rates to increase. At the same time, increased biofilm thickness changes its permeability and has the potential to cause significant clogging, increasing contact area and residence and reaction times at the biofilm surface, which in some cases has been shown to accelerate biogeochemical turnover (Battin et al. 2007), or even change the type of chemical reactions, inducing shifts from aerobic to anaerobic conditions or limiting biogeochemical processing at the interface (Treese et al. 2009).

Improving the understanding of the functioning of ecohydrological interfaces across spatial and temporal scales will require the acknowledgement that traditional hierarchical classification schemes where physical boundary conditions and hydrological behaviour control thermodynamic processes and biogeochemistry, which then define the biological template or ecological niche, are not suitable to adequately describe the complex interactions between biological, biogeochemical, and hydrological processes at ecohydrological interfaces. As previously discussed, biological activity can be a major driver of the spatial and temporal organization of ecohydrological interface functions and often actually shape the physicochemical template. It is essential to fully acknowledge this complexity of multidirectional interactions also in experimental and conceptual model designs as oversimplification of cause–impact relationships will not yield the required understanding of what drives the organizational principles of ecohydrological interface functions.

1.3.4 How do the Impacts of Hotspots and Hot Moments at Ecohydrological Interfaces Upscale to Ecosystem Ecohydrological, Biogeochemical, and Ecological Processes?

Our capacity to quantify and predict the large-scale and long-term importance of hotspots and hot moments at ecohydrological interfaces is hampered by our limited understanding of how mechanisms structuring ecohydrological interfaces and their processes scale in space and time (Krause et al. 2011b; Pinay et al. 2015). The effects of interface hotspot activity have been observed at scales ranging from micro-scales, such as biofilms, to intermediate scales of stream reaches (Lautz and Fanelli 2008; Trauth et al. 2015), and conceptual frameworks have been developed to explain interface process dynamics (Fisher et al. 1998; Harms and Grimm 2008; McClain et al. 2003; Pinay et al. 2015). For example, there is evidence that hyporheic zone processes can have implications for the whole stream network (Gomez-Velez et al. 2015; Harvey and Gooseff 2015; Kiel and Cardenas 2014; Zarnetske et al. 2015), with hyporheic nitrification and denitrification in headwater streams altering the nitrogen load in rivers (Alexander et al. 2007). Although hotspot activity has been shown to be at least temporarily significant at small local scales, its larger-scale importance for energy transfer or biogeochemical turnover in entire river networks or catchments is still widely debated. This partly results from the fact that processes specific to ecohydrological interfaces have often been studied by coupling conceptual models of different ecosystem types (e.g. coupling groundwater and surface water models: Markstrom et al. 2008; Yuan et al. 2011) or land–surface schemes and atmospheric models with hydrological models (e.g. Maxwell and Miller 2005). In both cases, ecohydrological interface conditions are at least partly defined as boundary conditions instead of integrating ecohydrological interface conditions and behaviour implicitly, a practice that restricts the way dynamic interface processes can be analysed across scales.

1.4 A Vision for Integrated Research at Ecohydrological Interfaces

The pressing challenges of global environmental change, such as increasing frequencies and magnitude of extreme events (Blöschl et al. 2015; Hall et al. 2014), call for improved understanding of their impacts from plot to regional scales, across ecosystem types, and beyond disciplinary boundaries. This will require advanced methods for multi-scale monitoring of highly dynamic ecosystem behaviour (Abbott et al. 2016; Blaen et al. 2016) in order to enhance the current understanding of quantitative implications and dynamic behaviour of ecohydrological interface processes for coupled water, matter, and energy fluxes and biogeochemical turnover. The most critical knowledge gaps outlined in this chapter include:

i) Inadequate conceptual frameworks for understanding how processes occurring at ecohydrological interfaces vary with scale and how and whether small-scale interface processes are manifested at large scales across complex landscapes; and
ii) Failure to transfer and integrate scale-dependent methods and knowledge of mechanisms controlling ecohydrological interface processes across disciplinary and ecosystem boundaries (Abbott et al. 2016; Hannah et al. 2007; Krause et al. 2014b, 2011b).

Interdisciplinary research strategies will need to move the research of ecohydrological interfaces from a descriptive to a mechanistic and predictive stage, extending the scope to a wider range of ecohydrological interfaces than explored in this chapter. Ecohydrological interfaces not only connect different environmental domains but also represent a research topic that requires and fosters novel linkage between traditionally distinct disciplines. The development of multi-scale conceptual models of ecohydrological interface functioning requires interdisciplinary thinking and integration of discipline-specific methods. Following this rationale, we propose the following "roadmap" to catalyse research advances.

Roadmap for Ecohydrological Interface Research:

I) Enhance capacities for multi-scale monitoring and modelling

Developing multi-scale conceptual models of ecohydrological interfaces will require advances in physical, microbial, biogeochemical, and ecological monitoring using innovative sensing and tracing technologies (Abbott et al. 2016; Blaen et al. 2016; Gonzáles-Pinzón et al. 2015). In turn, the application of these technologies will require new methods for managing big data sets, and advanced tools for spatial and time-series analysis. Recent advances in distributed sensor networks such as fibre-optic distributed temperature sensing (Selker et al. 2006), thermal IR imagery (Pfister et al. 2010), and high-frequency in situ sensors, analysers, and imagers (Jordan et al. 2007; Neal et al. 2012; Reidenbach et al. 2010) provide capacity for improved resolution and frequency in monitoring exchange fluxes across ecohydrological interfaces in real time (Blaen et al. 2016; Grant and Marusic 2011; Krause et al. 2015). Technology exchange among disciplines has the potential to advance process monitoring beyond current observations of average, compartmental system behaviour, including identification and quantification of hot moments and hotspots. Using terrestrial diatoms to detect the rapid onset and cessation of flow path connectivity in the hillslope–riparian zone–stream continuum (Martínez-Carreras, 2015; Pfister et al. 2009) is a promising example of such a cross-disciplinary approach.

II) Improve conceptual understanding of interface processes and their interactions

In addition to improving monitoring capacity, the resulting discipline and system-specific knowledge needs to be integrated to improve understanding of the scale-dependent processes and mechanisms that lead to the development of bioreactive hotspots and hot moments (Soulsby et al. 2008). For instance, the application of process understanding gained in groundwater–vadose zone or groundwater–surface water interfaces to other ecohydrological interfaces, ecosystem types, and disciplines will support the development of an integrated conceptual framework for ecohydrological interfaces. Promising examples include:

- The linking of spatial patterns and behaviour of anecic earthworm populations to the generation of preferential flow pathways through macropores, which in turn affects pesticide infiltration (Palm et al. 2013; van Schaik et al. 2014).
- Investigations of biogeochemical hotspots developing around chironomid burrow walls with fluxes of pore water infiltrating from the adjacent sediment and active ventilation of water from the tube into the surrounding sediment (Baranov et al. 2016; Roskosch et al. 2012).
- The extension of boundary-layer research from atmosphere–forest and atmosphere–soil interfaces (Finnegan 2000) to a variety of surface water–benthic interfaces (e.g. Larned et al. 2011; Nepf 2012; Nikora 2010).
- Novel approaches for analysing process dynamics at plant–soil interfaces including plant root endosphere and rhizosphere (Vandenkoornhuyse et al. 2015).

III) Quantify the impact of interface hotspots and hot moments at regional scales

To improve the prediction and quantification of landscape-scale water, matter, and energy fluxes and biogeochemical cycling, quantitative model frameworks need to incorporate improved mechanistic understanding of the space–time organization of ecohydrological interface activity. This may be achieved by advancing conceptual modelling frameworks that integrate traditionally separate model domains. In addition, there is great potential for enhanced interdisciplinary knowledge exchange by transferring subject-specific theory across disciplinary boundaries and testing its validity at ecohydrological interfaces (Abbott et al. 2016). For instance, concepts linking the spatial organization and hydrological functioning of intermediate-scale catchments (Zehe et al. 2014) provide potential for being applied to advance the process understanding of the functioning of ecohydrological interfaces. In this catchment-scale example, a hierarchy of functional units (i.e. co-evolving elementary functional units) has been shown to control catchment functioning, ultimately resulting in spatially organized landscapes (Zehe et al. 2014). In other applications, the blurring of system boundaries and adoption of flow path approaches has been advocated to more realistically scale up to larger landscapes (Fisher et al. 1998, 2004; Kolbe et al. 2016). Adaptations of such concepts may have great potential to improve large-scale quantification of ecohydrological interface activity.

IV) Manage ecohydrological interfaces to enhance ecosystems' services and increase resilience to environmental change

Ecosystem services provided by ecohydrological interfaces need to be restored and their resilience to future environmental perturbations improved, in order to better manage the adjacent ecosystems (Harvey and Gooseff 2015; Hester and Gooseff 2011; Kasahara and Hill 2008). For example, multiple restoration measures have been trialled with the goal of enhancing hyporheic exchange fluxes across groundwater–surface water interfaces, e.g. constructed channel structures (Crispell and Endreny 2009) and bedforms (Kasahara and Hill 2006), altered streambed hydraulic conductivity (Ward et al. 2011), planting (Gurnell 2014), and woody debris installation (Krause et al. 2014a). A key challenge remains to identify drivers that can be manipulated or managed at relevant scales. New high-frequency and high-resolution data obtained from novel distributed sensor networks can help to improve the understanding of dominant controls of ecohydrological interface reactivity (Blaen et al. 2016; Krause et al. 2015). Such an understanding is required to design potential engineering and management measures to restore, maintain, or enhance processes of ecohydrological interfaces. Explicit consideration of the dynamics of processes at ecohydrological interfaces also has the potential to improve management and risk assessment frameworks. Specifically, managing ecohydrological interfaces may permit their efficient use and promote their moderating impact and remediation potential, for example, by enhancing nutrient retention or removal at hyporheic or riparian interfaces.

1.5 Conclusions

This chapter has elaborated our view that to better understand the functioning of ecosystems, their component subsystems, and their interactions, it is important to explicitly account for the dynamics of processes occurring at ecohydrological interfaces. This implies consideration and analysis of ecohydrological interfaces in their own right, as entities with unique functioning and inherent, often complex, spatial patterns and temporal dynamics of physical, biogeochemical, and ecohydrological properties. Ecohydrological interfaces often occur at boundaries and ecotones, but they are not boundaries per se. They may appear and disappear, having a large or a small role in determining larger-scale processes that vary over space and time. An improved understanding of the wider landscape interactions between connected ecosystems will only be possible if current ecosystem and landscape concepts incorporate the processes that occur at ecohydrological interfaces.

The analysis of the actual causes of dynamic ecohydrological interface reactivity, including reasons for non-linear behaviour such as hotspots and hot moments, requires intensification of interdisciplinary research and enhanced capacity for high-frequency/resolution monitoring to adequately capture non-linear process dynamics as they occur. Combining technological and conceptual advancement from different disciplines can help us to understand the behaviour of non-linear ecohydrological interfaces. This will advance our understanding and conceptual frameworks of ecosystem processes, from their current, often disciplinary, descriptions of patterns and dynamics of ecosystems as segregated entities to dynamic systems with interconnected processes and interferences that are substantially controlled by the conditions at system and subsystem interfaces.

We recognize that we have provided insight into the importance of processes at a subset of the universe of ecohydrological interfaces, with a bias towards examples of ecohydrological interfaces involving freshwater ecosystems such as groundwater–surface water and benthic–pelagic interfaces. Further interdisciplinary research is needed to develop new strategies for extending and integrating this process understanding to other types of ecohydrological interfaces in more terrestrial ecosystems, such as plant–atmosphere, soil–plant, or microbe–plant interfaces.

Acknowledgements

"The chapter is based on the previously published paper Krause, S., et al. (2017), Ecohydrological interfaces as hot spots of ecosystem processes, Water Resour. Res., 53, 6359–6376, doi: 10.1002/2016WR019516 https://doi.org/10.1002/2016WR019516." (for Chapter 1)

References

Abbott, B.W., Baranov, V., Mendoza-Lera, C. et al. (2016). Using multi-tracer inference to move beyond single-catchment ecohydrology. *Earth Science Reviews* https://doi.org/10.1016/j.earscirev.2016.06.014.

Albayrak, I., Nikora, V., Miler, O. et al. (2014). Flow–plant interactions at leaf, stem and shoot scales: drag, turbulence, and biomechanics. *Aquatic Sciences* 76 (2): 269–294.

Alexander, M. and Scow, K.M. (1989). Kinetics of biodegradation in soil. In: *Reactions and Movements of Organic Chemicals in Soils, Soil Science Society of America* (ed. B.L. Sawhney and K. Brown), 243–269. Madison: American Society of Agronomy.

Alexander, R.B., Boyer, E.W., Smith, R.A. et al. (2007). The role of headwater streams in downstream water quality. *Journal of the American Water Resources Association* 43: 41–59.

Argerich, A., Haggerty, R., Martí, E. et al. (2011). Quantification of metabolically active transient storage (MATS) in two reaches with contrasting transient storage and ecosystem respiration. *Journal of Geophysical Research* 116.

Assouline, S., Narkis, K., and Or, D. (2010). Evaporation from partially covered water surfaces. *Water Resources Research* 46 (10).

Atkinson, C.L. and Vaughn, C.C. (2015). Biogeochemical hotspots: temporal and spatial scaling of the impact of freshwater mussels on ecosystem function. *Freshwater Biology* 60 (3): 563–574.

Baham, J. and Sposito, G. (1994). Adsorption of dissolved organic carbon extracted from sewage sludge on montmorillonite and kaolinite in the presence of metal ions. *Journal of Environmental Quality* 23: 147–153.

Baker, M.A., Dahm, C.N., and Valett, H.M. (2000). Anoxia, anaerobic metabolism, and biogeochemistry of the stream-water-groundwater interface. In: *Streams and Groundwaters* (ed. J.B. Jones and P.J. Mulholland), 259–283. San Diego: Academic Press.

Baranov, V., Lewandowski, J., Romeijn, P. et al. (2016). Effects of bioirrigation of non-biting midges (Diptera: chironomidae) on lake sediment respiration. *Scientific Reports* 6: 27329. https://doi.org/10.1038/srep27329.

Battin, T.J., Besemer, K., Bengtsson, M.M. et al. (2016). The ecology and biogeochemistry of stream biofilms. *Nature Reviews. Microbiology* 14 (4): 251–263.

Battin, T.J., Kaplan, L.A., Newbold, J.D. et al. (2003). Contributions of microbial biofilms to ecosystem processes in stream mesocosms. *Nature* 426: 439–442.

Battin, T.J., Sloan, W.T., Kjelleberg, S. et al. (2007). Microbial landscapes: new paths to biofilm research. *Nature Reviews Microbiology* 5 (1): 76–81.

Belnap, J., Hawkes, C.V., and Firestone, M.K. (2003). Boundaries in miniature: two examples from soil. *Bioscience* 53: 739–749.

Bernhardt, E.S., Blaszczak, J.R., Ficken, C.D. et al. (2017). Control points in ecosystems: moving beyond the hot spot hot moment concept. *Ecosystems* https://doi.org/10.1007/s10021-016-0103-y.

Blaen, P., Khamis, K., Lloyd, C.E.M. et al. (2016). Real-time monitoring of nutrients and dissolved organic matter in rivers: adaptive sampling strategies, technological challenges and future directions. *Science of the Total Environment* 569–570: 647–660. https://doi.org/10.1016/j.scitotenv.2016.06.116.

Blöschl, G., Gaál, L., Hall, J. et al. (2015). Increasing river floods: fiction or reality? *Wiley Interdisciplinary Reviews: Water* https://doi.org/10.1002/wat2.1079.

Boano, F., Demaria, A., Revelli, R. et al. (2010). Biogeochemical zonation due to intrameander hyporheic flow. *Water Resources Research* 46: W02511.

Boano, F., Harvey, J.W., Marion, A. et al. (2014). Hyporheic flow and transport processes: mechanisms, models, and biogeochemical implications. *Reviews of Geophysics* 52: 603–679. https://doi.org/10.1002/2012RG000417.

Botter, G., Bertuzzo, E., and Rinaldo, A. (2011). Catchment residence and travel time distributions: the master equation. *Geophysical Research Letters* 38.

Bourg, A.C.M. and Bertin, C. (1994). Seasonal and spatial trends in manganese solubility in an alluvial aquifer. *Environmental Science & Technology* 28: 868–876.

Briggs, M., Lautz, L., McKenzie, J. et al. (2012). Using high resolution distributed temperature sensing to quantify spatial and temporal variability in vertical hyporheic flux. *Water Resources Research* 48: W02527.

Briggs, M.A., Day-Lewis, F.D., Zarnetske, J.P. et al. (2015). A mechanistic explanation for the development of hyporheic redox microzones. *Geophysical Research Letters* 42: https://doi.org/10.1002/2015GL064200.

Briggs, M.A., Lautz, L.K., and Hare, D.K. (2014). Residence time control on hot moments of net nitrate production and uptake in the hyporheic zone. *Hydrological Processes* 28: 3741–3751.

Brodersen, K., Nielsen, D., Ralph, P. et al. (2014). A split flow chamber with artificial sediment to examine the below-ground microenvironment of aquatic macrophytes. *Marine Biology* 161 (12): 2921–2930.

Brunke, M. and Gonser, T. (1997). The ecological significance of exchange processes between rivers and groundwater. *Freshwater Biology* 37 (1): 1–33.

Cadenasso, M.L., Pickett, S.T.A., Weathers, K.C. et al. (2003). A framework for a theory of ecological boundaries. *Bioscience* 53 (8): 750–758.

Caraco, N., Cole, J., Findlay, S. et al. (2006). Vascular plants as engineers of oxygen in aquatic systems. *Bioscience* 56 (3): 219–225.

Cardenas, M.B. (2015). Hyporheic zone hydrologic science: a historical account of its emergence and a prospectus. *Water Resources Research* 51: 3601–3616. https://doi.org/10.1002/2015WR017028.

Cardenas, M.B. and Wilson, J.L. (2006). The influence of ambient groundwater discharge on exchange zones induced by current-bedform interactions. *Journal of Hydrology* 331 (1–2): 103–109.

Cardenas, M.B. and Wilson, J.L. (2007). Dunes, turbulent eddies, and interfacial exchange with permeable sediments. *Water Resources Research* 43: W08412.

Cardenas, M.B., Wilson, J.L., and Zlotnik, V.A. (2004). Impact of heterogeneity bedform configuration, and channel curvature on hyporheic exchange. *Water Resources Research* 40.

Clinton, S.M., Grimm, N.B., and Fisher, S.G. (1996). Response of a hyporheic invertebrate assemblage to drying disturbance in a desert stream. *Journal of the North American Benthological Society* 15 (4): 700–712.

Coco, G., Thrush, S.F., Green, M.O. et al. (2006). Feedbacks between bivalve density, flow, and suspended sediment concentration on patch stable states. *Ecology* 87: 2862–2870.

Collins, S.L., Belnap, J., Grimm, N.B. et al. (2014). A multiscale, hierachical model of pulse dynamics in aridland ecosystems. *Annual Reviews of Ecology, Evolution, and Systematics* 45: 397–419.

Cornelisen, C. and Thomas, F. (2009). Prediction and validation of flow-dependent uptake of ammonium over a seagrass-hardbottom community in Florida Bay. *Marine Ecology - Progress Series* 386: 71–81.

Crispell, J.K. and Endreny, T.A. (2009). Hyporheic exchange flow around constructed in-channel structures and implications for restoration design. *Hydrological Processes* 23 (8): 1158–1168.

Crump, B.C., Amaral-Zettler, L.A., and Kling, G.W. (2012). Microbial diversity in arctic freshwaters is structured by inoculation of microbes from soils. *ISME Journal* 6 (9): 1629–1639.

de Moraes, P.C., Franco, D.C., Pellizari, V.H. et al. (2014). Effect of plankton-derived organic matter on the microbial community of coastal marine sediments. *Journal of Experimental Marine Biology and Ecology* 461: 257–266.

Dijkstra, J.T. and Uittenbogaard, R.E. (2010). Modeling the interaction between flow and highly flexible aquatic vegetation. *Water Resources Research* 46: W12547.

Finnegan, J.J. (2000). Turbulence in plant canopies. *Annual Review of Fluid Mechanics* 32: 519–571.

Fisher, S.G., Grimm, N.B., Marti, E. et al. (1998). Material spiraling in stream corridors: a telescoping ecosystem model. *Ecosystems* 1 (1): 19–34.

Fisher, S.G., Sponseller, R.A., and Heffernan, J.B. (2004). Horizons in stream biogeochemistry: flowpaths to progress. *Ecology* 85: 2369–2379.

Folkard, A.M. (2005). Hydrodynamics of model *Posidonia oceanica* patches in shallow water. *Limnology and Oceanography* 50: 1592–1600.

Frei, S., Knorr, K.-H., Peiffer, S. et al. (2012). Surface micro-topography can cause hot-spots of biogeochemical activity in wetland systems. *Journal of Geophysical Research – Biogeosciences* 117.

Freitas, J.G., Rivett, M.O., Roche, R.S. et al. (2015). Heterogeneous hyporheic zone dechlorination of a TCE groundwater plume discharging to an urban river reach. *Science of the Total Environment* 505: 236–252.

Fromin, N., Pinay, G., Montuelle, B. et al. (2010). Impact of seasonal sediment dessication and rewetting on microbial processes involved in greenhouse gas emissions. *Ecohydrology* 3: 339–348.

Gomez-Velez, J.D., Harvey, J.W., Cardenas, M.B. et al. (2015). Denitrification in the Mississippi River network controlled by flow through river bedforms. *Nature Geoscience* 8: 941–945. https://doi.org/10.1038/ngeo2567.

Gomez-Velez, J.D., Krause, S., and Wilson, J.L. (2014). Effect of low-permeability layers on spatial patterns of hyporheic exchange and groundwater upwelling. *Water Resources Research* 50: 5196–5215.

Gomez-Velez, J.D., Wilson, J.L., and Cardenas, M.B. (2012). Residence time distributions in sinuosity-driven hyporheic zones and their biogeochemical effects. *Water Resources Research* 48: W09533.

González-Pinzón, R., Ward, A.S., Hatch, C.E. et al. (2015). A field comparison of multiple techniques to quantify surface water - groundwater interactions. *Freshwater Science* 34 (1): https://doi.org/10.1086/679738.

Grant, S.B. and Marusic, I. (2011). Crossing turbulent boundaries: interfacial flux in environmental flows. *Environmental Science & Technology* 45 (17): 7107–7113.

Grimm, N.B. (1988a). Role of macroinvertebrates in nitrogen dynamics of a desert stream. *Ecology* 69 (6): 1884–1893.

Grimm, N.B. (1988b). Feeding dynamics, nitrogen budgets, and ecosystem role of a desert stream omnivore, Agosia chrysogaster (Pisces: cyprinidae). *Environmental Biology of Fishes* 21: 143–152.

Gurnell, A. (2014). Plants as river system engineers. *Earth Surface Processes and Landforms* 39 (1): 4–25.

Haggerty, R., Martí, E., Argerich, A. et al. (2009). Resazurin as a “smart” tracer for quantifying metabolically active transient storage in stream ecosystems. *Journal of Geophysical Research* 114.

Haggerty, R., Ribot, M., Singer, G.A. et al. (2014). Ecosystem respiration increases with biofilm growth and bed forms: flume measurements with resazurin. *Journal of Geophysical Research-Biogeosciences* 119: 2220–2230.

Hall, J., Arheimer, B., Borga, M. et al. (2014). Understanding flood regime changes in Europe: a state of the art assessment. *Hydrology and Earth System Sciences* 18: 2735–2772. https://doi.org/10.5194/hess-18-2735-2014.

Hannah, D.M., Sadler, J.P., and Wood, J.P. (2007). Hydroecology and ecohydrology: a potential route forward? *Hydrological Processes* 21: 3385–3390.

Hanzel, J., Myrold, D., Sessitsch, A. et al. (2013). Microbial ecology of biogeochemical interfaces - diversity, structure, and function of microhabitats in soil. *Fems Microbiology Ecology* 86 (1): 1–2.

Harms, T.K. and Grimm, N.B. (2008). Hot spots and hot moments of carbon and nitrogen dynamics in a semi-arid riparian zone. *Journal of Geophysical Research–Biogeosciences* 113.

Harvey, J. and Gooseff, M. (2015). River corridor science: hydrologic exchange and ecological consequences from bedforms to basins. *Water Resources Research* 51: 6893–6922. https://doi.org/10.1002/2015WR017617.

Harvey, J.W., Böhlke, J.K., Voytek, M.A. et al. (2013). Hyporheic zone denitrification: controls on effective reaction depth and contribution to whole-stream mass balance. *Water Resources Research* 49: 6298–6316. https://doi.org/10.1002.

Harvey, J.W., Drummond, J.D., Martin, R.L. et al. (2012). Hydrogeomorphology of the hyporheic zone: stream solute and fine particle interactions with a dynamic streambed. *Journal of Geophysical Research – Biogeosciences* 117: G00N11. https://doi.org/10.1029/2012JG002043.

Harvey, J.W. and Fuller, C.C. (1998). Effect of enhanced manganese oxidation in the hyporheic zone on basin-scale geochemical mass balance. *Water Resources Research* 34 (4): 623–636.

Hedin, L. O., von Fischer, J. C., Ostrom, N. E. et al. (1998). Thermodynamic constraints on nitrogen transformations and other biogeochemical processes at soil–stream interfaces. *Ecology* 79: 684–703. https://doi.org/10.1890/0012-9658(1998)079[0684:TCONAO]2.0.CO;2.

Hester, E.T. and Gooseff, M.N. (2011). Hyporheic restoration in streams and rivers. In: *Stream Restoration in Dynamic Fluvial Systems* (ed. A. Simon, S.J. Bennett, and J.M. Castro), 167–187. Washington, D. C., USA: American Geophysical Union.

Hölker, F., Vanni, M.J., Kuiper, J.J. et al. (2015). Tube-dwelling invertebrates: tiny ecosystem engineers have large effects in lake ecosystems. *Ecological Mongraphs* 85: 333–351.

Huang, I., Rominger, J., and Nepf, H. (2011). The motion of kelp blades and the surface renewal model. *Limnology and Oceanography* 56 (4): 1453–1462.

Jordan, P., Arnscheidt, J., McGrogan, H. et al. (2007). Characterising phosphorus transfers in rural catchments using a continuous bank-side analyser. *Hydrology and Earth System Sciences* 11 (1): 372–381.

Jumars, P.A., Eckaman, J.E., and Koch, E. (2001). Macroscopic animals and plants in benthic flows. In: *The Benthic Boundary Layer* (ed. B.P. Boudreau and B.B. Jorgensen), 320–347. Oxford: Oxford University Press.

Kasahara, T. and Hill, A.R. (2006). Hyporheic exchange flows induced by constructed riffles and steps in lowland streams in southern Ontario. *Canada Hydrological Processes* 20 (20): 4287–4305.

Kasahara, T. and Hill, A.R. (2008). Modeling the effects of lowland stream restoration projects on stream-subsurface water exchange. *Ecological Engineering* 32 (4): 310–319.

Käser, D., Binley, A., Krause, S. et al. (2014a). Prospective modelling of 3-D hyporheic exchange based on high-resolution topography and stream elevation. *Hydrological Processes* 28 (4): 2579–2594.

Kennedy, C.D., Genereux, D.P., Corbett, D.R. et al. (2009). Spatial and temporal dynamics of coupled groundwater and nitrogen fluxes through a streambed in an agricultural watershed. *Water Resources Research* 45.

Kiel, B.A. and Cardenas, M.B. (2014). Lateral hyporheic exchange throughout the Mississippi River network. *Nature Geoscience* 7 (6): 413–417.

Klaus, J., Zehe, E., Elsner, M. et al. (2014). Controls of event-based pesticide leaching in natural soils: a systematic study based on replicated field scale irrigation experiments. *Journal of Hydrology* 512: 528–539.

Kolbe, T., Marçais, J., Thomas, Z. et al. (2016). Coupling 3D groundwater modeling with CFC-based age dating to classify local groundwater circulation in an unconfined crystalline aquifer. *Journal of Hydrology* 543 (A): 31–46. https://doi.org/10.1016/j.jhydrol.2016.05.020.

Krause, S., Boano, F., Cuthbert, M.O. et al. (2014b). Understanding process dynamics at aquifer-surfacewater interfaces: an introduction to the special section on new modeling approaches and novel experimental technologies. *Water Resources Research* 50: 1847–1855. https://doi.org/10.1002/2013WR014755.

Krause, S., Hannah, D.M., Fleckenstein, J.H. et al. (2011b). Inter-disciplinary perspectives on processes in the hyporheic zone. *Ecohydrology Journal* 4 (4): 481–499.

Krause, S., Hannah, D.M., Wood, P.J. et al. (2011a). Hydrology and Ecology interfaces: processes and interactions in wetland, riparian and groundwater-based ecosystems. *Ecohydrology Journal* 4 (4): 476–480.

Krause, S., Klaar, M.J., Hannah, D.M. et al. (2014a). The potential of large woody debris to alter biogeochemical processes and ecosystem services in lowland rivers. *WIREs Water* 1: 263–275.

Krause, S., Lewandowski, J., Dahm, C.N. et al. (2015). Frontiers in real-time ecohydrology – a paradigm shift in understanding complex environmental systems. *Ecohydrology* 8: 529–537.

Krause, S., Tecklenburg, C., Munz, M., and Naden, E. (2013). Streambed nitrogen cycling beyond the hyporheic zone: flow controls on horizontal patterns and depth distribution of nitrate and dissolved oxygen in the up-welling groundwater of a lowland river. *Journal of Geophysical Research (Biogeosciences)* 118 (1): 54–67.

Kreyling, J., Jentsch, A., and Beier, C. (2014). Beyond realism in climate change experiments: gradient approaches identify thresholds and tipping points. *Ecology Letters* 17: 125–e1.

Kumar, K., Dasgupta, C.N., Nayak, B. et al. (2011). Development of suitable photobioreactors for CO_2 sequestration addressing global warming using green algae and cyanobacteria. *Bioresource Technology* 102 (8): 4945–4953.

Larned, S.T., Nikora, V.I., and Biggs, B.J. (2004). Mass-transfer-limited nitrogen and phosphorus uptake by stream periphyton: a conceptual model and experimental evidence. *Limnology and Oceanography* 49: 1992–2000.

Larned, S.T., Packman, A.I., Plew, D.R. et al. (2011). Interactions between the mat-forming alga *Didymosphenia geminata* and its hydrodynamic environment. *Limnology and Oceanography: Fluids and Environments* 1 (1): 4–22.

Lautz, L.K. and Fanelli, R.M. (2008). Seasonal biogeochemical hotspots in the streambed around restoration structures. *Biogeochemistry* 91 (1): 85–104.

Layman, C.A., Allgeier, J.E., Yeager, L.A. et al. (2013). Thresholds of ecosystem response to nutrient enrichment from fish aggregations. *Ecology* 94 (2): 530–536.

Levi, P.S. and Tank, J.L. (2013). Nonnative Pacific salmon alter hot spots of sediment nitrification in Great Lakes tributaries. *Journal of Geophysical Research: Biogeosciences* 118 (2): 436–444.

Lewandowski, J., Laskov, C., and Hupfer, M. (2007). The relationship between *Chironomus plumosus* burrows and the spatial distribution of pore-water phosphate, iron and ammonium in lake sediments. *Freshwater Biology* 52: 331–343.

Li, H. and Jiao, J.J. (2005). One-dimensional airflow in unsaturated zone induced by periodic water table fluctuation. *Water Resources Research* 41.

Malzone, J.M., Lowry, C.S., and Ward, A.S. (2016). Response of the hyporheic zone to transient groundwater fluctuations on the annual and storm event time scales. *Water Resources Research*. https://doi.org/10.1002/2015WR018056.

Markstrom, S.L., Niswonger, R.G., Regan, R.S. et al. (2008). GSFLOW-coupled ground-water and surface-water FLOW model based on the integration of the Precipitation-Runoff Modeling System (PRMS) and the Modular Ground-Water Flow Model (MODFLOW-2005). *U.S. Geological Survey Techniques and Methods* 6-D1.

Martínez-Carreras, N., Wetzel, C.E., Frentress, J. et al. (2015). Hydrological connectivity inferred from diatom transport through the riparian-stream system. *Hydrology and Earth System Sciences* 19: 3133–3151. https://doi.org/10.5194/hess-19-3133-2015.

Maxwell, R.M. and Miller, N.L. (2005). Development of a coupled land surface and groundwater model. *Journal of Hydrometeorology* 6 (3): 233–247.

McClain, M.E., Boyer, E.W., Dent, L. et al. (2003). Biogeochemical hot spots and hot moments at the interface of terrestrial and aquatic ecosystems. *Ecosystems* 6: 301–312.

Morrice, J.A., Dahm, C.N., Valett, H.M. et al. (2000). Terminal electron accepting processes in the alluvial sediments of a headwater stream. *Journal of the North American Benthological Society* 19: 593–608.

Moslemi, J.M., Snider, S.B., MacNeill, K. et al. (2012). Impacts of an invasive snail (Tarebia granifera) on nutrient cycling in tropical streams: the role of riparian deforestation in Trinidad, West Indies (snail invasion and tropical streams). *PLoS ONE* 7 (6).

Naiman, R.J. and Decamps, H. (1997). The ecology of interfaces - riparian zones. *Annual Review of Ecology and Systematics* 28: 621–658.

Naiman, R.J., Decamps, H., Pastor, J. et al. (1988). The potential importance of boundaries of fluvial ecosystems. *Journal of the North American Benthological Society* 7: 289–306.

Naiman, R.J., Pinay, G., Johnston, C.A. et al. (1994). Beaver-induced influences on the long term characteristics of boreal forest drainage networks. *Ecology* 75 (4): 905–921.

Neal, C., Reynolds, B., Rowland, P. et al. (2012). High-frequency water quality time series in precipitation and streamflow: from fragmentary signals to scientific challenge. *Science of the Total Environment* 434: 3–12.

Nepf, H.M. (2012). Flow and transport in regions with aquatic vegetation. *Annual Review of Fluid Mechanics* 44: 123–142.

Nikora, V. (2010). Hydrodynamics of aquatic ecosystems: an interface between ecology, biomechanics and environmental fluid mechanics. *River Research and Applications* 26 (4): 367–384.

Nishihara, G.N. and Ackerman, J.D. (2009). Diffusive boundary layers do not limit the photosynthesis of the aquatic macrophyte, *Vallisneria americana*, at moderate flows and saturating light levels. *Limnology and Oceanography* 54: 1874–1882.

Nishizaki, M.T. and Carrington, E. (2014). The effect of water temperature and flow on respiration in barnacles: patterns of mass transfer versus kinetic limitation. *Journal of Experimental Biology* 217: 2101–2109.

Palm, J., van Schaik, N.L.M.B., and Schröder, B. (2013). Modelling distribution patterns of anecic, epigeic and endogeic earthworms at catchment-scale in agro-ecosystems. *Pedobiologia* 56 (1): 23–31.

Patrick, C.J. (2014). Macroinvertebrate communities of ecotones between the boundaries of streams, wetlands, and lakes. *Fundamental and Applied Limnology* 185 (3–4): 223–233.

Peipoch, M., Gacia, E., Bastias, E. et al. (2016). Small-scale heterogeneity of microbial N uptake in streams and its implications at the ecosystem level. *Ecology* 97: 1329–1344.

Perelo, L.W. (2010). Review: in situ and bioremediation of organic pollutants in aquatic sediments. *Journal of Hazardous Materials* 177 (1): 81–89.

Pfister, L., McDonnell, J.J., Hissler, C. et al. (2010). Ground-based thermal imagery as a simple, practical tool for mapping saturated area connectivity and dynamics. *Hydrological Processes* 24: 3123–3132.

Pfister, L., McDonnell, J.J., Wrede, S. et al. (2009). The rivers are alive: on the potential for diatoms as a tracer of water source and hydrological connectivity. *Hydrological Processes* 23: 2841–2845.

Pinay, G., Pfeiffer, S., de Dreuzy, J.-R. et al. (2015). Upscaling nitrogen removal capacity from hot spot to the landscape. *Ecosystems* 18 (6): 1101–1120.

Poungparn, S., Komiyama, A., Sangteian, T. et al. (2012). High primary productivity under submerged soil raises the net ecosystem productivity of a secondary mangrove forest in eastern Thailand. *Journal of Tropical Ecology* 28 (3): 303–306.

Puth, L. and Wilson, K.A. (2001). Boundaries and corridors as a continuum of ecological flow control: lessons from rivers and streams. *Conservation Biology* 15: 21–30.

Reidenbach, M.A., Limm, M., Hondzo, M. et al. (2010). Effects of bed roughness on boundary layer mixing and mass flux across the sediment-water interface. *Water Resources Research* 46 (7).

Reidenbach, M.A., Monismith, S.G., Koseff, J.R. et al. (2006). Boundary layer turbulence and flow structure over a fringing coral reef. *Limnology and Oceanography* 51: 1956–1968.

Ren, J. and Packman, A.I. (2004). Stream-subsurface exchange of zinc in the presence of silica and kaolinite colloids. *Environmental Science & Technology* 38 (24): 6571–6581.

Roskosch, A., Hette, N., Hupfer, M. et al. (2012). Alteration of *Chironomus plumosus* ventilation activity and bioirrigation-mediated benthic fluxes by changes in temperature, oxygen concentration, and seasonal variations. *Freshwater Science* 31: 269–281.

Sanford, L.P. and Crawford, S.M. (2000). Mass transfer versus kinetic control of uptake across solid- water boundaries. *Limnology & Oceanography* 45 (5): 1180–1186.

Sawyer, A.H., Cardenas, M.B., and Buttles, J. (2011). Hyporheic exchange due to channel-spanning logs. *Water Resources Research* 47: W08502.

Schade, J.D., Fisher, S.G., Grimm, N.B. et al. (2001). The influence of a riparian shrub on nitrogen cycling in a Sonoran Desert stream. *Ecology* 82: 3363–3376.

Schelker, J., Grabs, T., Bishop, K. et al. (2013). Drivers of increased organic carbon concentrations in stream water following forest disturbance: separating effects of changes in flow pathways and soil warming. *Journal of Geophysical Research: Biogeosciences* 118 (4).

Selker, J.S., Thevenaz, L., Huwald, H. et al. (2006). Distributed fiber-optic temperature sensing for hydrologic systems. *Water Resources Research* 42 (12): W12202.

Shahraeeni, E., Lehmann, P., and Or, D. (2012). Coupling of evaporative fluxes from drying porous surfaces with air boundary layer characteristics of evaporation from discrete pores. *Water Resources Research* 48 (9).

Singer, G., Besemer, K., Schmitt-Kopplin, P. et al. (2010). Physical heterogeneity increases biofilm resource use and its molecular diversity in stream mesocosms. *Plos One* 5: e9988.

Sizmur, T., Canário, J., Edmonds, S. et al. (2013). The polychaete worm Nereis diversicolor increases mercury lability and methylation in intertidal mudflats. *Environmental Toxicology and Chemistry* 32 (8): 1888–1895.

Soulsby, C., Neal, C., Laudon, H. et al. (2008). Catchment data for process conceptualization: simply not enough? *Hydrological Processes* 22: 2057–2061.

Statzner, B., Lamouroux, N., Nikora, V. et al. (2006). The debate about drag and reconfiguration of freshwater macrophytes: comparing results obtained by three recently discussed approaches. *Freshwater Biology* 51: 2173–2183.

Strayer, D.L., Power, M.E., Fagan, W.F. et al. (2003). A classification of ecological boundaries. *Bioscience* 53 (8): 723–729.

Stubbington, R. (2012). The hyporheic zone as an invertebrate refuge: a review of variability in space, time, taxa and behaviour. *Marine and Freshwater Research* 63 (4): 293–311.

Thibodeaux, L.J. and Boyle, J.D. (1987). Bedform-generated convective-transport in bottom sediment. *Nature* 325: 341–343.

Tonina, D. and Buffington, J.M. (2007). Hyporheic exchange in gravel bed rivers with pool-riffle morphology: laboratory experiments and three-dimensional modeling. *Water Resources Research* 43: W01421.

Trauth, N., Schmidt, C., Vieweg, M. et al. (2015). Hydraulic controls of in-stream gravel bar hyporheic exchange and reactions. *Water Resources Research* 51.

Treese, S., Meixner, T., and Hogan, J.F. (2009). Clogging of an effluent dominated semiarid river: a conceptual model of stream-aquifer interactions. *JAWRA Journal of the American Water Resources Association* 45 (4): 1047–1062.

van Schaik, L., Palm, J., Klaus, J. et al. (2014). Linking spatial earthworm distribution to macropore numbers and hydrological effectiveness. *Ecohydrology* 7 (2).

Vandenkoornhuyse, P., Quaiser, A., Duhamel, M. et al. (2015). The importance of the microbiome of the plant holobiont. *New Phytologist* 206: 1196–1206.

Vanni, M.J. (2002). Nutrient cycling by animals in freshwater ecosystems. *Annual Review of Ecology and Systematics* 33.

Ward, A.S., Gooseff, M.N., and Johnson, P.A. (2011). How can subsurface modifications to hydraulic conductivity be designed as stream restoration structures? Analysis of Vaux's conceptual models to enhance hyporheic exchange. *Water Resources Research* 47 (8).

Warren, D., Judd, K., Bade, D. et al. (2013). Effects of wood removal on stream habitat and nitrate uptake in two northeastern US headwater streams. *Hydrobiologia* 717 (1): 119–131.

Xing, Y., Xie, P., Yang, H. et al. (2006). The change of gaseous carbon fluxes following the switch of dominant producers from macrophytes to algae in a shallow subtropical lake of China. *Atmospheric Environment* 40 (40): 8034–8043.

Yarrow, M.M. and Marin, V.H. (2007). Toward conceptual cohesiveness: a historical analysis of the theory and utility of ecological boundaries and transition zones. *Ecosystems* 10 (3): 462–476.

Yuan, L.-R., Xin, P., Kong, J. et al. (2011). A coupled model for simulating surface water and groundwater interactions in coastal wetlands. *Hydrological Processes* 25: 3533–3546.

Zarnetske, J.P., Haggerty, R., and Wondzell, S.M. (2015). Coupling multiscale observations to evaluate hyporheic nitrate removal at the reach scale. *Freshwater Science* 34 (1): 172–186.

Zarnetske, J.P., Haggerty, R., Wondzell, S.M. et al. (2011a). Labile dissolved organic carbon supply limits hyporheic denitrification. *Journal of Geophysical Research* 116.

Zarnetske, J.P., Haggerty, R., Wondzell, S.M. et al. (2011b). Dynamics of nitrate production and removal as a function of residence time in the hyporheic zone. *Journal of Geophysical Research* 116.

Zehe, E., Ehret, U., Pfister, L. et al. (2014). HESS opinions: functional units: a novel framework to explore the link between spatial organization and hydrological functioning of intermediate scale catchments. *Hydrology and Earth System Sciences* 11: 3249–3313.

Zhang, Q., Katul, G.G., Oren, R. et al. (2015). The hysteresis response of soil CO_2 concentration and soil respiration to soil temperature. *Journal of Geophysical Research-Biogeosciences* 120: 1605–1618. https://doi.org/10.1002/2015JG003047.

2

Biological Activity as a Trigger of Enhanced Ecohydrological Interface Activity

Julian Klaus[1], *Viktor Baranov*[2], *Jörg Lewandowski*[3], *Anne Zangerlé*[4], *and Loes van Schaik*[5]

[1] *Department of Geography, University of Bonn, Bonn, Germany*
[2] *LMU Munich Biocenter, Großhaderner Str. 2, 82152 Planegg-Martinsried, Germany*
[3] *Lebniz Institute of Freshwater Ecology and Inland Fisheries, Berlin, Germany*
[4] *Ministry of Agriculture, Viticulture and Rural Development, Luxembourg, Luxembourg*
[5] *Department of Environmental Sciences, Wageningen University and Research, The Netherlands*

2.1 Introduction

In this chapter we will provide some examples of the influence of biological activity on ecohydrological processes at interfaces, for terrestrial and aquatic ecosystems. We will mainly focus on bioturbation-related processes; however, we will also briefly discuss other cases relevant for ecohydrology.

Ecosystem engineers are organisms, which change the availability of resources for other species by altering the biotic and abiotic environment. In doing so, they create and maintain habitats (Jones et al. 1996; Wright and Jones 2006). It is hardly possible to find organisms that do not influence their environment. However, species are only considered as ecosystem engineers if their impacts on habitats are disproportionally large in comparison with their biomass (Meysman et al. 2006). According to Meysman et al. (2006) "in modern ecological theory, bioturbation is now recognized as an archetypal example of 'ecosystem engineering', altering interface processes, modifying geochemical gradients, redistributing food resources, viruses, bacteria, resting stages and eggs".

Bioturbation as a field of study has been attracting scientific minds for over a century now. No one less than Charles Darwin studied bioturbation by earthworms (Darwin 1881; Meysman et al. 2006). According to a recent definition of Kristensen et al. (2012) bioturbation comprises "all transport processes carried out by animals that directly or indirectly affect sediment matrices. These processes include both particle reworking and burrow ventilation".

While the activity of larger animals is commonly recognizable to some extent (Section 2.2), also smaller animals strongly influence ecohydrological interface activity in terrestrial and aquatic environments (Sections 2.3 and 2.4). This is especially true when their abundance and bioturbation activity is high. In the following we give examples for terrestrial and aquatic environments.

Ecohydrological Interfaces, First Edition. Edited by Stefan Krause, David M. Hannah, and Nancy B. Grimm.

2.2 Larger Species

Biological activity of various species can have a strong effect on hydrology. This can be a direct effect, such as the digging activity of pocket gophers (*Thomomys bottae*) (Hakonson 1999) or the creation of alligator holes in marsh land, or an indirect effect. An example for the latter can be found in the work of Beschta and Ripple (2008). They showed how a truncated trophic cascade acted on the stream–riverbank interfaces. Beschta and Ripple (2008) could link the extirpation of wolves to the change of the riparian plant community by increasing elk populations; eventually this led to increased riverbank erosion and channel widening.

Alligators (*Alligator mississippiensis*) are another example for enhanced interface activity. Alligator holes in different landscapes, such as marshes in the Everglades, Florida, commonly lead to changes in plant community composition and structure (Palmer and Mazzotti 2004), and provide nest sites for other reptiles, such as turtles (Enge et al. 2000). Brandt et al. (2010) showed that the abundance of alligator holes in the Everglades is twice as high in drier areas than in wetter areas. In addition, such alligator holes exhibit different hydrological behaviour than the surrounding landscape, providing aquatic refugia under a range of hydrologic conditions (Liu et al. 2013).

The activity of beavers is a clear and visible example of how animal activity impacts landscape hydrology and creates new (and modifies) existing interfaces (water–soil, water–atmosphere, and stream–groundwater). Westbrook et al. (2006) showed that beavers by dam construction strongly influence hydrological conditions during low flow and flood peaks by elevating the groundwater level, extending and prolonging inundations during floods, and forcing water to flow around the dams as surface flow or groundwater seepage. Beaver activity also influences biogeochemical cycles. Painter et al. (2015) found that methyl mercury (MeHg) is significantly elevated in streamflow and in predatory invertebrates below beaver dams. Correll et al. (2000) found that the establishment of a beaver pond in a small headwater catchment reduced annual discharge of water, total-N, total-P, dissolved silicate, total organic carbon, and total suspended sediment. Other work found that the recovery of beavers led to increasing methane emission from the created wetlands compared to the situation in 1900, when beavers were nearly extinct (Whitfield et al. 2015).

2.3 Earthworms

Earthworms, through their burrowing activities, strongly influence soil structure by producing soil aggregates and more or less continuous macropores. They strongly enhance infiltration (Emmerling et al. 2015; Shipitalo and Butt 1999; Willoughby and Kladivko 2002), and water (Lee and Foster 1991; Klaus et al. 2013; van Schaik et al. 2014) and solute transport (Edwards et al. 1990; Kladivko et al. 1999; Klaus et al. 2014) in the soil. Shipitalo and Butt (1999) measured an infiltration rate up to 1000 ml water per minute, which is orders of magnitude higher than the soil matrix. Furthermore, earthworm burrows greatly contribute to soil aeration (Kretzschmar 1978).

The bioturbating activity of earthworms was already discovered and studied by Darwin as mentioned previously (Darwin 1881). Earthworms have been recognized as a very good example of ecosystem engineers (Jouquet et al. 2006; Lavelle et al. 1997). In temperate regions earthworm species can be grouped into three main ecological types, anecic, endogeic, and epigeic earthworms, according to differences in their feeding and burrowing behaviour (Bouché 1977). Endogeic earthworms have a more or less omnidirectional burrowing behaviour in the top soil, and mainly influence the diffuse transport of water and solutes there. On the contrary, deep-burrowing anecic earthworm species produce permanent continuous vertical macropores that can lead to rapid transport of water and solutes to deeper soil depths (Ernst et al. 2009), down to the bedrock (Paton et al. 1995). In a temperate climate the upper 15 cm of soil can be turned over by earthworm activity every 10 to 20 years (Edwards and Bohlen 1996). Thus, earthworms alter the hydrological regime of soils, but in the same time their occurrence and burrowing activity depends on the moisture regime of the soils (Edwards and Bohlen 1996; Evans and Guild 1947; Lavelle 1988), resulting in some kind of feedback loop (Jouquet et al. 2006).

Understanding the effects of earthworms on water and solute distribution and transport in soils requires taking into account the complexity of biological processes in the drilosphere. The drilosphere is defined by the burrows that earthworms create through digging. Digging results in a soil compaction of 4–8 mm soil layer surrounding the burrows (Brown et al. 2000; Lee 1985). Soil ingested by earthworms is excreted in the form of casts and coatings along the burrow walls. These coatings form an interface that is enriched in organic matter, microbes, and often in clay minerals (Brown et al. 2000). Earthworms ingest and simultaneously enrich drilosphere soil by the addition of saliva and intestinal mucus each time the earthworm passes through a burrow (Barois and Lavelle 1986; Brown et al. 2000). As a consequence, soil microbial communities benefit from earthworm activity and are highly concentrated at the surfaces of burrow walls and within the surrounding concentric soil layer (Brown et al. 2000; Bundt et al. 2001; Lee 1985; Vinther et al. 1999). The compaction of the drilosphere (Rogasik et al. 2014), the increased clay content and microbial biomass, as well as the organic matter composition of the macropore coatings, which can result in water repellency (Cammeraat 1992; Leue et al. 2010), influences the hydrological interactions between macropores and soil matrix (Leue et al. 2010; van Schaik et al. 2014). Generally, the macropore coating has a lower hydraulic conductivity than the bulk soil, reducing the flux between macropore and soil matrix (Gerke and Köhne 2002). The increased level of organic matter and clay particles in the macropore coating has a strong impact on solute adsorption by supplying more sorption sites than bulk soil (Binet et al. 2006; Lavelle et al. 2007; Stehouwer et al. 1994). Furthermore, the role of the drilosphere as "hot spots" of microbial activity (Bundt et al. 2001) can increase degradation of organic contaminants such as pesticides (Bolduan and Zehe 2006; Pivetz and Steenhuis 1995). Bolduan and Zehe (2006) showed the remarkable effect of this "hot spot" activity on degradation of the pesticide isoproturon. Degradation times at the burrow endings (>1 m depth) were in the same range as in the topsoil, while there was no measurable degradation (within 30 days) in a few centimetres distance from the burrows.

Earthworms have selective feeding habits and thus ingest preferred organic items (Bouché and Kretzschmar 1974; Curry and Schmidt 2007). Therfore, organic matter in the casts and burrows they produce varies both quantitatively and qualitatively among species and ecological types (Zangerlé et al. 2016). Different crop residues as a food source for earthworms on agricultural fields can change the organic matter composition of the coatings, which in turn also influences the amount of leached pesticides by different levels of adsorption (Farenhorst et al. 2001). As a consequence, different earthworm species and ecological groups and the available food source influence the water flux between macropores and soil matrix. Badawi et al. (2013) showed that hydraulically active macropores had a high density of pesticide degraders (e.g. microorganism from the *Streptomyces* and *Micromonospora* genus (Fuentes et al. 2010)) and pesticide mineralization activity in the coatings. However, impacts of the coatings on the infiltration from macropores to the soil matrix are relatively short effects since with age the porosity of the burrow walls increases and water repellency decreases (Leue et al. 2010).

While earthworm burrows are potential preferential flow paths, the occurrence of preferential flow in an area depends on their spatiotemporal distribution and on the precipitation regime (McGrath et al. 2008). The spatial distribution and temporal dynamics of the biopore networks are determined by the rate of burrowing and disturbance. The burrowing rate depends on the species (Felten and Emmerling 2009; Lee and Foster 1991) and seasonality, i.e. soil moisture content and temperature (Bouché 1972; Evans and Guild 1947). The main earthworm activity is in spring and autumn when soil moisture contents are optimal (Gates 1961). In summer- and wintertime earthworms often plug the surface burrow openings with leaves or soil to avoid desiccation or frost as they are very sensitive to drought (Garnsey 1994) and frost (Holmstrup 2003). Earthworms occur in almost all soils and climates of the world. Their densities, however, depend on many environmental factors such as soil moisture and temperature regime, texture, pH, and land use (Edwards and Bohlen 1996). Therefore, earthworm densities are highly variable in space as well as time. Summarized in a review by Spurgeon et al. (2013), the average earthworm abundances ± standard deviation were 56 ± 71 ind. m^{-2} in arable land versus 341 ± 402 ind. m^{-2} in grassland and 315 ± 431 ind. m^{-2} in woodland. The earthworm and macropore densities are related to each other as shown by Pérès et al. (2010) and van Schaik et al. (2014).

The long-term persistence of burrow networks, and thus the ecohydrological interface present in the soil, depends on different external disturbances: (i) natural processes such as physical disintegration, earthworm activity (Blanchart et al. 1997), and plant root growth (Milleret et al. 2009), or (ii) by soil management. The rate of natural decay of earthworm macropores is species-dependent (Capowiez et al. 2014). Generally, the macropores built by endogeic or epigeic earthworms are discontinuous voids, which are short-lived as the earthworms only pass through once. The macropores built by anecic earthworms, however, are comparatively stable as the anecic earthworms have compacted macropore walls due to the permanent use of the same burrow (Rogasik et al. 2014); these can even persist over many decades in deeper soil layers (Hagedorn and Bundt 2002). An intensive colonization of macropores by roots has often been observed in ageing earthworm burrows since plant roots take advantage of the nutrients available in the highly enriched coatings on the burrow walls (Decäens et al. 2001). In highly compacted soil, burrows facilitate the penetration of roots in deeper soil layers (Brown 2000; Darwin 1881), which in turn accelerates the physical degradation of burrows. In agricultural fields and in managed forests, the burrow networks in the topsoil are regularly destroyed by soil management.

2.4 Chironomus Plumosus Larvae

Macrozoobenthos exert manifold effects on biogeochemical processes and food web dynamics in shallow lakes, especially at the lake–sediment interface (Hölker et al. 2015). As the most diverse and abundant freshwater ecosystem engineers, larvae of non-biting midges (Diptera, Chironomidae) are a great example of bioturbating animals at ecohydrological interfaces (Ferrington 2008). For example, the well-known highly abundant *Chironomus plumosus* larvae pump water through their U-shaped burrows (diameter 1.5–2 mm, reaching up to 20 cm deep in the sediment (Roskosch et al. 2012)) to acquire oxygen for respiration and plankton food (Walshe 1947). The pumping periods are regularly interrupted by non-pumping periods, during which larvae spin conical mucus nets in their burrows (Roskosch et al. 2011). These nets and the entrapped particles are eaten by the larvae.

The filter feeding activity is the reason for the high ventilation rates of *C. plumosus* (60 ml ind.$^{-1}$ h^{-1}, flow velocity 15 mm s^{-1} (Roskosch et al. 2011)), which by far exceed respiratory requirements (Roskosch et al. 2011, 2010). Thus, ventilation rates are equal to filtration rates. *C. plumosus* typically occur at densities of 1000 ind. m^{-2} at the sediment–water interface of lakes, though values of up to 100 000 ind. m^{-2} have been reported (McLachlan 1977). *C. plumosus* is a very efficient filter feeder that reduces particle densities in the burrow outlet fluid (Hölker et al. 2015; Morad et al. 2010).

In shallow polymictic lakes, such as, for example, Lake Müggelsee in Berlin (mean water depth 5 m, mean summer densities of 700 ind. m^{-2}), average ventilation rates of the individual larvae (60 ml ind.$^{-1}$ h^{-1}, see above) result in ventilation rates of the entire population of 1 m^3 m^{-2} d^{-1}, which is equivalent to pumping the whole water column of the lake every 5 days through the chironomid tubes (Roskosch et al. 2010).

This is in the same order of magnitude as the filtration rate of daphnids in the same lake (0.3–1.0 m^3 m^{-2} d^{-1} (Hölker et al. 2015)). Thus, an organism present at the sediment–water interface, i.e. *C. plumosus*, may play an important, yet poorly understood role for phytoplankton succession in one of the adjacent compartments, the lake water body, and compete for food resources with filter feeding pelagic zooplankton, i.e. daphnids (Figure 2.1). Until recently, daphnids have been typically considered the main herbivore in lake food webs (e.g. Sterner 1989) but Hölker et al. (2015) could disprove this thesis.

Organisms such as *C. plumosus* increase the area of the biogeochemically highly relevant oxic–anoxic interface by building and ventilating their burrows. Ventilation and bioirrigation affect the exchange of dissolved compound across the sediment–water interface (Lewandowski et al. 2007; Stief et al. 2010), the spatio-temporal dynamics of oxygen, and redox conditions in the sediment surrounding the burrows (Polerecky et al. 2006; Roskosch et al. 2011). Redox oscillation can severely increase the metabolic rates of tube-associated sediment bacteria and stimulate organic matter degradation (Aller 1994). This releases phosphate and ammonium bound in organic detritus. Furthermore, fluctuating redox conditions allow for specific microbial processes, e.g. polyphosphate storage (Hupfer et al. 2007). Bacterial communities in chironomid tubes are well adapted to fluctuating redox conditions and nutrient supply. Periodic changes in basic environmental parameters due to ventilation affect the composition and activity of the bacterial community, e.g. in walls of crustaceans' burrows (Bertics and Ziebis 2009).

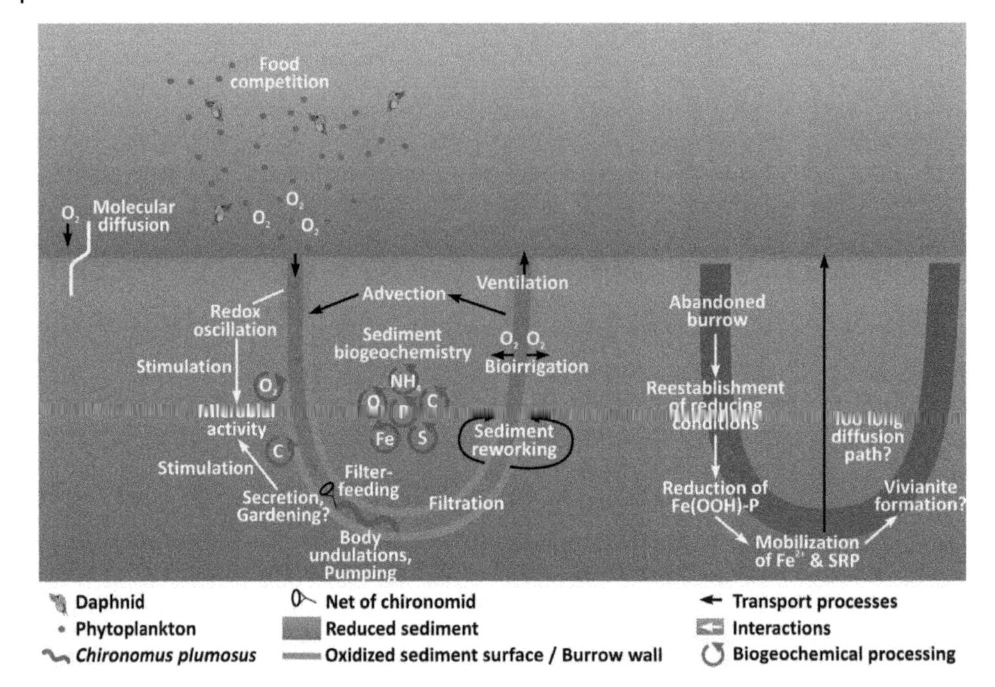

Figure 2.1 Mechanisms by which tube-dwelling Chironomus plumosus larvae may control water quality and trophic status. (Figure taken from Hölker et al., 2015 / John Wiley & Sons.)

An altered redox milieu impacts phosphorous biogeochemistry in the sediment and – depending on the complex interactions of elemental cycles known from other redoxclines – increases or decreases nutrient fluxes across the sediment–water interface due to ventilation and bioirrigation (e.g. Andersson et al. 1988; Lewandowski et al. 2007). Contradictory results for phosphate fluxes are caused by different sediment compositions such as the iron content (Lewandowski et al. 2007), which controls phosphate release processes.

It was proven that chironomids (and other benthic animal's activity) might cause more than just a local effect and influences the biogeochemistry of whole lakes or other large water bodies (Hölker et al. 2015). There is also emerging evidence that freshwater macrozoobenthos bioturbation might promote enhanced greenhouse gas (mainly methane) emissions from lakes (Kajan and Frenzel 1999).

2.5 Concluding Remarks

Within this chapter it was shown how biological organisms influence physical and chemical processes at ecohydrological interfaces. This chapter focused on the effects of macrofauna, while there is further clear evidence of the effect of microorganisms or plants. The effect of animal activity was linked to direct effects by bioturbation or indirect effects over a trophic cascade. These activities impact a range of environmental processes: the nutrient and solute loads and concentrations in streamflow, creation of habitat or refugia for fauna and flora, changes in gas emission from water bodies, enhanced soil aeration, and changes in hydrological flowpaths.

Today, the role that life plays in the formation of landscape is recognized (Dietrich and Perron 2006), but the detailed importance compared to physical processes remains poorly understood in both landscapes and seascapes (Meysman et al. 2006). Further, Meysman et al. (2006) stated that "from an evolutionary perspective, recent investigations provide evidence that bioturbation had a key role in the evolution of metazoan life at the end of the Precambrian Era".

References

Aller, R.C. (1994). Bioturbation and remineralization of sedimentary organic matter: effects of redox oscillation. *Chemical Geology* 114: 331–345.

Andersson, G., Graneli, W., and Stenson, J. (1988). The influence of animals on phosphorus cycling in lake ecosystems. *Hydrobiologia* 170: 267–284.

Badawi, N., Johnsen, A.R., Brandt, K.K. et al. (2013). Hydraulically active biopores stimulate pesticide mineralization in agricultural subsoil. *Soil Biology & Biochemistry* 57: 533–541.

Barois, I. and Lavelle, P. (1986). Changes in respiration rate and some physicochemical properties of a tropical soil during transit through pontoscolex-corethrurus (glossoscolecidae, oligochaeta). *Soil Biology & Biochemistry* 18: 539–541.

Bertics, V.J. and Ziebis, W. (2009). Biodiversity of benthic microbial communities in bioturbated coastal sediments is controlled by geochemical microniches. *ISME Journal* 3: 1269–1285.

Beschta, R.L. and Ripple, W.J. (2008). Wolves, trophic cascades, and rivers in the Olympic National Park, USA. *Ecohydrology* 1 (2): 118–130.

Binet, F., Kersanté, A., Munier-Lamy, C. et al. (2006). Lumbricid macrofauna alter atrazine mineralization and sorption in a silt loam soil. *Soil Biology and Biochemistry* 38 (6): 1255–1263.

Blanchart, E., Lavelle, P., Braudeau, E. et al. (1997). Regulation of soil structure by geophagous earthworm activities in humid savannas of Cote d'Ivoire. *Soil Biology and Biochemistry* 29 (3): 431–439.

Bolduan, R. and Zehe, E. (2006). Degradation of isoproturon in earthworm macropores and subsoil matrix-a field study. *Journal of Plant Nutrition and Soil Science* 169 (1): 87–94.

Bouché, M.B. (1972). Lombriciens de france. Écologie et Systématique. *Ann. Zool. Ecol. Anim.* 72 (HS): 671.

Bouché, M.B. (1977). Strategies lombriciennes. *Ecological Bulletins* 1: 122–132.

Bouche, M.B. and Kretzschmar, A. (1974). Functions of earthworms. 2. Methodologic studies of analysis of ingested soil (study of population of rcp-165-pbi station). *Revue D Ecologie Et De Biologie Du Sol.* 72: 127–139.

Brandt, L.A., Campbell, M.R., and Mazzotti, F.J. (2010). Spatial distribution of alligator holes in the central Everglades. *Southeastern Naturalist* 9 (3): 487–496.

Brown, G.G., Barois, I., and Lavelle, P. (2000). Regulation of soil organic matter dynamics and microbial activityin the drilosphere and the role of interactionswith other edaphic functional domains. *European Journal of Soil Biology* 36 (3): 177–198.

Bundt, M., Widmer, F., Pesaro, M. et al. (2001). Preferential flow paths: biological 'hot spots' in soils. *Soil Biology and Biochemistry* 33 (6): 729–738.

Cammeraat, L.H. (1992). *Hydro-geomorfologic processes in a small forested sub-catchment: preferred flow-paths of water.* Doctoral Thesis, University of Amsterdam, Faculty of Environmental Sciences, Amsterdam.

Capowiez, Y., Bottinelli, N., and Jouquet, P. (2014). Quantitative estimates of burrow construction and destruction, by anecic and endogeic earthworms in repacked soil cores. *Applied Soil Ecology* 74: 46–50.

Correll, D.L., Jordan, T.E., and Weller, D.E. (2000). Beaver pond biogeochemical effects in the Maryland Coastal Plain. *Biogeochemistry* 49 (3): 217–239.

Curry, J.P. and Schmidt, O. (2007). The feeding ecology of earthworms – a review. *Pedobiologia* 50: 463–477.

Darwin, C. (1881). *The Formation of Vegetable Mould through the Action of Worms, with Observations on Their Habits*. London: John Murray.

Decaens, T. and Rossi, J.P. (2001). Spatio-temporal structure of earthworm community and soil heterogeneity in a tropical pasture. *Ecography* 24: 671–682.

Dietrich, W.E. and Perron, J.T. (2006). The search for a topographic signature of life. *Nature* 439 (7075): 411–418.

Edwards, C.A. and Bohlen, P.J. (1996). *Biology and Ecology of Earthworms*, 3e. Springer Science & Business Media.

Edwards, W.M., Shipitalo, M.J., Owens, L.B., and Norton, L.D. (1990). Effect of lumbricus terrestris l. Burrows on hydrology of continuous no-till corn fields. *Geoderma* 46: 73–84.

Emmerling, C., Rassier, K.M., and Schneider, R. (2015). A simple and effective method for linking field investigations of earthworms and water infiltration rate into soil at pedon-scale. *Journal of Plant Nutrition and Soil Science* 178: 841–847.

Enge, K.M., Percival, H.F., Rice, K.G. et al. (2000). Summer nesting of turtles in alligator nests in Florida. *Journal of Herpetology* 34: 497–503.

Ernst, G., Felten, D., Vohland, M., and Emmerling, C. (2009). Impact of ecologically different earthworm species on soil water characteristics. *European Journal of Soil Biology* 45: 207–213.

Evans, A.C. and Guild, W.J.M. (1947). Studies on the relationships between earthworms and soil fertility. 1. Biological studies in the field. *Annals of Applied Biology* 34: 307–330.

Farenhorst, A., Topp, E., Bowman, B.T. et al. (2001). Sorption of atrazine and metolachlor by burrowlinings developed in soils with different crop residues at the surface. *Journal of Environmental Science and Health Part B* 36: 389–396.

Felten, D. and Emmerling, C. (2009). Earthworm burrowing behaviour in 2d terraria with single- and multi-species assemblages. *Biology and Fertility of Soils* 45: 789–797.

Ferrington, L.C., Jr. (2008). Global diversity of non-biting midges (Chironomidae; Insecta-Diptera) in freshwater. *Hydrobiologia* 595: 447–455.

Fuentes, M.S., Benimeli, C.S., Cuozzo, S.A., and Amoroso, M.J. (2010). Isolation of pesticide-degrading actinomycetes from a contaminated site: bacterial growth, removal and dechlorination of organochlorine pesticides. *International Biodeterioration & Biodegradation* 64 (6): 434–441.

Garnsey, R. (1994). Seasonal activity and estivation of lumbricid earthworms in the midlands of tasmania. *Soil Research* 32: 1355–1367.

Gates, G.E. (1961). Ecology of some earthworms with special reference to seasonal activity. *American Midland Naturalist* 66: 61–86.

Gerke, H.H. and Köhne, J.M. (2002). Estimating hydraulic properties of soil aggregate skins from sorptivity and water retention. *Soil Science Society America Journal* 66: 26–36.

Hagedorn, F. and Bundt, M. (2002). The age of preferential flow paths. *Geoderma* 108: 119–132.

Hakonson, T. E. (1999). The effects of pocket gopher burrowing on water balance and erosion from landfill covers. *J. Environ. Quality* 28 (2): 659–665.

Hölker, F., Vanni, M.J., Kuiper, J.J. et al. (2015). Tube-dwelling invertebrates: tiny ecosystem engineers have large effects in lake ecosystems. *Ecological Monographs* 85: 333–351.

Holmstrup, M. (2003). Overwintering adaptations in earthworms. *Pedobiologia* 47: 504–510.

Hupfer, M., Gloess, S., and Grossart, H.P. (2007). Polyphosphate-accumulating microorganisms in aquatic sediments. *Aquatic Microbial Ecology* 47: 299–311.

Jones, C.G., Lawton, J.H., and Shachak, M. (1996). Organisms as ecosystem engineers. In: *Ecosystem Management* (ed. F.B. Samson and F.L. Knopf), 130–147. New York: Springer.

Jouquet, P., Dauber, J., Lagerlöf, J. et al. (2006). Soil invertebrates as ecosystem engineers: intended and accidental effects on soil and feedback loops. *Applied Soil Ecology* 32: 153–164.

Kajan, R. and Frenzel, P. (1999). The effect of chironomid larvae on production, oxidation and fluxes of methane in a flooded rice soil. *FEMS Microbiology Ecology* 28: 121–129.

Kladivko, E.J., Grochulska, J., Turco, R.F. et al. (1999). Pesticide and nitrate transport into subsurface tile drains of different spacings. *Journal of Environmental Quality* 28: 997–1004.

Klaus, J., Zehe, E., Elsner, M. et al. (2013). Macropore flow of old water revisited: experimental insights from a tile-drained hillslope. *Hydrology and Earth System Sciences* 17 (1): 103–118.

Klaus, J., Zehe, E., Elsner, M. et al. (2014). Controls of event-based pesticide leaching in natural soils: a systematic study based on replicated field scale irrigation experiments. *Journal of Hydrology* 512: 528–539.

Kretzschmar, A. (1978). Ecological quantification of burrow systems of earthworms - techniques and 1st estimations. *Pedobiologia* 18: 31–38.

Kristensen, E., Penha-Lopes, G., Delefosse, M. et al. (2012). What is bioturbation? The need for a precise definition for fauna in aquatic sciences. *Marine Ecology Progress Series* 446: 285–302.

Lavelle, P. (1988). Earthworm activities and the soil system. *Biology and Fertility of Soils* 6: 237–251.

Lavelle, P., Barot, S., Blouin, M. et al. (2007). Earthworms as key actors in self-organized soil systems. In: *Ecosystem Engineers: Plants to Protists* (ed. K. Cuddington, J.E. Byers, W.G. Wilson, and A. Hastings), JKK. 77–106. Elsevier.

Lavelle, P., Bignell, D., Lepage, M. et al. (1997). Soil function in a changing world: the role of invertebrate ecosystem engineers. *European Journal of Soil Biology* 33 (4): 159–193.

Lee, K.E. (1985). *Earthworms: Their Ecology and Relationships with Soils and Land Use.* Sydney: Academic Press.

Lee, K.E. and Foster, R.C. (1991). Soil fauna and soil structure. *Australian Journal of Soil Research* 29: 745–775.

Leue, M., Ellerbrock, R.H., and Gerke, H.H. (2010). Drift mapping of organic matter composition at intact soil aggregate surfaces. *Vadose Zone Journal* 9: 317–324.

Lewandowski, J., Laskov, C., and Hupfer, M. (2007). The relationship between *Chironomus plumosus* burrows and the spatial distribution of pore-water phosphate, iron and ammonium in lake sediments. *Freshwater Biology* 52: 331–343.

Liu, Z., Brandt, L.A., Ogurcak, D.E., and Mazzotti, F.J. (2013). Morphometric and hydrologic characteristics of alligator holes in Everglades National Park, Florida from 1994 to 2007. *Ecohydrology* 6 (2): 275–286.

McGrath, G.S., Hinz, C., and Sivapalan, M. (2008). Modelling the impact of within-storm variability of rainfall on the loading of solutes to preferential flow pathways. *European Journal of Soil Science* 59 (1): 24–33.

McLachlan, A.J. (1977). Some effects of tube shape on feeding of *Chironomus plumosus* (Diptera-Chironomidae). *Journal of Animal Ecology* 46: 139–146.

Meysman, F.J.R., Middelburg, J.J., and Heip, C.H.R. (2006). Bioturbation: a fresh look at Darwin's last idea. *Trends in Ecology & Evolution* 21: 688–695.

Milleret, R., Le Bayon, R.C., and Gobat, J.M. (2009). Root, mycorrhiza and earthworm interactions: their effects on soil structuring processes, plant and soil nutrient concentration and plant biomass. *Plant and Soil* 316: 1–12.

Morad, M.R., Khalili, A., Roskosch, A., and Lewandowski, J. (2010). Quantification of pumping rate of *Chironomus plumosus* larvae in natural burrows. *Aquatic Ecology* 44: 143–153.

Painter, K.J., Westbrook, C.J., Hall, B.D. et al. (2015). Effects of in-channel beaver impoundments on mercury bioaccumulation in Rocky Mountain stream food webs. *Ecosphere* 6 (10): 1–17.

Palmer, M.L. and Mazzotti, F.J. (2004). Structure of Everglades alligator holes. *Wetlands* 24 (1): 115–122.

Paton, T.R., Humphreys, G.S., and Mitchell, P.B. (1995). *Soils: A New Global View*, 213 PP. London: UCL Press Limited.

Pérès, G., Bellido, A., Curmi, P. et al. (2010). Relationships between earthworm communities and burrow numbers under different land use systems. *Pedobiologia* 54 (1): 37–44.

Pivetz, B.E. and Steenhuis, T.S. (1995). Biodegradation and bioremediation - soil matrix and macropore biodegradation of 2,4-d. *Journal of Environmental Quality* 24: 564–570.

Polerecky, L., Volkenborn, N., and Stief, P. (2006). High temporal resolution oxygen imaging in bioirrigated sediments. *Environmental Science & Technology* 40: 5763–5769.

Rogasik, H., Schrader, S., Onasch, I. et al. (2014). Micro-scale dry bulk density variation around earthworm (lumbricus terrestris l.) burrows based on x-ray computed tomography. *Geoderma* 213: 471–477.

Roskosch, A., Hette, N., Hupfer, M., and Lewandowski, J. (2012). Alteration of *Chironomus plumosus* ventilation activity and bioirrigation-mediated benthic fluxes by changes in temperature, oxygen concentration, and seasonal variations. *Freshwater Science* 31: 269–281.

Roskosch, A., Hupfer, M., Nützmann, G., and Lewandowski, J. (2011). Measurement techniques for quantification of pumping activity of invertebrates in small burrows. *Fundamental and Applied Limnology* 178: 89–110.

Roskosch, A., Morad, M.R., Khalili, A., and Lewandowski, J. (2010). Bioirrigation by *Chironomus plumosus*: advective flow investigated by particle image velocimetry. *Journal of the North American Benthological Society* 29: 789–802.

Shipitalo, M.J. and Butt, K.R. (1999). Occupancy and geometrical properties of lumbricus terrestris l-burrows affecting infiltration. *Pedobiologia* 43: 782–794.

Spurgeon, D.J., Keith, A.M., Schmidt, O. et al. (2013). Land-use and land-management change: relationships with earthworm and fungi communities and soil structural properties. *BMC Ecology* 13 (1): 1.

Stehouwer, R.C., Dick, W.A., and Traina, S.J. (1994). Sorption and retention of herbicides in vertically oriented earthworm and artificial burrows. *Journal of Environmental Quality* 23: 286–292.

Sterner, R.W. (1989). The role of grazers in phytoplankton succession. In: *Plankton Ecology* (ed. U. Sommer), 107–170. Brook/Springer Series in Contemporary Bioscience.

Stief, P., Polerecky, L., Poulsen, M., and Schramm, A. (2010). Control of nitrous oxide emission from *Chironomus plumosus* larvae by nitrate and temperature. *Limnology and Oceanography* 55: 872–884.

van Schaik, L., Palm, J., Klaus, J. et al. (2014). Linking spatial earthworm distribution to macropore numbers and hydrological effectiveness. *Ecohydrology* 7: 401–408.

Vinther, F.P., Eiland, F., Lind, A.M., and Elsgaard, L. (1999). Microbial biomass and numbers of denitrifiers related to macropore channels in agricultural and forest soils. *Soil Biology & Biochemistry* 31: 608–611.

Walshe, B.M. (1947). Feeding mechanism of *Chironomus* larvae. *Nature* 160: 474.

Westbrook, C.J., Cooper, D.J., and Baker, B.W. (2006). Beaver dams and overbank floods influence groundwater–surface water interactions of a Rocky Mountain riparian area. *Water Resources Research* 42 (6).

Whitfield, C.J., Baulch, H.M., Chun, K.P., and Westbrook, C.J. (2015). Beaver-mediated methane emission: the effects of population growth in Eurasia and the Americas. *Ambio* 44 (1): 7–15.

Willoughby, G.L. and Kladivko, E.J. (2002). Water infiltration rates following reintroduction of lumbricus terrestris into no-till fields. *Journal of Soil and Water Conservation* 57: 82–88.

Wright, J.P. and Jones, C.G. (2006). The concept of organisms as ecosystem engineers ten years on: progress, limitations, and challenges. *BioScience* 56: 203–209.

Zangerlé, A., Hissler, C., Van Schaik, L., and McKey, D. (2016). Identification of earthworm burrow origins by near infrared spectroscopy: combining results from field sites and laboratory microcosms. *Soil & Tillage Research* 155: 280–288.

Section 2

3

The Four Interfaces' Components of Riparian Zones

Gilles Pinay[1], S. Bernal[2], Jake Diamond[3], Hanieh Seyedhashemi[3], Benjamin Abbott[4], and Florentina Moatar[3]

[1] *Environnement, Ville & Sociétés – UMR CNRS, Lyon, 5600, France*
[2] *CSIC, Blanes, Spain*
[3] *RiverLy, INRAe, Villeurbanne, France*
[4] *Brigham Young University, Provo, Utah, USA*

3.1 Introduction

Over the past 40 years, riparian corridors have been held up as an iconic interface between land and water. Yet, for several centuries, these riverine environments have been subject to intense anthropogenic disturbance such as urban development, clear-cutting for roads and agriculture, diking, straightening, and dredging. In the early 1980s, ecologists and biogeochemists demonstrated the importance of these vegetated ribbons alongside stream networks. Stream ecologists underlined the key role of riparian zones in structuring and fuelling riverine food webs (Vannote et al. 1980). In the meantime, from a more terrestrial perspective, biogeochemists unveiled the role of riparian forests in buffering stream nitrate input from upslope (Peterjohn and Correll 1984). These pluridisciplinary approaches have improved our understanding of the multiple functions of riparian ecosystems and led to the formalization of a more mechanistic definition of the ecotone concept (Weaver 1960). Hence, Holland et al. (1991) proposed to define an ecotone as a "zone of transition between adjacent ecological systems, having a set of characteristics uniquely defined by space and time scales, and by the strength of the interactions between adjacent ecological systems".

The growing understanding of riparian zones as part of the stream ecosystem builds on a perspective that Hynes advocated in his seminal paper *The Stream and its Valley* (Hynes 1975). Yet, this interface vision focused at the landscape scale has hidden a more complex biogeochemical functioning of riparian ecosystems that extends beyond a single interface between land and water, and which has led to some management simplifications and misinterpretations. For instance, several studies in the 1990s showed that the nitrate buffering capacity of the riparian zone was proportional to its width (see Mayer et al. 2007 for a review). These observations led to the establishment of minimal widths for riparian buffer strips, which are already used as a guideline for best management practices by many state environmental agencies (Lee et al. 2004). However, the potential for reducing

Ecohydrological Interfaces, First Edition. Edited by Stefan Krause, David M. Hannah, and Nancy B. Grimm.

nitrate loads is highly variable among riparian zones. Later studies demonstrated that this buffer capacity is a function of water table elevation and nitrate residence time within the system (Hefting et al. 2004; Ocampo et al. 2006). Although riparian zone efficiency to remove nitrate was once thought to be independent to its location along the stream network, it is now clear that contact length between terrestrial upslope and riparian zone is the key driver for nitrate removal. Therefore, riparian zones adjacent to small streams, which comprise the greatest proportion of stream length, should have management priority (Pinay et al. 2015).

While nitrate can be permanently removed from the riparian zone as an inert gas to the atmosphere through microbial denitrification, there is no such mechanism for phosphorus. Although phosphorus may be temporarily stored in plants and soils, it is inevitably released back to subsurface waters under waterlogged conditions; riparian zones are thus typically inefficient in reducing phosphorus loads to streams (Nair et al. 2015). Climatic conditions can also strongly influence the potential for riparian zone denitrification by controlling groundwater table depth and moisture in surface soil layers (Butturini et al. 2003; Pinay et al. 2018; Poblador et al. 2017). Hence, widespread ideas are usually taken as *totems* during the management and restoration of riparian ecosystems. For instance, the pervasive potential of riparian zones as green nutrient filters, or "the wider, the better", do not take into account the complex interplay between topography, hydrology, soil, and vegetation along spatial gradients, and how these interactions and climatic patterns ultimately influence riparian biogeochemical processes.

In fact, riparian zones are constituted of four interfaces with different biogeochemical drivers and effects on stream water quality and function (Figure 3.1). The interface between: (i) upslope and wetland, (ii) wetland and stream surface water, (iii) stream surface water and groundwater, and (iv) stream surface water and the atmosphere. The location and the intrinsic characteristics of each of these interfaces sustain and/or constrain different physical and biogeochemical processes, which ultimately dictate whether riparian ecosystems act either as sources or sinks of the different elements considered. The following chapter will consider each of these four interfaces and their main physical, hydrological, and biological controls.

3.2 The Boundary between Upslope and the Riparian Wetland

Three seminal papers published at about the same time demonstrated the role of riparian zones as buffers of diffuse nitrate fluxes from upslope to their draining stream (Jacobs and Gilliam 1985; Lowrance et al. 1984; Peterjohn and Correll 1984). They developed a similar nitrogen mass balance approach between subsurface nitrate inputs from upslope through the riparian zone's wetland. They found that nitrate was disappearing within a few metres of its transit through the upslope wetland interface. They inferred that this buffering capacity was under the control of plant nitrogen uptake and microbial denitrification process. It was confirmed a few years later that eventually heterotrophic denitrification was the mechanism by which nitrate was removed as a gas (dinitrogen N_2 and nitrous oxide N_2O) from the ecosystem (Pinay and Décamps 1988; Pinay and Labroue 1986). Indeed, the

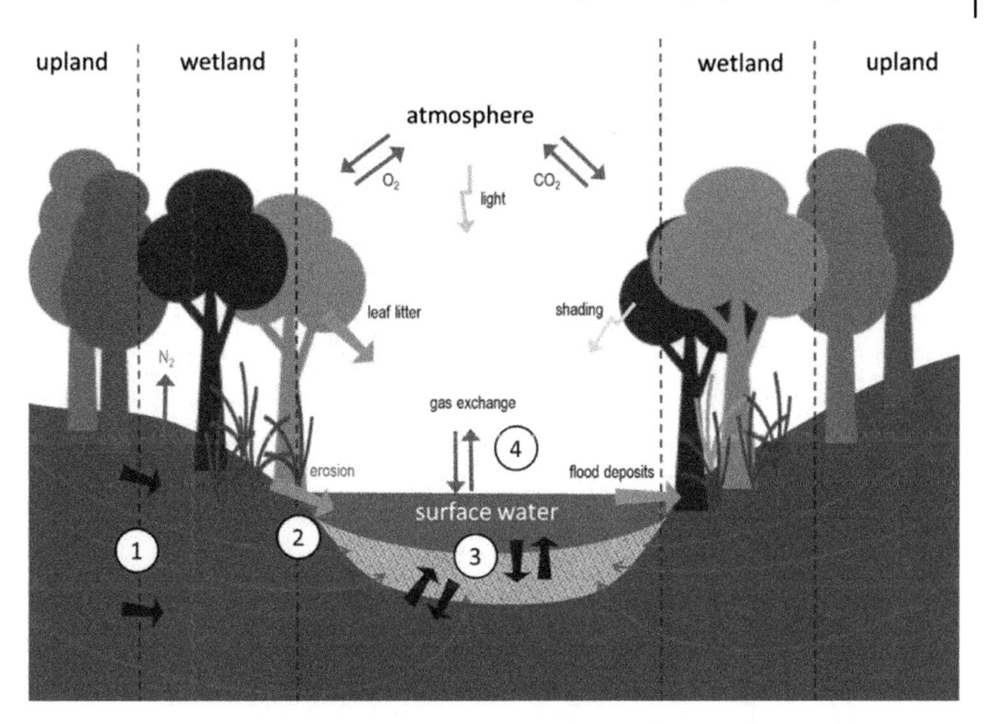

Figure 3.1 Schematic representation of the four interfaces that can be distinguished within the riparian ecotone. (1) The interface between upslope and wetland where N-rich subsurface waters from the hillslope can intersect C-rich riparian soils and promote denitrification and N gas emissions. (2) The interface between wetland and stream surface water, subjected to large sediment exchange during soil erosion and flood deposition processes, and where leaf litter inputs fuel stream metabolism. (3) The interface between stream surface water and groundwater, where vertical hydrological exchange controls biogeochemical processes within the hyporheic zone. (4) The interface between stream surface water and the atmosphere, control point of gas exchange fluxes and light inputs to the stream.

upslope boundary presents the necessary conditions for denitrification to occur (Knowles 1982), i.e. anaerobiosis caused by soil water-logging conditions, the presence of bioavailable organic matter from riparian forest leaves and roots exudates, and nitrate input from upslope subsurface flow. Wetlands classically present both soil anaerobic conditions and organic matter accumulation (Brinson 1993). Yet, denitrification would occur only if the third ingredient, i.e. nitrate, is added (Figure 3.2a). This has the implicit assumption that the wetland environment has to be connected to a source of nitrate for denitrification to occur. Hence, it might sound paradoxical in a world overloaded with nitrate, but the factor limiting denitrification in most wetlands and riparian zones is nitrate inputs (Figure 3.2b). This finding inspired the concept of the *biogeochemical hot spot*, a location showing disproportionally high reaction rates relative to the surrounding area, as it occurs in the upland–wetland interface of many riparian zones (McClain et al. 2003).

Once nitrate supply is met, the second factor limiting denitrification is the residence time of nitrate in the riparian zone. If subsurface flow is faster than denitrification reaction rate, then nitrate reduction will not be complete, and a portion will escape towards the stream (Sabater et al. 2003). In this context, Ocampo et al. (2006) used the Damköhler ratio,

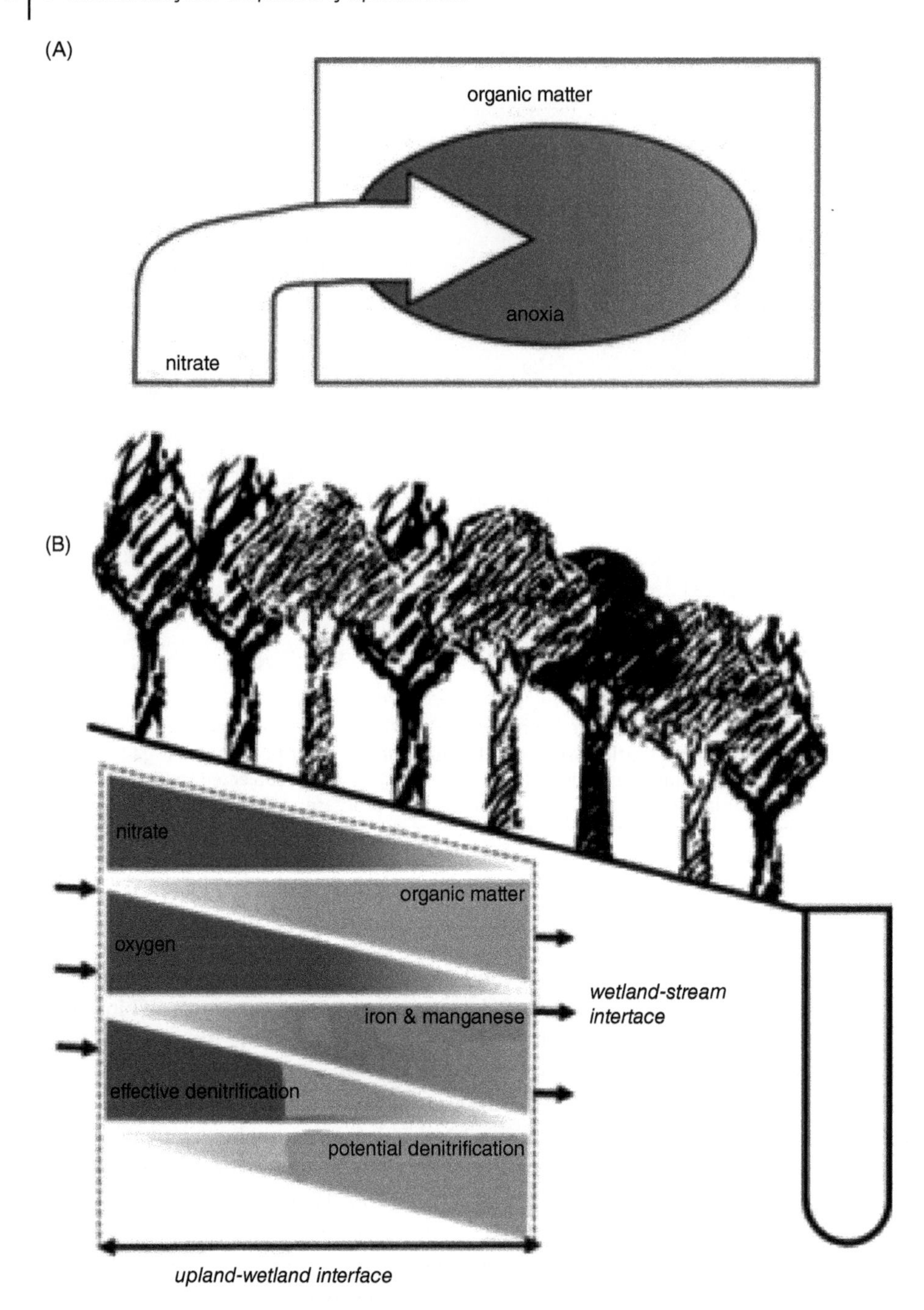

Figure 3.2 (a) Illustration of a biogeochemical hot spot for denitrification where nitrate flux intersects an anaerobic organic carbon-rich soil pool; and (b) major biogeochemical gradients along the upland–wetland interface and hypothetical associated effective and potential denitrification. Adapted from Pinay (1986) and McClain et al. (2003).

i.e. rate of nitrate input to rate of denitrification, to characterize and estimate the nitrate buffering capacity of riparian zones. They showed that the balance between transport and reaction is essential for determining the proportion of nitrate denitrified in riparian zones. Based on those findings, they proposed to use the Damköhler ratio as an indicator to define how wide a riparian should be, or to identify the most efficient riparian buffers within the landscape. In practice, riparian subsurface flows are so heterogeneous that heavy hydrogeological monitoring is needed to predict the local residence time of nitrate, which limits the possibility to use this type of approach in environmental management plans. Nonetheless, this research shed light on why riparian zone width is a poor surrogate of subsurface travel time, and thus, an insufficient criterion to warrant high nitrate retention in riparian zones. While predicting the nitrate buffering capacity of riparian zones at the local scale is highly uncertain, the likelihood of denitrification occurrence clearly increases with increasing the length of the upslope–wetland interface at the landscape scale. In other words, the higher the length of contact between upslopc and riparian zonc, thc higher the potential for denitrification. The second management lesson to learn is that preserving and restoring riparian zones along small stream orders would be more efficient to remove upslope nitrate input than preserving large riparian areas at particular locations. This is because the former strategy will contribute to elongate the upslope–riparian wetland interface, while the nitrate retention efficiency of the latter will mostly rely on the local features of the selected riparian zone.

Heterotrophic denitrification is perhaps the biogeochemical process that naturally contributes the most to the permanent removal of reactive nitrogen from the biosphere. Yet, it should be mentioned that denitrification can also contribute to nitrous oxide emission, a potent greenhouse gas, since it is an intermediary product of the chain of reduction reactions from nitrate to nitrogen gas. Nitrous oxide emission can be particularly important at the beginning of soil rewetting during rainfall or a flooding event because of limited activation of N_2O-reductase enzyme by denitrifying bacteria (Firestone et al. 1980). Additional denitrification pathways have been identified such as Anammox – the microbial oxidation of ammonium to nitrogen under anaerobic conditions (Jetten 2001), nitrifier denitrification (Wrage et al. 2001), and autotrophic denitrification, which uses inorganic substrates as electron donors. The existence of these different biogeochemical pathways highlights the complexity of soil microbial communities and their large potential to respire oxidized nitrogen compounds. Increased attention has also been given to the interaction between the nitrogen and sulphur cycles (Burgin and Hamilton 2008), and also to carbon–nitrogen interactions, with a wide range of research focused on the type and form of carbon needed to drive denitrification and the sustainability of these electron donor sources (Newcomer et al. 2012). Research on the interaction between the nitrogen and iron cycles was also undertaken, proposing the ferrous wheel concept with the oxidation of ammonium to nitrite using ferric iron as an electron donor, and further nitrite reduction into nitrogen gas through classic bacterially-mediated denitrification (Clément et al. 2005). The ferrous wheel concept widened our understanding of the nitrogen processing within riparian zones by demonstrating that alternative biogeochemical pathways, independent of organic carbon sources, can occur and substantially contribute to the gaseous purge of nitrate from riparian soils.

3.3 The Boundary Between Wetland and Surface Water

The lateral interface between surface water and riparian wetland is one of the most dynamic and complex in river corridors. These environments facilitate reciprocal exchanges of energy, water, nutrients, and organisms across the terrestrial-aquatic gradient (Baxter et al. 2005; Nakano and Murakami 2001). Small changes in water table elevation can restructure the extent and reverse the direction of these exchanges (Raymond et al. 2016; Zarnetske et al. 2018), which further depend on seasonal variation in organismal phenology (Fellman et al. 2009; Gücker et al. 2016; Saurer et al. 2014). Consequently, understanding the wetland–surface water interface requires multi-scale monitoring and interdisciplinary approaches (Abbott et al. 2017; Lee-Cullin et al. 2018).

One of the most important functions of the wetland–surface water interface is export of terrestrial carbon to surface water networks, where it exists as dissolved (DOM) and particulate organic matter (POM). The input of POM from riparian zones increases reach hydraulic heterogeneity, slowing flow and increasing in-channel transient storage, leading to greater opportunities for microbial processing (Ensign and Doyle 2005; Roberts et al. 2007a). DOM is a fundamental component of the global carbon cycle (Lupon et al. 2020; Wologo et al. 2021), and forms the base of many stream food webs (Acuña et al. 2007). While sources of DOM vary, much of it derives from riparian leaf litter and soil inputs that provide the dominant fuel for in-stream respiration and secondary production, especially in small and intermediate-size streams (Ledesma et al. 2015; Zarnetske et al. 2018). Stream ecologists integrated the importance of riparian leaf litter inputs for stream food webs and biogeochemical processing into classical conceptual models such as the River Continuum Concept (Vannote et al. 1980) and the nutrient spiralling theory (Newbold et al. 1981, 1983). Reach-scale experiments have supported these conceptual models, empirically demonstrating that streams rapidly process allochthonous organic matter and nutrients from adjacent wetlands (Brookshire et al. 2005; Peterson et al. 2001). Moreover, studies conducting mass balance approaches have shown that in-stream processes largely modify riparian groundwater inputs of nutrients, and that, far from being recalcitrant, dissolved organic carbon from lateral groundwater inputs increase heterotrophic activity in some streams (Lupon et al. 2019).

A paradigm shift in the potent biogeochemical processing capacity of streams has replaced earlier views that carbon and nutrient inputs across the riparian interface (Figure 3.3) were conservatively advected downstream (Brookshire et al. 2009; Goodale et al. 2009; Siegenthaler and Sarmiento 1993). Indeed, the conceptualization of streams as inert pipes connecting terrestrial and marine environments is rapidly shifting towards a perception of these ecosystems as biogeochemical funnels and bioreactors. Global estimates suggest that approximately 75% of the carbon imported by freshwater ecosystems, including both particulate and dissolved forms, is outgassed to the atmosphere as carbon dioxide or sequestered in sediments (Cole et al. 2007; Raymond et al. 2013). Likewise, stream networks remove substantial proportions of nitrogen inputs (Alexander et al. 2000), predominately through denitrification (Seitzinger et al. 2006), where removal efficiency and pathway depend strongly on nitrogen and carbon stream concentrations (Burgin and Hamilton 2007; Mulholland et al. 2008). Critically, the concentrations of necessary reactants and environmental conditions required for in-stream nitrogen removal are controlled by the degree of connectivity along wetland–stream surface water interfaces.

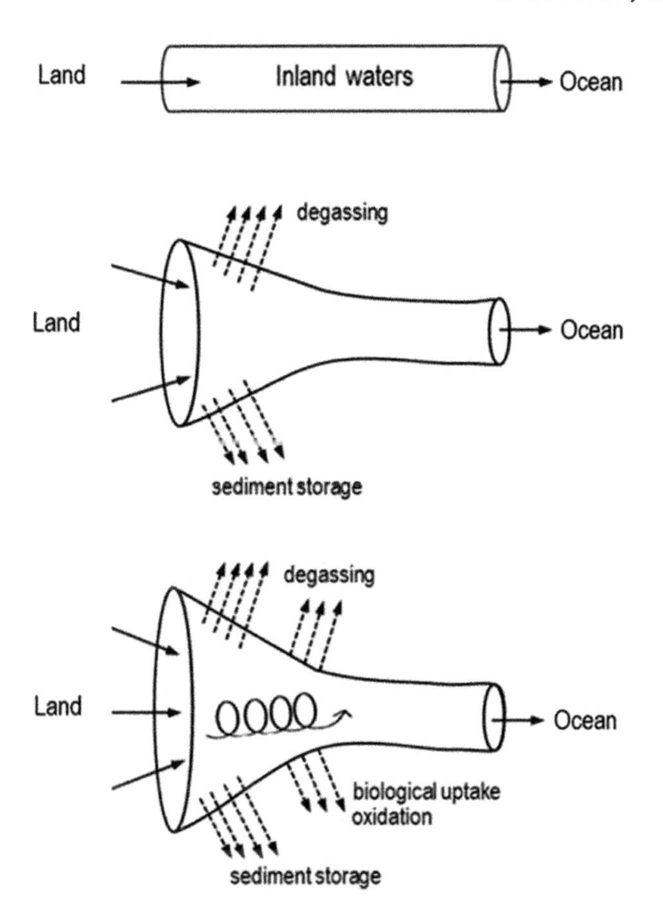

Figure 3.3 Perceptions of dissolved organic carbon in stream ecosystems. Adapted from Cole et al. (2007).

River systems and their riparian zones are dynamically linked longitudinally, laterally, and vertically by hydrologic and geomorphic processes (Ward 1989). The intensity of these processes varies from headwater to the river mouth with dominant erosional processes in the upper part and depositional processes in the lower reaches (Sullivan et al. 1987). River floodplains are important sinks for storing sediments and associated nutrients mobilized from upstream catchments during flood events. These transfers of energy, biotic, and abiotic matter in the floodplains are under the control of flood duration, frequency, and magnitude that create a mosaic of geomorphic surfaces influencing the spatial pattern and successional development of riparian vegetation (Roberts and Ludwig 1991; Salo et al. 1986). Flood characteristics control also the nutrient cycling intensity of floodplain soils and their impact on stream nutrient fluxes (Brinson et al. 1984; Mulholland 1992). Flooding directly affects nutrient cycling in alluvial soils by controlling the duration of oxic and anoxic phases (Ponnamperuma 1972; Tabacchi et al. 1998) (Figure 3.4).

For instance, it was found that the net nitrogen mineralization rate was four times greater in a spring-flooded marsh than in a non-flooded one, and that alternate aerobic and anaerobic conditions enhance organic matter decomposition and nitrogen loss through denitrification in flooded soils (Groffman and Tiedje 1988; Reddy and Patrick 1975).

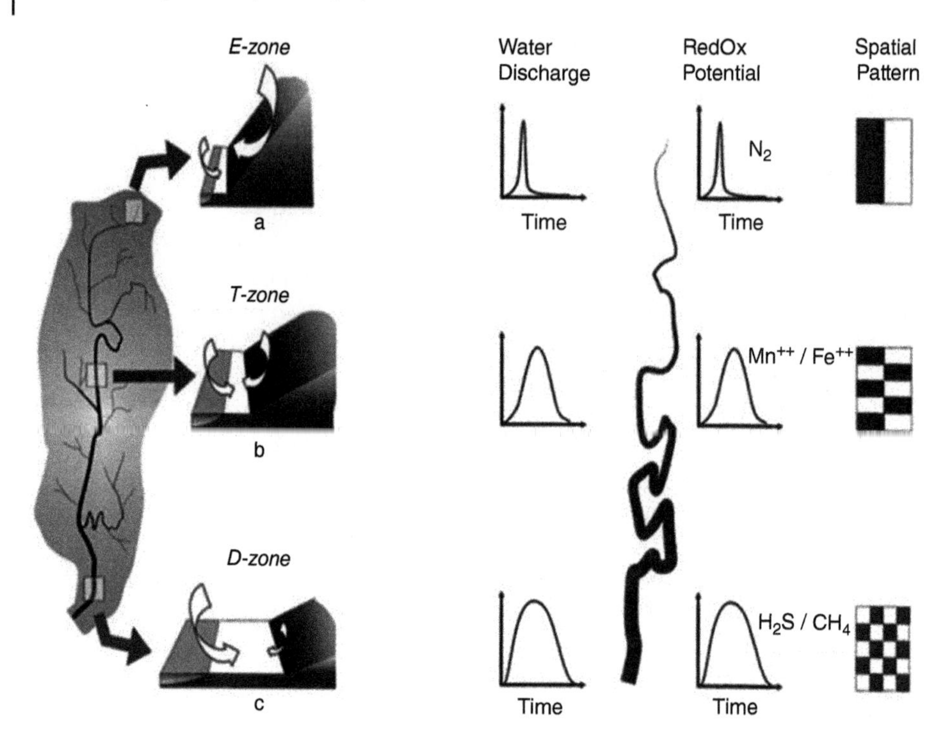

Figure 3.4 Illustration of the erosional (E), transitional (T), and depositional (D) zones along an idealized stream network. Differences in topographic and geomorphic features between these three zones result in different flood characteristics and spatial heterogeneity of soil properties and associated biogeochemical processes. Adapted from Tabacchi et al. (1998).

Flooding duration is controlled by local topography; low areas are flooded more often and longer than higher ones, producing large variations in biogeochemical patterns at a metre scale. Flooding also indirectly affects nutrient cycling in floodplain soils by influencing the soil structure and texture through sediment deposits. Hence, floodplain and stream channel geomorphic and hydrologic processes influence the sorting of flood sediment deposits on a grain size basis creating a mosaic of soils of different textures.

In a study realized on the Garonne River floodplain, Pinay et al. (2000) found that the floodplains' soil grain size could be a good proxy to estimate the likelihood of denitrification activity. Below a threshold of 65% of silt and clay content, the floodplain soils did not present any significant denitrification rates. Above that threshold denitrification increased significantly (Figure 3.5).

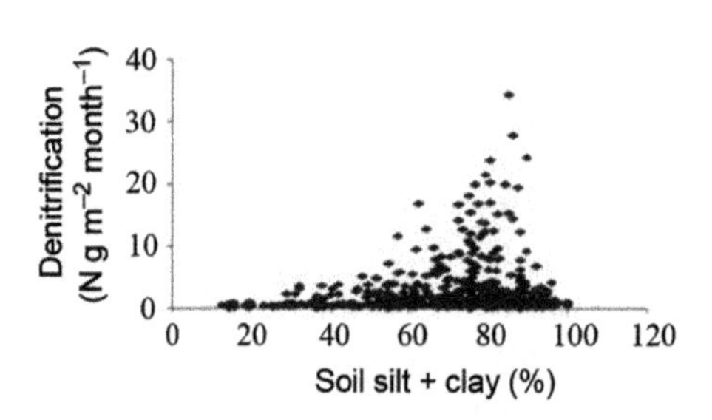

Figure 3.5 Relationship between the floodplain's soil grain size and denitrification in the Garonne River floodplain. Adapted from Pinay et al. (2000).

This relationship between soil grain size and denitrification was confirmed in another pan European study (Pinay et al. 2007). These results confirm the importance of texture on soil nitrogen cycling processes and suggest that soil grain size could be a good proxy at the reach scale to determine where denitrification occurs in a floodplain.

3.4 The Boundary between Surface and Groundwater

The riparian zone extends beneath the hyporheic zone, a highly dynamic subsurface region that encompasses both the stream channel and the riparian zone itself. The hyporheic zone is defined as the portion of sediments surrounding the stream that continuously exchange water and solutes with the stream (Boano et al. 2014). By definition, the hyporheic zone implies water exchange at relatively small spatial scales, typically from centimetres to metres, and it is characterized by slowly moving waters, which increase the hydrological opportunity for biogeochemical interactions (Battin et al. 2008). Yet, the area of influence of the hyporheic zone can be highly variable across sites depending on the hydromorphological characteristics of the river corridor as well as on a seasonal scale with changing climatic conditions (Harvey and Bencala 1993; Wondzell and Swanson 1996). Groundwater–streamwater exchange can also occur at larger spatial scales, from hundreds to thousands of metres, and this continuous gain and loss of water has large implications for the biogeochemical processing of solutes along river corridors (Covino and McGlynn 2007). In fact, the transition between riparian groundwater, hyporheic zone, and surface stream water can be understood as a continuum of hydrological and biogeochemical conditions, precluding delineation of a physical boundary between these water bodies (Figure 3.6).

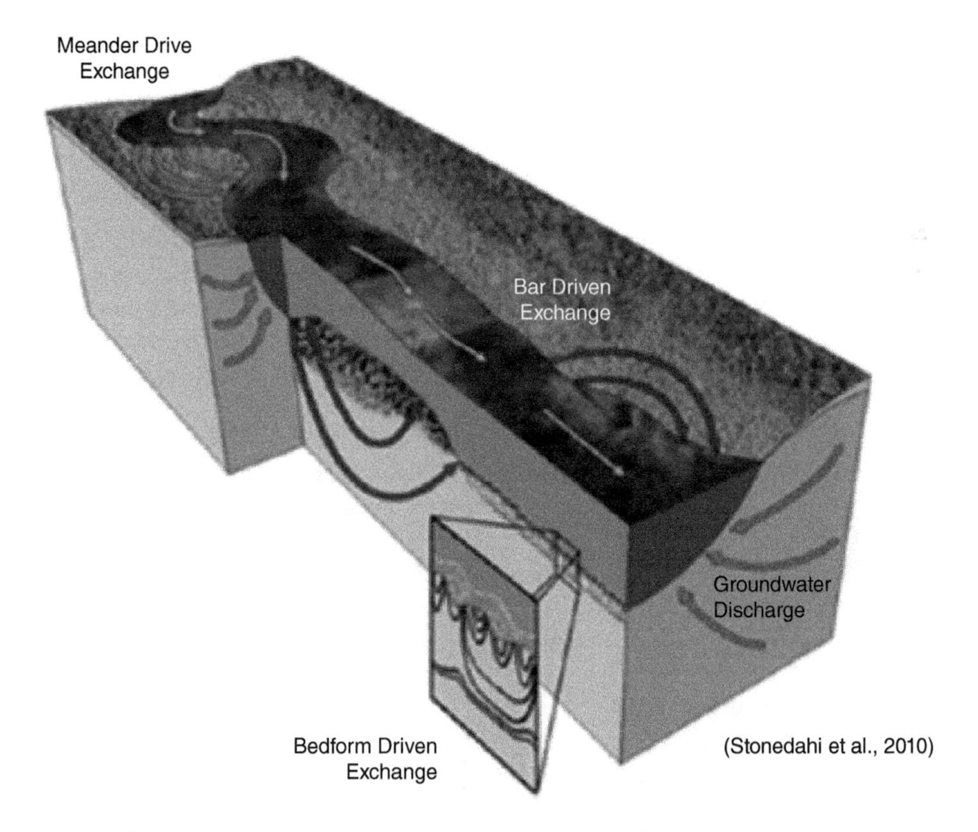

Figure 3.6 Hyporheic flows. Adapted from Stonedahl et al. (2010).

Hydraulic gradients usually promote the movement of water from hillslopes towards the stream, especially in headwaters. Thus, it is often assumed that groundwater preferentially flows from the riparian zone towards the stream. However, rather than being unidirectional, the hydrological exchange between riparian groundwater and stream water is highly dynamic over space and time depending on topography and watershed structure (Covino et al. 2011; Jencso et al. 2009). Further, there can be large mixing of groundwater and stream water in the riparian zone, especially in large alluvial valleys such as reported for the Garonne River in southwest France (Pinay et al. 1998). In semi-arid areas, water losses from the stream towards the riparian zone are commonly observed during low flow periods when hydraulic pressure from hillslope groundwater is small (Shade et al. 2005). Riparian tree evapotranspiration can further control water exchange fluxes at the stream–riparian interface by dropping down groundwater tables and favouring the movement of stream water towards the riparian zone at both daily and seasonal scales (Lupon et al. 2016; Wondzell et al. 2010). This ecohydrological process can exert a strong control on stream discharge in semi-arid areas, favouring stream channel desiccation and the lateral infiltration of stream water up to dozens of metres within the riparian zone, especially during periods with low hydrological connectivity (Bernal et al. 2013; Butturini et al. 2003). This is why planting riparian trees along buffer strips without considering local climate conditions can be a counterproductive management strategy in some cases.

Vertical and lateral hydrological exchange at the stream–riparian interface is accompanied by the exchange of dissolved organic matter and nutrients. The mixing of electron donors and acceptors along thermodynamic gradients from the riparian zone towards the near-stream area can promote intense microbial activity, and host high rates of biogeochemical processes within the hyporheic zone. For instance, Hedin et al. (1998) reported that denitrification at the riparian zone of the Smith Creek, a first-order stream in Michigan (USA), was mostly constrained to a very localized area where horizontal shallow subsurface flow rich in dissolved organic carbon interacts with nitrate-rich vertical upwelling of deep subsurface waters in the near-stream zone. They proposed that those areas of hydrological interaction may act as "control points" for fluxes of nitrogen and other nutrients at the soil–stream interface. In this line of thought, Zarnetske et al. (2011) proposed that thermodynamic gradients, from oxic to reduced conditions, organize along hyporheic flowpaths, so that the prevalence of either nitrification or denitrification is a function of the residence time of water and solutes in the hyporheic zone. There are also beautiful examples illustrating the transfer of water and solutes from the stream towards the riparian groundwater. Pinay et al. (2009) showed that nitrate mineralized from the carcasses of Pacific salmons in a small stream in Alaska (USA) was rapidly taken up by biota along hyporheic flowpaths from the stream towards the riparian zone. In Sycamore Creek (AZ, USA), riparian trees were enriched with the ^{15}N previously added to the adjacent stream, highlighting that solutes in the stream water column travel towards the riparian zone (Shade et al. 2005). Overall, these studies highlight the strong potential for biogeochemical interactions at the surface water–groundwater interface, which requires bidirectional fluxes between surface water and groundwater. The management lesson to learn from these studies is that maintaining and recovering kilometres of free flowing waters along the stream network contributes to naturalize hydrological exchange fluxes between hyporheic and surface waters, and thus, promote ecological and biogeochemical functions at this terrestrial–aquatic interface.

3.5 The Boundary between Stream and Atmosphere

A hierarchy of vertical interfaces controls the exchange of energy and materials between the stream and the atmosphere. Successive attenuation of solar energy by canopy shading, reflectance from the water surface, and absorption in the water column results in a small fraction of light energy entering ecosystems being converted to chemical energy via photosynthesis (Kirk 1994). Hence, the light regime – the timing, mode, and magnitude of photosynthetically active radiation (PAR) available for gross primary production (GPP) – for stream autotrophs differs from terrestrial autotrophs. Stream ecologists hypothesized that light regimes, and therefore GPP, would vary according to riparian shading (Minshall 1978), and that shading patterns were manifest in predictable network patterns of longitudinal increases in channel depth and width (Vannote et al. 1980). Subsequent efforts confirmed that indeed, light regime is the dominant driver of lotic GPP variation across biomes (Mulholland et al. 2001) and over time (Roberts et al. 2007b). However, complexities from riparian phenology (Hill and Dimick 2002) and variation in water column attenuation due to suspended sediment and coloured organic compounds (Davies-Colley and Nagels 2008; Davies-Colley and Smith 2001; Julian et al. 2008) limit the universal predictive capacity of atmospheric PAR for GPP, especially when autotrophs are attached to the stream benthos (Bernhardt et al. 2018). Still, recent improvements in modelling the true light availability for stream benthic autotrophs are increasing our predictive abilities (Julian et al. 2008; Kirk et al. 2020), and are further highlighting how the linked interfaces among atmosphere, riparian vegetation, and stream act in concert to control ecosystem productivity and function.

Stream water temperature represents the balance of energetic fluxes across the surface water–groundwater and surface water–atmosphere interfaces (Hannah and Garner 2015). In headwater streams, groundwater influence limits equilibrium with the atmosphere especially during the summer period (Edinger et al. 1968; Hrachowitz et al. 2010; Kelleher et al. 2012). In reducing solar radiation, riparian vegetation further increases the relative importance of streambed heat flux from groundwater. This is especially noticeable during summer when the net solar radiation is the dominant heat flux input (Hannah et al. 2008; Malcolm et al. 2008). Therefore, riparian forest shading reduces summer maximum daily stream water temperatures (Johnson 2004; Moore et al. 2005). The cooling effect of vegetation has been reported mainly for small streams (Garner et al. 2014; Johnson and Wilby 2015; Moore et al. 2005). It may be possible to maintain daily maximum water temperatures close to 20°C with shade levels of 70% in headwaters (Rutherford et al. 1997). For conditions at midday in July, about 74% decrease can be measured in net energy gain from an open reach under full sun to a full shaded reach (Johnson 2004). The effectiveness of riparian shading depends mainly on channel orientation, canopy density, and within-reach residence times (Garner et al. 2017). For instance, Garner et al. (2017) demonstrated that for reaches under high flow velocity and 30% vegetation density, stream temperatures varied by up to 0.8°C and 2.7 °C for mean and maximum daily temperatures, respectively. Under the same vegetation cover with a low velocity, temperatures varied by 2.7° and 4.3 °C for mean and maximum daily temperatures, respectively. They also showed that an increase in canopy density (from 10 to 90%) could decrease both maximum (≥3.0 °C) and mean (≥1.6 °C) daily temperatures. Riparian clear-cutting can increase maximum daily water temperature by up to 8 °C (Gomi et al. 2006; Johnson and Jones 2000), and can gradually return to pre-harvest temperature after a 15-year regrowth period (Johnson and Jones 2000).

Johnson and Wilby (2015) showed that approximately 0.5 km of complete shade is necessary to off-set stream temperature by 1 °C at midday in July in headwaters; whereas at 25 km downstream, 1.1 km of shade is required.

As rivers increase in size, their thermal regimes become dominated by inertial effects and are increasingly insensitive to the height and phenology of surrounding riparian vegetation (Caissie 2006). Yet, few studies have already modelled riparian shading by riparian canopy at any time and location for a given stream depending on stream azimuth and river width, canopy height, and overhang (Li et al. 2012). Integration of such model and dynamics of LIDAR-based riparian vegetation into a physical process-based thermal model showed a decrease in maximum daily water temperature ranging from −3 °C (upstream) to −1.3 °C (downstream) for medium streams (distance from source for 40 km to 300 km) as well as small streams (Loicq et al. 2018).

Despite this buffering effect of the riparian zone on temperature maximum in headwaters, stream water thermal regimes are generally synchronous with air temperature (Beaufort et al. 2020a). Yet, anthropogenic impoundments such as large dams, small reservoirs, and ponds can influence downstream thermal regimes (Beaufort et al. 2020a; Seyedhashemi et al. 2020). In this context, the riparian vegetation cooling effect can mitigate ponds' heating impact on stream thermal regimes (Maxted et al. 2005; Seyedhashemi et al. 2020).

3.6 Impact of Global Change on the Functioning of These Interfaces: Future Challenges and Opportunities

Climate change is one of the greatest challenges facing humanity, and the informed management of natural resources, particularly riparian interfaces, will be a critical component of our response. For example, increasing temperatures – particularly decreased winter minima and increased summer maxima – will shift the hydrological balance, likely reducing snowpack and storage, and increasing potential evapotranspiration. These effects will likely be regionally specific, necessitating continued local monitoring and integration of broad scientific understanding with local stakeholder knowledge. For example, in Europe, increased precipitation is likely to occur in winter in the northern areas, while southern Europe is expected to suffer a decrease in precipitation (recent references, Habets et al. 2013). Compounding the potential for reduced water supply, water demand is expected to increase due to growing irrigation needs and an expanding energy sector that requires water for cooling (Abbott et al. 2019).

As climate change shifts the spatiotemporal balance of energy inputs and water availability at the riparian interface, mismatches between biological and biogeochemical processes may arise. For example, hydrological connectivity of the river network may fragment, decreasing biodiversity due to phenological dependence of organisms on saturating or flowing conditions. Likewise, riparian soil saturation, a requirement for denitrification that limits landscape nitrate loads to streams, may reduce in extent. Land cover and land-use change factors can amplify these mismatches. Agriculture intensification leads to higher/chronic nutrients and pesticides leakages, and the timing of inputs is likely to occur when potential removal rates in the riparian zone are minimized (e.g. in the dormant season, Van Meter et al. 2020). It is therefore critical to maintain, conserve, and re-establish hydrological and habitat connectivity across the riparian interface to support the biodiverse ecosystems on which we depend.

Critically, the how-to of this important effort, e.g. by preserving and expanding riparian vegetation, is an emerging issue (Dufour et al. 2019), and how we use our knowledge of riparian interfaces to interface with stakeholders and decision-makers remains a difficult, but not insurmountable, challenge. We can use our position of expertise to suggest policy measures, for example, an "ecological re-allotment programme" to restructure the landscape. We are also in dire need of a rehabilitation of public perception of and relationship to riparian corridors; these environments have always been a focus of community life (transportation, recreation, food gathering, and water provisioning). We can conduct outreach programmes that focus on sustainable agriculture and wise application of fertilizers via the RRRR approach (Right type, Right amount, Right place, and Right time).

We can still do more to understand and support our view that riparian zones are critical interfaces in the landscape, with (in)valuable services provided to us at little-to-no cost. While we know that long hydrological travel times through subsurface hot spots of biogeochemical reactivity are the most efficient for removing undesirable solutes from reaching streams, we still lack a landscape perspective on how to design riparian interfaces that maximize this connectivity. How can we improve the drainage and imperviousness of agricultural soils to reduce the rapid bypass of nutrients and pesticides to the stream network? What is the ideal landscape arrangement of hedges, agricultural fields, and riparian zones that maximizes nutrient removal? Global changes bring enormous challenges to environmental scientists, but also unveil grand opportunities to learn, engage, and transform how humans interface with riparian zones.

References

Abbott, B.W., Bishop, K., Zarnetske, J.P. et al. (2019). Human domination of the global water cycle excluded from depictions and perceptions. *Nature Geoscience*. https://doi.org/10.1038/s41561-019-0374-y.

Abbott, B.W., Gruau, G., Zarnetske, J.P. et al. (2017). Structure and synchrony of water quality in headwater stream networks. *Ecology Letters*. https://doi.org/10.1111/ele.12897.

Acuña, V., Giorgi, A., Muñoz, I. et al. (2007). Meteorological and riparian influences on organic matter dynamics in a forested Mediterranean stream. *Journal North American Benthological Society* 26 (1): 54–69.

Alexander, R.B., Smith, R.A., and Schwarz, G.E. (2000). Effect of stream channel size on the delivery of nitrogen to the Gulf of Mexico. *Nature* 403 (6771): 758–761.

Battin, T.J., Kaplan, L.A., Findlay, S. et al. (2008). Biophysical controls on organic carbon fluxes in fluvial networks. *Nature Geosciences* 1: 95–100.

Baxter, C.V., Fausch, K.D., and Saunders, W.C. (2005). Tangled webs: reciprocal flows of invertebrate prey link streams and riparian zones. *Freshwater Biology* 50 (2): 201–220. https://doi.org/10.1111/j.1365-2427.2004.01328.x.

Beaufort, A., Moatar, F., Sauquet, E. et al. (2020a). Influence of landscape and hydrological factors on stream–air temperature relationships at regional scale. *Hydrological Processes* 34 (3): 583–597. https://doi.org/10.1002/hyp.13608.

Beaufort, A., Moatar, F., Sauquet, E. et al. (2020b). Influence of landscape and hydrological factors on stream–air temperature relationships at regional scale. *Hydrological Processes* 34 (3): 583–597.

Bernal, S., von Schiller, D., Sabater, F., and Martí, E. (2013). Hydrological extremes modulate nutrient dynamics in Mediterranean climate streams across different spatial scales. *Hydrobiologia* 719 (1): 31–42.

Bernhardt, E.S., Heffernan, J.B., Grimm, N.B. et al. (2018). The metabolic regimes of flowing waters. *Limnology & Oceanography* 63 (S1): S99–S118.

Boano, F., Harvey, J.W., Marion, A. et al. (2014). Hyporheic flow and transport processes: mechanisms, models and biogeochemical implications. *Reviews of Geophysics* 52: 603–679. https://doi.org/10.1002/2012RG000417.

Brinson, M.M. (1993). Changes in the functioning of wetlands along environmental gradients. *Wetlands* 13 (2): 65–74.

Brinson, M.M., Bradshaw, H.D., and Kane, E.S. (1984). Nutrient assimilative capacity of an alluvial flood plain swamp. *Journal of Applied Ecology* 21: 1041–1057.

Brookshire, E.N.J., Valett, H.M., and Gerber, S.G. (2009). Maintenance of terrestrial nutrient loss signatures during instream transport. *Ecology* 90: 293–299.

Brookshire, E.N.J., Valett, H.M., Thomas, S.A., and Webster, J.R. (2005). Coupled cycling of dissolved organic nitrogen and carbon in a forest stream. *Ecology* 86 (9): 2487–2496.

Burgin, A.J. and Hamilton, S.K. (2007). Have we overemphasized the role of denitrification in aquatic ecosystems? A review of nitrate removal pathways. *Frontiers in Ecology and the Environment* 5 (2): 89–96.

Burgin, A.J. and Hamilton, S.K. (2008). NO_3-driven SO_4^{2}-production in freshwater ecosystems: implications for N and S cycling. *Ecosystems* 11: 908–922.

Butturini, A., Bernal, S., Nin, E. et al. (2003). Influences of the stream groundwater hydrology on nitrate concentration in unsaturated riparian area bounded by an intermittent Mediterranean stream. *Water Resources Research* 39 (4): 1110. https://doi.org/10.1029/2001WR001260.

Caissie, D. (2006). The thermal regime of rivers: a review. *Freshwater Biology* 51 (8): 1389–1406. https://doi.org/10.1111/j.1365-2427.2006.01597.x.

Clément, J.C., Shrestha, J., Ehrenfeld, J.G., and Jaffe, P.R. (2005). Ammonium oxidation coupled to dissimilatory reduction of iron under anaerobic conditions in wetland soils. *Soil Biology & Biochemistry* 37 (12): 2323–2328.

Cole, J.J., Prairie, Y.T., Caraco, N.F. et al. (2007). Plumbing the global carbon cycle: integrating inland waters into the terrestrial carbon budget. *Ecosystems* 10: 171–184.

Covino, T.P. and McGlynn, B.L. (2007). Stream gains and losses across a mountain-to-valley transition: impacts on watershed hydrology and stream water chemistry. *Water Resources Research* 43: W10431. https://doi.org/10.1029/2006WR005544.

Covino, T.P., McGlynn, B.L., Mallard, J. et al. (2011). Stream-groundwater exchange and hydrologic turover at the network scale. *Water Resources Research* 47: W12521. https://doi.org/10.1029/2011WR010942.

Davies-Colley, R.J. and Nagels, J.W. (2008). Predicting light penetration into river waters. *Journal of Geophysical Research* 113: G03028.

Davies-Colley, R.J. and Smith, D.G. (2001). Turbidity, suspended sediment, and water clarity: a review. *Journal of the American Water Resources Association* 37: 1085–1101.

Dufour, S., Rodríguez-González, P.M., and Laslier, M. (2019). Tracing the scientific trajectory of riparian vegetation studies: main topics, approaches and needs in a globally changing world. *Science of the Total Environment* 653: 1168–1185.

Edinger, J.E., Duttweiler, D.W., and Geyer, J.C. (1968). The response of water temperatures to meteorological conditions. *Water Resources Research* 4 (5): 1137–1143. https://doi.org/10.1029/WR004i005p01137.

Ensign, S.H. and Doyle, A.W. (2005). In-channel transient storage and associated nutrient retention: evidence from experimental manipulations. *Limnology & Oceanography* 50 (6): 1740–1751.

Fellman, J.B., Hood, E., Edwards, R.T., and Jones, J.B. (2009). Uptake of allochthonous dissolved organic matter from soil and salmon in coastal temperate rainforest streams. *Ecosystems* 12 (5): 747–759. https://doi.org/10.1007/s10021-009-9254-4.

Firestone, M.K., Firestone, R.B., and Tiedje, J.M. (1980). Nitrous oxide from soil denitrification. Factors controlling biological production. *Science* 208 (4445): 749–751.

Garner, G., Malcolm, I.A., Sadler, J.P., and Hannah, D.M. (2014). What causes cooling water temperature gradients in a forested stream reach? *Hydrology and Earth System Sciences* 18 (12): 5361–5376. https://doi.org/10.5194/hess-18-5361-2014.

Garner, G., Malcolm, I.A., Sadler, J.P., and Hannah, D.M. (2017). The role of riparian vegetation density, channel orientation and water velocity in determining river temperature dynamics. *Journal of Hydrology* 553: 471–485. https://doi.org/10.1016/j.jhydrol.2017.03.024.

Gomi, T., Moore, R.D., and Dhakal, A.S. (2006). Headwater stream temperature response to clear-cut harvesting with different riparian treatments, coastal British Columbia, Canada. *Water Resources Research* 42 (8): https://doi.org/10.1029/2005WR004162.

Goodale, C.L., Thomas, S.A., Fredriksen, G. et al. (2009). Unusual seasonal patterns and inferred processes of nitrogen retention in forested headwaters of the Upper Susquehanna River. *Biogeochemistry* 93: 197–2018.

Groffman, P.M. and Tiedje, J.M. (1988). Denitrification hysteresis during wetting and drying cycles in soils. *Soil Science Society of America Journal* 52: 1626–1629.

Gücker, B., Silva, R.C.S., Graeber, D. et al. (2016). Dissolved nutrient exports from natural and human-impacted neotropical catchments. *Global Ecology and Biogeography* 25 (4): 378–390. https://doi.org/10.1111/geb.12417.

Habets, F., Boé, J., Déqué, M. et al. (2013). Impact of climate change on the hydrogeology of two basins in northern France. *Climatic Change* 121 (4): 771–785.

Hannah, D.M. and Garner, G. (2015). River water temperature in the United Kingdom: changes over the 20th century and possible changes over the 21st century. *Progress in Physical Geography* 39 (1): 68–92. https://doi.org/10.1177%2F0309133314550669.

Hannah, D.M., Malcolm, I.A., Soulsby, C., and Youngson, A.F. (2008). A comparison of forest and moorland stream microclimate, heat exchanges and thermal dynamics. *Hydrological Processes* 22 (7): 919–940. https://doi.org/10.1002/hyp.7003.

Harvey, J.W. and Bencala, K.E. (1993). The effect of streambed topography on surface-subsurface water exchange in mountain catchments. *Water Resources Research* 29 (1): 89–98. https://doi.org/10.1029/92WR01960.

Hedin, L.O., von Fisher, J.C., Ostrom, N.E. et al. (1998). Thermodynamic constraints on nitrogen transformations and other biogeochemical processes at soil-stream interfaces. *Ecology* 79 (2): 684–703.

Hefting, M., Clément, J.C., Dowrick, D. et al. (2004). Water table elevation controls on soil nitrogen cycling in riparian wetlands along a European climatic gradient. *Biogeochemistry* 67: 113–134.

Hill, W.R. and Dimick, S.M. (2002). Effects of riparian leaf dynamics on periphyton photosynthesis and light utilisation efficiency. *Freshwater Biology* 47 (7): 1245–1256.

Holland, M.M., Risser, P.G., and Naiman, R.J. (1991). *Ecotones: The Role of Landscape Boundaries in the Management and Restoration of Changing Environments*. New York: Chapman and Hall.

Hrachowitz, M., Soulsby, C., Imholt, C. et al. (2010). Thermal regimes in a large upland salmon river: a simple model to identify the influence of landscape controls and climate change on maximum temperatures. *Hydrological Processes* 24 (23): 3374–3391. https://doi.org/10.1002/hyp.7756.

Hynes, H.B.N. (1975). The stream and its valley. *Verhandlungen Internationale Vereinigung Theoretische and Angewandte Limnologie* 19: 1–15.

Jacobs, T.C. and Gilliam, J.W. (1985). Riparian losses of nitrate from agricultural drainage waters. *Journal of Environmental Quality* 14 (4): 472–478.

Jencso, K.G., McGlynn, B.L., Gooseff, M.N. et al. (2009). Hydrological connectivity between landscapes and streams: transferring reach and plot scale understanding to the catchment scale. *Water Resources Research* 45: W04428. https://doi.org/10.1029/2008WR007225.

Jetten, M.S.M. (2001). New pathways for ammonia conversion in soil and aquatic systems. *Plant and Soil* 230: 9–19.

Johnson, M.F. and Wilby, R.L. (2015). Seeing the landscape for the trees: metrics to guide riparian shade management in river catchments. *Water Resources Research* 51 (5): 3754–3769. https://doi.org/10.1002/2014WR016802.

Johnson, S.L. (2004). Factors influencing stream temperatures in small streams: substrate effects and a shading experiment. *Canadian Journal of Fisheries and Aquatic Sciences* 61 (6): 913–923. https://doi.org/10.1139/f04-040.

Johnson, S.L. and Jones, J.A. (2000). Stream temperature responses to forest harvest and debris flows in western Cascades, Oregon. *Canadian Journal of Fisheries and Aquatic Sciences* 57 (S2): 30–39. https://doi.org/10.1139/f00-109.

Julian, J.P., Doyle, M.W., and Stanley, E.H. (2008). Empirical modeling of light availability in rivers. *Journal of Geophysical Research: Biogeosciences* 113 (G3).

Kelleher, C., Wagener, T., Gooseff, M. et al. (2012). Investigating controls on the thermal sensitivity of Pennsylvania streams. *Hydrological Processes* 26 (5): 771–785. https://doi.org/10.1002/hyp.8186.

Kirk, J.T.O. (1994). *Light and Photosynthesis in Aquatic Ecosystems*. Cambridge University Press.

Kirk, L., Hensley, R.T., Savoy, P. et al. (2020). Estimating benthic light regimes improves predictions of primary production and constrains light-use efficiency in streams and rivers. *Ecosystems* 24: 1–15.

Knowles, R. (1982). Denitrification. *Microbiological Reviews* 46: 43–70.

Ledesma, J.L., Grabs, T., Bishop, K.H. et al. (2015). Potential for long-term transfer of dissolved organic carbon from riparian zones to streams in boreal catchments. *Global Change Biology* 21 (8): 2963–2979. https://doi.org/10.1111/gcb.12872.

Lee, P., Smyth, C., and Boutin, S. (2004). Quantitative review of riparian buffer width guidelines from Canada and the United States. *Journal of Environmental Management* 70: 165–180.

Lee-Cullin, J.A., Zarnetske, J.P., Ruhala, S.S., and Plont, S. (2018). Toward measuring biogeochemistry within the stream-groundwater interface at the network scale: an initial assessment of two spatial sampling strategies. *Limnology & Oceanography: Methods* 16 (11): 722–733. https://doi.org/10.1002/lom3.10277.

Li, G., Jackson, C.R., and Kraseski, K.A. (2012). Modeled riparian stream shading: agreement with field measurements and sensitivity to riparian conditions. *Journal of Hydrology* 428: 142–151. https://doi.org/10.1016/j.jhydrol.2012.01.032.

Loicq, P., Moatar, F., Jullian, Y. et al. (2018). Improving representation of riparian vegetation shading in a regional stream temperature model using LiDAR data. *Science of the Total Environment* 624: 480–490. https://doi.org/10.1016/j.scitotenv.2017.12.129.

Lowrance, R.R., Todd, R.L., and Asmussen, L.E. (1984). Nutrient cycling in an agricultural watershed: I. Phreatic movement. *Journal of Environmental Quality* 13 (1): 22–27.

Lupon, A., Bernal, S., Poblador, S. et al. (2016). The influence of riparian evapotranspiration on stream hydrology and nitrogen retention in a subhumid Mediterranean catchment. *Hydrology and Earth System Sciences* 20: 3831–3824. https://doi.org/10.5194/hess-2016-56.

Lupon, A., Catalán, N., Martí, E., and Bernal, S. (2020). Influence of dissolved organic matter sources on in-stream net dissolved organic carbon uptake in a Mediterranean stream. *Water* 12 (6): 1722. https://doi.org/10.3390/w12061722.

Lupon, A., Denfeld, B.A., Laudon, H. et al. (2019). Groundwater inflows control patterns and sources of greenhouse gas emissions from streams. *Limnology & Oceanography* 64: 1545–1557.

Malcolm, I.A., Soulsby, C., Hannah, D.M. et al. (2008). The influence of riparian woodland on stream temperatures: implications for the performance of juvenile salmonids. *Hydrological Processes* 22 (7): 968–979. https://doi.org/10.1002/hyp.6996.

Maxted, J.R., McCready, C.H., and Scarsbrook, M.R. (2005). Effects of small ponds on stream water quality and macroinvertebrate communities. *New Zealand Journal of Marine and Freshwater Research* 39 (5): 1069–1084. https://doi.org/10.1080/00288330.2005.9517376.

Mayer, P.M., Reynolds, S.K., McCutchen, M.D., and Canfield, T.J. (2007). Meta-analysis of nitrogen removal in riparian buffers. *Journal of Environmental Quality* 36: 1172–1180. https://doi.org/10.2134/jeq2006.0462.

McClain, M.E., Boyer, E.W., Dent, C.L. et al. (2003). Biogeochemical hot spots and hot moments at the interface of terrestrial and aquatic ecosystems. *Ecosystems* 6: 301–312.

Minshall, G.W. (1978). Autotrophy in stream ecosystems. *BioScience* 28 (12): 767–771.

Moore, D.R., Spittlehouse, D.L., and Story, A. (2005). Riparian microclimate and stream temperature response to forest harvesting: a review. *Journal of the American Water Resources Association* 41 (4): 813–834. https://doi.org/10.1111/j.1752-1688.2005.tb03772.x.

Mulholland, P.J. (1992). Regulation of nutrient concentrations in a temperate forest stream: roles of upland, riparian, and instream processes. *Limnology & Oceanography* 37 (7): 1512–1526.

Mulholland, P.J., Fellows, C.S., Tank, J.L. et al. (2001). Inter-biome comparison of factors controlling stream metabolism. *Freshwater Biology* 46 (11): 1503–1517.

Mulholland, P.J., Helton, A.M., Poole, G.C. et al. (2008). Stream denitrification across biomes and its response to anthropogenic nitrate loading. *Nature* 452 (7184): 202–205.

Nair, V.D., Clark, M.W., and Reddy, K.R. (2015). Evaluation of legacy phosphorus storage and release from wetland soils. *Journal of Environmental Quality* 44 (6): 1956–1964.

Nakano, S. and Murakami, M. (2001). Reciprocal subsidies: dynamic interdependence between terrestrial and aquatic food webs. *Proceedings of the National Academy of Sciences* 98 (1): 166–170. https://doi.org/10.1073/pnas.98.1.166.

Newbold, J.D., Elwood, J.W., O'Neill, R.V., and Sheldon, A.L. (1983). Phosphorus dynamics in a woodland stream ecosystem – a study of nutrient spiraling. *Ecology* 64: 1249–1265. https://doi.org/10.2307/1937833.

Newbold, J.D., Elwood, J.W., O'Neill, R.V., and Van Winkle, W. (1981). Measuring nutrient spiralling in streams. *Canadian Journal of Fisheries and Aquatic Sciences* 38: 860–863. https://doi.org/10.1139/f81-114.

Newcomer, T.A., Kaushal, S.S., Mayer, P.M. et al. (2012). Influence of natural and novel organic carbon sources on denitrification in forest, degraded urban, and restored streams. *Ecological Monographs* 82 (4): 449–466.

Ocampo, C.J., Oldham, C.E., and Sivapalan, M. (2006). Nitrate attenuation in agricultural catchments: shifting balances between transport and reaction. *Water Resources Research* 42: W01408. https://doi.org/10.1029/2004WR003773.

Peterjohn, W.T. and Correll, D.L. (1984). Nutrient dynamics in an agricultural watershed: observations on the role of a riparian forest. *Ecology* 65 (5): 1466–1475.

Peterson, B.J., Wollheim, W.M., Mulholland, P.J. et al. (2001). Control of nitrogen export from watersheds by headwater streams. *Science* 292: 86–90.

Pinay, G. (1986). *Relations sol - nappe dans les bois riverains de la Garonne. Etude de la dénitrification*. 200. France: Université C. Bernard, Lyon I.

Pinay, G., Bernal, S., Abbott, B.W. et al. (2018). Riparian corridors: a new conceptual framework for assessing nitrogen buffering across biomes. *Frontiers in Environmental Science* https://doi.org/10.3389/fenvs.2018.00047.

Pinay, G., Black, V.J., Planty-Tabacchi, A.M. et al. (2000). Geomorphic control of denitrification in large river floodplain soils. *Biogeochemistry* 50: 163–182.

Pinay, G. and Décamps, H. (1988). The role of riparian woods in regulating nitrogen fluxes between the alluvial aquifer and surface water: a conceptual model. *Regulated Rivers* 2: 507–516.

Pinay, G. et al. (2015). Upscaling nitrogen removal capacity from hot spot to the landscape. *Ecosystems* 18 (6): 1101–1120.

Pinay, G., Gumiero, B., Tabacchi, E. et al. (2007). Patterns of denitrification rates in European alluvial soils under various hydrological regimes. *Freshwater Biology* 52: 252–266.

Pinay, G. and Labroue, L. (1986). Une station d'épuration naturelle des nitrates transportés par les nappes alluviales: l'aulnaie glutineuse. *Comptes Rendus de l'Académie des Sciences de Paris* 302 III (17): 629–632.

Pinay, G., O'Keefe, T.C., Edwards, R.T., and Naiman, R.J. (2009). Nitrate removal in the hyporheic zone of a salmon river in Alaska. *River Research and Applications* 25: 367–375.

Pinay, G., Ruffinoni, C., Wondzell, S., and Gazelle, F. (1998). Change in groundwater nitrate concentration in a large river floodplain: denitrification, uptake or mixing? *Journal of North American Benthological Society* 17 (2): 179–189.

Poblador, S., Lupon, A., Sabaté, S., and Sabater, F. (2017). Soil water content drives spatiotemporal patterns of CO_2 and N_2O emissions from a Mediterranean riparian forest soil. *Biogeosciences* 14: 4195–4208. https://doi.org/10.5194/bg-14-4195-2017.

Ponnamperuma, F.N. (1972). The chemistry of submerged soils. *Advances in Agronomy* 24: 29–96.

Raymond, P.A., Hartmann, J., Lauerwald, R. et al. (2013). Global carbon dioxide emissions from inland waters. *Nature* 503: 355–359.

Raymond, P.A., Saiers, J.E., and Sobczak, W.V. (2016). Hydrological and biogeochemical controls on watershed dissolved organic matter transport: pulse-shunt concept. *Ecology* 97 (1): 5–16. https://doi.org/10.1890/14-1684.1.

Reddy, K.R. and Patrick, W.H., Jr. (1975). Effect of alternate aerobic and anaerobic conditions on redox potential, organic matter decomposition and nitrogen loss in a flooded soil. *Soil Biology and Biochemistry* 7: 87–94.

Roberts, B.J., Mulholland, P.J., and Hill, W.R. (2007a). Multiple scales of temporal variability in ecosystem metabolism rates: results from 2 years of continuous monitoring in a forested headwater stream. *Ecosystems* 10 (4): 588–606.

Roberts, B.J., Mulholland, P.J., and Houser, J.N. (2007b). Effects of upland disturbance and instream restoration on hydrodynamics and ammonium uptake in headwater streams. *Journal of the North American Benthological Society* 26 (1): 38–53.

Roberts, J. and Ludwig, J.A. (1991). Riparian vegetation along current-exposure gradients in floodplain wetlands of the River Murray, Australia. *Journal of Ecology* 79: 117–127.

Rutherford, J.C., Blackett, S., Blackett, C. et al. (1997). Predicting the effects of shade on water temperature in small streams. *New Zealand Journal of Marine and Freshwater Research* 31 (5): 707–721. https://doi.org/10.1080/00288330.1997.9516801.

Sabater, S., Butturini, A., Clément, J.C. et al. (2003). Nitrogen removal by riparian buffers under various N loads along a European climatic gradient: patterns and factors of variation. *Ecosystems* 6: 20–30.

Salo, J., Kalliola, R., Häkkinen, J. et al. (1986). River dynamics and the diversity of Amazon lowland forest. *Nature* 332: 254–258.

Saurer, M., Spahni, R., Frank, D.C. et al. (2014). Spatial variability and temporal trends in water-use efficiency of European forests. *Global Change Biology* 20 (12): 3700–3712. https://doi.org/10.1111/gcb.12717.

Seitzinger, S., Harrison, J.A., Böhlke, J.K. et al. (2006). Denitrification across landscapes and waterscapes: a synthesis. *Ecological Applications* 16 (6): 2064–2090.

Seyedhashemi, H., Moatar, F., Vidal, J.P. et al. (2020). Thermal signatures identify the influence of dams and ponds on stream temperature at the regional scale. *Science of the Total Environment* 142667. https://doi.org/10.1016/j.scitotenv.2020.142667.

Shade, J.D., Welter, J.R., Martí, E., and Grimm, N.B. (2005). Hydrological exchange and N uptake by riparian vegetation in an arid-land stream. *Journal of North American Benthological Society* 24 (1): 19–28.

Siegenthaler, U. and Sarmiento, J.L. (1993). Atmospheric carbon dioxide and the ocean. *Nature* 365: 119–125.

Stonedahl, S. H. et al. (2010). A multiscale model for integrating hyporheic exchange from ripples to meanders. *Water Resources Research* 46: https://doi.org/10.1029/2009WR008865.

Sullivan, K.T., Lisle, C.A., Dollof, G.E., and Reid, I.M. (1987). Stream channels: the links between forests and fishes. In: *Streamside Management: Forestry and Fishery Interactions* (ed. E.O. Salo and T.W. Cundy), Seattle: University of Washington, Institute of Forest Resources. Contribution No. 57.

Tabacchi, E., Correll, D.L., Hauer, R. et al. (1998). Development, maintenance and role of riparian vegetation in the river landscape. *Freshwater Biology* 40 (1): 1–21.

Van Meter, K.J., Chowdhury, S., Byrnes, D.K., and Basu, N.B. (2020). Biogeochemical asynchrony: ecosystem drivers of seasonal concentration regimes across the Great Lakes Basin. *Limnology & Oceanography* 65 (4): 848–862.

Vannote, R.L., Minshall, G.W., Cummins, K.W. et al. (1980). The river continuum concept. *Canadian Journal of Fisheries and Aquatic Sciences* 37: 130–137.

Ward, J.V. (1989). The four dimentional nature of lotic ecosystems. *Journal of North American Benthological Society* 8 (1): 2–8.

Weaver, J.E. (1960). Flood plain vegetation of the central Missouri Valley and contacts of woodland with prairie. *Ecological Monographs* 30: 37–64.

Wologo, E., Shakil, S., Zolkos, S. et al. (2021). Stream dissolved organic matter in permafrost regions shows surprising compositional similarities but negative priming and nutrient effects. *Global Biogeochemical Cycles* 35 (1): e2020GB006719. https://doi.org/10.1029/2020GB006719.

Wondzell, S.M., Gooseff, M.N., and McGlynn, B.L. (2010). An analysis of alternative conceptual models relatins hyporheic exchange flow to diel fluctuations in discharge during baseflow recession. *Hydrological Processes* 24: 686–694.

Wondzell, S.M. and Swanson, F.J. (1996). Seasonal and storm dynamics of the hyporheic zone of a 4th-order mountain stream. 2. Nitrogen cycling. *Journal of North American Benthological Society* 15: 20–34. https://doi.org/10.2307/1467430.

Wrage, N., Velthof, G.L., van Beusichem, M.L., and Oenema, O. (2001). Role of nitrifier denitrification in the production of nitrous oxide. *Soil Biology and Biochemistry* 33 (12–13): 1723–1732.

Zarnetske, J.P., Bouda, M., Abbott, B.W. et al. (2018). Generality of hydrologic transport limitation of watershed organic carbon flux across ecoregions of the United States. *Geophysical Research Letters* 45 (21): 11,702–11,711. https://doi.org/10.1029/2018GL080005.

Zarnetske, J.P., Haggerty, R., Wondzell, S.M., and Baker, M.A. (2011). Dynamics of nitrate production and removal as a function of residence time in the hyporheic zone. *Journal of Geophysical Research* 116: G01025. https://doi.org/10.1029/2010JG001356.

4

Organizational Principles of Hyporheic Exchange Flow and Biogeochemical Cycling in River Networks across Scales

Stefan Krause[1], Benjamin W. Abbott[2], Viktor Baranov[3], Susana Bernal[4], Phillip Blaen[1], Thibault Datry[5], Jennifer Drummond[4], Jan H. Fleckenstein[6], Jesus Gomez Velez[7], David M. Hannah[1], Julia L. A. Knapp[8], Marie Kurz[9], Jörg Lewandowski[10], Eugènia Martí[4], Clara Mendoza-Lera[11], Alexander Milner[1], Aaron Packman[12], Gilles Pinay[13], Adam S. Ward[14], and Jay P. Zarnetzke[15]

[1] *University of Birmingham, Birmingham, B15 2TT, UK*
[2] *Brigham Young University, Provo, UT, USA*
[3] *Ludwig Maximillian University Munich, Munich, Germany*
[4] *Center for Advanced Studies of Blanes, Blanes, Spain*
[5] *Institut National de Recherche en Sciences et Technologies pour L'Environnement et L'Agriculture – IRSTEA, Paris, France*
[6] *Helmholtz Center for Environmental Research – UFZ, Leipzig, Germany*
[7] *Vanderbilt University, Nashville, TN, USA*
[8] *Durham University, Durham, UK*
[9] *Drexel University, Philadelphia, PA, USA*
[10] *Leibniz-Institute of Freshwater Ecology and Inland Fisheries, Berlin, Germany*
[11] *University of Koblenz-Landau, Mainz, Germany*
[12] *Northwestern University, Evanston, IL, USA*
[13] *Centre National de la Recherche Scientifique – CNRS, Paris, France*
[14] *Indiana University Bloomington, Bloomington, IN, USA*
[15] *Michigan State University, East Lansing, MI, USA*

4.1 Introduction – The Need for Identifying Landscape Organizational Principles of Hyporheic Zone Functioning

Over the past three decades, there has been a paradigm shift in groundwater–surface water research from defining rivers and aquifers as discrete entities towards understanding them as an integral part of the stream-catchment continuum (Bencala 1993, 2000; Bencala et al. 2011; Boulton and Hancock 2006; Brunke and Gonser 1997; Fleckenstein et al. 2010; Harvey and Gooseff 2015; Jencso et al. 2009; Winter 1998, 1999). This shift started with the early works of hydrobiologists defining the hyporheos, or hyporheic zone, as a new habitat (Angelier 1953; Karaman 1935; Orghidan 1959). From an interdisciplinary perspective, the hyporheic zone may be defined as the interface between aquifers and streams where the flow of surface water into streambed sediments and river banks promotes interactions with streambed porewater and potentially groundwater before re-emerging to the stream (Bencala 2000; Boano et al. 2014; Cardenas 2015; Gooseff 2010; Krause et al. 2011b; Ward 2016).

Ecohydrological Interfaces, First Edition. Edited by Stefan Krause, David M. Hannah, and Nancy B. Grimm.

Interdisciplinary research focused on the hydrological, biogeochemical, and ecological functioning of hyporheic zones has aimed to understand the extent to which this ecohydrological interface (Krause et al. 2017) influences surface water and groundwater flows, water quality, and ecological status of freshwaters (Boano et al. 2014; Boulton 2007; Boulton et al. 2010, 1998; Brunke and Gonser 1997; Buffington and Tonina 2009; Graham et al. 2019; Harvey and Gooseff 2015). Hyporheic zones both structure and connect surface and subsurface environments, providing key ecosystem services, including stream flow modulation (Costigan et al. 2016; Fleckenstein et al. 2006; Malzone et al. 2016; Villeneuve et al. 2015), moderation of thermal conditions in streams and groundwater (e.g. Arrigoni et al. 2008; Burkholder et al. 2008; Folegot et al. 2018a; Hannah et al. 2009; Krause et al., 2011a), enhanced metabolism and biogeochemical cycling (Bardini et al. 2012; DelVecchia et al. 2016; Fellows et al. 2001; Gomez-Velez et al. 2015; Knapp et al. 2017; Krause et al. 2009, 2013; Pinay et al. 2009; Schmadel et al. 2016a; Trauth et al. 2015; Valett et al. 1997), and contaminant transformation (Freitas et al. 2015; Gandy et al. 2007; Jaeger et al. 2019; Lawrence et al. 2013; Liu et al. 2019; Packman and Brooks 2001; Posselt et al. 2020; Schaper et al. 2019; Weatherill et al. 2018, 2014). With their distinct physicochemical conditions, hyporheic zones provide unique habitats that serve as refugia for a wide range of species (Boulton et al. 2010; Datry and Larned 2008; Datry et al. 2008; Folegot et al. 2018b; Hancock et al. 2005; Stanford and Ward 1988). Clearly, hyporheic zones, and their hydrological interactions with surface water and groundwater, are important for the resilience of freshwater ecosystems to hydrological extremes, landscape development, and global environmental changes including hydrological and climate modifications (Dole-Olivier 2011; Dole-Olivier et al. 1997; Fisher et al. 1998; Lewandowski et al. 2019; Nelson et al. 2020; Stubbington et al. 2009).

Several landscape-scale conceptualizations of river corridors highlight the relevance of connectivity between groundwater and surface water (Bencala et al. 2011, 2011; Buffington and Tonina 2009; Harvey and Gooseff 2015; Jencso et al. 2009; Malard et al. 2002; Pinay et al. 2015; Winter 1998). Evidence from experimental and modelling studies reveals the impact of stream flow velocity and bedform geometry on exchange flow (Cardenas and Wilson 2007a; Cardenas et al. 2004; Harman et al. 2016; Marzadri et al. 2010; Trauth et al. 2013), residence time distributions (Cardenas et al. 2008; Hester and Doyle 2008), and biogeochemical cycling (Pinay et al. 2009; Zarnetske et al. 2011a, 2011b) in the hyporheic zone. Previous research has demonstrated the functional significance of hyporheic zone ecosystem services (Boulton et al. 1998; Brunke and Gonser 1997; Stanford and Ward 1993) and has explored the drivers and controls of specific hyporheic zone processes and their relevance at multiple scales and in different landscape contexts (Buffington and Tonina 2009). However, the landscape-wide organizational principles of the drivers and controls of key hyporheic functions in space and time remain elusive (Harvey and Gooseff 2015; Krause et al., 2014a; Krause et al. 2017; Lee-Cullin et al. 2018; Pinay et al. 2015; Stonedahl et al. 2010). The longstanding focus on conceptualizing the principles of hyporheic zone functioning primarily at the local and reach scale is understandable, given the difficulty investigating this environment. However, this local focus constrains our understanding of the influence of the hyporheic zone on relevant hydrological, biogeochemical, and ecological processes at larger spatial scales, particularly at the catchment scale. Additionally, as studies of temporal dynamics in drivers of hyporheic exchange are now emerging (Hester et al. 2019; Kaufman et al. 2017; Malzone et al. 2016; Singh et al. 2019; Wu et al. 2018a), we have a new opportunity to describe and predict hyporheic zone functioning through space and time (Boano et al. 2014; Cardenas 2015; Gooseff 2010; Harvey and Gooseff 2015; Krause et al. 2011b, 2017; Lee-Cullin et al. 2018; Ward 2016; Ward and Packman 2019).

We propose that current limitations of upscaling conceptual models of hyporheic zone hydrological and biogeochemical functioning towards a landscape perspective can be overcome by:

i) Better integration and synthesis of complexity from field observations across different scales (and beyond small headwater streams) with systematic modelling and controlled laboratory studies; and
ii) Rigorous testing of assumptions of drivers and controls of hyporheic process dynamics at their specific scale before extrapolating process knowledge from small-scale studies to the landscape context.

Technological advances in sensor and tracer technologies have yielded very detailed data from field investigations, enabling quantifications of hyporheic residence time distributions (Marçais et al. 2018; Rinaldo et al. 2015) and resulting influences on biogeochemical processes under site-specific conditions and hydro-geomorphic settings (González-Pinzón et al. 2015; Harvey et al. 2013; Krause et al. 2013, 2014a, 2014b; Zarnetske et al. 2011a). The resulting mechanistic process knowledge helps one to understand hyporheic zone functioning under those site-specific conditions. However, the transferability of process understanding to other sites and conditions is still limited because the broader context of the drivers and controls of hyporheic exchange and biogeochemical reactivity are complex and difficult to observe, and the dominant underlying mechanisms that interact in their situation-specific control of hyporheic exchange flow processes have not been investigated in sufficient detail to enable cross-site comparisons or upscale projections. Moreover, different field studies reveal that a wide range of site-specific conditions control the relative importance of drivers and controls of hyporheic zone processes at particular locations and scales (Endreny and Lautz 2012; Jones and Holmes 1996; Krause et al. 2012b; Munz et al. 2011; Ward and Packman 2019). These conditions complicate further generalizations, such as the potential relevance of small-scale low-conductivity structures in streambed sediments for larger-scale patterns of hyporheic zone processes (Bardini et al. 2013; Gomez-Velez et al. 2014; Laube et al. 2018; Sawyer 2015; Sawyer and Cardenas 2009). We suggest that these limiations can be overcome by accounting for the wider landscape controls of the broad range of encountered site-specific variability in streambed properties. Generalizing and transferring process understanding and concepts across river systems and spatial scales beyond the specific study area is difficult, but the focus to date on local understanding has limited the possibilities for advancing our understanding of hyporheic zone functioning within the wider river network and landscape context.

A substantial amount of our existing theory and understanding of hyporheic zone processes has been based on systematic studies designed to advance beyond limitations of individual system observations and to analyse the dynamics of hyporheic zone processes across a range of conditions. In particular, systematic modelling (Bardini et al. 2012; Boano et al. 2010a, 2010b; Cardenas et al. 2004; Gomez-Velez et al. 2014) and controlled laboratory studies in flumes (Arnon et al. 2009; Fox et al. 2014, 2016; Salehin et al. 2004; Thibodeaux and Boyle 1987) have revealed key mechanisms controlling hyporheic exchange fluxes and their associated residence time and ecological function. However, we frequently fail to relate core principles identified in these controlled studies to observations of more complex dynamics and patterns at a river network scale. This failure suggests multiple knowledge gaps that prevent us from effectively linking the design and underlying assumptions of many of our systematic modelling studies to the actual governing mechanisms of those process dynamics and their spatiotemporal variability that can be observed in situ. As an example, current conceptual models propose that hyporheic residence times and the relationship between residence and reaction times (as expressed by the non-dimensional Damköhler number; Marzadri et al. 2012; Pinay et al. 2015; Zarnetske et al. 2012) act as a primary control on the fate of reactive solutes in the hyporheic zone. Longer residence time in the hyporheic zone results in a shift from aerobic to anaerobic metabolic pathways, including denitrification, sulphur reduction, and methanogenesis (Briggs et al. 2014; Pinay et al. 2009; Trauth et al. 2014, 2015; Zarnetske et al. 2011a). However, despite promising advances in representing spatial variability in physical sediment properties (Tonina et al. 2016) and improved in situ measurements (Bray and Dunne 2017; Ryan and Boufadel 2007), field observations frequently reveal hyporheic carbon, nitrogen, and oxygen concentration patterns that are inconsistent with the assumption that bulk hyporheic residence time controls biogeochemical reactions and turnover rates. In this sense, considering the spatial variability in sediment biogeochemical reactivity resulting from the structural controls such as the patterns of deposited sediments and autochthonous reagents (e.g. terrestrial organic carbon; Krause et al. 2013), microbial community structure and activity may help understanding the observed patterns.

There is a similar disconnection between empirical observations and conceptual models for the effects of streambed structural heterogeneity on hyporheic exchange, residence time distributions, and nutrient cycling (Bardini et al. 2012, 2013; Cardenas et al. 2004; Laube et al. 2018; Sawyer and Cardenas 2009; Tonina et al. 2016). This divergence often occurs due to the significant spatial heterogeneity of hyporheic exchange (Genereux et al. 2008) as well as biogeochemical properties, nutrient concentrations, and turnover rates observed in hyporheic zone laboratory and field studies across multiple scales (Hou et al. 2017; Krause et al. 2013; Marion et al. 2008; Packman et al. 2006; Salehin et al. 2004). Consequently, the conceptual boundaries set for many model studies might be based on assumptions which do not necessarily represent the most relevant processes governing the respective real-world context.

We suggest intensifying our efforts on improving the transferability of findings required to overcome fragmentation in process understanding and to increase our capacity to conduct, interpret, and conceptualize field observations across river network and landscape scales. Useful strategies include the development of standardized methodologies for collecting comparable hyporheic zone data and understanding of the drivers of their landscape organizing principles (Barthel and Banzhaf 2016; González-Pinzón et al. 2015; Krause et al. 2011b; Lee-Cullin et al. 2018), consistent descriptions of metadata to enable synthesis efforts, and organized synoptic field sampling to assess global patterns in exchange processes and the resultant ecosystem services.

There is a critical need for integrating and advancing existing conceptual approaches to identify landscape organizational principles of hyporheic exchange flow and biogeochemical processing in order to better contextualize and understand the role of hyporheic zone functioning in river networks across both spatial and temporal scales. The principal aim of this chapter is to provide a comprehensive analysis and synthesis of the interactions between important drivers and controls of hyporheic exchange and biogeochemical cycling and how they vary across scales, integrating results from a wide range of case studies that go beyond current conceptual model frameworks. In Section 4.2, we therefore discuss the interactions of different local-to-regional controls and drivers of hyporheic zone processes such as hydrodynamic and hydrostatic drivers of hyporheic exchange, sediment hydraulic conductivity, the role of autochthonous organic matter sources, and feedbacks between hydrological exchange and ecological processes in the streambed. We explore the implications of these interactions for biogeochemical cycling in the landscape context. Emerging from this discussion, we identify existing knowledge gaps and mismatches between empirical observations and current concepts and theories. In Section 4.3, we integrate conceptualizations of organizational principles of hyporheic exchange and biogeochemical cycling from reach to catchment scale. We expect that increasing awareness and embracing the landscape organizing principles of hyporheic zones will advance the future of research at groundwater–surface water interfaces.

4.2 Drivers and Controls of Hyporheic Exchange Flow: Unravelling Spatiotemporal Complexity and Their Implications for Biogeochemical Cycling

Mechanistic understanding of hyporheic exchange has advanced significantly in recent years with a large body of field-based, laboratory (flumes), and numerical model investigations. These studies have revealed how hydrostatic and hydrodynamic drivers of hyporheic exchange are controlled by regional flow acting on local head gradients and patterns of stream flow velocity, channel morphology, and flow turbulence (Boano et al. 2006, 2007; Bottacin-Busolin and Marion 2010; Buffington and Tonina 2009; Cardenas and Wilson 2007a; Cardenas et al. 2004; Fox et al. 2014; Hester and Doyle 2008). Despite this significant progress, the relative importance of different individual drivers and controls of hyporheic exchange, their scale-specific and context-dependent relevance, and the principles that explain their spatial organization in river networks and landscapes are still under debate (Boano et al. 2014; Gomez-Velez et al. 2014; Krause et al. 2014b; Stonedahl et al. 2010; Tonina and Buffington 2011; Ward 2016). This lack of consensus drives us to unravel the importance of both spatial variability (e.g. sediment hydraulic conductivity, autochthonous organic matter) and temporal dynamics (e.g. in stream flow and stage) as drivers and controls of hyporheic exchange. Moreover, we embrace the idea that this fluvial structural variability strongly influences spatial patterns and temporal dynamics of biogeochemical cycling in hyporheic zones, and ultimately in river networks. In the following sections we discuss – and at times challenge – accepted conceptualizations of drivers and controls of hyporheic exchange and biogeochemical cycling by presenting evidence from field, laboratory, and modelling experiments that do not always fit or may even contradict the application of current concepts and theory.

4.2.1 Interactive Effects of Hydrodynamic and Hydrostatic Drivers of Hyporheic Exchange Flow

Deconvolution of the combined effects of multiple geomorphic drivers is essential for quantifying scale-dependencies of hyporheic exchange, residence time distributions and subsequently, biogeochemical transformation rates. Hyporheic exchange and associated hyporheic residence time distributions vary and interact across orders of mangitude in spatial and temporal scales (Figure 4.1). These exchanges range from relatively short-term (seconds–minutes) and small spatial scale (mm) dynamics to long-term (weeks–years) and large-scale (km) patterns, such as inter-meander flow of several hundreds of metres and beyond (Boano et al. 2014; Krause et al. 2011b; Stonedahl et al. 2010; Wondzell 2011).

The drivers of hyporheic exchange, forcing surface water to down-well into the streambed and reside in the hyporheic zone before re-emerging to the river (Figure 4.2a), include hydrodynamic and hydrostatic forcings. Hydrodynamic drivers include stream flow turbulence (Boano et al. 2011; Cardenas and Wilson 2007a; Roche et al. 2018, 2019; Trauth et al. 2013) and advective pumping induced by bedforms such as ripples, dunes, steps, and pool-riffle sequences (Boano et al. 2014, 2007; Bottacin-Busolin and Marion 2010; Buffington and Tonina 2009; Cardenas et al. 2004; Elliott and Brooks 1997a; Singh et al. 2019; Storey et al. 2003; Trauth et al. 2014), or by flow obstacles such as weirs, boulders, woody debris, and streambed engineering, or restoration structures (Briggs et al. 2013; Kasahara and Hill 2006; Krause et al. 2014b; Wondzell et al., 2009b; Zhou and Endreny 2013). Hydrodynamic drivers usually create small and intermediate scale hyporheic exchange at scales ranging from 0.1–100 m (Figure 4.1). Concurrent hydrostatic forcing is induced by elevation head gradients between surface water and groundwaters or between different surface waters such as river branches or confluences. Hydrostatic gradients are primarily controlled by the morphology of individual features, or larger river structures (e.g. multi-thread channels), and regional groundwater flows. Hydrostatic drivers cause exchange via mechanisms including bank exchange, inter-meander flow, and large-scale groundwater circulation cells ranging from tens of metres to several kilometres (Boano et al. 2006, 2010a, 2010c; Gooseff et al. 2006; Munz et al. 2011; Pinay et al. 1998) (Figure 4.1).

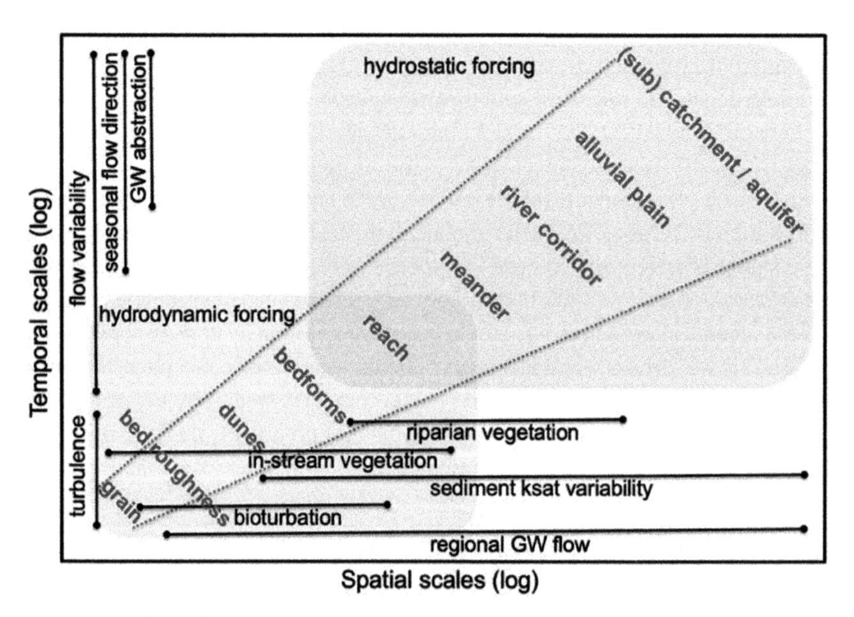

Figure 4.1 Spatial and temporal scales, overlaps and interactions of example physical and biological drivers and controls of hyporheic exchange flow and hyporheic residence time distributions (adapted partly from Boano et al. 2014) with consequences for streambed solute mixing and biogeochemical cycling *(the processes depicted are selected to demonstrate scale overlaps and are not claiming to be exhaustive; sediment ksat = sediment hydraulic conductivity).*

The majority of hydrological studies on flow in the hyporheic zone have conceptualized hyporheic exchange as a result of either small-scale (streambed feature scale) or large-scale (catchment-scale) drivers (Figure 4.1). Few studies have so far attempted to quantify the impact of interactions, both potentially overlapping or counter-acting, between hydrodynamic and hydrostatic forces across spatial and temporal scales (Figure 4.1). Experimental findings include suppression of local hyporheic exchange by regional groundwater up-welling (Angermann et al. 2012; Krause et al. 2011a, 2009, 2013), which has been systematically investigated in a range of conceptual models of bedform-induced hyporheic exchange impacted by groundwater up-welling and/or ambient lateral groundwater flow (Boano et al. 2008, 2009; Cardenas and Wilson 2006, 2007b; Storey et al. 2003; Trauth and Fleckenstein 2017; Trauth et al. 2013, 2015; Wu et al. 2018a) (Figure 4.2b). However, the impact of regionally losing conditions that potentially expand the hyporheic zone and enhance hyporheic exchange and broaden residence time distributions (Figure 4.2c) has been less examined (De Falco et al. 2016; Fox et al. 2014; Preziosi-Ribero et al. 2020). In particular, the combined influence of hydrodynamic and hydrostatic forcings on hyporheic exchange and biogeochemical cycling under losing conditions still needs to be established in detail (Trauth et al. 2015).

Systematic analyses of hyporheic exchange and hyporheic residence time distributions have predominantly investigated the impact of singular features and successions thereof (Bardini et al. 2013; Boano et al. 2007; Bottacin-Busolin and Marion 2010; Cardenas and Wilson 2006; Cardenas et al. 2008; Elliott and Brooks 1997b; Herzog et al. 2019). Previous research has provided increased evidence of the co-existence of the integrated, and often nested, impacts of different geomorphic structures on hyporheic exchange, such as the overlapping effects of ripples along pool-riffle structures nested in an inter-meander flow system (Azizian et al. 2017; Kasahara and Wondzell 2003; Poole et al. 2008; Stonedahl et al. 2012, 2010). The complexity of overlapping geomorphic drivers and controls of hyporheic exchange (Figure 4.1) requires advanced observation that explicitly focuses on understanding and conceptualization of hierarchical, interacting geomorphological drivers, which are commonly analysed separately. Such an integrated approach allows systematic exploration of the conditions under which either the impacts of small-scale processes are expressed at larger scales, or the conditions under which the effects of small-scale processes are overwhelmed by larger-scale drivers (Herzog et al. 2019; Krause et al. 2017; Stonedahl et al. 2010).

4.2.2 Potential Influence of Heterogeneous Substrate Hydraulic Conductivity on Hyporheic Exchange Flow

The nested influence of hydrostatic and hydrodynamic forcings of hyporheic exchange is modified by the spatial patterns and temporal dynamics of substrate hydraulic conductivity (Conant 2004; Genereux et al. 2008; Hester et al. 2017, 2019; Stewardson et al. 2016). Controlled flume experiments (Fox et al. 2014; Salehin et al. 2004) and field studies (Genereux et al. 2008; Krause et al. 2013; Weatherill et al. 2014) confirm that even small-scale spatial variability of sediment hydraulic conductivity can have the potential to substantially impact hyporheic exchange and residence time distributions because of preferential flow through higher-conductivity pathways. Only a limited number of numerical modelling studies consider spatial heterogeneity in streambed properties, such as hydraulic conductivity (Bardini et al. 2013; Cardenas et al. 2004; Gomez-Velez et al. 2014; Irvine

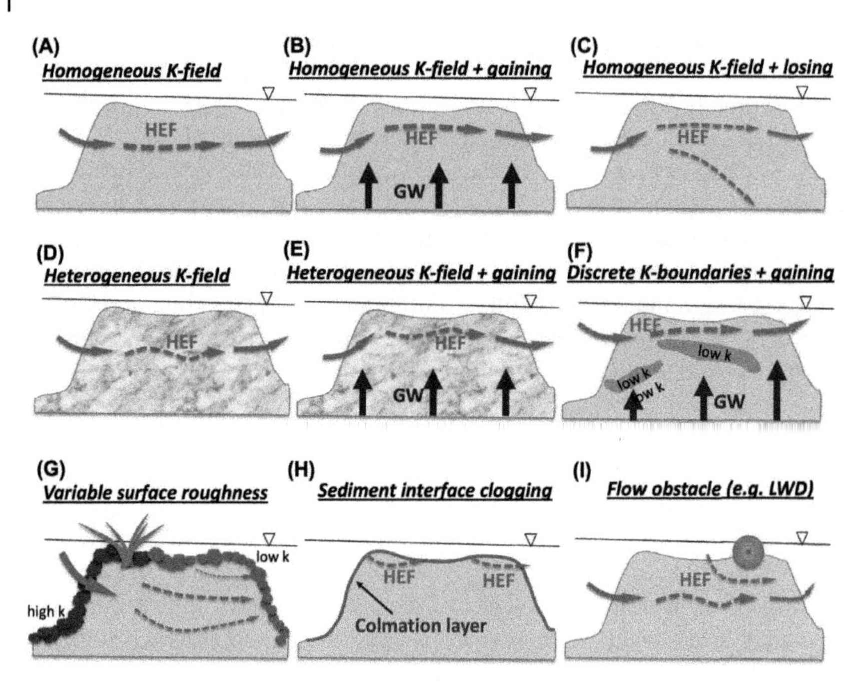

Figure 4.2 Drivers and controls of hyporheic exchange flow (HEF) through stream bedforms in different landscape contexts with predominantly surface hydrology driven HEF and homogeneous streambed sediments (A), interacting surface water and groundwater (GW) forcings on HEF and homogeneous streambed sediments under gaining (B) and losing (C) conditions, dominant surface water forcing in spatially heterogeneous sediments without (D) and with (E) GW up-welling, under consideration of highly heterogeneous sediment hydraulic conductivities with discrete sediment boundaries (F), under the influence of variable streambed surface roughness through sediment properties and vegetation (G), sediment clogging (H), and stream flow obstacles (I) such as large woody debris (LWD).

et al. 2012; Laube et al. 2018; Poole et al. 2006; Salehin et al. 2004; Sawyer and Cardenas 2009) (Figure 4.2d). With few exceptions (Bray and Dunne 2017; Cardenas et al. 2004; Tonina et al. 2016), hydraulic conductivity in numerical modelling studies is conceptualized to be spatially homogeneous, despite abundant field evidence of the importance of river bed heterogeneity in physical properties, such as hydraulic conductivity, particularly in mid-stream or lowland sections of high-order streams (Chen 2004; Genereux et al. 2008; Krause et al. 2012a, 2013; Leek et al. 2009; Mendoza-Lera and Datry 2017; Mendoza-Lera et al. 2019; Sebok et al. 2015; Stewardson et al. 2016; Wondzell et al. 2009a). This common assumption highlights a possible bias in model parameterization towards conditions that have been more frequently observed in low-order headwater streams or in alluvial rivers with moderate to low hydraulic gradients and not constrained vertically or laterally.

Initial modelling studies that aimed to quantify streambed heterogeneity impacts on hyporheic exchange, residence time distributions, and biogeochemical cycling considered some limited spatial variability of streambed hydraulic conductivity with patterns often characterized by assumed correlation lengths (Bardini et al. 2013; Cardenas et al. 2004; Salehin et al. 2004; Sawyer and Cardenas 2009). Such studies could be extended towards analysis of how this structural variability affects the interactions of groundwater up-welling and hyporheic exchange (Figure 4.2e). This variability has been observed and simulated independently, but not included together in integrated multi-scale models. With a few exceptions (Gomez-Velez et al. 2014; Laube et al. 2018; Y. Zhou et al. 2014b), previous conceptual modelling studies do not consider the effects of many orders of magnitude differences in channel morphology and hydraulic conductivity found in situ (Chen 2004; Conant 2004; Fox et al. 2014; Genereux et al. 2008; Krause et al. 2013; Nowinski et al. 2011; Weatherill et al. 2014) (Figure 4.2f). These limitations propagate to conclusions that have suggested only limited impacts of streambed structural heterogeneity on residence time distributions and biogeochemical cycling in the hyporheic zone (Bardini et al. 2013; Laube et al. 2018). In addition, it is crucial to determine how streambed sediment structures and hydraulic conductivity patterns are controlled by ecological drivers such as interactions between aquatic vegetation and streambed sediments (Baranov et al. 2017; Jones et al. 2012, 2008; Ullah et al. 2014) (Figure 4.2g) causing sediment clogging and by particle deposition and biofilm growth (Brunke 1999; Nogaro et al. 2010; Rode et al. 2015) (Figure 4.2h), bioengineers causing bioturbation (Baranov et al. 2016; Mendoza-Lera and Mutz 2013), and the impact of flow obstacles such as large woody debris (Gippel 1995; Krause et al. 2014b; Sawyer et al. 2012; Shelley et al. 2017) (Figure 4.2i). These processes are critical to engineering streambeds for purposes such as nutrient removal and river restoration, which involve using spatial heterogeneity to control fluxes and residence times to achieve desired outcomes (Herzog et al. 2018; Vaux 1968; Ward et al. 2011).

4.2.3 Multi-scale Interactions of Lateral and Vertical Drivers of Hyporheic Exchange in the River Corridor

Similar principles to those identified for in-channel controls on hyporheic exchange also apply to interactions between groundwater and surface water across multiple scales in river corridors (Figure 4.1) (Boano et al., 2010c; Revelli et al. 2008; Stonedahl et al. 2010). For instance, stream sinuosity is a dominant control of inter-meander subsurface flow (Figure 4.3a), in addition to stream flow velocity and sediment hydraulic conductivity (Boano et al. 2010a, 2006; Pescimoro et al. 2019). Regional groundwater up-welling (Figure 4.3b) and down-welling (Figure 4.3c) interact with local channel morphology to control patterns of surface–groundwater exchange (Balbarini et al. 2017). Resulting inter-meander flow has been shown to control residence time distributions; and thus, redox zonation and nutrient turnover in sediments (Boano et al., 2010c; Dwivedi et al. 2017). However to date from field studies, we rarely consider the vast spatial heterogeneity of hydraulic and hydrogeological properties of sediments between the river channel, the meanders, and floodplains (Figure 4.3d) (Bersezio et al. 2007; Bridge et al. 1995, 1995; Dara et al. 2019), such as preferential flow through subsurface palaeo-channels (Lowell et al. 2009; Słowik 2014; Stanford and Ward 1993).

Hyporheic exchange is occurring across a range of scales being controlled by a variety of the processes discussed previously (Boano et al. 2014; Harvey and Gooseff 2015; Krause et al. 2014b, 2017; Magliozzi et al. 2018; Poole et al. 2008; Stonedahl et al. 2010). However, increased efforts are required to integrate and compare the respective context-specific importance of multi-scale interactions between groundwater and surface water. Recent model-based attempts to quantify the relative importance of bedform-driven vs.

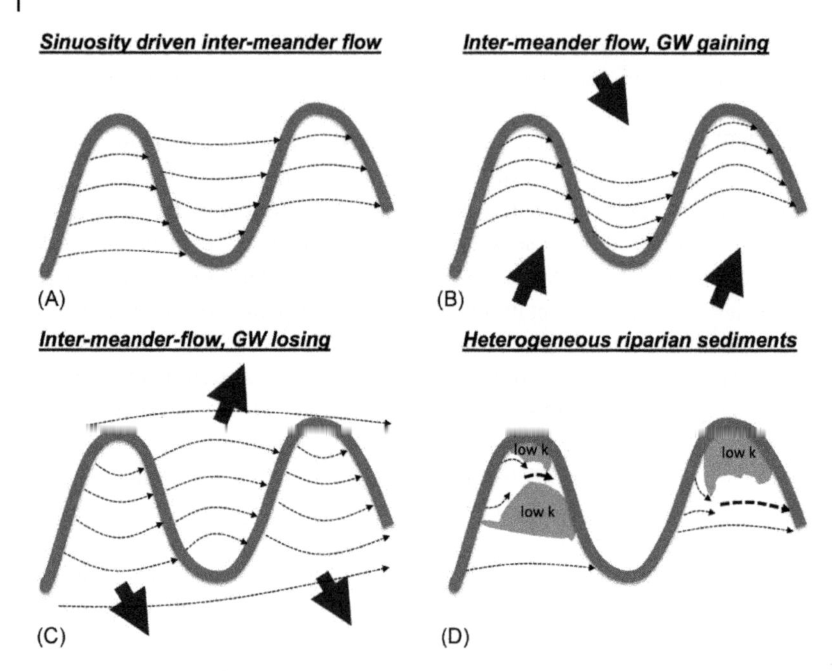

Figure 4.3 Inter-meander flow of surface water through alluvial sediment driven by stream flow velocity and sinuosity (A), under the influence of regional groundwater (GW) flow with gaining (B) and losing (C) conditions as well as spatial heterogeneity in riparian sediment conditions including palaeo-channels as fast conduits, and preferential flow paths and stagnation zones caused by spatially variable hydraulic conductivity patterns (D).

meander-driven exchange between surface water and groundwater for nitrogen processing in river networks provide a promising path forward (Gomez-Velez et al. 2015). However, these results still require field validation and as yet, do not account for spatial heterogeneity in sediment hydraulic conductivity, biogeochemical properties, nor landscape context, known to control both small-scale hyporheic exchange and large-scale groundwater flow.

4.2.4 Dynamic Hydrological Forcing of Hyporheic Exchange Flow

Recent experimental and modelling based research has started to explore the impacts of transience (non-steady conditions) in hydrostatic and hydrodynamic forcing on hyporheic exchange and residence time distributions (Boano et al. 2007, 2013), with a particular focus on extreme flow conditions during freshets and flood scenarios (Malzone et al. 2016; Schmadel et al. 2016a, 2016b; Singh et al. 2019; Trauth and Fleckenstein 2017; Ward et al. 2018) and temporally variable groundwater flow (Trauth et al. 2014; Wu et al. 2018a).

An increasing number of field observations and controlled laboratory experiments provide evidence of the ecological and biogeochemical implications of temporally dynamic hyporheic exchange and hyporheic residence time distributions. However, a systematic analysis of the influence of stream flow dynamics on hyporheic exchange and residence time distributions in different landscape contexts is long overdue. (Datry and Larned 2008; Dole-Olivier et al. 1997; Malcolm et al. 2009; Schmadel et al. 2016a). Recent laboratory and numerical modelling studies have started to more systematically explore the variable importance of different drivers on the temporal dynamics of hyporheic exchange (Kaufman et al. 2017; Singh et al. 2019; Wu et al. 2018b). Such studies still need to be extended to explore the impact of episodic high-flow events that mobilize sediments, yielding spatial and temporal erosion, and deposition dynamics, and subsequent non-stationary patterns of bed morphology, sediment structure, and hydraulic conductivity as seen in field and flume experiments (Ahmerkamp et al. 2015; Kessler et al. 2015; Packman and Brooks 2001). Despite small bedforms, such as ripples originating directly from sediment movement and changing in time, most flume studies, conceptual models, and modelling exercises simplify reality and assume stationary bedforms. Some flume studies are now considering bedform migration (Ahmerkamp et al. 2015; Kessler et al. 2015; Wolke et al. 2019) with far-reaching hydrological and biogeochemical implications. Systematic analyses of the wide spectrum of flow transience across different river types, combined with regional groundwater flow interactions, will reveal the degree to which short-term and long-term changes in stream flow alter hyporheic exchange, hyporheic residence time distributions, and related ecological and biogeochemical processes.

4.2.5 Spatiotemporal Variability of Hydrological Opportunity and Biogeochemical Reactivity in the Hyporheic Zone

The magnitude and array of hyporheic biogeochemical processes, associated with transitions between aerobic and anaerobic respiration, are a function of the hydrological opportunity for metabolism, defined by the influx and residence time of reactants, nutrients, and metabolic substrates in local environments with specific reactivity. These processes are set by the concentrations of reactants and the frequency of their spatial coincidence (Battin et al. 2008; Marcé et al. 2018; Reeder et al. 2018a), microbial community dynamics (e.g. recruitment, growth, and activity), and the hyporheic biogeochemical reactivity (Krause et al. 2017; McClain et al. 2003). The supply and mixing of reactants (including buried autochthonous streambed organic matter), the residence time distributions, and the bulk average reaction rates, are controlled directly by hyporheic exchange, which transports solutes and fine particles from the surface water into and through the streambed sediments, and hence controls both distributions and rates of reactions within porewater.

Most current conceptual models of streambed biogeochemical cycling assume surface water solute concentrations and hyporheic exchange flow-driven residence times in streambed sediments to be the primary (and often only) controls of biogeochemical reactions and rates in the hyporheic zone (Bardini et al. 2012; Boano et al. 2014; Hester and Doyle 2008; Marzadri et al. 2012; Zarnetske et al. 2011a, 2012). In this context, the hyporheic zone is conceptualized as a single, homogeneous chemical reactor that receives reactants (e.g. dissolved organic carbon, nutrients, and dissolved oxygen) exclusively via hyporheic exchange from the surface water (Figure 4.4a). Biogeochemical reactions and rates are thus dependent on the turnover of solutes in the hyporheic zone (Bardini et al. 2013; Boano et al. 2014; Briggs et al. 2014; Trauth et al. 2014; Zarnetske et al. 2011a). With the exception of recent modelling work that allows for variation in reaction rates with sediment depth (Aubeneau et al. 2015; Caruso et al. 2017; Li et al. 2017) and heterogeneities in the physical pore network structure of hyporheic sediments (Briggs et al. 2015; Sawyer 2015), biogeochemical reaction rates are typically considered independent from the location of the chemical reaction taking place in the hyporheic zone. Consequently, the efficiency of biogeochemical turnover in the hyporheic zone is often assumed to be limited by the availability of reactants transported by hyporheic exchange from the surface into the streambed (Aubeneau et al. 2015; Bardini et al. 2013; Li et al. 2017; Trauth et al. 2015; Zarnetske et al. 2011b). Such conditions, where types of reactions and rates are solely a function of surface water concentrations of reactants and their hyporheic exchange flow-controlled hyporheic residence time, have been observed in the field. For example, carbon respiration in the hyporheic zone of oligotrophic headwater streams has been found to depend on surface water inputs and hyporheic travel time, with respiration shifting from aerobic to anaerobic conditions along hyporheic flow paths (Zarnetske et al. 2011b). As a result, denitrification is reliant on both sufficient residence time to consume sufficient dissolved oxygen from the infiltrating surface water and bioavailable organic carbon remaining as an electron donor (Holmes et al. 1994; Jones and Holmes 1996; Ocampo et al. 2006; Zarnetske et al. 2011b). Hyporheic zones support nutrient retention in river corridors, with hyporheic metabolism reducing concentrations of organic carbon, bioavailable inorganic nitrogen, and dissolved oxygen in the hyporheic water before it subsequently returns to the stream (Figure 4.4a) (Gomez-Velez et al. 2015; Krause et al. 2011b, 2014b; Li et al. 2017; Pinay et al. 2009; Poole et al. 2008; Wondzell 2011).

Such conceptualizations of spatially homogeneous hyporheic reactivity are certainly useful to simplify estimates of solute turnover in the hyporheic zone as the ratio of residence time and biogeochemical reaction time, expressed by the dimensionless Damköhler number (Marzadri et al. 2012; Zarnetske et al. 2011a). This approach represents a potentially powerful methodology for spatial upscaling (Pinay et al. 2015; Reeder et al. 2018a, 2018b). However, if the goal is to nest biogeochemical function of hyporheic zones at landscape scales, the Damköhler number approach has further potential to be enhanced in its ability to upscale the hyporheic biogeochemical function to river network scales (Marzadri et al. 2021, 2017). It currently does not capture the full range of coupling between biogeochemical processes and hyporheic exchange, abiotic and biotic heterogeneities, or the scale dependency resulting from decreasing hyporheic exchange flow (HEF) rates and concentrations of exchanged solutes with depth in the hyporheic zone.

Water residence time may be the main control of hyporheic biogeochemical cycling for many oligotrophic and relatively homogeneous, low-order headwater streams with limited variability in sediment texture (such as in Pinay et al. 2009; Zarnetske et al. 2011a, 2011b). However, this concept is frequently contradicted by field observations in other small streams (Drummond et al. 2016; Marcé et al. 2018) as well as in more complex lowland rivers, particularly in agricultural areas with enriched nutrient conditions (Abbott et al. 2018b; Frei et al. 2020; Krause et al. 2009, 2013; Sawyer 2015). The assumption of an otherwise "empty" and inert, homogeneous streambed reactor charged by hyporheic exchange driven solute inputs from surface water (Bardini et al. 2012; Boano et al. 2014; Trauth et al. 2014, 2015; Zarnetske et al. 2011a) is not applicable when streambed sediments also contain autochthonous reactants, as both the dissolved form and particulate organic matter (Figure 4.4b, c). In this case, the encountered diversity of turnover rates and reaction types cannot be solely explained by hyporheic exchange controls of reaction times and surface water solute inputs, as these processes are strongly influenced by the concentrations of

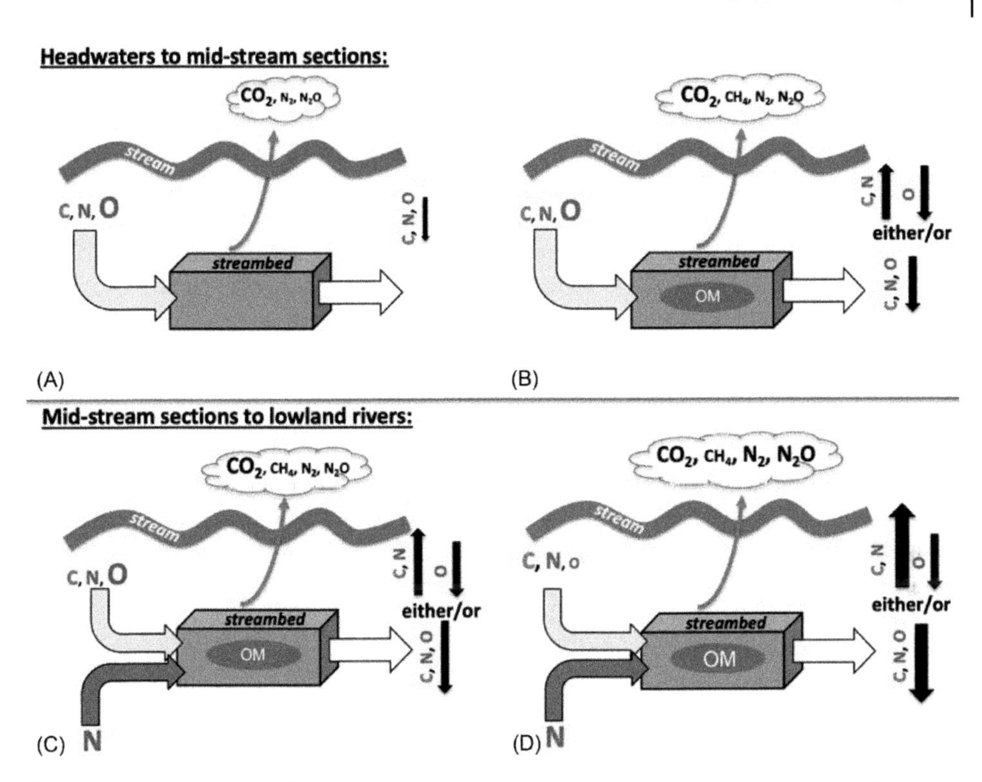

Figure 4.4 Conceptual models of the hyporheic zone as a biogeochemical reactor: Variability in the mixing of different reactive solute sources (carbon (C) and nitrogen (N) fractions, and free oxygen (O)), impacts on stream(bed) metabolism and downstream effects of hyporheic processing, for single water source (surface water) dominated systems (A+B) and systems with reactive solute contributions from multiple water sources mixing in the hyporheic zone (C+D), with (B, C, D) and without (A) relevance of autochthonous streambed organic matter (OM). Clouds representing gaseous losses of metabolites and symbol sizes indicating relative proportions of their concentrations with arrow sizes and directions indicating minor or major increases or reductions in downstream concentrations.

bioavailable dissolved organic matter and mineralization rates of particulate organic matter in the sediment (Corson-Rikert et al. 2016; Krause et al. 2009, 2013; Reeder et al. 2018a, 2018a; Reeder et al. 2018b; Trimmer et al. 2012). Dissolved and particulate organic matter concentrations have significant spatial variability within the sediment (Datry et al. 2017; Drummond et al. 2017, 2018; Krause et al. 2009, 2013; Shelley et al. 2017). The spatial patterns of organic matter distributions in the streambed often coincide with the spatial organization of physical sediment properties that result from the fluvial depositional history of the river (Drummond et al. 2017, 2018; Larsen and Harvey 2017; Larsen et al. 2015). High organic matter content is generally associated with low hydraulic conductivity strata of organic sediments, while highly permeable mineral sediments are often characterized by low organic matter content (Pinay et al. 2000, 1995). The relationship between physical and biogeochemical sediment controls provide additional and perhaps underutilized predictive capacity to explain observed heterogeneity in hyporheic zone biogeochemical reactivity, as a function of interacting sediment conductivity, residence time, and reactivity patterns. Concordantly, field studies using hydrometabolic tracers have indicated that the entire hyporheic zone is not metabolically active contributing to ecosystem respiration and biogeochemical cycling (Argerich et al. 2011), though the locations and timescales associated with transformation are only beginning to be understood (Ward et al. 2019b). This finding has been corroborated by particle-tracking and pore-network models showing that hyporheic zone biogeochemical turnover can be largely driven by the residence time of water in hyporheic bioactive layers or redox micro-zones (Briggs et al. 2015; Li et al. 2020).

As a consequence of the heterogenous distributions of residence times (and related hydrological opportunities) and sediment biogeochemical reactivities, reactions of the interlinked carbon and nitrogen cycle are often more complex. For example, contrasting concentrations of dissolved oxygen have resulted in comparable rates of microbial carbon processing due to compensation by the composition of the microbial community (Risse-Buhl et al. 2017). Previous conceptualizations of residence time control of biogeochemical turnover in hyporheic zones also widely ignore other nitrogen transformation processes evidenced in the field such as dissimilatory nitrate reduction to ammonium (DNRA) and anaerobic oxidation of ammonium (Anammox) (Lansdown et al. 2015, 2016; Trimmer et al. 2012, 2015; S. Zhou et al. 2014a). Depending on the relative importance of the contribution from buried streambed autochthonous organic matter to hyporheic biogeochemical cycling, hyporheic exchange might result in either a reduction or an increase of in-stream loading of carbon and nitrogen (Figure 4.4b). Further, hyporheic exchange could decrease in-stream carbon and nitrogen loading due to transport towards the aquifer in the case of losing conditions (Figure 4.4c).

In many cases, up-welling groundwater may contribute reactive solutes to the streambed (Figure 4.4d), which is particularly relevant for legacy pollutants, such as nitrate contamination in groundwater, which in many lowland agricultural catchments represent the main nitrogen source (Basu et al. 2010; Bochet et al. 2020; Frei et al. 2020; Withers et al. 2014), and industrial contaminants such as chlorinated solvents in urban areas (Rivett et al. 2012; Weatherill et al. 2018). The concurrence of spatially variable up-welling of solutes from groundwater and temporally dynamic down-welling of surface water pollutants has frequently been observed to create complex patterns of reactions in the hyporheic zone (Liu et al. 2019; Shelley et al. 2017; Weatherill et al. 2014) that go far beyond the current concepts of hyporheic exchange and residence time controls on streambed biogeochemical cycling. In fact, the observed impacts of groundwater solute contributions and autochthonous sediment organic matter have been shown to produce a hot spot of biogeochemical transformation in the hyporheic zone (Krause et al. 2009, 2013), which does not match current conceptualizations. In particular, the net effect of hyporheic zone biogeochemical cycling on nitrate removal might be underestimated given that model-based quantifications do not consider the interactions of multiple solute pathways into the streambed sediments, which may already contain standing stocks of bioavailable organic matter.

To capture these effects, process conceptualizations currently used in numerical models should be extended to improve identification and representation of the dominant process dynamics across multiple scales, considering solute mixing from different sources, including buried autochthonous streambed organic matter (Figure 4.4b–d). This approach will require a dialogue to incorporate the frequently observed behaviours that have been identified as being relevant in the field into existing numerical models in parsimonious approaches where parameters remain tractable and identifiable within acceptable confidence bounds. Many of the existing models should have the capability to account for these additional processes if parameterized adequately. However, new numerical frameworks are also needed to better capture multi-scale interactions, process interactions, and feedbacks that change system conditions.

4.2.6 The Missing Link? Ecological Controls on Hyporheic Exchange Flow and Biogeochemical Cycling

Previous interdisciplinary research has mainly focused on the analysis and quantification of hyporheic exchange and biogeochemical cycling impacts on aquatic ecosystem functioning (Figure 4.5) (Boulton et al. 2010, 1998; Boulton and Hancock 2006; Hancock et al. 2005). This research highlights that hyporheic exchange and streambed biogeochemical processes create a unique ecological niche (Brunke and Gonser 1997; Stanford and Ward 1993; Stubbington et al. 2009) that potentially acts as refuge during extreme conditions (Folegot et al. 2018b) enhancing biodiversity and ecosystem resilience to environmental change (Kurz et al. 2017). However, our understanding of the impacts are in the reverse direction, where ecological processes can influence hyporheic exchange and biogeochemical cycling is in its infancy (Figure 4.5) (Buxton et al. 2015). Initial work on the functioning of microbial biofilms established how dynamic growth of benthic and hyporheic biofilms affects turbulent flow in the stream channel and consequently modifies turbulence-driven hyporheic exchange (Nikora 2010; Roche et al. 2017). Biofilms also cause bioclogging of streambed sediments (Caruso et al. 2017; Newcomer et al. 2016; Roy Chowdhury et al. 2020) (Figure 4.2h), where complex biofilm communities on the streambed surface and within sediment pores reduce the effective porosity of the streambed substrate and subsequently hyporheic exchange (Battin and Sengschmitt 1999; Mendoza-Lera and Mutz 2013; Newcomer et al. 2016; Roche et al. 2017) (Figure 4.2h). At larger scales, the dynamic growth of submerged macrophytes has been shown to strongly modify turbulent flow patterns in the channel and enhance the trapping of fine sediment (Drummond et al. 2014; Liu et al. 2018; Liu and Nepf 2016; Sand-jensen 1998) (Figure 4.2g), directly impacting hyporheic exchange and hyporheic biogeochemical processes (Salehin et al. 2003; Ullah et al. 2014). The presence of macrophytes alters flow paths and residence time distributions, providing additional substrate and input of organic carbon as well as enhancing nutrient uptake during the growth phase (Baranov et al. 2017; Nikolakopoulou et al. 2018; Ribot et al. 2019).

In addition to these microbial- and plant-induced influences on the hyporheic hydrology and biogeochemistry, there is increasing evidence that some invertebrate bioturbator species and ecosystem engineers modify hyporheic hydrological and biogeochemical conditions. These ecosystem engineers alter the conditions of their habitat to augment their ecological needs, with direct and indirect influences on the dynamics of both hyporheic exchange and hyporheic biogeochemical process (Hölker et al. 2015). For example, burrowing chironomid larvae pump significant amounts of surface water through their U-shaped sediment burrows, directly affecting hyporheic exchange by actively transporting greater volumes of water and solute mass to deeper sediments. This process influences sediment metabolism and biogeochemical cycles (Mermillod-Blondin 2011; Mermillod-Blondin et al. 2004; Nogaro et al. 2009) with potentially significant impacts on greenhouse

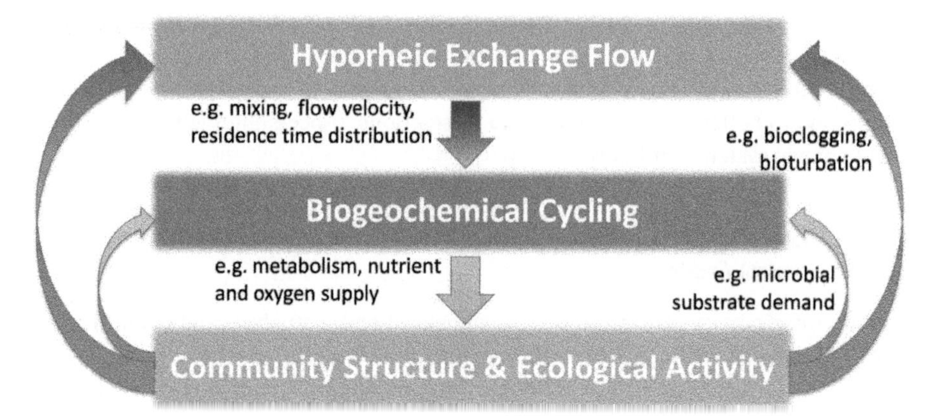

Figure 4.5 Multi-directional interactions between physical controls of hyporheic exchange flow, streambed biogeochemical cycling, and biological community structure and ecological (metabolic) functioning (red arrows), including ecological feedbacks on streambed biogeochemistry, e.g., microbial demand for substrate shifting porewater from oxic to anoxic (aerobic to anaerobic metabolic pathways) as well as hyporheic exchange (e.g., bioturbation, bioclogging, and ecosystem engineering).

gas production and sequestration (Baranov et al., 2016, 2017). These findings highlight the urgent need to extend analyses towards frequently observed behaviour of other species, such as the burrowing of crayfish in streambeds during hydrological extremes or the active movement of *Gammarus pulex* (freshwater shrimp) and other hyporheic invertebrates triggered by thermal and hydrological stress (DiStefano et al. 2009; Statzner et al. 2000; Vander Vorste et al. 2016). The activities of these invertebrates are likely to alter sediment structure and thus, hydraulic conductivity and hyporheic exchange. Many vertebrates, such as fish or freshwater mammals (Janzen and Westbrook 2011; Shurin et al. 2020) also directly affect hyporheic exchange and streambed biogeochemical conditions. For instance, when fish select a gravel spawning habitat with their preferred hyporheic exchange and biogeochemical conditions, these conditions may be affected also by their spawning activities (Baxter and Hauer 2000; Buxton et al. 2015; Harrison et al. 2019; Malcolm et al. 2004, 2009; Moir and Pasternack 2010; Tonina and Buffington 2009). In Columbia, one of the world's largest mammals, the non-native hippopotamus (*Hyppopothamus amphibious*), acts as an ecosystem engineer affecting hyporheic exchange and impacting hydrological habitats (Shurin et al. 2020). This activity is considered valuable to fill an important ecosystem function as megaherbivores amphibious ecosystem engineers became extinct in South America at the end of the Pleistocene (MacPhee and Schouten 2019).

Given the observed ecological impacts on hyporheic exchange and biogeochemical processes in the hyporheic zone (Figure 4.5), future research requires in-depth attempts to integrate the advancing knowledge of ecological controls into conceptual and quantitative models of hyporheic hydrological and biogeochemical process dynamics (Hannah et al. 2007; Krause et al. 2011b, 2017). We are only beginning to understand the magnitude of ecological controls and their impact on non-stationarity and temporal dynamics of hyporheic processes, caused, for instance, by time-variable biological processes. Hence, a fully coupled approach is required that considers ecosystem process response to changes of both hydrological and biogeochemical habitat conditions, as well as the impact of biological activity on physical and chemical properties of the hyporheic zone (Figure 4.5).

4.3 A Landscape Perspective of Organizational Principles of Hyporheic Exchange Fluxes and Biogeochemical Cycling Along the Catchment Continuum

Our synthesis of field observations and modelling studies highlights complex interactions among a broad diversity of drivers and controls of hyporheic exchange, biogeochemical cycling, and biological activity. This synthesis furthermore provides evidence that hyporheic exchange, hyporheic biogeochemical cycling, and hyporheic ecosystems are highly organized and spatially structured in fluvial landscapes. Understanding the underlying organizational principles of these systems is key for enabling transferability and generalization of knowledge to predict the landscape-wide significance of hyporheic exchange and hyporheic biogeochemical cycling on water balance, nutrient dynamics, reactive contaminant transport, and ecosystem functioning at catchment scale.

The majority of experimental or modelling studies to date have focused on individual stream reaches and then scaled up observations, with only a few experimental studies attempting to quantify hyporheic exchange along a river continuum using a river network approach (Gootman et al. 2020; Lee-Cullin et al. 2018; Ward et al. 2019b; Wondzell 2011). Here, we build on previous conceptual models of landscape organizational principles (Boulton et al. 1998; Boulton and Hancock 2006; Buffington and Tonina 2009; Frissell et al. 1986; Helton et al. 2011; Malard et al. 2002) to synthesize and conceptualize the spatial and temporal organization of different drivers and controls of hyporheic exchange and hyporheic biogeochemical cycling along a river network continuum from first order headwaters to lowland streams (Figure 4.6a). We propose advances to existing landscape-scale conceptualizations of hyporheic exchange that account not only for interactions among different drivers and controls of hyporheic flow, biogeochemical cycling, and ecology, but also their spatially nested co-existence (Figure 4.1).

Integrating the drivers of hyporheic exchange and hyporheic biogeochemical cycling into a catchment context requires using landscape organizational principles developed in hydrology, geomorphology, and ecology to explain hyporheic exchange patterns (Boano et al. 2014; Magliozzi et al. 2019, 2018). For example, basic principles of sediment transport and storage along river networks indicate a general down-stream reduction in channel slope, lateral channel confinement, sediment grain size, and channel roughness coupled with an increase in streambed organic matter from headwater to lowland streams (Figure 4.6). This longitudinal change in the characteristics of the streambed sediments results in a downstream shift in hyporheic exchange and often coincides with an increase in groundwater contributions (Figure 4.3). Decreasing hydraulic gradients causes a downstream reduction in driving forces for vertical hyporheic exchange, coinciding with deeper fluvial and alluvial deposits, leading to longer and deeper hyporheic flow paths and slower hyporheic flow velocities in finer-grained sediments with lower permeability, which results in increased hyporheic residence times (Figure 4.6a). At the same time, river meandering in low-gradient mid-stream sections results in enhanced river corridor connectivity, longer lateral hyporheic exchange

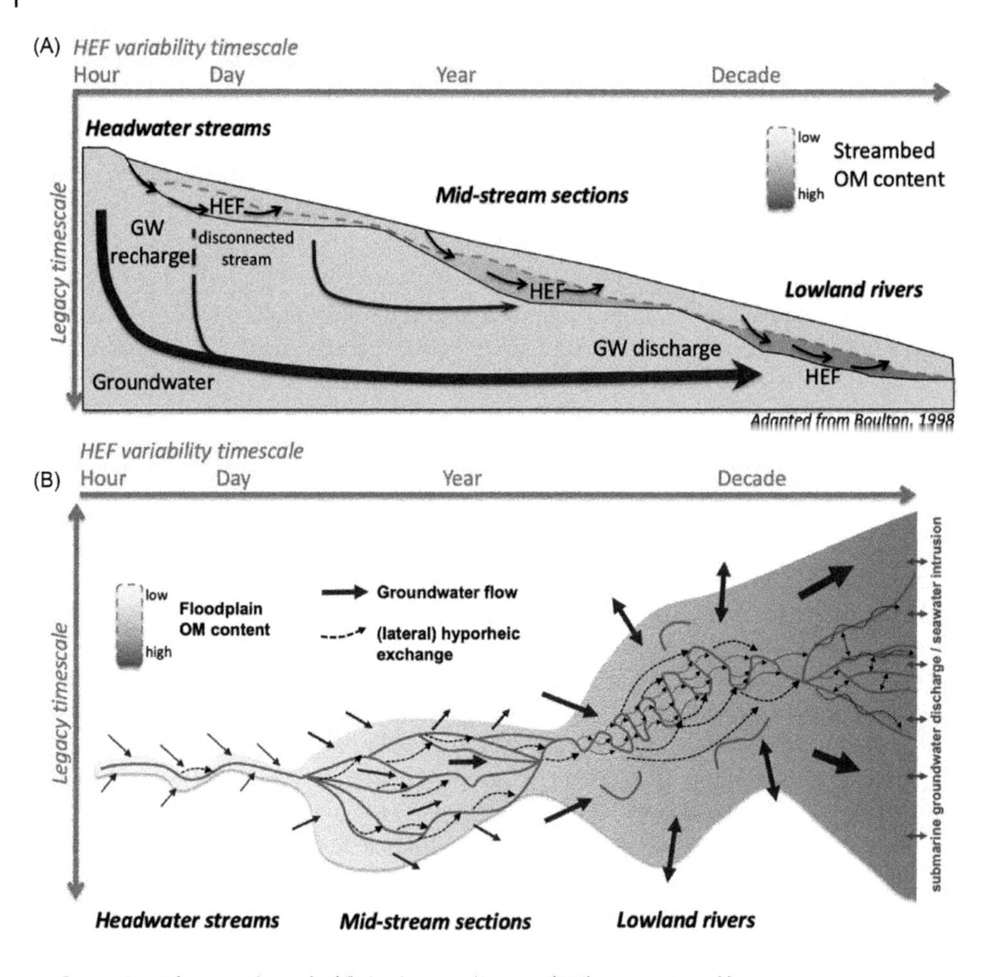

Figure 4.6 Landscape-scale organizational principles of vertical hyporheic exchange flow (HEF) from headwater streams to lowland rivers (A – vertical profile) and of lateral inter-meander and parafluvial flow (B – plane view) as a function of surface hydrological drivers and multi-scale groundwater flow controls, and distribution of physical and chemical sediment properties such as spatial variability in sediment hydraulic conductivity, hyporheic residence time (RT), or streambed organic matter (OM) content as drivers of biogeochemical processing *(visualization of downstream tendencies are indicating general trends only and do not imply linear changes).*

flow path, and increased hyporheic and riparian residence times and flow permanence (Figure 4.6b). Depending on the dominant geomorphodynamic processes, these changes and flow path transitions can be highly non-linear, yielding sharp thresholds at regions of known geomorphic transitions such as between mountain ranges, foothills, valleys, and lowlands (Marzadri et al. 2017; Wondzell 2011). Abrupt changes in multiple factors, such as the transition from steep coarse-bedded and constrained mountain rivers into finer-grained and less-constrained lowland rivers is expected to yield sharp transitions in hyporheic exchange (Figure 4.6a, b), but these patterns have not been systematically investigated for a range of fluvial system conditions.

The wider application of fluvial sedimentology principles (Dara et al. 2019) and understanding of alluvial depositional history (Słowik 2014) provides further and perhaps underutilized predictive capacity for the spatial distributions of sediment properties in river valleys and their impact on hydrologic connectivity between streams and groundwaters, including hyporheic exchange, residence time distributions, and biogeochemical reactivity in river channel and riparian sediments (Figure 4.6b). The potential for combining model-based information of fluvial sediment transport to predict river valley and streambed sediment stratigraphy as controls of hyporheic exchange and hyporheic biogeochemical processes is currently untapped, leaving a great underutilized potential for achieving step changes in the understanding of hyporheic zone processes across large catchments.

This landscape perspective emphasizes that local hyporheic exchange dynamics are strongly modulated by larger-scale patterns of topography, biogeography, and groundwater circulation (Figure 4.6). In this context, hyporheic interactions can be considered a local, near-surface manifestation of larger spatial scale and longer temporal scale surface–groundwater circulation patterns. Similarly, landscape patterns of terrestrial ecosystems, primary production, and organic matter inputs both drive and condition hyporheic microbial activity and biogeochemistry. Looking forward, the accumulated advances in the knowledge of surface–groundwater systems outlined previously, together with new capability in sensing, simulation, and data science, provide the potential to unify understanding of local drivers of ecological processes and their interactions with larger aquatic and terrestrial ecosystems.

The intensity and distribution of groundwater up-welling and the associated delivery of legacy pollutants, such as nitrate or chlorinated solvents, into hyporheic zones are likely to increase from headwaters (often with less agriculture and urbanization) to more intensively managed and impacted downstream lowland ecosystems (Figure 4.6). Similarly, the flow permanence of river channels increases in downstream direction, with many headwater streams being prone to drying and flow cessation (Benstead and Leigh 2012; Boulton et al. 2017) and largely unknown impacts of dry phases on hyporheic zone functioning (Boulton et al. 2017; Datry et al. 2017; Datry and Larned 2008). An overall downstream increase is assumed in the complexity of interactions of different hydrodynamic, sedimentological, and biogeochemical drivers, including distributions of sediment structure and properties, groundwater up-welling and solute contributions as well as patterns of autochthonous organic matter content in streambed sediments. On the other hand, spatial variability in stream chemistry, including pollutants, typically decreases moving downstream in river networks, suggesting a homogenization arising from the averaging of different signals and attenuation of discrete sources (Abbott et al. 2018a; Creed et al. 2015; Dupas et al. 2019).

We advocate for hyporheic research to embrace a wider landscape perspective when interpreting local observations, and to avoid applying principles derived predominantly from small headwater streams throughout the river network continuum. Further, the current fragmentary approaches can lead to inaccuracies in system-level understanding and management of the hyporheic zone at catchment scale. Therefore, studies considering a greater diversity in the ecological conditions of the hyporheic zone are needed. Arid and semi-arid systems have fundamentally different hydrology and biogeochemistry (Fisher et al. 1998; Harms and Grimm 2008). Still, many of the conceptualizations considered in this chapter predominantly reflect patterns under hydrologically gaining conditions with net groundwater to surface water flux, typical of temperate regions.

Acknowledging interactions between different drivers and controls of hyporheic exchange and hyporheic biogeochemical cycling in a landscape context provides a pathway towards more accurate representations of governing processes in conceptual hyporheic zone models. This does not necessarily need to lead to an increase in complexity for site-specific models but supports the development of parsimonious approaches where the selection of representative processes is justified by understanding the most important hyporheic exchange controls at each location in a wider landscape context. We emphasize here that the general patterns illustrated in Figure 4.6 represent an overall expectation based on current understanding of watershed structure, both geophysical and ecological. Hyporheic hydrology and biogeochemistry at any site in the landscape can vary substantially from the general expectation, necessitating careful consideration of both local- and landscape-scale drivers. New observational approaches are needed that are capable of capturing a wider range of environmental conditions in hyporheic zones across river networks (Krause et al. 2011b; Lee-Cullin et al. 2018; Ward and Packman 2019). To ensure the contribution of hyporheic zone research to efficiently manage the interface between aquifers and rivers (Krause et al. 2014b, 2017; Lewandowski et al. 2019), future research will need to test how those landscape principles either hold or need to be adapted across catchments, including heavily anthropogenically modified and polluted urban streams that are currently under-represented in hyporheic investigations (Lawrence et al. 2013; Schaper et al. 2019, 2018).

The proposed advancements of process conceptualizations across scales also highlight the need for intensifying efforts to improve mechanistic process understanding through interdisciplinary research and knowledge exchange in emerging areas of hyporheic research. This includes providing evidence of the biological, physical, and biogeochemical process interactions (i.e. how the physical environment controls habitat functioning), and also of how biological behaviour is a feedback to hyporheic physical and biogeochemical conditions. Therefore, it will be critical to establish landscape organizational principles of the abundance and activity of hyporheic and benthic fauna, including bioturbators, along the river continuum and as a consequence of changes in the type of sediments and accumulation of organic matter in streambeds.

To improve promising recent attempts into predictions of larger-scale implications of hyporheic exchange and hyporheic biogeochemical processing, as well as to quantify the resilience of hyporheic functioning to global environmental change, future research also needs to specifically address the flow-dependent mobility of streambed sediments, and its impact on hyporheic zone processes and ecological functioning. With climate change projections suggesting that many rivers are likely to experience an increase in extreme hydrological events, it will be important to advance the understanding of hyporheic processes, in particular under conditions of increased flow intermittence and sediment mobilizing flow events. The key for success in both hyporheic science and river ecosystem management lies in understanding the interactions of physical, biogeochemical, and ecological processes and how they vary across scales, as well as integrating knowledge between (field and lab) experimental and modelling approaches. We hope that the framework we propose here will stimulate discussions and open opportunities for further integrating existing and new process knowledge across scales within the landscape context.

Acknowledgment

"The chapter is based on the previously published paper Krause, S., et al. (2022). Organizational principles of hyporheic exchange flow and biogeochemical cycling in river networks across scales. Water Resources Research, 58, e2021WR029771. https://doi.org/10.1029/2021WR029771." (for Chapter 4)

References

Abbott, B.W., Gruau, G., Zarnetske, J.P. et al. (2018a). Unexpected spatial stability of water chemistry in headwater stream networks. *Ecol. Lett.* 21: 296–308. https://doi.org/10.1111/ele.12897.

Abbott, B.W., Moatar, F., Gauthier, O. et al. (2018b). Trends and seasonality of river nutrients in agricultural catchments: 18 years of weekly citizen science in France. *Sci. Total Environ.* 624: 845–858. https://doi.org/10.1016/j.scitotenv.2017.12.176.

Ahmerkamp, S., Winter, C., Janssen, F. et al. (2015). The impact of bedform migration on benthic oxygen fluxes: bedform migration and benthic fluxes. *J. Geophys. Res. Biogeosciences* 120: 2229–2242. https://doi.org/10.1002/2015JG003106.

Angelier, E. (1953). Recherches écologiques et biogéographiques sur la faune des sabeles sumergés. *Arch. Zool. Exp. Gen* 37–161: 37–161.

Angermann, L., Krause, S., and Lewandowski, J. (2012). Application of heat pulse injections for investigating shallow hyporheic flow in a lowland river: heat pulse injections for investigating hyporheic flow. *Water Resour. Res.* 48: n/a–n/a. https://doi.org/10.1029/2012WR012564.

Argerich, A., Haggerty, R., Martí, E. et al. (2011). Quantification of metabolically active transient storage (MATS) in two reaches with contrasting transient storage and ecosystem respiration. *J. Geophys. Res.* 116: G03034. https://doi.org/10.1029/2010JG001379.

Arnon, S., Marx, L.P., Searcy, K.E., and Packman, A.I. (2009). Effects of overlying velocity, particle size, and biofilm growth on stream-subsurface exchange of particles. *Hydrol. Process.* n/a–n/a. https://doi.org/10.1002/hyp.7490.

Arrigoni, A.S., Poole, G.C., Mertes, L.A.K. et al. (2008). Buffered, lagged, or cooled? Disentangling hyporheic influences on temperature cycles in stream channels: hyporheic influence on stream temperature. *Water Resour. Res.* 44. https://doi.org/10.1029/2007WR006480.

Aubeneau, A.F., Drummond, J.D., Schumer, R. et al. (2015). Effects of benthic and hyporheic reactive transport on breakthrough curves. *Freshw. Sci.* 34: 301–315. https://doi.org/10.1086/680037.

Azizian, M., Boano, F., Cook, P.L.M. et al. (2017). Ambient groundwater flow diminishes nitrate processing in the hyporheic zone of streams: ambient groundwater and stream N-cycling. *Water Resour. Res.* 53: 3941–3967. https://doi.org/10.1002/2016WR020048.

Balbarini, N., Boon, W.M., Nicolajsen, E. et al. (2017). A 3-D numerical model of the influence of meanders on groundwater discharge to a gaining stream in an unconfined sandy aquifer. *J. Hydrol.* 552: 168–181. https://doi.org/10.1016/j.jhydrol.2017.06.042.

Baranov, V., Lewandowski, J., Romeijn, P. et al. (2016). Effects of bioirrigation of non-biting midges (Diptera: chironomidae) on lake sediment respiration. *Sci. Rep.* 6. https://doi.org/10.1038/srep27329.

Baranov, V., Milošević, D., Kurz, M.J. et al. (2017). Helophyte impacts on the response of hyporheic invertebrate communities to inundation events in intermittent streams. *Ecohydrology* 10: e1857. https://doi.org/10.1002/eco.1857.

Bardini, L., Boano, F., Cardenas, M.B. et al. (2012). Nutrient cycling in bedform induced hyporheic zones. *Geochim. Cosmochim. Acta* 84: 47–61. https://doi.org/10.1016/j.gca.2012.01.025.

Bardini, L., Boano, F., Cardenas, M.B. et al. (2013). Small-scale permeability heterogeneity has negligible effects on nutrient cycling in streambeds: effect of small-scale heterogeneity. *Geophys. Res. Lett.* 40: 1118–1122. https://doi.org/10.1002/grl.50224.

Barthel, R. and Banzhaf, S. (2016). Groundwater and surface water interaction at the regional-scale – a review with focus on regional integrated models. *Water Resour. Manag.* 30: 1–32. https://doi.org/10.1007/s11269-015-1163-z.

Basu, N.B., Destouni, G., Jawitz, J.W. et al. (2010). Nutrient loads exported from managed catchments reveal emergent biogeochemical stationarity. *Geophys. Res. Lett.* 37: n/a–n/a. https://doi.org/10.1029/2010GL045168.

Battin, T.J., Kaplan, L.A., Findlay, S. et al. (2008). Biophysical controls on organic carbon fluxes in fluvial networks. *Nat. Geosci.* 1: 95–100. https://doi.org/10.1038/ngeo101.

Battin, T.J. and Sengschmitt, D. (1999). Linking sediment biofilms, hydrodynamics, and river bed clogging: evidence from a large river. *Microb. Ecol.* 37: 185–196. https://doi.org/10.1007/s002489900142.

Baxter, C.V. and Hauer, F.R. (2000). Geomorphology, hyporheic exchange, and selection of spawning habitat by bull trout (*Salvelinus confluentus*). *Can. J. Fish. Aquat. Sci.* 57: 1470–1481. https://doi.org/10.1139/f00-056.

Bencala, K.E. (1993). A perspective on stream-catchment connections. *J. North Am. Benthol. Soc.* 12: 44–47. https://doi.org/10.2307/1467684.

Bencala, K.E. (2000). Hyporheic zone hydrological processes. *Hydrol. Process.* 14: 2797–2798. https://doi.org/10.1002/1099-1085(20001030)14:15<2797::AID-HYP402>3.0.CO;2-6.

Bencala, K.E., Gooseff, M.N., and Kimball, B.A. (2011). Rethinking hyporheic flow and transient storage to advance understanding of stream-catchment connections. *Water Resour. Res.* 47. https://doi.org/10.1029/2010WR010066.

Benstead, J.P. and Leigh, D.S. (2012). An expanded role for river networks. *Nat. Geosci.* 5: 678–679. https://doi.org/10.1038/ngeo1593.

Bersezio, R., Giudici, M., and Mele, M. (2007). Combining sedimentological and geophysical data for high-resolution 3-D mapping of fluvial architectural elements in the Quaternary Po plain (Italy). *Sediment. Geol.* 202: 230–248. https://doi.org/10.1016/j.sedgeo.2007.05.002.

Boano, F., Camporeale, C., and Revelli, R. (2010a). A linear model for the coupled surface-subsurface flow in a meandering stream: surface-subsurface flow in a meandering stream. *Water Resour. Res.* 46. https://doi.org/10.1029/2009WR008317.

Boano, F., Camporeale, C., Revelli, R., and Ridolfi, L. (2006). Sinuosity-driven hyporheic exchange in meandering rivers: hyporheic exchange in meandering rivers. *Geophys. Res. Lett.* 33: n/a–n/a. https://doi.org/10.1029/2006GL027630.

Boano, F., Demaria, A., Revelli, R., and Ridolfi, L. (2010c). Biogeochemical zonation due to intrameander hyporheic flow: intrameander biogeochemical zonation. *Water Resour. Res.* 46. https://doi.org/10.1029/2008WR007583.

Boano, F., Harvey, J.W., Marion, A. et al. (2014). Hyporheic flow and transport processes: mechanisms, models, and biogeochemical implications: hyporheic flow and transport processes. *Rev. Geophys.* 52: 603–679. https://doi.org/10.1002/2012RG000417.

Boano, F., Revelli, R., and Ridolfi, L. (2007). Bedform-induced hyporheic exchange with unsteady flows. *Adv. Water Resour.* 30: 148–156. https://doi.org/10.1016/j.advwatres.2006.03.004.

Boano, F., Revelli, R., and Ridolfi, L. (2008). Reduction of the hyporheic zone volume due to the stream-aquifer interaction. *Geophys. Res. Lett.* 35. https://doi.org/10.1029/2008GL033554.

Boano, F., Revelli, R., and Ridolfi, L. (2009). Quantifying the impact of groundwater discharge on the surface-subsurface exchange. *Hydrol. Process.* 23: 2108–2116. https://doi.org/10.1002/hyp.7278.

Boano, F., Revelli, R., and Ridolfi, L. (2010b). Effect of streamflow stochasticity on bedform-driven hyporheic exchange. *Adv. Water Resour.* 33: 1367–1374. https://doi.org/10.1016/j.advwatres.2010.03.005.

Boano, F., Revelli, R., and Ridolfi, L. (2011). Water and solute exchange through flat streambeds induced by large turbulent eddies. *J. Hydrol.* 402: 290–296. https://doi.org/10.1016/j.jhydrol.2011.03.023.

Boano, F., Revelli, R., and Ridolfi, L. (2013). Modeling hyporheic exchange with unsteady stream discharge and bedform dynamics: unsteady hyporheic exchange with moving bed forms. *Water Resour. Res.* 49: 4089–4099. https://doi.org/10.1002/wrcr.20322.

Bochet, O., Bethencourt, L., Dufresne, A. et al. (2020). Iron-oxidizer hotspots formed by intermittent oxic–anoxic fluid mixing in fractured rocks. *Nat. Geosci.* 13: 149–155. https://doi.org/10.1038/s41561-019-0509-1.

Bottacin-Busolin, A. and Marion, A. (2010). Combined role of advective pumping and mechanical dispersion on time scales of bed form-induced hyporheic exchange: bed form-induced hyporheic exchange. *Water Resour. Res.* 46. https://doi.org/10.1029/2009WR008892.

Boulton, A.J. (2007). Hyporheic rehabilitation in rivers: restoring vertical connectivity. *Freshw. Biol.* 52: 632–650. https://doi.org/10.1111/j.1365-2427.2006.01710.x.

Boulton, A.J., Datry, T., Kasahara, T. et al. (2010). Ecology and management of the hyporheic zone: stream–groundwater interactions of running waters and their floodplains. *J. North Am. Benthol. Soc.* 29: 26–40. https://doi.org/10.1899/08-017.1.

Boulton, A.J., Findlay, S., Marmonier, P. et al. (1998). The functional significance of the hyporheic zone in streams and rivers. *Annu. Rev. Ecol. Syst.* 29: 59–81. https://doi.org/10.1146/annurev.ecolsys.29.1.59.

Boulton, A.J. and Hancock, P.J. (2006). Rivers as groundwater-dependent ecosystems: a review of degrees of dependency, riverine processes and management implications. *Aust. J. Bot.* 54: 133. https://doi.org/10.1071/BT05074.

Boulton, A.J., Rolls, R.J., Jaeger, K.L., and Datry, T. (2017). Hydrological connectivity in intermittent rivers and ephemeral streams. In: *Intermittent Rivers and Ephemeral Streams* (ed. T. Datry, N. Bonada, and A.J. Boulton), 79–108. Elsevier. https://doi.org/10.1016/B978-0-12-803835-2.00004-8.

Bray, E.N. and Dunne, T. (2017). Subsurface flow in lowland river gravel bars: subsurface flow gravel bars. *Water Resour. Res.* 53: 7773–7797. https://doi.org/10.1002/2016WR019514.

Bridge, J.S., Alexander, J., Collier, R.E.L. et al. (1995). Ground-penetrating radar and coring used to study the large-scale structure of point-bar deposits in three dimensions. *Sedimentology* 42: 839–852. https://doi.org/10.1111/j.1365-3091.1995.tb00413.x.

Briggs, M.A., Day-Lewis, F.D., Zarnetske, J.P., and Harvey, J.W. (2015). A physical explanation for the development of redox microzones in hyporheic flow. *Geophys. Res. Lett.* 42: 4402–4410. https://doi.org/10.1002/2015GL064200.

Briggs, M.A., Lautz, L.K., and Hare, D.K. (2014). Residence time control on hot moments of net nitrate production and uptake in the hyporheic zone: residence time control on temporal hyporheic nitrate cycling. *Hydrol. Process.* 28: 3741–3751. https://doi.org/10.1002/hyp.9921.

Briggs, M.A., Lautz, L.K., Hare, D.K., and González-Pinzón, R. (2013). Relating hyporheic fluxes, residence times, and redox-sensitive biogeochemical processes upstream of beaver dams. *Freshw. Sci.* 32: 622–641. https://doi.org/10.1899/12-110.1.

Brunke, M. (1999). Colmation and depth filtration within streambeds: retention of particles in hyporheic interstices. *Int. Rev. Hydrobiol.* 84: 99–117. https://doi.org/10.1002/iroh.199900014.

Brunke, M. and Gonser, T. (1997). The ecological significance of exchange processes between rivers and groundwater. *Freshw. Biol.* 37: 1–33. https://doi.org/10.1046/j.1365-2427.1997.00143.x.

Buffington, J.M. and Tonina, D. (2009). Hyporheic exchange in mountain rivers II: effects of channel morphology on mechanics, scales, and rates of exchange. *Geogr. Compass* 3: 1038–1062. https://doi.org/10.1111/j.1749-8198.2009.00225.x.

Burkholder, B.K., Grant, G.E., Haggerty, R. et al. (2008). Influence of hyporheic flow and geomorphology on temperature of a large, gravel-bed river, Clackamas River, Oregon, USA. *Hydrol. Process.* 22: 941–953. https://doi.org/10.1002/hyp.6984.

Buxton, T.H., Buffington, J.M., Tonina, D. et al. (2015). Modeling the influence of salmon spawning on hyporheic exchange of marine-derived nutrients in gravel stream beds. *Can. J. Fish. Aquat. Sci.* 72: 1146–1158. https://doi.org/10.1139/cjfas-2014-0413.

Cardenas, M.B. (2015). Hyporheic zone hydrologic science: a historical account of its emergence and a prospectus: hyporheic zone hydrologic science: a historical account. *Water Resour. Res.* 51: 3601–3616. https://doi.org/10.1002/2015WR017028.

Cardenas, M.B. and Wilson, J.L. (2006). The influence of ambient groundwater discharge on exchange zones induced by current–bedform interactions. *J. Hydrol.* 331: 103–109. https://doi.org/10.1016/j.jhydrol.2006.05.012.

Cardenas, M.B. and Wilson, J.L. (2007a). Dunes, turbulent eddies, and interfacial exchange with permeable sediments: dunes, eddies, and interfacial exchange. *Water Resour. Res.* 43. https://doi.org/10.1029/2006WR005787.

Cardenas, M.B. and Wilson, J.L. (2007b). Exchange across a sediment–water interface with ambient groundwater discharge. *J. Hydrol.* 346: 69–80. https://doi.org/10.1016/j.jhydrol.2007.08.019.

Cardenas, M.B., Wilson, J.L., and Haggerty, R. (2008). Residence time of bedform-driven hyporheic exchange. *Adv. Water Resour.* 31: 1382–1386. https://doi.org/10.1016/j.advwatres.2008.07.006.

Cardenas, M.B., Wilson, J.L., and Zlotnik, V.A. (2004). Impact of heterogeneity, bed forms, and stream curvature on subchannel hyporheic exchange: modeling study of hyporheic exchange. *Water Resour. Res.* 40. https://doi.org/10.1029/2004WR003008.

Caruso, A., Boano, F., Ridolfi, L. et al. (2017). Biofilm-induced bioclogging produces sharp interfaces in hyporheic flow, redox conditions, and microbial community structure: bioclogging-induced biochemical zonation. *Geophys. Res. Lett.* 44: 4917–4925. https://doi.org/10.1002/2017GL073651.

Chen, X. (2004). Streambed hydraulic conductivity for rivers in South-Central Nebraska. *J. Am. Water Resour. Assoc.* 40: 561–573. https://doi.org/10.1111/j.1752-1688.2004.tb04443.x.

Conant, B. (2004). Delineating and quantifying ground water discharge zones using streambed temperatures. *Ground Water* 42: 243–257. https://doi.org/10.1111/j.1745-6584.2004.tb02671.x.

Corson-Rikert, H.A., Wondzell, S.M., Haggerty, R., and Santelmann, M.V. (2016). Carbon dynamics in the hyporheic zone of a headwater mountain stream in the Cascade Mountains, Oregon: HZ carbon dynamics in headwater stream. *Water Resour. Res.* 52: 7556 7576. https://doi.org/10.1002/2016WR019303.

Costigan, K.H., Jaeger, K.L., Goss, C.W. et al. (2016). Understanding controls on flow permanence in intermittent rivers to aid ecological research: integrating meteorology, geology and land cover: integrating science to understand flow intermittence. *Ecohydrology* 9: 1141–1153. https://doi.org/10.1002/eco.1712.

Creed, I.F., McKnight, D.M., Pellerin, B.A. et al. (2015). The river as a chemostat: fresh perspectives on dissolved organic matter flowing down the river continuum. *Can. J. Fish. Aquat. Sci.* 72: 1272–1285. https://doi.org/10.1139/cjfas-2014-0400.

Dara, R., Kettridge, N., Rivett, M.O. et al. (2019). Identification of floodplain and riverbed sediment heterogeneity in a meandering UK lowland stream by ground penetrating radar. *J. Appl. Geophys.* 171: 103863. https://doi.org/10.1016/j.jappgeo.2019.103863.

Datry, T., Bonada, N., and Boulton, A.J. (Eds.) (2017). *Intermittent Rivers and Ephemeral Streams: Ecology and Management.* London; San Diego, CA: Academic Press, an imprint of Elsevier.

Datry, T. and Larned, S.T. (2008). River flow controls ecological processes and invertebrate assemblages in subsurface flowpaths of an ephemeral river reach. *Can. J. Fish. Aquat. Sci.* 65: 1532–1544. https://doi.org/10.1139/F08-075.

Datry, T., Scarsbrook, M., Larned, S., and Fenwick, G. (2008). Lateral and longitudinal patterns within the stygoscape of an alluvial river corridor. *Fundam. Appl. Limnol. Arch. Für Hydrobiol.* 171: 335–347. https://doi.org/10.1127/1863-9135/2008/0171-0335.

De Falco, N., Boano, F., and Arnon, S. (2016). Biodegradation of labile dissolved organic carbon under losing and gaining streamflow conditions simulated in a laboratory flume: DOC uptake in losing and gaining streams. *Limnol. Oceanogr.* 61: 1839–1852. https://doi.org/10.1002/lno.10344.

DelVecchia, A.G., Stanford, J.A., and Xu, X. (2016). Ancient and methane-derived carbon subsidizes contemporary food webs. *Nat. Commun.* 7: 13163. https://doi.org/10.1038/ncomms13163.

DiStefano, R.J., Magoulick, D.D., Imhoff, E.M., and Larson, E.R. (2009). Imperiled crayfishes use hyporheic zone during seasonal drying of an intermittent stream. *J. North Am. Benthol. Soc.* 28: 142–152. https://doi.org/10.1899/08-072.1.

Dole-Olivier, M.-J. (2011). The hyporheic refuge hypothesis reconsidered: a review of hydrological aspects. *Mar. Freshw. Res.* 62: 1281. https://doi.org/10.1071/MF11084.

Dole-Olivier, M.-J., Marmonier, P., and Beffy, J.-L. (1997). Response of invertebrates to lotic disturbance: is the hyporheic zone a patchy refugium? *Freshw. Biol* 37: 257–276. https://doi.org/10.1046/j.1365-2427.1997.00140.x.

Drummond, J.D., Bernal, S., von Schiller, D., and Martí, E. (2016). Linking in-stream nutrient uptake to hydrologic retention in two headwater streams. *Freshw. Sci.* 35: 1176–1188. https://doi.org/10.1086/688599.

Drummond, J.D., Davies-Colley, R.J., Stott, R. et al. (2014). Retention and remobilization dynamics of fine particles and microorganisms in pastoral streams. *Water Res* 66: 459–472. https://doi.org/10.1016/j.watres.2014.08.025.

Drummond, J.D., Larsen, L.G., González-Pinzón, R. et al. (2017). Fine particle retention within stream storage areas at base flow and in response to a storm event: particle retention stream storage areas. *Water Resour. Res.* 53: 5690–5705. https://doi.org/10.1002/2016WR020202.

Drummond, J.D., Larsen, L.G., González-Pinzón, R. et al. (2018). Less fine particle retention in a restored versus unrestored urban stream: balance between hyporheic exchange, resuspension, and immobilization. *J. Geophys. Res. Biogeosciences* 123: 1425–1439. https://doi.org/10.1029/2017JG004212.

Dupas, R., Minaudo, C., and Abbott, B.W. (2019). Stability of spatial patterns in water chemistry across temperate ecoregions. *Environ. Res. Lett.* 14: 074015. https://doi.org/10.1088/1748-9326/ab24f4.

Dwivedi, D., Steefel, I.C., Arora, B., and Bisht, G. (2017). Impact of intra-meander hyporheic flow on nitrogen cycling. *Procedia Earth Planet. Sci.* 17: 404–407. https://doi.org/10.1016/j.proeps.2016.12.102.

Elliott, A.H. and Brooks, N.H. (1997a). Transfer of nonsorbing solutes to a streambed with bed forms: laboratory experiments. *Water Resour. Res.* 33: 137–151. https://doi.org/10.1029/96WR02783.

Elliott, A.H. and Brooks, N.H. (1997b). Transfer of nonsorbing solutes to a streambed with bed forms: theory. *Water Resour. Res.* 33: 123–136. https://doi.org/10.1029/96WR02784.

Endreny, T.A. and Lautz, L.K. (2012). Comment on 'Munz M, Krause S, Tecklenburg C, Binley A. Reducing monitoring gaps at the aquifer-river interface by modelling groundwater-surfacewater exchange flow patterns. Hydrological processes. DOI: 10.1002/hyp.8080': comments on 'Reducing monitoring gaps at the aquifer-river interface'. *Hydrol. Process* 26: 1586–1588. https://doi.org/10.1002/hyp.8410.

Fellows, C.S., Valett, M.H., and Dahm, C.N. (2001). Whole-stream metabolism in two montane streams: contribution of the hyporheic zone. *Limnol. Oceanogr.* 46: 523–531. https://doi.org/10.4319/lo.2001.46.3.0523.

Fisher, S.G., Grimm, N.B., Martí, E. et al. (1998). Material spiraling in stream corridors: a telescoping ecosystem model. *Ecosystems* 1: 19–34. https://doi.org/10.1007/s100219900003.

Fleckenstein, J.H., Krause, S., Hannah, D.M., and Boano, F. (2010). Groundwater-surface water interactions: new methods and models to improve understanding of processes and dynamics. *Adv. Water Resour.* 33: 1291–1295. https://doi.org/10.1016/j.advwatres.2010.09.011.

Fleckenstein, J.H., Niswonger, R.G., and Fogg, G.E. (2006). River-aquifer interactions, geologic heterogeneity, and low-flow management. *Ground Water* 44: 837–852. https://doi.org/10.1111/j.1745-6584.2006.00190.x.

Folegot, S., Hannah, D.M., Dugdale, S.J. et al. (2018a). Low flow controls on stream thermal dynamics. *Limnologica* 68: 157–167. https://doi.org/10.1016/j.limno.2017.08.003.

Folegot, S., Krause, S., Mons, R. et al. (2018b). Mesocosm experiments reveal the direction of groundwater-surface water exchange alters the hyporheic refuge capacity under warming scenarios. *Freshw. Biol.* 63: 165–177. https://doi.org/10.1111/fwb.13049.

Fox, A., Boano, F., and Arnon, S. (2014). Impact of losing and gaining streamflow conditions on hyporheic exchange fluxes induced by dune-shaped bed forms. *Water Resour. Res.* 50: 1895–1907. https://doi.org/10.1002/2013WR014668.

Fox, A., Laube, G., Schmidt, C. et al. (2016). The effect of losing and gaining flow conditions on hyporheic exchange in heterogeneous streambeds: hyporheic exchange in heterogeneous streambeds. *Water Resour. Res.* 52: 7460–7477. https://doi.org/10.1002/2016WR018677.

Frei, R.J., Abbott, B.W., Dupas, R. et al. (2020). Predicting nutrient incontinence in the anthropocene at watershed scales. *Front. Environ. Sci.* 7: 200. https://doi.org/10.3389/fenvs.2019.00200.

Freitas, J.G., Rivett, M.O., Roche, R.S. et al., 2015. Heterogeneous hyporheic zone dechlorination of a TCE groundwater plume discharging to an urban river reach. *Sci. Total Environ.* 505, 236–252. https://doi.org/10.1016/j.scitotenv.2014.09.083.

Frissell, C.A., Liss, W.J., Warren, C.E., and Hurley, M.D. (1986). A hierarchical framework for stream habitat classification: viewing streams in a watershed context. *Environ. Manage.* 10: 199–214. https://doi.org/10.1007/BF01867358.

Gandy, C.J., Smith, J.W.N., and Jarvis, A.P. (2007). Attenuation of mining-derived pollutants in the hyporheic zone: a review. *Sci. Total Environ.* 373: 435–446. https://doi.org/10.1016/j.scitotenv.2006.11.004.

Genereux, D.P., Leahy, S., Mitasova, H. et al. (2008). Spatial and temporal variability of streambed hydraulic conductivity in West Bear Creek, North Carolina, USA. *J. Hydrol.* 358: 332–353. https://doi.org/10.1016/j.jhydrol.2008.06.017.

Gippel, C.J. (1995). Environmental hydraulics of large woody debris in streams and rivers. *J. Environ. Eng.* 121: 388–395. https://doi.org/10.1061/(ASCE)0733-9372(1995)121:5(388).

Gomez-Velez, J.D., Harvey, J.W., Cardenas, M.B., and Kiel, B. (2015). Denitrification in the Mississippi River network controlled by flow through river bedforms. *Nat. Geosci.* 8: 941–945. https://doi.org/10.1038/ngeo2567.

Gomez-Velez, J.D., Krause, S., and Wilson, J.L. (2014). Effect of low-permeability layers on spatial patterns of hyporheic exchange and groundwater upwelling. *Water Resour. Res.* 50: 5196–5215. https://doi.org/10.1002/2013WR015054.

González-Pinzón, R., Ward, A.S., Hatch, C.E. et al. (2015). A field comparison of multiple techniques to quantify groundwater–surface-water interactions. *Freshw. Sci.* 34: 139–160. https://doi.org/10.1086/679738.

Gooseff, M.N. (2010). Defining hyporheic zones - advancing our conceptual and operational definitions of where stream water and groundwater meet: defining hyporheic zones. *Geogr. Compass* 4: 945–955. https://doi.org/10.1111/j.1749-8198.2010.00364.x.

Gooseff, M.N., Anderson, J.K., Wondzell, S.M. et al. (2006). A modelling study of hyporheic exchange pattern and the sequence, size, and spacing of stream bedforms in mountain stream networks, Oregon, USA. *Hydrol. Process.* 20: 2443–2457. https://doi.org/10.1002/hyp.6349.

Gootman, K.S., González-Pinzón, R., Knapp, J.L.A. et al. (2020). Spatiotemporal variability in transport and reactive processes across a 1st - 5th order fluvial network. *Water Resour. Res.* e2019WR026303. https://doi.org/10.1029/2019WR026303.

Graham, E.B., Stegen, J.C., Huang, M. et al. (2019). Subsurface biogeochemistry is a missing link between ecology and hydrology in dam-impacted river corridors. *Sci. Total Environ.* 657: 435–445. https://doi.org/10.1016/j.scitotenv.2018.11.414.

Hancock, P.J., Boulton, A.J., and Humphreys, W.F. (2005). Aquifers and hyporheic zones: towards an ecological understanding of groundwater. *Hydrogeol. J.* 13: 98–111. https://doi.org/10.1007/s10040-004-0421-6.

Hannah, D.M., Malcolm, I.A., and Bradley, C. (2009). Seasonal hyporheic temperature dynamics over riffle bedforms. *Hydrol. Process.* 23: 2178–2194. https://doi.org/10.1002/hyp.7256.

Hannah, D.M., Sadler, J.P., and Wood, P.J. (2007). Hydroecology and ecohydrology: a potential route forward? *Hydrol. Process* 21: 3385–3390. https://doi.org/10.1002/hyp.6888.

Harman, C.J., Ward, A.S., and Ball, A. (2016). How does reach-scale stream-hyporheic transport vary with discharge? Insights from rSAS analysis of sequential tracer injections in a headwater mountain stream: time-variable stream transport. *Water Resour. Res.* 52: 7130–7150. https://doi.org/10.1002/2016WR018832.

Harms, T.K. and Grimm, N.B. (2008). Hot spots and hot moments of carbon and nitrogen dynamics in a semiarid riparian zone: riparian hot spots and hot moments. *J. Geophys. Res. Biogeosciences* 113: n/a–n/a. https://doi.org/10.1029/2007JG000588.

Harrison, L.R., Bray, E., Overstreet, B. et al. (2019). Physical controls on Salmon Redd site selection in restored reaches of a regulated, gravel-bed river. *Water Resour. Res.* 55: 8942–8966. https://doi.org/10.1029/2018WR024428.

Harvey, J. and Gooseff, M. (2015). River corridor science: hydrologic exchange and ecological consequences from bedforms to basins: river corridors from bedforms to basins. *Water Resour. Res.* 51: 6893–6922. https://doi.org/10.1002/2015WR017617.

Harvey, J.W., Böhlke, J.K., Voytek, M.A. et al. (2013). Hyporheic zone denitrification: controls on effective reaction depth and contribution to whole-stream mass balance: scaling hyporheic flow controls on stream denitrification. *Water Resour. Res.* 49: 6298–6316. https://doi.org/10.1002/wrcr.20492.

Helton, A.M., Poole, G.C., Meyer, J.L. et al. (2011). Thinking outside the channel: modeling nitrogen cycling in networked river ecosystems. *Front. Ecol. Environ.* 9: 229–238. https://doi.org/10.1890/080211.

Herzog, S.P., Higgins, C.P., Singha, K., and McCray, J.E. (2018). Performance of engineered streambeds for inducing hyporheic transient storage and attenuation of resazurin. *Environ. Sci. Technol.* 52: 10627–10636. https://doi.org/10.1021/acs.est.8b01145.

Herzog, S.P., Ward, A.S., and Wondzell, S.M. (2019). Multiscale feature-feature interactions control patterns of hyporheic exchange in a simulated headwater mountain stream. *Water Resour. Res.* 55: 10976–10992. https://doi.org/10.1029/2019WR025763.

Hester, E.T., Cardenas, M.B., Haggerty, R., and Apte, S.V. (2017). The importance and challenge of hyporheic mixing: hyporheic mixing. *Water Resour. Res.* 53: 3565–3575. https://doi.org/10.1002/2016WR020005.

Hester, E.T. and Doyle, M.W. (2008). In-stream geomorphic structures as drivers of hyporheic exchange: in-stream structures and hyporheic exchange. *Water Resour. Res.* 44. https://doi.org/10.1029/2006WR005810.

Hester, E.T., Eastes, L.A., and Widdowson, M.A. (2019). Effect of surface water stage fluctuation on mixing-dependent hyporheic denitrification in riverbed dunes. *Water Resour. Res.* 2018WR024198. https://doi.org/10.1029/2018WR024198.

Hölker, F., Vanni, M.J., Kuiper, J.J. et al. (2015). Tube-dwelling invertebrates: tiny ecosystem engineers have large effects in lake ecosystems. *Ecol. Monogr.* 85: 333–351. https://doi.org/10.1890/14-1160.1.

Holmes, R.M., Fisher, S.G., and Grimm, N.B. (1994). Parafluvial nitrogen dynamics in a desert stream ecosystem. *J. North Am. Benthol. Soc.* 13: 468–478. https://doi.org/10.2307/1467844.

Hou, Z., Nelson, W.C., Stegen, J.C. et al. (2017). Geochemical and microbial community attributes in relation to hyporheic zone geological facies. *Sci. Rep.* 7: 12006. https://doi.org/10.1038/s41598-017-12275-w.

Irvine, D.J., Brunner, P., Franssen, H.-J.H., and Simmons, C.T. (2012). Heterogeneous or homogeneous? Implications of simplifying heterogeneous streambeds in models of losing streams. *J. Hydrol.* 424–425: 16–23. https://doi.org/10.1016/j.jhydrol.2011.11.051.

Jaeger, A., Posselt, M., Betterle, A. et al. (2019). Spatial and temporal variability in attenuation of polar organic micropollutants in an urban lowland stream. *Environ. Sci. Technol.* 53: 2383–2395. https://doi.org/10.1021/acs.est.8b05488.

Janzen, K. and Westbrook, C.J. (2011). Hyporheic flows along a channelled peatland: influence of beaver dams. *Can. Water Resour. J. Rev. Can. Ressour. Hydr.* 36: 331–347. https://doi.org/10.4296/cwrj3604846.

Jencso, K.G., McGlynn, B.L., Gooseff, M.N. et al. (2009). Hydrologic connectivity between landscapes and streams: transferring reach- and plot-scale understanding to the catchment scale: connectivity between landscapes and streams. *Water Resour. Res.* 45. https://doi.org/10.1029/2008WR007225.

Jones, J.B. and Holmes, R.M. (1996). Surface-subsurface interactions in stream ecosystems. *Trends Ecol. Evol.* 11: 239–242. https://doi.org/10.1016/0169-5347(96).

Jones, J.I., Collins, A.L., Naden, P.S., and Sear, D.A. (2012). The relationship between fine sediment and macrophytes in rivers: fine sediment and macrophytes. *River Res. Appl.* 28: 1006–1018. https://doi.org/10.1002/rra.1486.

Jones, K.L., Poole, G.C., Woessner, W.W. et al. (2008). Geomorphology, hydrology, and aquatic vegetation drive seasonal hyporheic flow patterns across a gravel-dominated floodplain. *Hydrol. Process.* 22: 2105–2113. https://doi.org/10.1002/hyp.6810.

Karaman, S. (1935). Die Fauna der unterirdischen Gewässer Jugoslaviens: mit 5 Abbildungen. *SIL Proc* 1922-2010 (7): 46–73. https://doi.org/10.1080/03680770.1935.11902405.

Kasahara, T. and Hill, A.R. (2006). Hyporheic exchange flows induced by constructed riffles and steps in lowland streams in southern Ontario, Canada. *Hydrol. Process.* 20: 4287–4305. https://doi.org/10.1002/hyp.6174.

Kasahara, T. and Wondzell, S.M. (2003). Geomorphic controls on hyporheic exchange flow in mountain streams: geomorphic controls on hyporheic exchange. *Water Resour. Res.* 39: SBH 3-1–SBH 3-14. https://doi.org/10.1029/2002WR001386.

Kaufman, M.H., Cardenas, M.B., Buttles, J. et al. (2017). Hyporheic hot moments: dissolved oxygen dynamics in the hyporheic zone in response to surface flow perturbations: dissolved oxygen dynamics in the hyporheic zone. *Water Resour. Res.* 53: 6642–6662. https://doi.org/10.1002/2016WR020296.

Kessler, A.J., Cardenas, M.B., and Cook, P.L.M. (2015). The negligible effect of bed form migration on denitrification in hyporheic zones of permeable sediments: denitrification under moving bedforms. *J. Geophys. Res. Biogeosciences* 120: 538–548. https://doi.org/10.1002/2014JG002852.

Knapp, J.L.A., González-Pinzón, R., Drummond, J.D. et al. (2017). Tracer-based characterization of hyporheic exchange and benthic biolayers in streams: hyporheic exchange and benthic biolayers. *Water Resour. Res.* 53: 1575–1594. https://doi.org/10.1002/2016WR019393.

Krause, S., Blume, T., and Cassidy, N.J. (2012a). Investigating patterns and controls of groundwater up-welling in a lowland river by combining fibre-optic distributed temperature sensing with observations of vertical hydraulic gradients. *Hydrol. Earth Syst. Sci.* 16: 1775–1792. https://doi.org/10.5194/hess-16-1775-2012.

Krause, S., Boano, F., Cuthbert, M.O. et al. (2014a). Understanding process dynamics at aquifer-surface water interfaces: an introduction to the special section on new modeling approaches and novel experimental technologies: introduction. *Water Resour. Res.* 50: 1847–1855. https://doi.org/10.1002/2013WR014755.

Krause, S., Hannah, D.M., and Blume, T. (2011a). Interstitial pore-water temperature dynamics across a pool-riffle-pool sequence. *Ecohydrology* 4: 549–563. https://doi.org/10.1002/eco.199.

Krause, S., Hannah, D.M., Fleckenstein, J.H. et al. (2011b). Inter-disciplinary perspectives on processes in the hyporheic zone. *Ecohydrology* 4: 481–499. https://doi.org/10.1002/eco.176.

Krause, S., Heathwaite, L., Binley, A., and Keenan, P. (2009). Nitrate concentration changes at the groundwater-surface water interface of a small Cumbrian river. *Hydrol. Process.* 23: 2195–2211. https://doi.org/10.1002/hyp.7213.

Krause, S., Klaar, M.J., Hannah, D.M. et al. (2014b). The potential of large woody debris to alter biogeochemical processes and ecosystem services in lowland rivers: potential of large woody debris to alter biogeochemical processes. *Wiley Interdiscip. Rev. Water* 1: 263–275. https://doi.org/10.1002/wat2.1019.

Krause, S., Lewandowski, J., Grimm, N.B. et al. (2017). Ecohydrological interfaces as hot spots of ecosystem processes: ecohydrological interfaces as hot spots. *Water Resour. Res.* 53: 6359–6376. https://doi.org/10.1002/2016WR019516.

Krause, S., Munz, M., Tecklenburg, C., and Binley, A. (2012b). The effect of groundwater forcing on hyporheic exchange: reply to comment on 'Munz M, Krause S, Tecklenburg C, Binley A. Reducing monitoring gaps at the aquifer-river interface by modelling groundwater-surfacewater exchange flow patterns. hydrological pro: the effect of groundwater forcing on hyporheic exchange. *Hydrol. Process.* 26: 1589–1592. https://doi.org/10.1002/hyp.9271.

Krause, S., Tecklenburg, C., Munz, M., and Naden, E. (2013). Streambed nitrogen cycling beyond the hyporheic zone: flow controls on horizontal patterns and depth distribution of nitrate and dissolved oxygen in the upwelling groundwater of a lowland river: nitrogen cycling beyond the hyporheic. *J. Geophys. Res. Biogeosciences* 118: 54–67. https://doi.org/10.1029/2012JG002122.

Kurz, M.J., Drummond, J.D., Martí, E. et al. (2017). Impacts of water level on metabolism and transient storage in vegetated lowland rivers: insights from a mesocosm study: water level impacts on metabolism. *J. Geophys. Res. Biogeosciences*. https://doi.org/10.1002/2016JG003695.

Lansdown, K., Heppell, C.M., Trimmer, M. et al. (2015). The interplay between transport and reaction rates as controls on nitrate attenuation in permeable, streambed sediments: nitrate removal in permeable sediments. *J. Geophys. Res. Biogeosciences* 120: 1093–1109. https://doi.org/10.1002/2014JG002874.

Lansdown, K., McKew, B.A., Whitby, C. et al. (2016). Importance and controls of anaerobic ammonium oxidation influenced by riverbed geology. *Nat. Geosci.* 9: 357–360. https://doi.org/10.1038/ngeo2684.

Larsen, L., Harvey, J., Skalak, K., and Goodman, M. (2015). Fluorescence-based source tracking of organic sediment in restored and unrestored urban streams: organic sediment source tracking. *Limnol. Oceanogr.* 60: 1439–1461. https://doi.org/10.1002/lno.10108.

Larsen, L.G. and Harvey, J.W. (2017). Disrupted carbon cycling in restored and unrestored urban streams: critical timescales and controls: disrupted carbon cycling in urban streams. *Limnol. Oceanogr.* 62: S160–S182. https://doi.org/10.1002/lno.10613.

Laube, G., Schmidt, C., and Fleckenstein, J.H. (2018). The systematic effect of streambed conductivity heterogeneity on hyporheic flux and residence time. *Adv. Water Resour.* 122: 60–69. https://doi.org/10.1016/j.advwatres.2018.10.003.

Lawrence, J.E., Skold, M.E., Hussain, F.A. et al. (2013). Hyporheic zone in urban streams: a review and opportunities for enhancing water quality and improving aquatic habitat by active management. *Environ. Eng. Sci.* 30: 480–501. https://doi.org/10.1089/ees.2012.0235.

Lee-Cullin, J.A., Zarnetske, J.P., Ruhala, S.S., and Plont, S. (2018). Toward measuring biogeochemistry within the stream-groundwater interface at the network scale: an initial assessment of two spatial sampling strategies: network scale stream-groundwater biogeochemistry. *Limnol. Oceanogr. Methods* 16: 722–733. https://doi.org/10.1002/lom3.10277.

Leek, R., Wu, J.Q., Wang, L. et al. (2009). Heterogeneous characteristics of streambed saturated hydraulic conductivity of the Touchet River, south eastern Washington, USA. *Hydrol. Process.* 23: 1236–1246. https://doi.org/10.1002/hyp.7258.

Lewandowski, J., Arnon, S., Banks, E. et al. (2019). Is the hyporheic zone relevant beyond the scientific community? *Water* 11: 2230. https://doi.org/10.3390/w11112230.

Li, A., Aubeneau, A.F., Bolster, D. et al. (2017). Covariation in patterns of turbulence-driven hyporheic flow and denitrification enhances reach-scale nitrogen removal: hyporheic flow-denitrification upscaling. *Water Resour. Res.* 53: 6927–6944. https://doi.org/10.1002/2016WR019949.

Li, A., Bernal, S., Kohler, B. et al. (2020). Residence time in hyporheic bioactive layers explains nitrate uptake in streams. *Water Resour. Res.* https://doi.org/10.1029/2020WR027646.

Liu, C., Hu, Z., Lei, J., and Nepf, H. (2018). Vortex structure and sediment deposition in the wake behind a finite patch of model submerged vegetation. *J. Hydraul. Eng.* 144: 04017065. https://doi.org/10.1061/(ASCE)HY.1943-7900.0001408.

Liu, C. and Nepf, H. (2016). Sediment deposition within and around a finite patch of model vegetation over a range of channel velocity. *Water Resour. Res.* 52: 600–612. https://doi.org/10.1002/2015WR018249.

Liu, Y., Zarfl, C., Basu, N.B., and Cirpka, O.A. (2019). Turnover and legacy of sediment-associated PAH in a baseflow-dominated river. *Sci. Total Environ.* 671: 754–764. https://doi.org/10.1016/j.scitotenv.2019.03.236.

Lowell, J.L., Gordon, N., Engstrom, D. et al. (2009). Habitat heterogeneity and associated microbial community structure in a small-scale floodplain hyporheic flow path. *Microb. Ecol.* 58: 611–620. https://doi.org/10.1007/s00248-009-9525-9.

MacPhee, R.D.E. and Schouten, P. (2019). *End of the Megafauna: The Fate of the World's Hugest, Fiercest, and Strangest Animals*, 1e. New York: W.W. Norton & Company.

Magliozzi, C., Coro, G., Grabowski, R.C. et al. (2019). A multiscale statistical method to identify potential areas of hyporheic exchange for river restoration planning. *Environ. Model. Softw.* 111: 311–323. https://doi.org/10.1016/j.envsoft.2018.09.006.

Magliozzi, C., Grabowski, R., Packman, A.I., and Krause, S. (2018). Toward a conceptual framework of hyporheic exchange across spatial scales (preprint). *Rivers and Lakes/Theory Development.* https://doi.org/10.5194/hess-2018-268.

Malard, F., Tockner, K., Dole-Olivier, M.-J., and Ward, J.V. (2002). A landscape perspective of surface-subsurface hydrological exchanges in river corridors. *Freshw. Biol.* 47: 621–640. https://doi.org/10.1046/j.1365-2427.2002.00906.x.

Malcolm, I.A., Soulsby, C., Youngson, A.F. et al. (2004). Hydrological influences on hyporheic water quality: implications for salmon egg survival. *Hydrol. Process.* 18: 1543–1560. https://doi.org/10.1002/hyp.1405.

Malcolm, I.A., Soulsby, C., Youngson, A.F., and Tetzlaff, D. (2009). Fine scale variability of hyporheic hydrochemistry in salmon spawning gravels with contrasting groundwater-surface water interactions. *Hydrogeol. J.* 17: 161–174. https://doi.org/10.1007/s10040-008-0339-5.

Malzone, J.M., Anseeuw, S.K., Lowry, C.S., and Allen-King, R. (2016). Temporal hyporheic zone response to water table fluctuations. *Groundwater* 54: 274–285. https://doi.org/10.1111/gwat.12352.

Marçais, J., Gauvain, A., Labasque, T. et al. (2018). Dating groundwater with dissolved silica and CFC concentrations in crystalline aquifers. *Sci. Total Environ.* 636: 260–272. https://doi.org/10.1016/j.scitotenv.2018.04.196.

Marcé, R., von Schiller, D., Aguilera, R. et al. (2018). Contribution of hydrologic opportunity and biogeochemical reactivity to the variability of nutrient retention in river networks. *Glob. Biogeochem. Cycles* 32: 376–388. https://doi.org/10.1002/2017GB005677.

Marion, A., Packman, A.I., Zaramella, M., and Bottacin-Busolin, A. (2008). Hyporheic flows in stratified beds: hyporheic flow in stratified beds. *Water Resour. Res.* 44. https://doi.org/10.1029/2007WR006079.

Marzadri, A., Amatulli, G., Tonina, D. et al. (2021). Global riverine nitrous oxide emissions: the role of small streams and large rivers. *Sci. Total Environ.* 776: 145148. https://doi.org/10.1016/j.scitotenv.2021.145148.

Marzadri, A., Dee, M.M., Tonina, D. et al. (2017). Role of surface and subsurface processes in scaling N_2O emissions along riverine networks. *Proc. Natl. Acad. Sci.* 114: 4330–4335. https://doi.org/10.1073/pnas.1617454114.

Marzadri, A., Tonina, D., and Bellin, A. (2012). Morphodynamic controls on redox conditions and on nitrogen dynamics within the hyporheic zone: application to gravel bed rivers with alternate-bar morphology: NR dynamics within the hyporheic zone. *J. Geophys. Res. Biogeosciences* 117: n/a–n/a. https://doi.org/10.1029/2012JG001966.

Marzadri, A., Tonina, D., Bellin, A. et al. (2010). Semianalytical analysis of hyporheic flow induced by alternate bars: hyporheic flow induced by alternate bars. *Water Resour. Res.* 46. https://doi.org/10.1029/2009WR008285.

McClain, E.M., Boyer, W.E., Dent, L.C. et al. (2003). Biogeochemical hot spots and hot moments at the interface of terrestrial and aquatic ecosystems. *Ecosystems* 6: 301–312. https://doi.org/10.1007/s10021-003-0161-9.

Mendoza-Lera, C. and Datry, T. (2017). Relating hydraulic conductivity and hyporheic zone biogeochemical processing to conserve and restore river ecosystem services. *Sci. Total Environ.* 579: 1815–1821. https://doi.org/10.1016/j.scitotenv.2016.11.166.

Mendoza-Lera, C. and Mutz, M. (2013). Microbial activity and sediment disturbance modulate the vertical water flux in sandy sediments. *Freshw. Sci.* 32: 26–38. https://doi.org/10.1899/11-165.1.

Mendoza-Lera, C., Ribot, M., Foulquier, A. et al. (2019). Exploring the role of hydraulic conductivity on the contribution of the hyporheic zone to in-stream nitrogen uptake. *Ecohydrology* 12. https://doi.org/10.1002/eco.2139.

Mermillod-Blondin, F. (2011). The functional significance of bioturbation and biodeposition on biogeochemical processes at the water–sediment interface in freshwater and marine ecosystems. *J. North Am. Benthol. Soc.* 30: 770–778. https://doi.org/10.1899/10-121.1.

Mermillod-Blondin, F., Gaudet, J.-P., Gerino, M. et al. (2004). Relative influence of bioturbation and predation on organic matter processing in river sediments: a microcosm experiment. *Freshw. Biol.* 49: 895–912. https://doi.org/10.1111/j.1365-2427.2004.01233.x.

Moir, H.J. and Pasternack, G.B. (2010). Substrate requirements of spawning Chinook salmon (*Oncorhynchus tshawytscha*) are dependent on local channel hydraulics: substrate requirements of spawning chinook salmon. *River Res. Appl.* 26: 456–468. https://doi.org/10.1002/rra.1292.

Munz, M., Krause, S., Tecklenburg, C., and Binley, A. (2011). Reducing monitoring gaps at the aquifer-river interface by modelling groundwater-surface water exchange flow patterns. *Hydrol. Process.* 25: 3547–3562. https://doi.org/10.1002/hyp.8080.

Nelson, W.C., Graham, E.B., Crump, A.R. et al. (2020). Distinct temporal diversity profiles for nitrogen cycling genes in a hyporheic microbiome. *PLOS ONE* 15: e0228165. https://doi.org/10.1371/journal.pone.0228165.

Newcomer, M.E., Hubbard, S.S., Fleckenstein, J.H. et al. (2016). Simulating bioclogging effects on dynamic riverbed permeability and infiltration: bioclogging, dynamic riverbed permeability, and infiltration. *Water Resour. Res.* 52: 2883–2900. https://doi.org/10.1002/2015WR018351.

Nikolakopoulou, M., Argerich, A., Drummond, J.D. et al. (2018). Emergent macrophyte root architecture controls subsurface solute transport. *Water Resour. Res.* 54: 5958–5972. https://doi.org/10.1029/2017WR022381.

Nikora, V. (2010). Hydrodynamics of aquatic ecosystems : an interface between ecology, biomechanics and environmental fluid mechanics. *River Res. Appl.* 26: 367–384. https://doi.org/10.1002/rra.1291.

Nogaro, G., Datry, T., Mermillod-Blondin, F. et al. (2010). Influence of streambed sediment clogging on microbial processes in the hyporheic zone: influence of clogging on microbial processes. *Freshw. Biol.* 55: 1288–1302. https://doi.org/10.1111/j.1365-2427.2009.02352.x.

Nogaro, G., Mermillod-Blondin, F., Valett, M.H. et al. (2009). Ecosystem engineering at the sediment–water interface: bioturbation and consumer-substrate interaction. *Oecologia* 161: 125–138. https://doi.org/10.1007/s00442-009-1365-2.

Nowinski, J.D., Cardenas, M.B., and Lightbody, A.F., 2011. Evolution of hydraulic conductivity in the floodplain of a meandering river due to hyporheic transport of fine materials: hydraulic conductivity in the floodplain. *Geophys. Res. Lett.* 38: https://doi.org/10.1029/2010GL045819.

Ocampo, C.J., Oldham, C.E., and Sivapalan, M., 2006. Nitrate attenuation in agricultural catchments: shifting balances between transport and reaction: shifting balances in nitrate attenuation. *Water Resour. Res.* 42. https://doi.org/10.1029/2004WR003773.

Orghidan, T. (1959). Ein neuer Lebensraum des unterirdischen Wassers: der hyporheische Biotop. *Arch. Hydrobiol.* 55: 392–414.

Packman, A.I. and Brooks, N.H. (2001). Hyporheic exchange of solutes and colloids with moving bed forms. *Water Resour. Res.* 37: 2591–2605. https://doi.org/10.1029/2001WR000477.

Packman, A.I., Marion, A., Zaramella, M. et al. (2006). Development of layered sediment structure and its effects on pore water transport and hyporheic exchange. *Water Air Soil Pollut. Focus* 6: 433–442. https://doi.org/10.1007/s11267-006-9057-y.

Pescimoro, E., Boano, F., Sawyer, A.H., and Soltanian, M.R. (2019). Modeling influence of sediment heterogeneity on nutrient cycling in streambeds. *Water Resour. Res.* 55: 4082–4095. https://doi.org/10.1029/2018WR024221.

Pinay, G., Black, V.J., Planty-Tabacchi, A.M. et al. (2000). Geomorphic control of denitrification in large river floodplain soils. *Biogeochemistry* 50: 163–182. https://doi.org/10.1023/A:.

Pinay, G., O'Keefe, T.C., Edwards, R.T., and Naiman, R.J. (2009). Nitrate removal in the hyporheic zone of a salmon river in Alaska. *River Res. Appl.* 25: 367–375. https://doi.org/10.1002/rra.1164.

Pinay, G., Peiffer, S., De Dreuzy, J.-R. et al. (2015). Upscaling nitrogen removal capacity from local hotspots to low stream orders' drainage basins. *Ecosystems* 18: 1101–1120. https://doi.org/10.1007/s10021-015-9878-5.

Pinay, G., Ruffinoni, C., and Fabre, A. (1995). Nitrogen cycling in two riparian forest soils under different geomorphic conditions. *Biogeochemistry* 30: 9–29. https://doi.org/10.1007/BF02181038.

Pinay, G., Ruffinoni, C., Wondzell, S., and Gazelle, F. (1998). Change in groundwater nitrate concentration in a large river floodplain: denitrification, uptake, or mixing? *J. North Am. Benthol. Soc.* 17: 179–189. https://doi.org/10.2307/1467961.

Poole, G.C., O'Daniel, S.J., Jones, K.L. et al. (2008). Hydrologic spiralling: the role of multiple interactive flow paths in stream ecosystems. *River Res. Appl.* 24: 1018–1031. https://doi.org/10.1002/rra.1099.

Poole, G.C., Stanford, J.A., Running, S.W., and Frissell, C.A. (2006). Multiscale geomorphic drivers of groundwater flow paths: subsurface hydrologic dynamics and hyporheic habitat diversity. *J. North Am. Benthol. Soc.* 25: 288–303. https://doi.org/10.1899/0887-3593(2006)25[288:MGDOGF]2.0.CO;2.

Posselt, M., Mechelke, J., Rutere, C. et al. (2020). Bacterial diversity controls transformation of wastewater-derived organic contaminants in river-simulating flumes. *Environ. Sci. Technol.* 54: 5467–5479. https://doi.org/10.1021/acs.est.9b06928.

Preziosi-Ribero, A., Packman, A.I., Escobar-Vargas, J.A. et al. (2020). Fine sediment deposition and filtration under losing and gaining flow conditions: a particle tracking model approach. *Water Resour. Res.* 56. https://doi.org/10.1029/2019WR026057.

Reeder, W.J., Quick, A.M., Farrell, T.B. et al. (2018a). Hyporheic source and sink of nitrous oxide. *Water Resour. Res.* 54: 5001–5016. https://doi.org/10.1029/2018WR022564.

Reeder, W.J., Quick, A.M., Farrell, T.B. et al. (2018b). Spatial and temporal dynamics of dissolved oxygen concentrations and bioactivity in the hyporheic zone. *Water Resour. Res.* 54: 2112–2128. https://doi.org/10.1002/2017WR021388.

Revelli, R., Boano, F., Camporeale, C., and Ridolfi, L. (2008). Intra-meander hyporheic flow in alluvial rivers: hyporheic exchange in meandering rivers. *Water Resour. Res.* 44. https://doi.org/10.1029/2008WR007081.

Ribot, M., Cochero, J., Vaessen, T.N. et al. (2019). Leachates from helophyte leaf-litter enhance nitrogen removal from wastewater treatment plant effluents. *Environ. Sci. Technol.* 53: 7613–7620. https://doi.org/10.1021/acs.est.8b07218.

Rinaldo, A., Benettin, P., Harman, C.J. et al. (2015). Storage selection functions: a coherent framework for quantifying how catchments store and release water and solutes: on storage selection functions. *Water Resour. Res.* 51: 4840–4847. https://doi.org/10.1002/2015WR017273.

Risse-Buhl, U., Mendoza-Lera, C., Norf, H. et al. (2017). Contrasting habitats but comparable microbial decomposition in the benthic and hyporheic zone. *Sci. Total Environ.* 605–606: 683–691. https://doi.org/10.1016/j.scitotenv.2017.06.203.

Rivett, M.O., Turner, R.J., Glibbery (Née Murcott), P., and Cuthbert, M.O. (2012). The legacy of chlorinated solvents in the Birmingham aquifer, UK: observations spanning three decades and the challenge of future urban groundwater development. *J. Contam. Hydrol.* 140–141, 107–123. https://doi.org/10.1016/j.jconhyd.2012.08.006.

Roche, K.R., Blois, G., Best, J.L. et al. (2018). Turbulence links momentum and solute exchange in coarse-grained streambeds. *Water Resour. Res.* 54: 3225–3242. https://doi.org/10.1029/2017WR021992.

Roche, K.R., Drummond, J.D., Boano, F. et al. (2017). Benthic biofilm controls on fine particle dynamics in streams: biofilm-particle dynamics. *Water Resour. Res.* 53: 222–236. https://doi.org/10.1002/2016WR019041.

Roche, K.R., Li, A., Bolster, D. et al. (2019). Effects of turbulent hyporheic mixing on reach-scale transport. *Water Resour. Res.* 2018WR023421. https://doi.org/10.1029/2018WR023421.

Rode, M., Hartwig, M., Wagenschein, D. et al. (2015). The importance of hyporheic zone processes on ecological functioning and solute transport of streams and rivers. In: *Ecosystem Services and River Basin Ecohydrology* (ed. L. Chicharo, F. Müller, and N. Fohrer), 57–82. Dordrecht: Springer Netherlands. https://doi.org/10.1007/978-94-017-9846-4_4.

Roy Chowdhury, S., Zarnetske, J.P., Phanikumar, M.S. et al. (2020). Formation criteria for hyporheic anoxic microzones: assessing interactions of hydraulics, nutrients, and biofilms. *Water Resour. Res.* 56. https://doi.org/10.1029/2019WR025971.

Ryan, R.J. and Boufadel, M.C. (2007). Evaluation of streambed hydraulic conductivity heterogeneity in an urban watershed. *Stoch. Environ. Res. Risk Assess.* 21: 309–316. https://doi.org/10.1007/s00477-006-0066-1.

Salehin, M., Packman, A.I., and Paradis, M. (2004). Hyporheic exchange with heterogeneous streambeds: laboratory experiments and modeling: hyporheic exchange with heterogeneous streambeds. *Water Resour. Res.* 40. https://doi.org/10.1029/2003WR002567.

Salehin, M., Packman, A.I., and Wörman, A. (2003). Comparison of transient storage in vegetated and unvegetated reaches of a small agricultural stream in Sweden: seasonal variation and anthropogenic manipulation. *Adv. Water Resour.* 26: 951–964. https://doi.org/10.1016/S0309-1708(03).

Sand-jensen, K. (1998). Influence of submerged macrophytes on sediment composition and near-bed flow in lowland streams. *Freshw. Biol.* 39: 663–679. https://doi.org/10.1046/j.1365-2427.1998.00316.x.

Sawyer, A.H. (2015). Enhanced removal of groundwater-borne nitrate in heterogeneous aquatic sediments. *Geophys. Res. Lett.* 42: 403–410. https://doi.org/10.1002/2014GL062234.

Sawyer, A.H., Bayani Cardenas, M., and Buttles, J. (2012). Hyporheic temperature dynamics and heat exchange near channel-spanning logs: hyporheic heat and LWD. *Water Resour. Res.* 48. https://doi.org/10.1029/2011WR011200.

Sawyer, A.H. and Cardenas, M.B. (2009). Hyporheic flow and residence time distributions in heterogeneous cross-bedded sediment: hyporheic flow in heterogeneous sediment. *Water Resour. Res.* 45. https://doi.org/10.1029/2008WR007632.

Schaper, J.L., Posselt, M., Bouchez, C. et al. (2019). Fate of trace organic compounds in the hyporheic zone: influence of retardation, the benthic biolayer, and organic carbon. *Environ. Sci. Technol.* 53: 4224–4234. https://doi.org/10.1021/acs.est.8b06231.

Schaper, J.L., Seher, W., Nützmann, G. et al. (2018). The fate of polar trace organic compounds in the hyporheic zone. *Water Res* 140: 158–166. https://doi.org/10.1016/j.watres.2018.04.040.

Schmadel, N.M., Ward, A.S., Kurz, M.J. et al. (2016a). Stream solute tracer timescales changing with discharge and reach length confound process interpretation: solute tracer timescales confound process interpretation. *Water Resour. Res.* 52: 3227–3245. https://doi.org/10.1002/2015WR018062.

Schmadel, N.M., Ward, A.S., Lowry, C.S., and Malzone, J.M. (2016b). Hyporheic exchange controlled by dynamic hydrologic boundary conditions: dynamic hyporheic exchange. *Geophys. Res. Lett.* 43: 4408–4417. https://doi.org/10.1002/2016GL068286.

Sebok, E., Duque, C., Engesgaard, P., and Boegh, E. (2015). Spatial variability in streambed hydraulic conductivity of contrasting stream morphologies: channel bend and straight channel: hydraulic conductivity of streambed sediments. *Hydrol. Process.* 29: 458–472. https://doi.org/10.1002/hyp.10170.

Shelley, F., Klaar, M., Krause, S., and Trimmer, M. (2017). Enhanced hyporheic exchange flow around woody debris does not increase nitrate reduction in a sandy streambed. *Biogeochemistry* 136: 353–372. https://doi.org/10.1007/s10533-017-0401-2.

Shurin, J.B., Aranguren-Riaño, N., Duque Negro, D. et al. (2020). Ecosystem effects of the world's largest invasive animal. *Ecology* 101. https://doi.org/10.1002/ecy.2991.

Singh, T., Wu, L., Gomez-Velez, J.D. et al. (2019). Dynamic hyporheic zones: exploring the role of peak flow events on bedform-induced hyporheic exchange. *Water Resour. Res.* 55: 218–235. https://doi.org/10.1029/2018WR022993.

Słowik, M. (2014). Holocene evolution of meander bends in lowland river valley formed in complex geological conditions (the Obra river, Poland). *Geogr. Ann. Ser. Phys. Geogr.* 96: 61–81. https://doi.org/10.1111/geoa.12029.

Stanford, J.A. and Ward, J.V. (1988). The hyporheic habitat of river ecosystems. *Nature* 335: 64–66. https://doi.org/10.1038/335064a0.

Stanford, J.A. and Ward, J.V. (1993). An ecosystem perspective of alluvial rivers: connectivity and the hyporheic corridor. *J. North Am. Benthol. Soc.* 12: 48–60. https://doi.org/10.2307/1467685.

Statzner, B., Fièvet, E., Champagne, J.-Y. et al. (2000). Crayfish as geomorphic agents and ecosystem engineers: biological behavior affects sand and gravel erosion in experimental streams. *Limnol. Oceanogr.* 45: 1030–1040. https://doi.org/10.4319/lo.2000.45.5.1030.

Stewardson, M.J., Datry, T., Lamouroux, N. et al. (2016). Variation in reach-scale hydraulic conductivity of streambeds. *Geomorphology* 259: 70–80. https://doi.org/10.1016/j.geomorph.2016.02.001.

Stonedahl, S.H., Harvey, J.W., Detty, J. et al. (2012). Physical controls and predictability of stream hyporheic flow evaluated with a multiscale model: evaluating hyporheic flow with a multiscale model. *Water Resour. Res.* 48. https://doi.org/10.1029/2011WR011582.

Stonedahl, S.H., Harvey, J.W., Wörman, A. et al. (2010). A multiscale model for integrating hyporheic exchange from ripples to meanders: a 3-D flow model for hyporheic exchange. *Water Resour. Res.* 46. https://doi.org/10.1029/2009WR008865.

Storey, R.G., Howard, K.W.F., and Williams, D.D. (2003). Factors controlling riffle-scale hyporheic exchange flows and their seasonal changes in a gaining stream: a three-dimensional groundwater flow model: factors controlling exchange flows. *Water Resour. Res.* 39. https://doi.org/10.1029/2002WR001367.

Stubbington, R., Wood, P.J., and Boulton, A.J. (2009). Low flow controls on benthic and hyporheic macroinvertebrate assemblages during supra-seasonal drought. *Hydrol. Process.* 23: 2252–2263. https://doi.org/10.1002/hyp.7290.

Thibodeaux, L.J. and Boyle, J.D. (1987). Bedform-generated convective transport in bottom sediment. *Nature* 325: 341–343. https://doi.org/10.1038/325341a0.

Tonina, D. and Buffington, J.M. (2009). A three-dimensional model for analyzing the effects of salmon redds on hyporheic exchange and egg pocket habitat. *Can. J. Fish. Aquat. Sci.* 66: 2157–2173. https://doi.org/10.1139/F09-146.

Tonina, D. and Buffington, J.M. (2011). Effects of stream discharge, alluvial depth and bar amplitude on hyporheic flow in pool-riffle channels: hyporheic flow in pool-riffle channels. *Water Resour. Res.* 47. https://doi.org/10.1029/2010WR009140.

Tonina, D., de Barros, F.P.J., Marzadri, A., and Bellin, A. (2016). Does streambed heterogeneity matter for hyporheic residence time distribution in sand-bedded streams? *Adv. Water Resour.* 96: 120–126. https://doi.org/10.1016/j.advwatres.2016.07.009.

Trauth, N. and Fleckenstein, J.H. (2017). Single discharge events increase reactive efficiency of the hyporheic zone: discharge events increase reactivity. *Water Resour. Res.* 53: 779–798. https://doi.org/10.1002/2016WR019488.

Trauth, N., Schmidt, C., Maier, U. et al. (2013). Coupled 3-D stream flow and hyporheic flow model under varying stream and ambient groundwater flow conditions in a pool-riffle system: coupled 3-D stream flow and hyporheic flow model. *Water Resour. Res.* 49: 5834–5850. https://doi.org/10.1002/wrcr.20442.

Trauth, N., Schmidt, C., Vieweg, M. et al. (2014). Hyporheic transport and biogeochemical reactions in pool-riffle systems under varying ambient groundwater flow conditions. *J. Geophys. Res. Biogeosciences* 119: 910–928. https://doi.org/10.1002/2013JG002586.

Trauth, N., Schmidt, C., Vieweg, M. et al. (2015). Hydraulic controls of in-stream gravel bar hyporheic exchange and reactions. *Water Resour. Res.* 51: 2243–2263. https://doi.org/10.1002/2014WR015857.

Trimmer, M., Grey, J., Heppell, C.M. et al. (2012). River bed carbon and nitrogen cycling: state of play and some new directions. *Sci. Total Environ.* 434: 143–158. https://doi.org/10.1016/j.scitotenv.2011.10.074.

Trimmer, M., Shelley, F.C., Purdy, K.J. et al. (2015). Riverbed methanotrophy sustained by high carbon conversion efficiency. *ISME J* 9: 2304–2314. https://doi.org/10.1038/ismej.2015.98.

Ullah, S., Zhang, H., Heathwaite, A.L. et al. (2014). Influence of emergent vegetation on nitrate cycling in sediments of a groundwater-fed river. *Biogeochemistry* 118: 121–134. https://doi.org/10.1007/s10533-013-9909-2.

Valett, H.M., Dahm, C.N., Campana, M.E. et al. (1997). Hydrologic influences on groundwater-surface water ecotones: heterogeneity in nutrient composition and retention. *J. North Am. Benthol. Soc.* 16: 239–247. https://doi.org/10.2307/1468254.

Vander Vorste, R., Mermillod-Blondin, F., Hervant, F. et al. (2017). Gammarus pulex (Crustacea: amphipoda) avoids increasing water temperature and intraspecific competition through vertical migration into the hyporheic zone: a mesocosm experiment. *Aquat. Sci.* 79: 45–55. https://doi.org/10.1007/s00027-016-0478-z.

Vander Vorste, R., Mermillod-Blondin, F., Hervant, F. et al. (2016). Increased depth to the water table during river drying decreases the resilience of *Gammarus pulex* and alters ecosystem function: increased depth to water table decreases resilience and alters function. *Ecohydrology* 9: 1177–1186. https://doi.org/10.1002/eco.1716.

Vaux, W.G. (1968). Intragravel flow and interchange of water in a streambed. *Fish. Bull.* 66: 479–489.

Villeneuve, S., Cook, P.G., Shanafield, M. et al. (2015). Groundwater recharge via infiltration through an ephemeral riverbed, central Australia. *J. Arid Environ.* 117: 47–58. https://doi.org/10.1016/j.jaridenv.2015.02.009.

Ward, A.S. (2016). The evolution and state of interdisciplinary hyporheic research: the evolution and state of interdisciplinary hyporheic research. *Wiley Interdiscip. Rev. Water* 3: 83–103. https://doi.org/10.1002/wat2.1120.

Ward, A.S., Gooseff, M.N., and Johnson, P.A. (2011). How can subsurface modifications to hydraulic conductivity be designed as stream restoration structures? Analysis of Vaux's conceptual models to enhance hyporheic exchange: vaux revisited-subsurface structure design. *Water Resour. Res.* 47. https://doi.org/10.1029/2010WR010028.

Ward, A.S., Kurz, M.J., Schmadel, N.M. et al. (2019a). Solute transport and transformation in an intermittent, headwater mountain stream with diurnal discharge fluctuations. *Water* 11: 2208. https://doi.org/10.3390/w11112208.

Ward, A.S. and Packman, A.I. (2019). Advancing our predictive understanding of river corridor exchange. *Wiley Interdiscip. Rev. Water* 6: e1327. https://doi.org/10.1002/wat2.1327.

Ward, A.S., Schmadel, N.M., and Wondzell, S.M. (2018). Time-variable transit time distributions in the hyporheic zone of a headwater mountain stream. *Water Resour. Res.* 54: 2017–2036. https://doi.org/10.1002/2017WR021502.

Ward, A.S., Wondzell, S.M., Schmadel, N.M. et al. (2019b). Spatial and temporal variation in river corridor exchange across a 5th-order mountain stream network. *Hydrol. Earth Syst. Sci.* 23: 5199–5225. https://doi.org/10.5194/hess-23-5199-2019.

Weatherill, J., Krause, S., Voyce, K. et al. (2014). Nested monitoring approaches to delineate groundwater trichloroethene discharge to a UK lowland stream at multiple spatial scales. *J. Contam. Hydrol.* 158: 38–54. https://doi.org/10.1016/j.jconhyd.2013.12.001.

Weatherill, J.J., Atashgahi, S., Schneidewind, U. et al. (2018). Natural attenuation of chlorinated ethenes in hyporheic zones: a review of key biogeochemical processes and in-situ transformation potential. *Water Res* 128: 362–382. https://doi.org/10.1016/j.watres.2017.10.059.

Winter, T.C. (Ed.), 1998. *Ground water and surface water: a single resource, U.S. Geological Survey circular*. U.S. Geological Survey, Denver, Colo.

Winter, T.C. (1999). Relation of streams, lakes, and wetlands to groundwater flow systems. *Hydrogeol. J.* 7: 28–45. https://doi.org/10.1007/s100400050178.

Withers, P., Neal, C., Jarvie, H., and Doody, D. (2014). Agriculture and eutrophication: Where do we go from here? *Sustainability* 6: 5853–5875. https://doi.org/10.3390/su6095853.

Wolke, P., Teitelbaum, Y., Deng, C. et al. (2019). Impact of bed form celerity on oxygen dynamics in the hyporheic zone. *Water* 12: 62. https://doi.org/10.3390/w12010062.

Wondzell, S.M. (2011). The role of the hyporheic zone across stream networks. *Hydrol. Process.* 25: 3525–3532. https://doi.org/10.1002/hyp.8119.

Wondzell, S.M., LaNier, J., and Haggerty, R. (2009a). Evaluation of alternative groundwater flow models for simulating hyporheic exchange in a small mountain stream. *J. Hydrol.* 364: 142–151. https://doi.org/10.1016/j.jhydrol.2008.10.011.

Wondzell, S.M., LaNier, J., Haggerty, R. et al., 2009b. Changes in hyporheic exchange flow following experimental wood removal in a small, low-gradient stream: hyporheic exchange and large wood removal. *Water Resour. Res.* 45. https://doi.org/10.1029/2008WR007214.

Wu, L., Singh, T., Gomez-Velez, J. et al. (2018a). Impact of dynamically changing discharge on hyporheic exchange processes under gaining and losing groundwater conditions. *Water Resour. Res.* 54. https://doi.org/10.1029/2018WR023185.

Wu, L., Singh, T., Gomez-Velez, J. et al. (2018b). Impact of dynamically changing discharge on hyporheic exchange processes under gaining and losing groundwater conditions. *Water Resour. Res.* 54. https://doi.org/10.1029/2018WR023185.

Zarnetske, J.P., Haggerty, R., Wondzell, S.M., and Baker, M.A. (2011a). Dynamics of nitrate production and removal as a function of residence time in the hyporheic zone. *J. Geophys. Res.* 116. https://doi.org/10.1029/2010JG001356.

Zarnetske, J.P., Haggerty, R., Wondzell, S.M., and Baker, M.A. (2011b). Labile dissolved organic carbon supply limits hyporheic denitrification. *J. Geophys. Res.* 116. https://doi.org/10.1029/2011JG001730.

Zarnetske, J.P., Haggerty, R., Wondzell, S.M. et al. (2012). Coupled transport and reaction kinetics control the nitrate source-sink function of hyporheic zones: hyporheic n source-sink controls. *Water Resour. Res.* 48. https://doi.org/10.1029/2012WR011894.

Zhou, S., Borjigin, S., Riya, S. et al. (2014a). The relationship between anammox and denitrification in the sediment of an inland river. *Sci. Total Environ.* 490: 1029–1036. https://doi.org/10.1016/j.scitotenv.2014.05.096.

Zhou, T. and Endreny, T.A. (2013). Reshaping of the hyporheic zone beneath river restoration structures: flume and hydrodynamic experiments: hyporheic exchange beneath river restoration structures. *Water Resour. Res.* 49: 5009–5020. https://doi.org/10.1002/wrcr.20384.

Zhou, Y., Ritzi, R.W., Soltanian, M.R., and Dominic, D.F. (2014b). The influence of streambed heterogeneity on hyporheic flow in gravelly rivers. *Groundwater* 52: 206–216. https://doi.org/10.1111/gwat.12048.

5

Groundwater–Lake Interfaces

Jörg Lewandowski[1], Donald O. Rosenberry[2], and Karin Meinikmann[3]

[1] *Leibniz Institute of Freshwater Ecology and Inland Fisheries, Berlin, Germany*
[2] *Geography Department, Humbold University Berlin, Germany*
[3] *Julius Kühn Institute of Ecological Chemistry, Plant Analysis and Stored Product Protection, Berlin, Germany*

5.1 General Principles

In comparison to the two other groundwater–surface water interfaces, i.e. groundwater–stream (Chapters 3 and 4) and groundwater–ocean (Chapter 6), there are fewer studies in the literature about the groundwater–lake interface; it is less well studied than other groundwater–surface water interfaces. Any and all flow of groundwater from the lakebed to the lake is called lacustrine groundwater discharge (LGD) (Lewandowski et al. 2013).

In general, LGD originates from groundwater that has been recharged in the subsurface catchment of a lake, i.e. in the area where hydraulic gradients cause groundwater flow to a lake. As groundwater flows towards and reaches the lake, it becomes LGD once it crosses the sediment–water interface (i.e. the lakebed). This process is usually focused in the near-shore margins of a lake. The reverse process of lake water infiltrating into the adjacent aquifer also occurs primarily along or near the shoreline.

LGD delivers dissolved compounds to a lake. These compounds originate either from water that recharges the aquifer, or they are released from the aquifer matrix via weathering of the aquifer porous medium. Some of them, e.g. nutrients and contaminants, are of special interest for lake geochemists and limnologists. When crossing the sediment–water interface, concentrations of groundwater-borne compounds might be altered by reactions within the interface, which can be caused by large amounts of easily degradable organic matter or intense redox processes. This complicates the quantifications of mass loads (constituent concentration × LGD volume) contributed by groundwater exfiltration to both lakes and rivers. The magnitude of constituent mass entering a lake depends on both LGD volume and the concentrations of the parameters of interest. Both may vary in space and time, which imposes even more complexity on investigations of LGD and related mass loads.

Eutrophication – the over-enrichment of water by nutrients such as nitrogen (N) and phosphorus (P) – is a major cause of water-quality impairment for freshwaters. Anthropogenic eutrophication has been attributed to intensive agriculture (fertilizer and manure), industrial activities, and population growth (Conley et al. 2009). Eutrophication of surface waters due to nutrient-rich LGD (diffuse source) is an emerging topic (Knights et al. 2017; Meinikmann et al. 2015; Nisbeth et al. 2019b, 2019a) because its relative impact on nutrient budgets increases with decreasing inputs of nutrients from point sources (Lewandowski et al. 2014).

As a basis for implementing effective management measures, water and nutrient budgets are required, and all input paths need to be quantified. Theoretically, a budget is quite simple: the sum of all inputs/sources minus the sum of all losses/sinks should equal changes in storage. Important input paths of water and nutrients to lakes include overland run-off, inflows of streams and rivers, precipitation onto the lake's water surface, dry atmospheric deposition (in the case of the nutrient budget), and LGD. Important losses are surface water outflows, evaporation from the open water surface (only water budget), and groundwater recharge (infiltration) through the lakebed. Although the latter might be ecologically relevant in some cases, the focus here is on groundwater discharge to the lake (i.e. LGD). Reliable quantifications of most components of the water and nutrient budget are a challenge especially with regard to temporal variations and spatial heterogeneities (Lewandowski et al. 2014). LGD and accompanying mass loads might, however, be the most complicated terms to quantify due to the spatial heterogeneity of LGD, large areas over which diffuse LGD occurs, difficult accessibility of deeper zones of the lake for measurements, etc. (Section 5.5, Lewandowski et al. 2015; Rosenberry et al. 2015).

5.2 Hydrology of Lacustrine Groundwater Discharge

The rate of groundwater flow to a lake depends on the hydraulic gradient and the hydraulic conductivity of the porous medium adjacent to the lake. In simple settings with uniform geology (homogeneous, isotropic), LGD may be focused along the near-shore margins. Several authors observed an exponential decrease of LGD with increasing distance from the shore line. This was found both in modelling studies and by in situ measurements (McBride and Pfannkuch 1975; Pfannkuch and Winter 1984). Flow paths of groundwater approaching the interface bend upwards (but will never cross each other) since groundwater will always flow in the direction of the largest gradient (McCobb et al. 2003). As a consequence, a vertical distribution of different groundwater compositions occurring in the aquifer is projected by the upwards bending flow paths more or less horizontally onto the lakebed, but is commonly compressed close to the shore and extended with increasing distance from the shore (Figure 5.1). Focusing of LGD to near-shore zones is conducive to measurements of LGD since near-shore areas are often shallow and, therefore, more easily accessible (e.g. by wading) (Lewandowski et al. 2014).

The focusing of LGD in near-shore areas is valid for aquifers that are more or less homogenous. Spatially restricted zones of high hydraulic conductivity in otherwise low-conductivity aquifers may serve as preferential flow paths resulting in locally increased LGD rates. This phenomenon is particularly common in lakes located in fractured-rock settings where highly localized LGD is focused in fractures. Lakes in contact with a more permeable, deeper aquifer may also have large groundwater discharge rates in deeper zones of the lakebed (Lewandowski et al. 2014).

Rosenberry et al. (2015) reviewed the hydrology of LGD. In addition to presenting general principles, the authors surveyed a large number of scientific studies about LGD and reported groundwater discharge rates found by the different authors. Figure 5.2 presents a

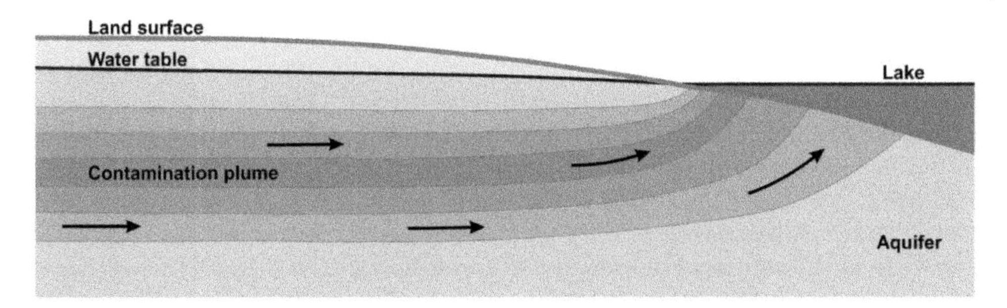

Figure 5.1 Schematic drawing of flow paths of a contamination plume approaching a lake.

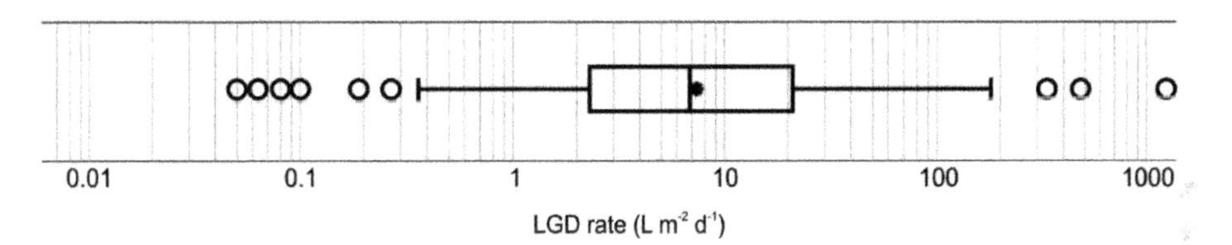

Figure 5.2 Boxplot of logarithmic data of lacustrine groundwater discharge rates reported in the literature (n = 109, data and details in Rosenberry et al. 2015, the upper and lower end of the box are the first and third quartiles, the band inside the box is the median, the filled dot the average, the end of the whiskers represent the lowest and highest values still within 1.5 × interquartile range of the lower/upper quartile, open circles: outliers).

boxplot summarizing the results shown in more detail in Rosenberry et al. (2015). Basically, the LGD rates varied over more than four orders of magnitude covering the range 0.05 to 1241 L m^{-2} d^{-1} with a median of 6.8 L m^{-2} d^{-1}, an arithmetic mean of the logarithmic data of 7.4 L m^{-2} d^{-1} (37.6 L m^{-2} d^{-1} for the non-logarithmic data) and lower and upper quartiles of 2.3 and 20.9 L m^{-2} d^{-1}, respectively.

In some settings the groundwater component dominates lake water budgets. For example, LGD represents 74% of all inflows to Williams Lake, Minnesota (LaBaugh et al. 1997), 94% to Mary Lake, Minnesota (Stets et al. 2010), 90% to Cliff Lake, Montana (Gurrieri and Furniss 2004), 70% to Lake Steißling, Germany (Gilfedder et al. 2018), 85% to Lake Annie, Florida (Sacks et al. 1998), and 77% to Lake Ammelshaimer (see (Petermann et al. 2018, based on water stable isotope measurements) . High LGD rates are reported for several lakes. For example, LGD was 477 L m^{-2} d^{-1} for Ashumet Pond, Massachusetts (McCobb et al. 2009), 155 L m^{-2} d^{-1} for Dickson Lake, Ontario (Ridgeway & Blanchfield 1998), and 138 L m^{-2} d^{-1} for Shingobee Lake, Minnesota (Rosenberry et al. 2000); all rates were determined from point measurements by using seepage meters (Section 5.4) (Lewandowski et al.2014).

5.3 Nitrogen and Phosphorus Biogeochemistry of Lacustrine Groundwater Discharge

Many natural systems have become artificially enriched in nitrogen (N) (Ibanhez et al. 2011), and nitrate contamination is a common problem worldwide (Delkash et al. 2018; Le Moal et al. 2019, Wakida and Lerner 2005). Nitrogen in aquifers often originates from mineral fertilizers (agriculture, urban lawns, and golf courses), atmospheric deposition, livestock, and sewage (Lewandowski et al. 2015; Shukla and Saxena 2020). Excess applications of fertilizers and manure have resulted in very high N concentrations in some aquifers. Acid rain and dry atmospheric N deposition have resulted in a ubiquitous N contamination of soils and subsequently of aquifers. Techniques for sewage disposal vary widely by region and culture; in many cases treated sewage with substantial N loads reaches the aquifer through decentralized sewage pits, sewage infiltration beds, on-site sewage treatment systems with drain fields, and leaky sewers (e.g. Lewandowski et al. 2014, Mester et al. 2019; Wakida and Lerner 2005).

Nitrogen in the aquifer occurs in the forms of nitrate, nitrite, ammonium, and dissolved organic nitrogen. Nitrate is highly mobile in oxic aquifers but is denitrified and converted to gaseous N_2 under anoxic conditions (Lewandowski et al. 2015). Ammonium is less mobile and occurs under anoxic conditions. Under oxic conditions, ammonium is nitrified to nitrate (Slomp and Van Cappellen 2004). Dissolved organic nitrogen is less well studied although it occurs in high concentrations in some aquifers. Eventually, the nitrogen species are transported from the subsurface catchment through the reactive aquifer–lake interface into the lake. Some N species (nitrate, nitrite) might be decreased at the interface due to reducing redox conditions, high microbial activity, and easily degradable organic matter. Other N species such as ammonium might even be increased as water passes across the interface (Lewandowski et al. 2015).

In general, high concentrations of nitrogen in groundwater approaching the groundwater–lake interface can be expected to result in high N loads to the lake. For example, N loads were 641 g m^{-2} yr^{-1} in Colgada Lake, Spain (Pina-Ochoa and Alvarez-Cobelas 2009) and 453 and 456 g m^{-2} yr^{-1} in two different gravel pit lakes, Austria (Muellegger et al. 2013; Weilhartner et al. 2012). Figure 5.3 shows a boxplot of the logarithmic distribution of LGD data for total N, which are taken from the international literature and are reviewed in Lewandowski et al. (2015). The range is 0 to 640 900 mg m^{-2} yr^{-1}, the median is 1080 mg m^{-2} yr^{-1}, and the arithmetic average of the logarithmic data is 1840 mg m^{-2} yr^{-1} (73 927 mg m^{-2} yr^{-1} for the non-logarithmic data). The lower and upper quartiles are 293 and 43 525 mg m^{-2} yr^{-1} (Figure 5.3).

Phosphorus (P) concentrations in pristine groundwater are usually but not always low (<50 μg L^{-1}) (Lewandowski et al. 2015). For example, Nisbeth et al. (2019a, 2019b) found

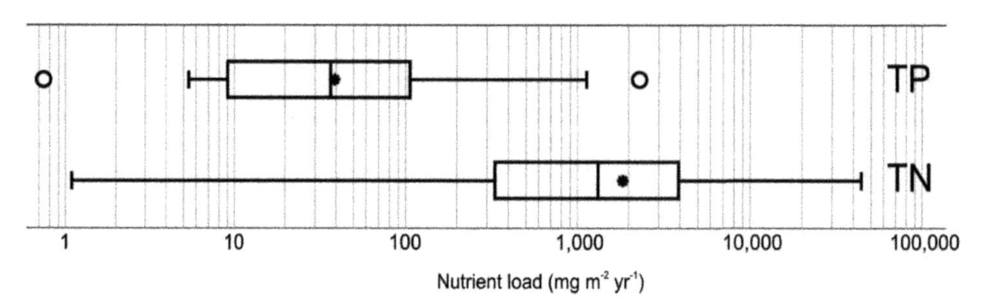

Figure 5.3 Boxplots of logarithmic data of groundwater-borne nutrient (phosphorus, n = 28 and nitrogen, n = 26) loads to lakes (lacustrine groundwater discharge) reported in the literature (data and details in Lewandowski et al. 2015; legend of boxplot symbols given in the caption of Figure 5.2).

that geogenic P might be a much more important P source in some instances than is generally assumed. Analogous to N, groundwater-P might be anthropogenically increased due to fertilizers, atmospheric deposition, manure, and sewage (500 to 5000 $\mu g\ L^{-1}$; Kunkel et al. 2005; Liao et al. 2019; Meinikmann et al. 2015; Wendland et al. 2005) often related to urban and/or industrial land use (e.g. Huang et al. 2020; Schilling et al. 2018; Stuart and Lapworth 2016). The mineralization of naturally present organic matter might also increase P concentrations. P itself occurs only in the oxidation state P(+V). However, its binding partners are redox sensitive and mobile under reducing conditions. Thus, reducing conditions favour the transport of P in the aquifer, but even under reducing conditions there is some retardation (approximately by a factor 10) compared to groundwater flow velocities (Lewandowski et al. 2015).

P in aquifers was considered immobile in the last century (Lewandowski et al. 2015). Thus, P concentrations in groundwater were assumed negligible, and consequently, P transfer via LGD received little attention (Cherkauer et al. 1992; Kilroy and Coxon 2005; Vanek 1987). P in groundwater was seldom measured with ecologically-adequate detection limits since the high ecological sensitivity of receiving water bodies to P was not the focus of hydrologists conducting groundwater investigations. However, some researchers have disproved the statement that P in aquifers is immobile and coined the statement "Phosphate does not migrate in groundwater – Better think again" (http://toxics.usgs.gov/highlights/phosphorous_migration.html, Access: 10 June 2020).

P is commonly considered the limiting nutrient in freshwater settings (Lewandowski et al. 2015). P concentrations in groundwater might be high, resulting in high P loads; e.g. 2.3 $g\ m^{-2}\ yr^{-1}$ in Lake Bysjön, Sweden (Vanek 1993), 1.1 $g\ m^{-2}\ yr^{-1}$ in Sparkling Lake, Wisconsin (Hargerthey & Kerfoot, 1998), and 0.4 $g\ m^{-2}\ yr^{-1}$ in Fishermans Cove of Ashumet Pond, Cape Cod, Massachusetts (McCobb et al. 2003). Reported P loads contributed by LGD generally are about two orders of magnitude lower than reported N loads (Lewandowski et al. 2015; Figure 5.3). Data for Figure 5.3 were taken from the international literature and were reviewed by Lewandowski et al. (2015). The range is 0.74 to 2292 $mg\ m^{-2}\ yr^{-1}$, the median is 36.8, and the arithmetic average of the logarithmic data is 39.0 $mg\ m^{-2}\ yr^{-1}$ (186.4 $mg\ m^{-2}\ yr^{-1}$ for the non-logarithmic data). The lower and upper quartiles are 9.14 and 107 $mg\ m^{-2}\ yr^{-1}$ (Figure 5.3).

5.4 Measurement Techniques

LGD can be identified and measured by several techniques that can be grouped into (1) local point measurements, (2) integrating methods, and (3) methods of pattern identification (Lewandowski et al. 2013, 2015; Rosenberry and LaBaugh 2008; Rosenberry et al. 2015).

Point methods:

a) Seepage meters are open-ended cylindrical vessels; the open end of the cylinder is commonly inserted 0.05 to 0.1 m into the sediment. An attached plastic bag collects the groundwater discharging within the area enclosed by the vessel. The LGD volume can be determined by the mass change of the plastic bag during the period of bag attachment. This is the only direct method to quantify seepage volumes. However, several difficulties should be considered when applying seepage meters since they might affect the results. The difficulties associated with using seepage meters are insertion problems, wave action, resistance of collection bag, and preferential flow paths (Rosenberry et al. 2008).

b) Temperature-depth profiles in the lakebed make use of natural temperature differences between the groundwater and lake water. LGD rates are calculated (1) from the curvature of the temperature depth profiles, which have an intense curvature at high discharge rates, or (2) from the phase shift in the sediment, which is small if flow is fast (Gordon et al. 2012; Schmidt et al. 2006).
c) Depth profiles of conservative ions such as chloride in the lakebed can be used if concentrations in the groundwater approaching the interface are significantly different from lake water concentrations. Ion-depth profiles are evaluated analogously to temperature depth profiles (Schuster et al. 2003).
d) Piezometers installed at a single location at multiple depths in the lakebed can indicate vertical hydraulic gradients; piezometers installed along a line perpendicular to the shoreline (transect, horizontal flow from catchment towards lake), or several groundwater observation wells distributed across the catchment (map of water table contour lines), can show hydraulic gradients along a horizontal axis, allowing LGD rates to be calculated based on Darcy's law. Estimates of hydraulic conductivity are required additionally for LGD analysis (Lewandowski et al. 2014).

Integrating methods:

a) Calculations of the mean annual groundwater recharge in the lake's subsurface catchment must equal the amount of mean annual LGD. In cases where groundwater discharges additionally to springs or streams before reaching the lake, the corresponding volumes have to be subtracted from annual LGD. The same applies to artificial groundwater abstractions, e.g. for drinking water supply.
b) Quantifying all of the easier-to-measure components of a lake water budget, and solving for the groundwater component as a residual, is a relatively simple concept. However, the errors of all components are included in the groundwater term and the method determines only the net groundwater component.
c) Mass balances of conservative ions such as chloride, stable water isotopes, or radon can also be used to refine estimates of LGD in settings where all other inputs and losses are known (Kazmierczak et al. 2016; Lewandowski et al. 2014). The two latter ones, especially, have recently become very popular in hydrological sciences. In fact, many studies use radon and water stable isotopes' complementary information to quantify lake water budgets and the role of LGD (e.g. Arnoux et al. 2017; Jiang 2020; Kebede and Zewdu 2019; Liao et al. 2018; Luo et al. 2018; Petermann et al. 2018).

Pattern identification:

Temperature differences between LGD and lake water are a useful natural tracer for pattern identification. Determining the most representative sites or those having particularly high LGD rates is a promising approach to guide spatially detailed point measurements. This kind of pre-information is useful since point measurements are usually very labour- and time-consuming. Furthermore, pattern identification might be applied for upscaling of point measurements (Blume et al. 2013).

a) Fibre-optic Distributed Temperature Sensing (FO-DTS, Chapter 8) detects localized temperature anomalies caused by LGD at the sediment–water interface (e.g. Blume et al. 2013; Liu et al. 2016; McCobb et al. 2018; Sebok et al. 2018). Several hundred metre to several kilometre long fibre-optic cables are deployed at the sediment–water interface, and temperatures are measured with a spatial resolution of 0.25 to 2 m along the cable. The broad areal coverage provided by the fibre-optic cable allows the detection of thermal anomalies associated with varying rates of LGD. Alternatively, if natural temperature differences between groundwater and surface water are insufficient, a heated FO-DTS cable can be used and spatially varying rates of heat dissipation along the cable can be related to varying rates of groundwater discharge (Kurth et al. 2015). This method is especially interesting when LGD is highly heterogeneous rather than diffusive and more or less homogenous.
b) If LGD is significantly warmer and thus less dense than the lake water, discharging groundwater will rise to the surface of the lake. In such cases, LGD might be detected as an area of temperature anomaly at the water surface by airborne thermal infrared (TIR) imagery (Lewandowski et al. 2013, Chapter 8). This method works particularly well in marine settings where fresh groundwater is substantially less dense than the much more saline surface water, whereas the use of TIR in freshwater settings to identify areas of LGD is much more difficult (Pöschke et al. 2015). Marruedo Arricibita et al. (2018a, 2018b) investigated the circumstances under which the warm groundwater signal can be reliably identified in lake settings.

5.5 Methods to Quantify Nutrient Fluxes

All of the aforementioned methods aim at quantifying discharging groundwater volumes. LGD-derived mass loads are commonly calculated by multiplying the volume (rate) of groundwater discharge by the concentration of the target parameter. Since both discharge rates and solute concentrations are often spatially quite heterogeneous, approaches for segmenting the area of interest are required (Meinikmann et al. 2013). For the determination of concentrations, a range of groundwater sampling strategies can be applied to each of such segments:

a) Near-shore groundwater wells are used to sample and analyse the groundwater approaching the aquifer–surface water interface. Usually the number of available near-shore wells is quite limited because the installation of groundwater wells by a truck-mounted drilling rig can be expensive.
b) Since near-shore groundwater levels are usually close to the land surface level, hand-drilled near-shore piezometers are also used to sample shallow groundwater. They can usually be installed in larger numbers to provide an overview of groundwater quality with a large spatial resolution.
c) Multi-level samplers are additionally used to determine the groundwater composition where high vertical resolution is required (Rivett et al. 2008).
d) Groundwater samples can be collected very close to the point of entry (i.e. the groundwater–lake interface) with piezometers and multi-level samplers installed in the lakebed. Due to the reactivity of the interface, these samples may better represent the actual quality of the water discharging across the sediment–water interface.

e) Finally, samples from seepage meters can be used to collect the discharging groundwater and determine its composition. This approach requires a minimized water residence time within the seepage cylinder and seepage-collection bag to avoid changes of the water composition between the time of actual groundwater discharge and the time of sample collection (Rosenberry and LaBaugh 2008).

5.6 Reasons for Neglecting Groundwater Discharge in the Recent Past

In spite of its importance (Figures 5.2 and 5.3), groundwater has often been disregarded in water and nutrient budgets of lakes due to a variety of reasons (Lewandowski et al. 2014, 2015; Rosenberry et al. 2015):

a) Groundwater and surface water have long been considered as separate entities and thus the impact of groundwater on surface water bodies was neglected.
b) Groundwater discharge is invisible (except in the case of springs) and thus people are less aware of groundwater discharge compared to surface inflows, overland flow, or precipitation.
c) The interface is difficult to access, especially in deep lakes and coastal oceans.
d) Due to the large extent of the interface, local LGD rates are usually small.
e) Although a lot of different measurement techniques for LGD were developed in the last decades, there is still a dearth of simple or convenient methodology.
f) Spatial heterogeneity of discharge rates and groundwater composition result in large numbers of measurements required for reliable determinations of the LGD component in mass budgets.
g) Temporal variability of discharge rates add to the tremendous challenge of reliable estimations of LGD fluxes.
h) The net groundwater component in lake settings is often small, leading to the assumption that it can be ignored – even if gross inflows and gross losses are large. In spite of the small net size, the groundwater component might be relevant in nutrient budgets. Solute concentrations in inflowing groundwater might be orders of magnitude higher than those in outflowing water. Thus, groundwater might be important in mass budgets of lakes even if it is not important in water budgets of that lake.
i) The interface is often chemically reactive. Thus, even if parameter concentrations determined close to the shore are low, they might differ from concentrations actually discharging into the lake since chemical processing might occur in the last decimetre of sediment before entering the surface water body.

5.7 Case Study Lake Arendsee

Lake Arendsee is located in north-eastern Germany. The lake covers an area of 5.1 km^2, has a maximum depth of 49 m, and a mean depth of 29 m. Excess P from an unknown source has led to pelagic total P concentrations of nearly 200 μg L^{-1} within the last years (Meinikmann et al. 2015), raising the question of the portion of LGD-derived P loads. The hydrogeological situation is spatially complex with hydrologically conductive sediments (saturated hydraulic conductivity k_{sat} from 0.3×10^{-4} to 6×10^{-4} m s^{-1}) being intersected by lenses of extremely low conductivity. This indicates some spatial heterogeneity also in LGD.

To identify infiltration and exfiltration zones along the lake shore, the size and shape of the subsurface catchment of the lake were determined from hydraulic-head contour lines (Meinikmann et al. 2013). To verify the deduced zones where groundwater discharges into the lake (LGD) and lake water recharges adjacent groundwater, stable water isotopes of near-shore groundwater were measured along the lake shore. The results showed only negligible deviations from interaction zones defined based on hydraulic-head contour lines (Meinikmann et al. 2014, Figure 5.4). It was concluded that LGD would enter the lake along the southern, south-western, and eastern shoreline.

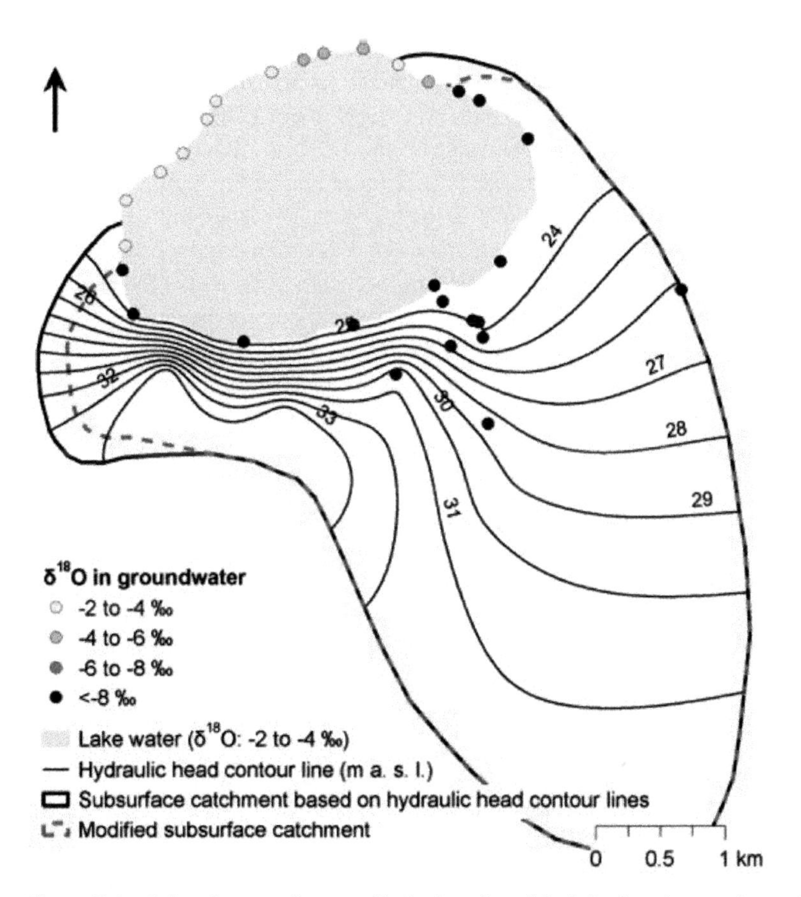

Figure 5.4 Subsurface catchment of Lake Arendsee (black line) and groundwater head contour lines based on water levels measured in several wells in and close to the catchment (Meinikmann et al. 2013). The blue dashed line represents a modification of the subsurface catchment, i.e. of in- and exfiltration zones along the shoreline, due to stable isotope composition of near-shore shallow groundwater (Meinikmann et al. 2014). Stable isotopes of oxygen of near-shore shallow groundwater at the northern and north-western shore show signatures similar to lake water, confirming lake water infiltration into the aquifer in this shoreline section. Signatures similar to the ones of groundwater in the catchment indicate groundwater is discharging into the lake (LGD) in the other shoreline sections.

Aiming at the assessment of local patterns of groundwater discharge, LGD rates were calculated from temperature-depth profiles (Chapter 8) measured in the lakebed along the discharge zone. Temperature measurements were conducted using probes consisting of 16 temperature sensors (NTC 10K, TDK EPCOS, Munich, Germany) with a distance of 7 cm to each other. With one sensor situated in the surface-water column and another placed at the sediment–water interface, the probe was able to measure sediment temperatures to a depth of 0.98 m. The mathematical procedure by Schmidt et al. (2006) was followed to calculate LGD rates from the profiles (Meinikmann et al. 2013). Measurement locations were spaced every 200 m along the shoreline. Since LGD is known to be focused in near-shore areas given uniform geology (McBride and Pfannkuch 1975), probes were placed at 0.5, 1, 2, and 4 m distance from the shoreline at each site. Resulting LGD rates ranged between 0 and 122 L m^{-2} d^{-1} (Figure 5.5a). The highest LGD rates occurred along a 2200 m section of the southern shoreline. At the western and eastern shorelines LGD rates decreased to almost zero at some sites (Figure 5.5a). The results are in agreement with what was expected based on hydraulic head contour lines and reveal major LGD along the 2200 m section of the southern shoreline ranging between 60 and 122 L m^{-2} d^{-1} (Figure 5.5b).

Looking at LGD within individual transects (examples shown in Figure 5.6), rates decreased with distance from the shore at 65% of the transects. However, this decrease did not follow a consistent pattern. Some transects showed a consequent exponential decrease with distance to shore indicating isotropic and homogeneous sediments (McBride and Pfannkuch 1975). In other transects the LGD rates decreased, but not explicitly exponentially (Figure 5.6). Furthermore, nine of the overall 26 transects did not follow the general pattern of decreasing LGD rates with increasing shore distance at all. LGD rates at these locations seemed to be of individual inconsistency as the two examples in Figure 5.6 demonstrate. The described diversity of LGD patterns covered the entire range of LGD rates measured at Lake Arendsee, i.e. spatial heterogeneity of LGD rates is independent of LGD intensity. This confirms that medium-scale patterns of hydraulic conductivity and/or hydraulic gradients might be locally obscured by features influencing LGD rates on smaller scales (e.g. transpiration of shore line vegetation).

LGD rates derived from temperature measurements confirmed hydraulic head contour lines in general. The spatial heterogeneity, however, could not be depicted adequately by hydraulic head contour lines alone (Figure 5.5b, Meinikmann et al. 2014). Since the approach using heat as a tracer has some uncertainties concerning absolute LGD rates, they were not upscaled to estimate an overall amount of LGD. Instead, they were applied as a weighting factor for LGD segmented volumes per shoreline section using the total groundwater recharge in the catchment (Meinikmann et al.2013): by definition, recharge to a subsurface catchment is equal to LGD if there are no springs or gaining streams and the system is in a steady state, which is usually the case on annual timescales (Section 5.4). Combining this whole-lake approach with local-scale point measurements of LGD will improve both the overall LGD amount and the local LGD rates that are the basis for a lake's mass budget.

Sampling at four groundwater observation sites with a total of 10 wells along the southern shoreline revealed high concentrations of soluble reactive phosphorus (SRP) in some of the wells, especially at site 3 (Figure 5.7a), where the shallower of the two wells had a mean SRP concentration of 1600 μg L^{-1} ($n = 33$, April 2010 to December 2013; Meinikmann et al. 2015).

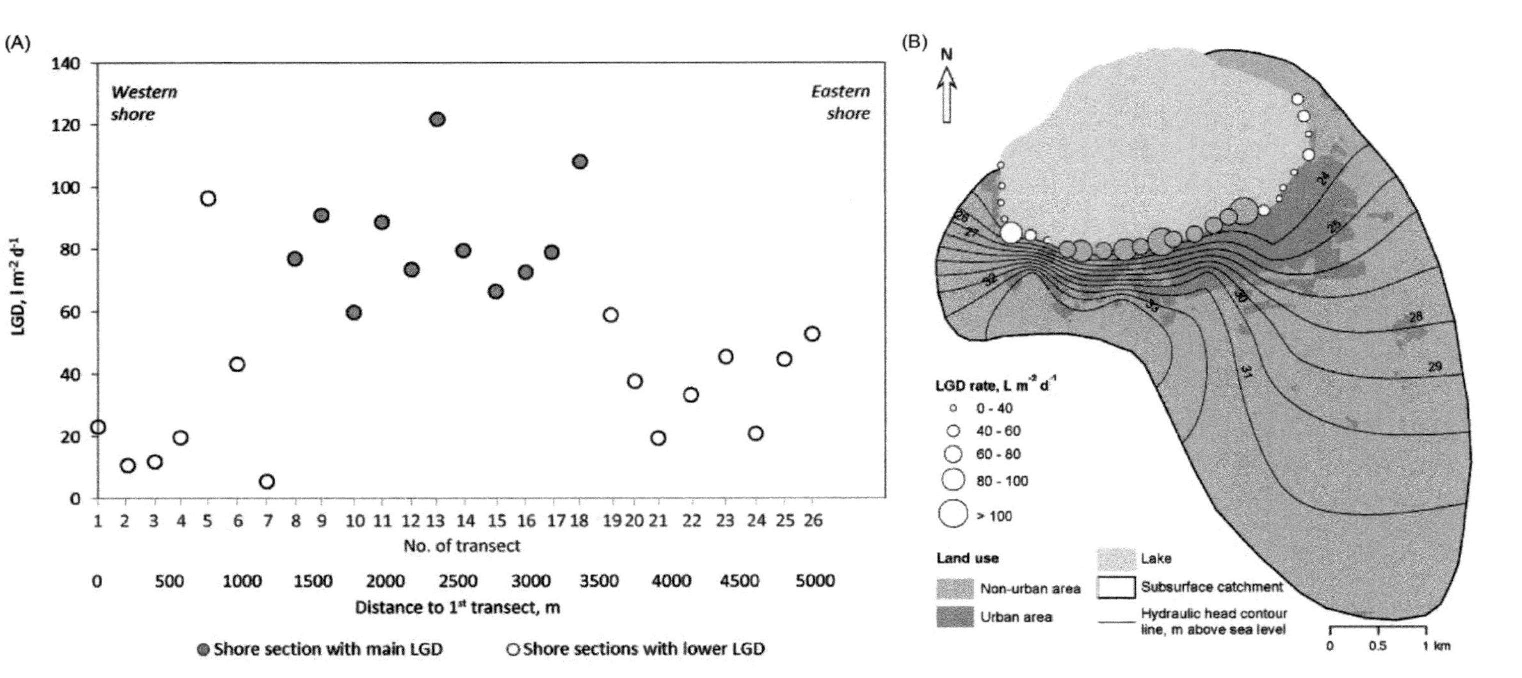

Figure 5.5 Maximum LGD rate of each transect calculated from temperature depth profiles of the lakebed taken in 0.5, 1, 2, and 4 m distance to the shoreline. (a) Diagram of LGD over distance along the shoreline to 1st western transect. (b) Map of Lake Arendsee illustrating LGD rates at different locations (Adapted from Meinikmann et al. 2013). Grey circles in (a) and (b) represent the reach of main LGD (2200 m along the southern shoreline). White circles in (a) and (b) indicate the LGD rates at the eastern and western shoreline, which are generally lower.

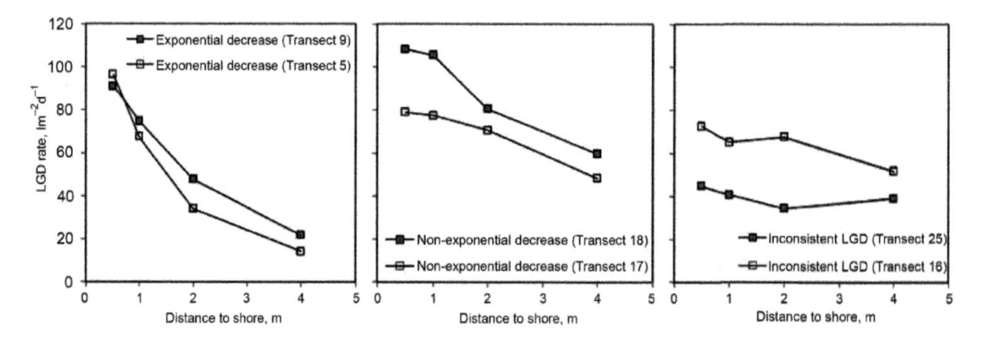

Figure 5.6 Examples of transects of LGD rates showing exponential decrease (left), non-exponential decrease (middle), as well as inconsistent LGD patterns (right). Numbers of transects are counted from the 1st western transect along the shoreline (Figure 5.5). Transect numbers refer to numbers in Figure 5.5a.

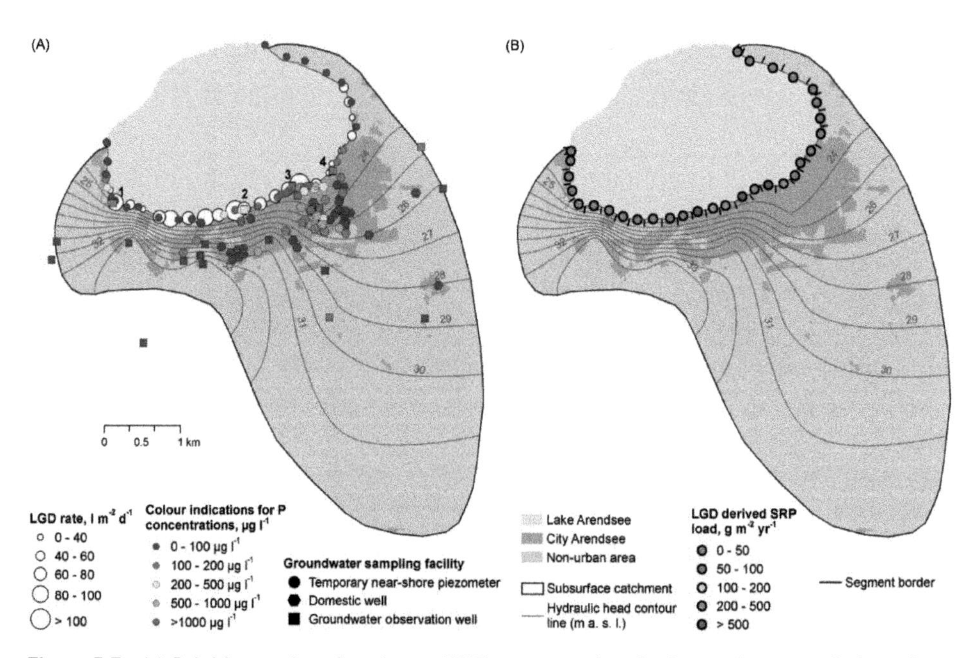

Figure 5.7 (a) Soluble reactive phosphorus (SRP) concentrations in the catchment and along the shoreline of Lake Arendsee (dots – groundwater observation wells and domestic wells; squares – temporary piezometers). Numbers 1 to 4 indicate near-shore groundwater observation wells. (b) LGD-derived SRP loads in different sections along the shoreline (Meinikmann et al. 2015/ with permission of Elsevier).

In order to track back a potential SRP plume, a number of additional groundwater observation wells were monitored for several years. They covered both the agricultural background of the lake's catchment as well as the settled area of the City of Arendsee, which is situated directly along the southern shoreline (Figure 5.7a). A time series of these investigations showed hardly any temporal variability of the groundwater SRP concentrations (Meinikmann et al. 2015). Therefore, it was postulated that a single sampling of groundwater at additional sites should be representative and allow for a more detailed picture of the spatial variability of nutrient concentrations. Thus, 44 hand-drilled shallow temporary piezometers were installed and sampled all along the shoreline section where exfiltration occurred (squares in Figure 5.7a). Furthermore, the citizens living in the catchment were asked to bring water from their private wells for chemical analyses (Figure 5.7a), resulting in 56 additional samples.

Mean SRP concentrations in the catchment and close to the lake shore varied by orders of magnitude. In the urban parts of the catchment, SRP concentrations of some groundwater samples were very high (up to 4500 $\mu g\ l^{-1}$). In contrast, SRP concentrations in the rural catchment were relatively low compared to concentrations in the urban catchment. Results of near-shore groundwater sampling revealed a reach of heavily contaminated groundwater along the south-eastern shoreline (Figure 5.7a). The reach is in the urban area of the City of Arendsee. SRP concentrations up to 4060 $\mu g\ L^{-1}$ were found here. In contrast, near-shore shallow groundwater in unsettled areas around the lake had low SRP concentrations. Near-shore shallow groundwater had SRP concentrations of more than 100 $\mu g\ L^{-1}$ at only two sites outside the urban area. Both sites are located at the western shoreline, where a settlement of vacation cottages is situated.

The intense investigation of the groundwater revealed a large degree of spatial heterogeneity in SRP concentrations in the urban area (Figure 5.7a). In several cases, low SRP concentrations occurred in close proximity (20 to 50 m) to severely contaminated sites. When evaluating such results, it has to be taken into account that screens of different groundwater wells are located in different depths of the aquifer and that groundwater plumes have a three-dimensional structure with strong horizontal and vertical concentration gradients (McCobb et al. 2003). At some sites in the catchment of Lake Arendsee, however, even groundwater samples collected from similar depths and in close proximity showed strongly differing SRP concentrations. Such small-scale spatial heterogeneity might be caused by differences in hydraulic conductivities. Sediments of high hydraulic conductivities surrounded by sediments of low hydraulic conductivities can act as preferential flow paths for groundwater and its constituents. Furthermore, sediments of low hydraulic conductivity do commonly have higher specific retention capacities leading to more efficient reduction of groundwater P concentrations compared to highly conductive sediments.

Despite the large number of available monitoring sites, it was not possible to unambiguously identify a single location or source of the groundwater contamination. The exclusive occurrence of severely contaminated groundwater in the settled areas of the City of Arendsee indicates that the contamination likely originates from urban sources rather than from agricultural ones. The high spatial variability of SRP concentrations indicates that there are several contamination sources (Meinikmann et al. 2015). Increased groundwater SRP concentrations in urban areas relative to agricultural ones have also been found by other authors (Holman et al. 2008; Qian et al. 2011; Roy & Malenica 2013). The most common sources for groundwater contamination beneath settlements are leakages from waste water facilities (Bishop et al. 1998; Katz et al. 2011; Rutsch et al. 2008; Schirmer et al. 2013; Wakida and Lerner 2005). Given the large spatial heterogeneity, the contamination likely is caused by several malfunctioning waste water facilities distributed over the City of Arendsee. Some acesulfame (artificial sweetener) measurements (data not shown) support this statement.

As described previously, LGD-derived SRP loads were calculated using a segmented approach that multiplied LGD volumes per shoreline section by the corresponding nutrient concentration measured in shallow piezometers. The combined SRP load from LGD was 830 kg yr^{-1}, 94% of which entered the lake along a 1.4 km long reach at the southern shoreline (Figure 5.7b). The determination of an overall P budget of the lake revealed that LGD-derived SRP loads account for about 53% of the overall P load to the lake, followed by atmospheric deposition (19%), water fowl (13%), drainage from agriculture (12%), and other sources (3%) (Meinikmann et al. 2015). Eutrophication of Lake Arendsee therefore appears to be substantially driven by excess P from groundwater discharge.

Substantial resources were required to provide the detailed results presented previously. Two other data-collection scenarios are compared with the full investigation to determine whether similar results could be achieved with a smaller investment (Table 5.1). Scenario 1 represents the studies as described previously in which both LGD rates and SRP concentrations are taken into account to determine groundwater-borne SRP loads (Meinikmann et al. 2015). Scenario 2 (Meinikmann et al. 2013) determined SRP concentrations at only four groundwater observation wells along the southern shoreline; results from temporary piezometers (Figure 5.7a) were not included. By upscaling those four values a much smaller LGD load was calculated (Table 5.1). The third scenario excludes spatial heterogeneity of LGD; it is based on the overall LGD volume derived from groundwater recharge calculations. LGD rates from temperature depth profiles are not taken into account. This results in equal LGD rates along the entire shoreline. Scenario 3 does include SRP concentrations segmented per shoreline section from near-shore piezometers. Presuming that Scenario 1 accounts for the exact load of SRP being discharged to the lake via LGD, Scenario 2 underestimates the load by about 50%. Excluding the spatial heterogeneity of LGD rates from SRP load calculations (Scenario 3) reduces the groundwater-borne SRP load to about 75% of the load of Scenario 1 (Table 5.1).

Table 5.1 Scenarios for LGD-derived SRP load determination. Scenario 1 includes spatial heterogeneity of both SRP concentrations (temporary piezometer samples) and LGD volumes (temperature depth profiles). Scenario 2 reduces the heterogeneity of SRP concentrations by using only the four SRP concentrations from the four near-shore groundwater observation wells (Figure 5.7a). Scenario 3 includes SRP heterogeneity from temporary piezometers but does not take into account temperature-derived LGD rates.

	Taking into account spatial heterogeneity of			
	LGD	**SRP**	**LGD-derived SRP load**	**% of Scenario 1-SRP load**
Scenario 1	Yes	Yes	830	100
Scenario 2	Yes	No	425	51
Scenario 3	No	Yes	610	73

5.8 Conclusions and Knowledge Gaps

The case of Lake Arendsee is an example of the potential importance of LGD for lake P budgets. The study shows that both LGD rates and concentrations can vary spatially to a large degree. Investigating these heterogeneities is resource intensive but incorporating heterogeneity with a segmented approach increases the accuracy of LGD-derived SRP loads (Table 5.1). The results indicate that SRP concentrations in groundwater were greatly increased by multiple anthropogenic sources. The groundwater was most heavily contaminated where the largest rates of LGD occurred, which resulted in very high P loads compared to other components of the lake's P budget. Although this might represent a "worst case", SRP concentrations of more than 4 mg L^{-1} in the groundwater might not be an exclusive phenomenon of Lake Arendsee but may also occur at other lakes where they contribute undetected to lake eutrophication via LGD. High SRP concentrations in groundwater become especially relevant when the contamination of the aquifer occurs close to a lake shore where the vadose zone is thin and travel times to the lake are relatively short. Degradation and sorption capacities are limited in such cases.

Quantification of volumes and mass loads from groundwater remains a challenge, particularly for large lakes (Naranjo et al. 2019). Private domestic wells and temporary piezometers are useful tools to improve the spatial information about the subsurface nutrient distribution (Meinikmann et al. 2015). Further development of measurement methods is required, e.g. to unambiguously identify sources of P contamination. Other open questions address the reactivity of the groundwater–lake interface. The importance of the biogeochemical processing of nutrients and whether that results in a delayed transport of nutrients to the receiving water bodies, or perhaps permanent removal, has yet to be determined. Given that contaminated groundwater at Lake Arendsee was only found in the urban area of the catchment, a focus of further research might be on lakes with urbanized areas in close proximity to shore sections where LGD occurs. Since groundwater P is usually not monitored on a regular basis, Lake Arendsee is probably not a special case of lake eutrophication by LGD. Other studies focusing on P loading by LGD have been presented by Nisbeth et al. (2019a, 2019b), Kidmose et al. (2013), Schindler (1974), Shaw et al. (1990), and Vanek (1993). Further effort should be spent on the role of leaking sewer facilities for groundwater quality in general. Although it could not be finally proved that the groundwater at Lake Arendsee was contaminated by leaking waste water, such leaks can be regarded as a probable source.

Acknowledgements

We thank Larry R. Rick Arnold, Erik Day, and Keith J. Lucey for their very detailed and helpful comments to improve our book chapter. Any use of trade, firm, or product names is for descriptive purposes only and does not imply endorsement by the U.S. Government.

References

Arnoux, M., Gibert-Brunet, E., Barbecot, F. et al. (2017). Interactions between groundwater and seasonally ice-covered lakes: using water stable isotopes and radon-222 multilayer mass balance models. *Hydrol. Process.* 31: 2566–2581.

Bishop, P.K., Misstear, B.D., White, M., and Harding, N.J. (1998). Impacts of sewers on groundwater quality. *J. Chart. Inst. Water E.* 12 (3): 216–223.

Blume, T., Krause, S., Meinikmann, K., and Lewandowski, J. (2013). Upscaling lacustrine groundwater discharge rates 1 by fiber-optic distributed temperature sensing. *Water Resourc. Res.* 49: 1–16.

Cherkauer, D.S., Mckereghan, P.F., and Schalch, L.H. (1992). Delivery of chloride and nitrate by ground water to the Great Lakes: case-study for the Door Peninsula, Wisconsin. *Ground Water* 30 (6): 885–894.

Conley, D.J., Paerl, H.W., Howarth, R.W. et al. (2009). Controlling eutrophication: nitrogen and phosphorus. *Science* 323 (5917): 1014–1015.

Delkash, M., Al-Faraj, F.A.M., and Scholz, M. (2018). Impacts of anthropogenic land use changes on nutrient concentrations in surface waterbodies: a review. *Clean-Soil Air Water* 46 (5).

Gilfedder, B., Peiffer, S., Pöschke, F., and Spirkaneder, A. (2018). Mapping and quantifying groundwater discharge to Lake Steißling and its influence on lake water chemistry (in German). *Grundwasser* 23 (2): 155–165.

Gordon, R.P., Lautz, L.K., Briggs, M.A., and McKenzie, J.M. (2012). Automated calculation of vertical pore-water flux from field temperature time series using the VFLUX method and computer program. *J. Hydrol.* 420–421: 142–158.

Gurrieri, J.T. and Furniss, G. (2004). Estimation of groundwater exchange in alpine lakes using non-steady mass-balance methods. *J. Hydrol.* 297 (1–4): 187–208.

Hagerthey, S.E. and Kerfoot, W.C. (1998). Groundwater flow influences the biomass and nutrient ratios of epibenthic algae in a north temperate seepage lake. *Limnol. Oceanogr.* 43 (6): 1227–1242.

Holman, I.P., Whelan, M.J., Howden, N.J.K. et al. (2008). Phosphorus in groundwater—an overlooked contributor to eutrophication? *Hydrol. Process.* 22 (26): 5121–5127.

Huang, G.X., Liu, C.Y., Zhang, Y., and Chen, Z.Y. (2020). Groundwater is important for the geochemical cycling of phosphorus in rapidly urbanized areas: a case study in the Pearl River Delta. *Environ. Pollut.* 260.

Ibanhez, J.S.P., Leote, C., and Rocha, C. (2011). Porewater nitrate profiles in sandy sediments hosting submarine groundwater discharge described by an advection-dispersion-reaction model. *Biogeochem* 103 (1–3): 159–180.

Jiang, J. (2020). Quantifying the influence of groundwater discharge induced by permafrost degradation on lake water budget in Qinghai-Tibet Plateau: using Rn-222 and stable isotopes. *J. Radioanal. Nucl. Ch.* 323: 1125–1134.

Katz, B.G., Eberts, S.M., and Kauffman, L.J. (2011). Using Cl/Br ratios and other indicators to assess potential impacts on groundwater quality from septic systems: a review and examples from principal aquifers in the United States. *J. Hydrol.* 397 (3–4): 151–166.

Kazmierczak, J., Müller, S., Nilsson, B. et al. (2016). Groundwater flow and heterogeneous discharge into a seepage lake: combined use of physical methods and hydrochemical tracers. *Water Resour. Res.* 52: 9109–9130.

Kebede, S. and Zewdu, S. (2019). Use of ^{222}Rn and δ^{18}O-δ^2H isotopes in detecting the origin of water and in quantifying groundwater inflow rates in an alarmingly growing Lake, Ethiopia. *Water* 11 (12).

Kidmose, J., Nilsson, B., Engesgaard, P. et al. (2013). Focused groundwater discharge of phosphorus to a eutrophic seepage lake (Lake Væng, Denmark): implications for lake ecological state and restoration. *Hydrogeol. J.* 21 (8): 1787–1802.

Kilroy, G. and Coxon, C. (2005). Temporal variability of phosphorus fractions in Irish karst springs. *Environ. Geol.* 47 (3): 421–430.

Knights, D., Parks, K.C., Sawyer, A.H. et al. (2017). Direct groundwater discharge and vulnerability to hidden nutrient loads along the Great Lakes coast of the United States. *J. Hydrol.* 554: 331–341.

Kunkel, R., Voigt, H.J., Wendland, F., and Hannappel, S. (2005). *Die natürliche, ubiquitär überprägte Grundwasserbeschaffenheit in Deutschland.* Schriften des Forschungszentrums Jülich, Forschungszentrum Jülich. 201.

Kurth, A.-M., Weber, C., and Schirmer, M. (2015). How effective is river restoration in re-establishing groundwater-surface water interactions? - A case study. *Hydrol. Earth Sys. Sci.* 19 (6): 2663–2672.

LaBaugh, J.W., Winter, T.C., Rosenberry, D.O. et al. (1997). Hydrological and chemical estimates of the water balance of a closed-basin lake in north central Minnesota. *Water Resour. Res.* 33 (12): 2799–2812.

Le Moal, M., Gascuel-Odoux, C., Menesguen, A. et al. (2019). Eutrophication: a new wine in an old bottle? *Sci. Total Environ.* 651: 1–11.

Lewandowski, J., Meinikmann, K., Nützmann, G., and Rosenberry, D.O. (2015). Groundwater discharge to lakes - the disregarded component in lake nutrient budgets. 2. Biogeochemical considerations. *Hydrol. Process.* 29: 2922–2955.

Lewandowski, J., Meinikmann, K., Pöschke, F. et al. (2014): From submarine to lacustrine groundwater discharge. In: *Complex interfaces under change: Sea–river–groundwater–lake: Proceedings of symposia HP2 and HP3 held during the IAHS-IAPSO-IASPEI Assembly, Gothenburg, Sweden*, 365 (ed. C. Cudennec, M. Kravchishina, J. Lewandowski et al.), 72–78. Oxfordshire: IAHS Publ.

Lewandowski, J., Meinikmann, K., Ruhtz, T. et al. (2013). Localization of lacustrine groundwater discharge (LGD) by airborne measurement of thermal infrared radiation. *Remote Sens. Environ.* 138: 119–125.

Liao, F., Wang, G.C., Shi, Z.M. et al. (2018). Estimation of groundwater discharge and associated chemical fluxes into Poyang Lake, China: approaches using stable isotopes (δD and δ^{18}O) and radon. *Hydrogeol. J.* 26: 1625–1638.

Liao, X.L., Nair, V.D., Canion, A. et al. (2019). Subsurface transport and potential risk of phosphorus to groundwater across different land uses in a karst springs basin, Florida, USA. *Geoderma* 338: 97–106.

Liu, C.K., Liu, J., Wang, X.S., and Zheng, C.M. (2016). Analysis of groundwater-lake interaction by distributed temperature sensing in Badain Jaran Desert, Northwest China. *Hydrol. Process.* 30: 1330–1341.

Luo, X., Kuang, X.X., Jiao, J.J. et al. (2018). Evaluation of lacustrine groundwater discharge, hydrologic partitioning, and nutrient budgets in a proglacial lake in the Qinghai-Tibet Plateau: using Rn-222 and stable isotopes. *Hydrol. Earth Syst. Sci.* 22: 5579–5598.

Marruedo Arricibita, A.I., Lewandowski, J., Krause, S. et al. (2018a). Mesocosm experiments identifying hotspots of groundwater upwelling in a water column by fibre optic distributed temperature sensing. *Hydrol. Process* 32: 185–199.

Marruedo Arricibita, A.I., Dugdale, S.J., Krause, S. et al. (2018b). Thermal infrared imaging for the detection of relatively warm lacustrine groundwater discharge at the surface of freshwater bodies. *J. Hydrol.* 562: 281–289.

McBride, M.S. and Pfannkuch, H.O. (1975). The distribution of seepage within lakebeds. *U.S. Geol. Survey J. Res.* 3 (5): 505–512.

McCobb, T.D., Briggs, M.A., LeBlanc, D.R. et al. (2018). Evaluating long-term patterns of decreasing groundwater discharge through a lake-bottom permeable reactive barrier. *J. Environ. Manage.* 220: 233–245.

McCobb, T.D., LeBlanc, D.R., and Massey, A.J. (2009). Monitoring the removal of phosphate from ground water discharging through a pond-bottom permeable reactive barrier. *Ground Water Monit. Remed.* 29 (2): 43–55.

McCobb, T.D., LeBlanc, D.R., Walter, D.A. et al. (2003). *Phosphorus in a Ground-water Contaminant Plume Discharging to Ashumet Pond, Cape Cod, Massachusetts, 1999.* US Geol. Survey Water Resources Investigations Report 02-4306.

Meinikmann, K., Hupfer, M., and Lewandowski, J. (2015). Phosphorus in groundwater discharge – a potential source for lake eutrophication. *J. Hydrol.* 524: 214–226.

Meinikmann, K., Nützmann, G., and Lewandowski, J. (2013). Lacustrine groundwater discharge: combined determination of volumes and spatial patterns. *J. Hydrol.* 502: 202–211.

Meinikmann, K., Nützmann, G., and Lewandowski, J. (2014) Empirical quantification of lacustrine groundwater discharge – different methods and their limitations. In: *Complex interfaces under change: Sea–river–groundwater–lake: Proceedings of symposia HP2 and HP3 held during the IAHS-IAPSO-IASPEI Assembly, Gothenburg, Sweden*, 365 (ed. C. Cudennec, M. Kravchishina, J. Lewandowski et al.), 85–90. Oxfordshire: IAHS Publ.

Mester, T., Balla, D., Karancsi, G. et al. (2019). Effects of nitrogen loading from domestic wastewater on groundwater quality. *Water SA* 45 (3): 349–358.

Muellegger, C., Weilhartner, A., Battin, T.J., and Hofmann, T. (2013). Positive and negative impacts of five Austrian gravel pit lakes on groundwater quality. *Sci. Tot. Environ.* 443: 14–23.

Naranjo, R.C., Niswonger, R.G., Smith, D. et al. (2019). Linkages between hydrology and seasonal variations of nutrients and periphyton in a large oligotrophic subalpine lake. *J. Hydrol.* 568: 877–890.

Nisbeth, C.S., Jessen, S., Bennike, O. et al. (2019b). Role of groundwater-borne geogenic phosphorus for the internal P release in shallow lakes. *Water* 11: 1783.

Nisbeth, C.S., Kidmose, J., Weckström, K. et al (2019a). Dissolved inorganic geogenic phosphorus load to a groundwater-fed lake: implications of terrestrial phosphorus cycling by groundwater. *Water* 11: 2213.

Petermann, E., Gibson, J.J., Knoller, K. et al. (2018). Determination of groundwater discharge rates and water residence time of groundwater-fed lakes by stable isotopes of water (O-18, H-2) and radon (Rn-222) mass balances. *Hydrol. Process.* 32 (6): 805–816.

Pfannkuch, H.O. and Winter, T.C. (1984). Effect of anisotropy and groundwater system geometry on seepage through lakebeds, 1. Analog and dimensional analysis. *J. Hydrol.* 75: 213–237.

Pina-Ochoa, E. and Alvarez-Cobelas, M. (2009). Seasonal nitrogen dynamics in a seepage lake receiving high nitrogen loads. *Mar. Freshw. Res.* 60 (5): 435–445.

Pöschke, F., Lewandowski, J., Engelhardt, C. et al. (2015). Upwelling of deep water during thermal stratification onset—A major mechanism of vertical transport in small temperate lakes in spring? *Water Resour. Res.* 51: https://doi.org/10.1002/2015WR017579.

Qian, J.Z., Wang, L.L., Zhan, H.B., and Chen, Z. (2011). Urban land-use effects on groundwater phosphate distribution in a shallow aquifer, Nanfei River basin. *China. Hydrogeol. J.* 19 (7): 1431–1442.

Ridgway, M.S. and Blanchfield, P.J. (1998). Brook trout spawning areas in lakes. *Ecol. Freshw. Fish.* 7: 140–145.

Rivett, M.O., Ellis, R., Greswell, R.B. et al. (2008). Cost-effective mini drive-point piezometers and multilevel samplers for monitoring the hyporheic zone. *Q. J. Eng. Geol. Hydrogeol.* 41: 49–60.

Rosenberry, D.O. and LaBaugh, J.W. (2008). *Field Techniques for Estimating Water Fluxes between Surface Water and Ground Water.* U.S. Geological Survey Techniques and Methods 4-D2.

Rosenberry, D.O., LaBaugh, J.W., and Hunt, R.J. (2008). Use of monitoring wells, portable piezometers, and seepage meters to quantify flow between surface water and ground water. In: *Field Techniques for Estimating Water Fluxes between Surface Water and Ground Water* (ed. D.O. Rosenberry and J.W. LaBaugh). 39–70. Denver: U.S. Geological Survey Techniques and Methods 4-D2.

Rosenberry, D.O., Lewandowski, J., Meinikmann, K., and Nützmann, G. (2015). Groundwater discharge to lakes – the disregarded component in lake nutrient budgets. 1. Hydrologic considerations. *Hydrol. Process.* 29: 2895–2921.

Rosenberry, D.O., Striegl, R.G., and Hudson, D.C. (2000). Plants as indicators of focused ground water discharge to a northern Minnesota lake. *Ground Water* 38: 296–303.

Roy, J.W. and Malenica, A. (2013). Nutrients and toxic contaminants in shallow groundwater along Lake Simcoe urban shorelines. *Inland Waters* 3: 125–138.

Rutsch, M., Rieckermann, J., Cullmann, J. et al. (2008). Towards a better understanding of sewer exfiltration. *Water Res* 42 (10–11): 2385–2394.

Sacks, L.A., Swancar, A., and Lee, T.M. (1998). *Estimating Ground-water Exchange with Lakes Using Water-budget and Chemical Mass-balance Approaches for ten Lakes in Ridge Areas of Polk and Highlands Counties.* Florida: U.S. Geological Survey Water-Resources Investigations Report 98-4133.

Schilling, K.E., Streeter, M.T., Bettis, E.A. et al. (2018). Groundwater monitoring at the watershed scale: an evaluation of recharge and nonpoint source pollutant loading in the Clear Creek Watershed, Iowa. *Hydrol. Process.* 32 (4): 562–575.

Schindler, D.W. (1974). Eutrophication and recovery in experimental lakes: implications for lake management. *Science* 184: 897–898.

Schirmer, M., Leschik, S., and Musolff, A. (2013). Current research in urban hydrogeology – a review. *Adv. Water Resour.* 51: 280–291.

Schmidt, C., Bayer-Raich, M., and Schirmer, M. (2006). Characterization of spatial heterogeneity of groundwater-stream water interactions using multiple depth streambed temperature measurements at the reach scale. *Hydrol. Earth Syst. Sci.* 10: 849–859.

Schuster, P.F., Reddy, M.M., LaBaugh, J.W. et al. (2003). Characterization of lake water and ground water movement in the littoral zone of Williams Lake, a closed-basin lake in north central Minnesota. *Hydrol. Process.* 17: 823–838.

Sebok, E., Karan, S., and Engesgaard, P. (2018). Using hydrogeophysical methods to assess the feasibility of lake bank filtration. *J. Hydrol.* 562: 423–434.

Shaw, R.D., Shaw, J.F.H., Fricker, H., and Prepas, E.E. (1990). An integrated approach to quantify groundwater transport of phosphorus to Narrow Lake, Alberta. *Limnol. Oceanogr.* 35 (4): 870–886.

Shukla, S. and Saxena, A. (2020). Sources and leaching of nitrate contamination in groundwater. *Curr Sci.* 118 (6): 883–891.

Slomp, C.P. and Van Cappellen, P. (2004). Nutrient inputs to the coastal ocean through submarine groundwater discharge: controls and potential impact. *J. Hydrol.* 295 (1–4): 64–86.

Stets, E.G., Winter, T.C., Rosenberry, D.O., and Striegl, R.G. (2010). Quantification of surface water and groundwater flows to open- and closed- basin lakes in a headwaters watershed using a descriptive oxygen stable isotope model. *Water Resour. Res.* 46: W03515. 03510.01029/02009WR007793.

Stuart, M.E. and Lapworth, D.J. (2016). Macronutrient status of UK groundwater: nitrogen, phosphorus and organic carbon. *Sci. Total Environ.* 572: 1543–1560.

Vanek, V. (1987). The interactions between lake and groundwater and their ecological significance. *Stygologia* 3: 1–23.

Vanek, V. (1993). Transport of groundwater-borne phosphorus to Lake Bysjon, south Sweden. *Hydrobiologia* 251 (1–3): 211–216.

Wakida, F.T. and Lerner, D.N. (2005). Non-agricultural sources of groundwater nitrate: a review and case study. *Water Res* 39 (1): 3–16.

Weilhartner, A., Muellegger, C., Kainz, M. et al. (2012). Gravel pit lake ecosystems reduce nitrate and phosphate concentrations in the outflowing groundwater. *Sci. Tot. Environ.* 420: 222–228.

Wendland, F., Hannappel, S., Kunkel, R. et al. (2005). A procedure to define natural groundwater conditions of groundwater bodies in Germany. *Wat. Sci. Technol.* 51 (3–4): 249–257.

6

Coastal–Groundwater Interfaces (Submarine Groundwater Discharge)

Michael E. Böttcher[1], Ulf Mallast[2], Gudrun Massmann[3], Nils Moosdorf[4], Mike Müller-Petke[5], and Hannelore Waska[6]

[1] *Geochemistry & Isotope Biogeochemistry, Leibniz Institute for Baltic Sea Research (IOW), Germany*
[2] *Geography and Geology Institute, University of Greifswald, Germany*
[3] *Department of Maritime Systems, Interdisciplinary Faculty, University of Rostock, Germany*
[4] *Leibniz Centre for Tropical Marine Research, Germany*
[5] *Leibniz, Institute for Applied Geophysics, Germany*
[6] *Institute of Chemistry and Biology of the Sea, Carl von Ossietzky University of Oldenburg Ammerländer Heerstraße, Germany*

6.1 General Principles

Submarine groundwater discharge (SGD) at the interface of land and sea is likely an important part of the global hydrological cycle and has started to attract the attention of a growing interdisciplinary scientific community. While before the year 2000 only a few papers about that topic were listed in the ISI Web of Science, by now about 100 publications per year address the topic. Submarine groundwater discharge has been defined as "direct groundwater outflow across the ocean–land interface into the ocean" (Church 1996), later refined to "any and all flow of water on continental margins from the seabed to the coastal ocean" (Burnett et al. 2003), consisting of fresh terrestrial groundwater of modern origin ("meteoric water"), connate water, and recirculated seawater (Figure 6.1). Although the majority of the SGD flux is derived from recirculated seawater, the term "groundwater discharge" often tends to be misleadingly reduced on its fresh terrestrial groundwater proportion. Additional confusion may also be caused by the use of different synonyms for this proportion which comprise "freshwater discharge", "submarine spring" (if discharge occurs in spatially focused form, such as in karst environments), "freshwater spring" (or "Vrulja") (Bögli 1980; d'Elia et al. 1981; Fleury et al. 2007; Kohout 1966; Milne 1897). The processes controlling the fluxes of SGD belong to topic 13 of the currently unsolved problems in hydrology (Blöschl et al. 2019).

Independent of its composition, SGD is of particular relevance for estuarine and coastal areas, since it provides a direct pathway for element and water flow at the connection between land and ocean. It could impact a variety of aspects in these ecosystems from

Ecohydrological Interfaces, First Edition. Edited by Stefan Krause, David M. Hannah, and Nancy B. Grimm.

possible water resource usage (Bakken et al. 2012) over local ecological consequences (e.g. Johannes 1980; Miller and Ullmann 2004) to global scale transport of nitrogen, phosphorous, metals, and anthropogenic contaminants (Beusen et al. 2013; Knee and Paytan 2011). Since SGD may be enriched in nutrients, dissolved inorganic carbon, methane, and metals (Cole et al. 2007; Jurasinski et al. 2018; Knee and Paytan 2011; Moore et al. 2011; Winde et al. 2014), it is linked to society-relevant issues, such as coastal eutrophication (excess loads of nutrients into aquatic systems) and the development of harmful algal blooms (red tides), the development of hypoxia (oxygen deficit due to enhanced primary production and organic matter re-mineralization), greenhouse effect (increasing concentrations of CO_2 and CH_4 in the atmosphere), as well as ocean acidification (changes in the carbonate system and proton activity of ocean surface water) to mention a few (Paerl, 1997; Hu et al. 2006; Santos et al. 2011; Winde et al. 2014; Macklin et al. 2014; Cyronak et al. 2013). The liberation of biological growth-promoting, inhibiting, or even toxic substances into the coastal environment may influence biological activity on all scales starting from microbial activity, the abundance and activity of meio- and macrofauna, up to human health issues (Knee and Paytan 2011; Kotwicki et al. 2014; Oehler et al. 2019a). Consequently, a variety of scientific fields have begun to recognize its potential relevance, while it is still mostly investigated by marine geochemists. As already mentioned in an Editorial Note by Kazemi (2008) and since SGD is a land–ocean interface, it is pertinent to involve more hydrogeologists into SGD research. However, the representation of different disciplines in SGD research increases and with it the emphasis on different complexities around this interface. The recognized controlling factors for SGD are sediment/aquifer properties (e.g. Russoniello et al. 2016), hydraulic heads (Moore 2010; Mulligan et al. 2006; Oberdorfer 2003), tidal amplitudes (e.g. Abarca et al. 2013; Robinson et al. 2007; Simmons 1992), wave amplitudes (e.g. Xin et al. 2010), and anthropogenic activity (irrigation or pumping of fresh water, modifications of surface water flow paths – to mention a few; Knee and Paytan 2011). Several of these factors are impacted by climate change. The current challenge is to localize and understand SGD in 4D: The quantitative role of SGD in the cycle of water and elements in time and space, and the processes that impact the fluxes between and interactions within reservoirs (e.g. Blöschl et al. 2019). Successful future SGD research will have to integrate different interdisciplinary experimental and theoretical approaches.

6.2 Hydro(geo)logy of Submarine Groundwater Discharge

Generally, hydraulic gradients in coastal aquifers point seawards, leading to the on- or offshore discharge of freshwater into the coastal sea (Figure 6.1). In the simplest case of a homogenous, non-tidal unconfined aquifer, the freshwater discharges at the mean water line. However, more complicated geological conditions may cause the water to discharge further out in the sea, for example, in karstic systems or where confined deeper aquifers are present (Figure 6.1).

In every confined or unconfined aquifer that is connected to the sea, an interface between the fresh groundwater and seawater exists and a wedge of denser saltwater underlies the discharging freshwater. In tidal systems the discharge zone, which is often also referred to as the subterranean estuary, is more complex. Tides and waves induce a saline

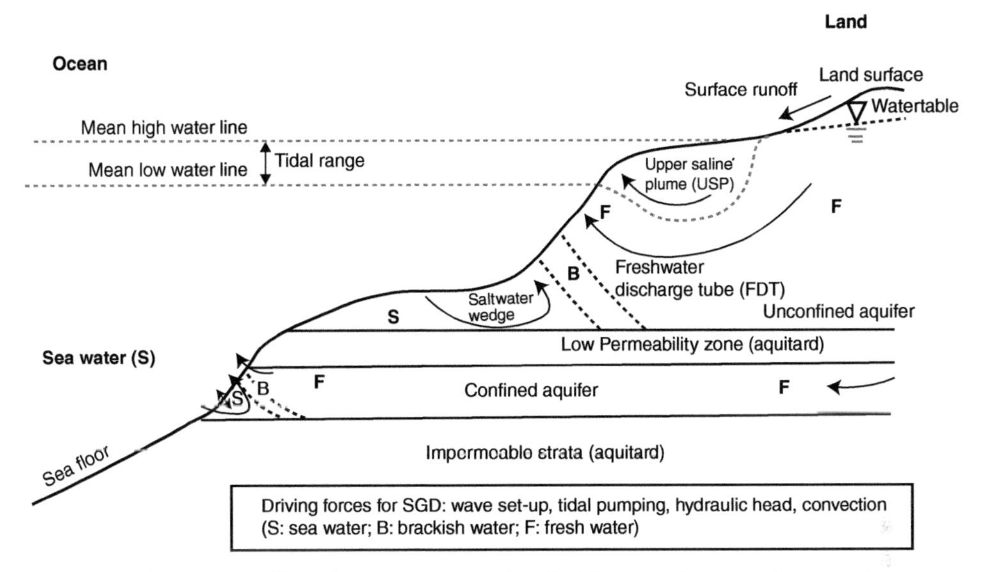

Figure 6.1 Scheme compiling the processes associated with submarine groundwater recharge (SGD) (Adapted from Swarzenski & Kindinger, 2003; Burnett et al, 2003; and Robinson et al., 2007). Arrows indicate fluid movement.

recirculation cell, often called upper saline plume (USP) in the shallow subsurface of the beach zone, overlying the discharging fresh groundwater (Figure 6.1). The fresh groundwater, being restricted by the classical saltwater wedge from below, is then forced through a "discharge tube" and exits near the low water tide mark. Below, very slow saline circulation by density- and current- driven convection occurs in the saltwater wedge (e.g. Michael et al. 2005; Robinson et al. 2007). In general, saltwater discharge rates are much larger and residence times considerably shorter in the USP compared to the wedge (weeks/years compared to decades/centuries; Beck et al. 2017; Robinson et al. 2007; Seidel et al. 2015). Freshwater discharge rates and age may be highly variable, depending on the local hydrogeology (Grünenbaum et al. 2020). Infrequent substantial saltwater intrusions into and movements within the coastal aquifers can occur, for instance, due to significant storm-events. These disturbed systems will be refreshened and reset afterwards over a certain time to their original steady-state conditions depending on the local hydrogeology. (Jenner et al., in prep.). However, while infrequent events are temporally limited, the whole system is furthermore under continuous long-term impact by sea-level change, possibly leading to substantial unidirectional system changes.

6.3 Nutrient, Carbon, and Metal Biogeochemistry of Submarine Groundwater Discharge

SGD may be highly enriched in radionuclides such as radon (Rn) and radium (Ra), macronutrients (phosphorous (P), silica (Si), nitrogen (N) species (nitrate, nitrite, ammonium)), dissolved carbon species (e.g. dissolved inorganic and organic carbon (DIC, DOC), methane, organic contaminants), metals as micronutrients and pollutants, and even microbes (Cyronak et al. 2013; Knee and Paytan 2011; Macklin et al., 2014; Santos et al. 2011; Slomp and van Cappellen 2004; Winde et al. 2014, and references therein). Therefore, SGD is generally considered as an important modifier for coastal biogeochemistry and plays a considerable part in the global element cycles, although the volumetric part in the overall water budget may sometimes be small when compared to direct surface water inflow (Knee and Paytan 2011; Moore 2010). The hydrochemical composition of the seepage depends on its origin (fresh or saline water), water–rock and water–sediment interactions, biological activity, and anthropogenic activity. The water isotopic composition is preserved in most settings and may be used for source identification and mixing calculations.

The sources of the macronutrients N, P, and Si and their transport behaviour differ in part substantially, but as limiting elements may increase coastal benthic and pelagic productivity. N species are essentially derived from both anthropogenic (sewage, animal farms; regional agricultural land use (fertilizers)) and less important natural sources (nitrogen fixation, mineralization of soil N, atmospheric sources), and the dominant dissolved species are controlled by the original speciation and the redox regime and associated microbial activity in the aquifer/sediment (Slomp and van Cappellen 2004; Zhang et al. 2012). Phosphate is derived from fertilizers, and waste and sewage (Slomp and van Cappellen 2004), and to a minor extent from the mineralization of organic matter. The high affinity of dissolved orthophosphate to mineral surfaces may limit the amount that is transported and finally released via SGD. However, recent research revealed that orthophosphate transport in aquifers is much more relevant than has been assumed for decades (Lewandowski et al. 2015). In contrast to N and P species, dissolved silica concentrations are mostly derived from water–aquifer rock interactions and controlled by the residence time in the aquifers (Davis 1964). The dissolution of diatom-derived biogenic silica, however, may further increase the concentration within the benthic mixing zone (Oehler et al. 2019b). Most or all of these processes lead to non-conservative behaviour of the nutrients that are impacted by the physical and mineralogical aquifer properties, biological activity, and residence times.

While the role of SGD as a source of nutrients is, in principle (although often not quantitatively), established, its contribution to regional and global inorganic and organic carbon budgets used to be much less known (Cole et al. 2007; Macklin et al. 2014). Analogous to the SGD water masses, dissolved organic carbon (DOC) in the subterranean estuary is generally of terrestrial (soil or buried peat layers) as well as marine origin (leachates from buried algal detritus and active microphytobenthos communities), with the former fraction being considered more refractory than the latter (Goñi and Gardner 2003; Santos et al. 2009). Consequently, proposed removal processes for terrestrial DOC in the subterranean estuary are more abiotic, such as sorption (Goñi and Gardner 2003) and coagulation with iron(oxi)hydroxides (Seidel et al. 2015), whereas marine DOC is seen as the "fuel" which drives microbial nutrient and trace metal transformation in the subterranean estuary (Santos et al. 2009; Seidel et al. 2015). Despite retention and remineralization processes in the subterranean estuary, DOC transport via SGD into the coastal ocean column can be substantial and even rival riverine inputs (Goñi and Gardner 2003; Kim et al. 2012; Santos et al. 2009). In areas with elevated fresh groundwater contributions, SGD is furthermore a substantial source of coloured dissolved organic matter (Kim et al. 2013; Nelson et al. 2015).

Methane, which is often observed with fresh water seepage, may partly be liberated to the atmosphere. In addition, the aerobic or anaerobic oxidation of methane may impact the DIC and alkalinity concentrations of coastal waters. The SGD-related changes in pH and concentrations of DIC have an impact on the CO_2 degassing potential (Macklin et al. 2014; Winde et al. 2014).

Like silica, dissolved metal concentrations in SGD (e.g. Cd, Cu, Mn, Pb, Zn, Fe, Mo) depend on the regional composition of the aquifers, but may also be highly enriched due to local and regional anthropogenic contaminations, e.g. from industrial outlets or mining activity (Knee and Paytan 2011). The transport behaviour in the aquifer and through the mixing zone depends strongly on particle–solute interactions, salinity, pH, ligand-availability, and the redox conditions. Complex and colloid formation, sorption on particle surfaces, ion exchange, and changes in speciation are some of the relevant processes that control the metal mobility, reactivity, and toxicity (Charette and Sholkovitz 2006). Karst systems, in particular, can be efficient pathways for the exfiltration of metal-enriched waters to coastal zones due to short aquifer residence times, an unfavourable relation of mineral surfaces to groundwater volumes, and high discharge rates (Knee and Paytan 2011). Also, microorganisms and plants may be impacted by the metal concentrations in the SGD feeding groundwater using selected trace metals as micronutrients (e.g. Fe, Mn, Ni, Zn, Mo) or contaminants (e.g. Pb, Se, Hg) (Salt et al. 1995). Trace metals may not only act as micronutrients, but have toxic effects, and may modify the sedimentary paleoceanographic record (Knee and Paytan 2011). In times of climate change and changing coastal eutrophic states, the importance of SGD-related metal fluxes into the coastal ocean may increase (Knee and Paytan 2011). In ultraoligotrophic waters surrounding oceanic islands, SGD has been suspected to spike phytoplankton productivity and contributes to the so-called "island mass effect" and subsequently the biological carbon pump (Gove et al. 2016). SGD is also a source of rare earth elements to the ocean, and may be the missing link in the global distribution of Nd, a key proxy for oceanic water-mass mixing (Johannesson and Burdige 2007; Johannesson et al. 2011; Paffrath et al. 2020).

6.4 Measurement Techniques

In contrast to the measurements of surface waters such as rivers and streams, the measurement of SGD on a volumetric and compositional base is still a major challenge because flow paths and possible impacting conditions (redox gradients, properties of surrounding substrate) are highly heterogeneous and hidden. Therefore, it is still difficult to estimate the volumetric and biogeochemical impact of SGD on local, regional, and global biogeochemical element budgets. In addition to spatial heterogeneity, the entrance of SGD into the coastal ocean is impacted by temporal dynamics such as tidal and seasonal cycles, and climate change. Therefore, the combination of different approaches is required to gain an understanding on the impact of SGD on the water and element balance of the coastal ocean on the local, regional, and global scale. One challenge is the proper characterization of the hydrochemical composition of the SGD component. The composition of coastal groundwaters derived from drilled wells, for instance, provides information about the regional overall input, but not about the chemical and biological processes that are expected at the small-scale interface of surface sediments or rocks. Particularly in areas with unconfined coastal aquifers, local processes defining the actual fluid composition entering the seawater reservoir have to be understood in great detail. Element concentrations in SGD in karst coastal environments are likely closest to the composition found in nearby groundwaters, but also impacted by the short-term interactions with surface waters. Whereas regional and global approaches allow for an estimate of the overall water and element balance, only local methods can give insight to the processes controlling SGD and groundwater-borne transport of compounds. The latter, however, is required for a fundamental system understanding and the build-up of the capacity to extrapolate present impacts into the future.

6.4.1 Methods Used at Local Scale

Prior to the quantification and sampling of SGD, it is necessary to detect and identify the discharge locations, which is often difficult. Amongst the few characteristic indicators for freshwater SGD that have been described in the literature are adaptions in benthic community structures, for instance, the absence of purely marine, sedentary species such as lugworms (Zipperle and Reise 2005). Also, the presence of geomorphological indicators such as pockmarks (Schlüter et al. 2004) may indicate discharge zones. A common tool to identify groundwater discharge locations in hydrological studies is the use of temperature differences during some periods of the year between surface water and groundwater, which can be made visible by thermal infrared (TIR) imagery. This method has also been successfully used to identify and map SGD on large scales, i.e. along coastlines, mostly by aerial surveys (see below). On a local scale, however, either unmanned aerial vehicles (UAVs) or ground-based handheld TIR can be used. UAV studies offer the advantage to be detailed on a spatial scale with ground resolutions <50 cm. However, they also comprise a spatial continuity (Lee et al. 2016) and the possibility to combine spatial and temporal scales and thus to observe an area of $\sim 10^2$ m^2 over a continuous time period of several minutes (Mallast and Siebert 2019). Handheld TIR approaches pose the option to be even more detailed, where object distances <2 m allow the observation of discharging springs in the order of 1–2 cm for a discrete moment of time (Röper et al. 2014) or on a continuous basis.

A field approach to measure local SGD directly is the application of seepage meters. Seepage meters are similar to a large-diameter bucket that is inserted upside down into the sediment. A tube with an inflatable bag is attached to the seepage meter to collect discharging groundwater or a measurement device measures the flow velocity in the tube (Lee 1977). Besides the ability to consider spatial and temporal patchiness at the local scale, it allows sample collection for the measurement of the hydrochemical composition. Apart from the classical manually operated Lee-type seepage meters (Lee 1977), more sophisticated designs allow for an automatization of direct flow measurements via heat transport, the application of dyes, or electromagnetic in-line flow meters (Duque et al. 2020; Rosenberry et al. 2013; Taniguchi et al. 2003). Seepage meters allow for an estimate of the heterogeneity of SGD at a given site, but multiple applications are required to upscale these results. Due to the direct measurement of the SGD composition and fluxes on small scales, however, biogeochemical and physical contributions to the adjacent coastal water column can be deduced. The characterization of a more complex dynamics in the hydrochemical composition requires the combination with multi in situ sensors or multi-sampling ports for discrete volume sampling.

Recently, both volume and hydrochemical composition of SGD from coastal sands were directly measured with stirred round benthic chambers with attached plastic bags, that allowed for an investigation of the impact of pore water advection caused by bottom water hydrodynamics on local SGD (Figure 6.2; Donis et al. 2017). Under the investigated conditions in Puck bay, southern Baltic Sea, the focused local upward flow of groundwater through permeable sands was high and not impacted by stirring.

Another direct local sampling approach comes from the application of push point and longer permanent pore water lances that allow the investigation of vertical pore water profiles on the decimetre to metre scale (e.g. Beck et al. 2007; Donis et al. 2017). In areas with high SGD flux and thereby the development of steep pore water gradients, also the application of ship-based sediment corers (multi corer, gravity core) may be successful, as long as the properties of the sediments allow for the recovery of undisturbed pore waters. The sediment core is sliced and centrifugation is used to separate pore water and sediment. The altered pressure (compared to the water depth from which the sediment core was retrieved) might cause a change of the sample by degassing and oxygen import might alter the redox milieu. An alternative approach for shallow sediments uses dialysis samplers (so-called peepers – for more details see Chapter 9 of this book), that, however, may change the hydrodynamic flow field in the surface sediment.

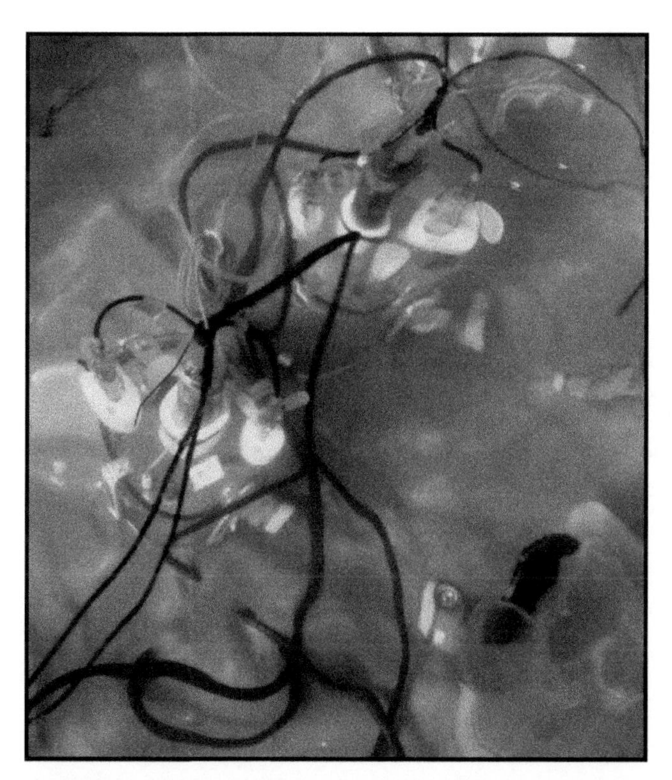

Figure 6.2 In situ application of stirred benthic chambers in the seepage-meter mode (Donis et al. 2017) in the coastal sands of Hel peninsula in the southern Baltic Sea (figure 6.5; D. Donis, 2011).

6.4.2 Methods Used at Regional Scale

At regional scale, discrete measurements need extension by spatially continuous information to depict heterogeneities of the land–ocean interface. Therefore, regionalizing methods are increasingly applied to provide spatially continuous information including the identification of possible heterogeneities on larger scales. The multiplicity of applied sensors, sensor systems, and measuring principles allows a variety of different spatial coverages, spatial resolutions (Figure 6.3), fields of application, aims, and penetration depths of which the most common methods are given below. Quasi-continuous measurements (No. 9 in Figure 6.3) are achieved through short interval measurements (seconds or minutes) of radon (e.g. using Durridge RAD-7), salinity, or temperature (e.g. with a CTD probe). The named sensors can be mounted onto a boat in a depth usually at 1 m water depth or deeper that slowly drives along the investigated coastline sections. Thus obtained spatial 1D resolutions depend on speed, defined interval periods, and number of mounted devices but range between 10^1–10^2 m while the linear spatial coverage is <30 km (Rocha et al. 2016; Schubert et al. 2014).

One of the most commonly used methods applied to quantify SGD volumes on a regional scale is the establishment of a mass balance using conservative SGD tracers such as radon (^{222}Rn) or the radium isotope quartet (^{223}Ra, ^{224}Ra, ^{226}Ra, and ^{228}Ra; Kwon et al. 2014; Moore et al. 2008; Santos et al. 2015; Swarzenski et al. 2006). Most of the aforementioned radionuclides are highly enriched in groundwater compared to surface waters (Moore et al. 2008). SGD volumes can be calculated with simple mass balance models using activities of the respective endmembers (offshore ocean and groundwater), the inventory of the impacted water body, and the local water residence time (which in turn can be determined

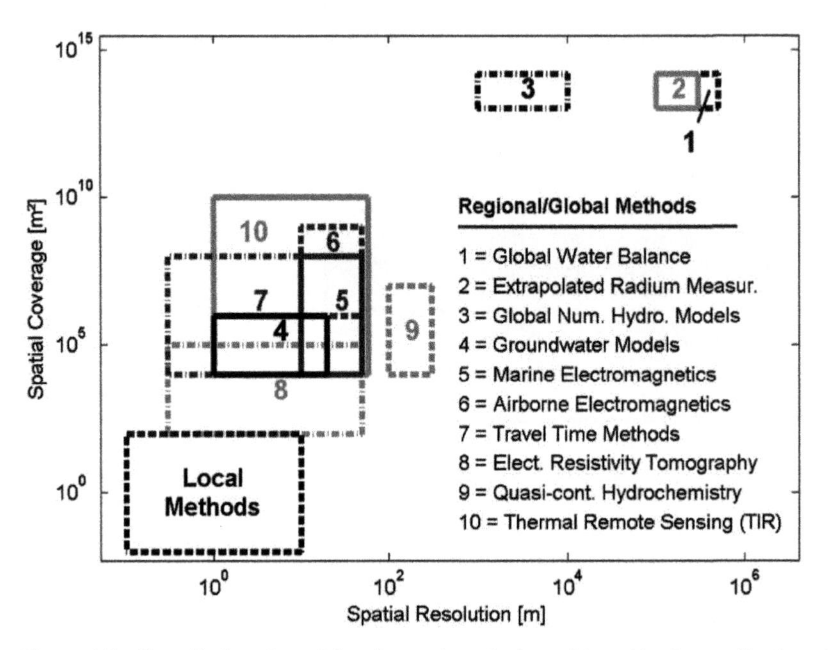

Figure 6.3 Compilation of spatial scales and resolution achieved by the application of different techniques in SGD research.

using $^{223/224}Ra$ decay models). The SGD-driven input signals of radionuclides are integrated over the water columns of semi-enclosed embayments and even marginal seas, allowing estimates on scales from several (Waska and Kim 2011) to thousands of kilometres (Moore et al. 2008; Rodellas et al. 2015).

Electric and electromagnetic methods (Nos. 5, 6, and 8 in Figure 6.3) reveal the subsurface conductivity structure in two or three dimensions. Thus, they are able to localize the fresh/saltwater interface due to its salinity contrast. Electrical resistivity tomography (ERT) uses steel electrodes pinned to the ground (either on land or at the sea bottom) injecting current into the ground measuring the resistance of the sediment (Stieglitz 2005; Swarzenski et al. 2006). A common profile length ranges from tens of metres up to hundreds of metres with depths of investigation down to some tens of metres and allows for imaging structures on the sub-metre scale. Recent developments include monitoring concepts to image temporal changes (e.g. Johnson et al. 2015). For instance, Morrow et al. (2010) and Sutter and Ingham (2017) used ERT imaging on a highly dynamic beach system and imaged a single saltwater circulation zone. Controlled source marine electromagnetics to image the shallow seafloor use a roughly 40 m long streamer towed in contact with the seafloor and consisting of a horizontal magnetic dipole and three receivers (Evans et al. 2007). Szpak et al. (2012) used the system to localize pockmarks off the northern coast of Ireland. Profiles can exceed the kilometre range as the streamer continuously moves at a speed of about 3 km/h. The depth of investigation is 20 m and structural variation of a few tens of metres can be detected. Airborne electromagnetics use systems towed beneath a helicopter carrying the electromagnetic field transmitter and receiver coils (Kirkegaard et al. 2011; Siemon et al. 2020; Steuer et al. 2009; Viezolli et al. 2010). Flight velocities range from 50–160 km/h, thus areas of some squared kilometres can be covered. Depending on the subsurface conductivity, investigation depth ranges to a hundred metres and lateral variations of some tens of metres can be resolved.

A set of methods is based on the travel time of reflected waves (No. 7 in Figure 6.3). Most prominent are multi-beam echosounder (to map sea bathymetry) and seismic reflections (to image the structure below sea-bottom). Hence, this allows for an acoustic/seismic identification of disturbances, so-called pockmarks, even in soft muddy sediments caused by fluid flow through sediments and associated (in particular methane) gas. Profiling speed, resolution, and penetration into the sediment depend very much on the selected system. For instance, echosounders resolve on the centimetre scale while seismic methods provide resolution of some decimetres combined with a penetration of some tens of metres.

One of the most commonly applied regional methods to map SGD is the usage of thermal infrared (TIR) remote sensing (No. 10 in Figure 6.3). The general principle relies on sensing emitted radiation in the thermal electromagnetic spectrum (8–14 μm) of a water area with thermal sensors mounted on various platforms (UAVs, kites, airplanes, satellites). Given a temperature contrast between discharging groundwater and the open water body and a positive buoyancy of the discharging groundwater, emitted radiation for those areas differs from ambient areas and displays as thermal anomalies on the sea surface. These anomalies serve as qualitative proxy for groundwater discharge occurrence on various spatial scales (10^1–10^9 m^2) and resolutions (10^{-1}–10^2 m) (e.g. Mejias et al. 2012; Kelly et al. 2013; Xing et al. 2016;; Schubert et al. 2014; Johnson et al. 2008; Röper et al. 2014). In addition, recent studies use anomaly sizes and extents to upscale quantitative discharge estimates from, for example, mass balances (Johnson et al. 2008) or in situ measurements (Kelly et al. 2013; Mallast et al. 2013; Tamborski et al. 2015), and hence combine discrete and continuous measurements.

This combination of discrete and continuous data is already applied in numerical groundwater models (# 4 in Figure 8.4) by using hydraulic conductivity, permeability, and porosity values as discrete information mostly obtained from well logs and pumping tests and information on elevation, slope, or geology as continuous data. Although "aquifer systems are usually heterogeneous, and it is difficult to obtain sufficient representative values" (Burnett et al. 2006), these models allow one to provide SGD magnitudes and to investigate associated processes for specific areas in a 2D/3D representation (e.g. Beck et al. 2017; Grünenbaum et al. 2020).

6.4.3 Methods Used at Global Scale

At global scale, SGD can be either quantified based on terrestrial water balances, by developing models based on assumed processes, or by extrapolation of regional/local studies. Terrestrial water balances (No. 1 in Figure 6.3) only represent the fresh groundwater component which is usually quantified as the part of the balance unaccounted for by river runoff (e.g. Garrels and Mackenzie 1971; Korzun 1977; Zektser et al. 1973). Unfortunately, the resulting numbers between 5% and 10% of river discharge estimates (Taniguchi et al. 2002) are well within the uncertainty ranges of other parameters of the water balances (Garrels and Mackenzie 1971). While early global models (No. 3 in Figure 6.3) agreed to this range (Beusen et al. 2013; Bouwman et al. 2006), the coarse resolution of the model framework (approx. 50 km grid size at the equator), which originally focused mostly on rivers, combined with the assumptions in the model approach, leads to substantial uncertainty of the results given the scale sensitivity of coastal processes. Recently developed models estimate the volumetric freshwater component of SGD to be less than 1% of river discharge but emphasize its particular abundance in tropical regions (Luijendijk et al. 2020; Zhou et al. 2019). This differs substantially from findings on local and regional scales that may be due to the role of coastal areas with no active SGD pathway in a global approach.

The only global study currently estimating total SGD, including recirculated seawater, combines a ^{228}Ra inverse model (No. 2 in Figure 6.3) to constrain the coastal sources of ^{228}Ra of which the SGD fraction was defined in a second step by a balance-approach (Kwon et al. 2014). Based on an ocean circulation model (DeVries and Primeau 2011), the coastal ^{228}Ra fluxes were identified against atmospheric inputs. The simulated integrated coastal ^{228}Ra fluxes were optimized with observed ^{228}Ra point measurements in SGD to differentiate between SGD, river, and sediment inputs. Similar methods were used to quantify the amount of SGD entering the Atlantic Ocean (Moore et al. 2008). However, at a global scale, not just coastal SGD, which most of the Ra studies represent, but also groundwater discharging further from the coast has to be addressed (e.g. Bakken et al. 2012). Even 200 km into the Atlantic (Manheim 1967), or 25 km off the coast of Taiwan (Lin et al. 2010), traces of fresh SGD were encountered in 500–1200 m depth, respectively, highlighting that global scale estimates of SGD should ideally go beyond extrapolating coastal studies.

6.5 Case Study (Local)

The contribution of SGD to the regional water budget of the Puck Bay (German: *Putziger Wiek*) in the southern Baltic Sea, was already recognized by pelagic salinity and stable isotope anomalies in the early 1990s (Jankowska et al. 1994), and the consequences for the coastal biogeochemistry were further investigated in detail within the BONUS+ project AMBER (Donis et al. 2017; Kotwicki et al. 2014; Szymczycha et al. 2012; Vogler et al. 2011). During several field campaigns the overall study area was characterized on a regional scale by ship-going CTD, acoustic and multi-corer methods for water column, sediment, and pore water properties (Figure 6.5) showing that SGD is escaping in a diffusive mode from muds of the central bay and is associated with pockmark structures and the presence of methane (Figures 6.4–6.6). In the following, it was shown that intense localized seep-type seepage also occurs from the sands at the southern boundaries of Hel peninsula (Figures 6.6–6.8). During several field campaigns with the application of short porewater lances (modified after Beck et al. 2007) and benthic chambers (figure 8.2) it was found that low-salinity groundwater escapes through these seeps from the permeable sands. Due to intense upward fluxes (volumetric seepage rates up to 187 L m^{-2} d^{-1}; Donis et al. 2017; Kotwicki et al. 2014) the salinity in the pore waters was lowered from 7 at the sediment–water interface to less than 0.5 at 20 cmbsf (figure 8.7; Donis et al. 2017; Kotwicki et al. 2014; Vogler et al. 2011).

The spatial SGD seepage distribution was highly heterogeneous and focused on a small area (Figure 6.6) and is assumed to be modulated by a wind-driven wave set-up (Massel 2001). The study was complemented by the hydrochemical and stable isotope characterization of regional groundwaters being possible sources of freshwater to the area. Binary mixing calculations indicate that groundwater escaping from the sands at Hel island is enriched compared

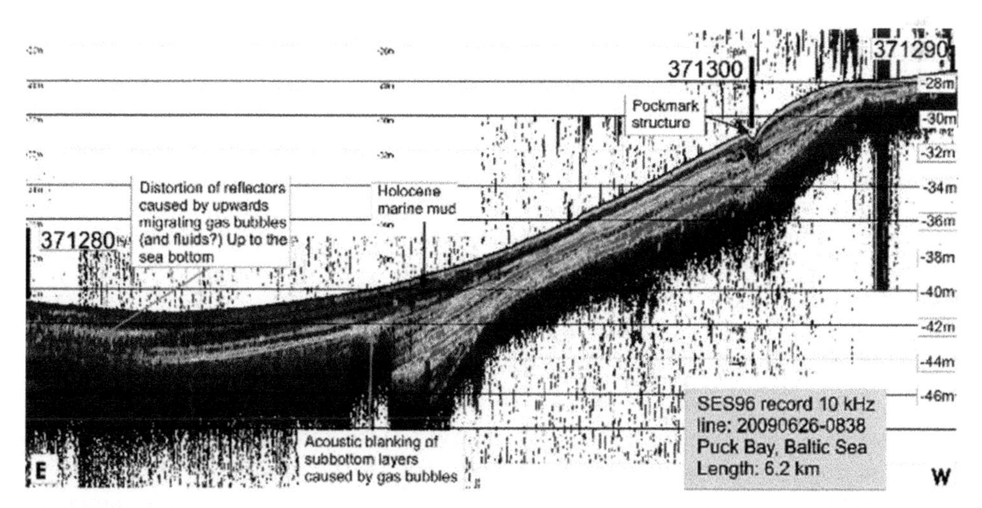

Figure 6.4 Acoustic transect AT2 in Outer Puck Bay, southern Baltic Sea (R. Endler, unpublished data, 2009). The pore water profile at Site 371330 (Figure 6.5) corresponds to a pockmark impacted sediment close to Site 371300. Note: Methane gas bubbles formed in supersaturated pore waters may act as shields, absorbing scattering acoustic energy and thereby hide deeper sediment structures.

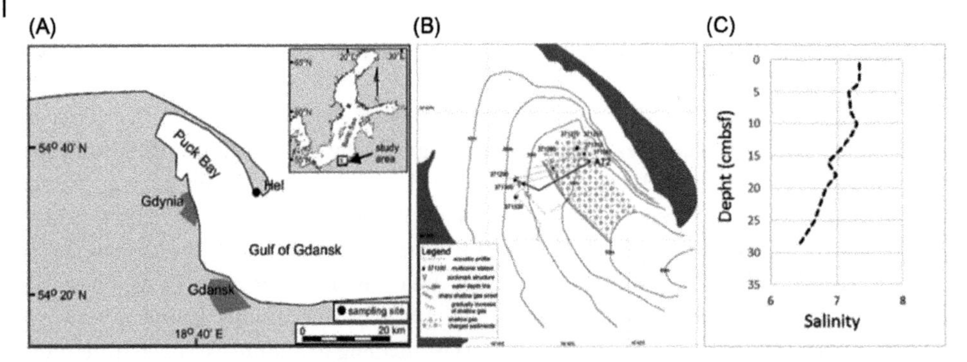

Figure 6.5 The study area Puck Bay at the southern coastline of the Baltic Sea, separated by Hel peninsula from the open Baltic Sea (A: Kotwicki et al., 2014 / with permission of Elsevier B: R. Endler, unpublished, 2009). The acoustic profile AT2 (in B) is shown in Figure 6.8. The salinity profile in porewater from a multicorer core (C) at Site 371330 was calculated from dissolved Na concentrations. Diffusive SGD was encountered by the analysis of pore water profiles in short sediment cores retrieved with a multi-corer in the central muddy part (Site 371330). Discrete seep-type locations were found in the coastal sands of Hel peninsula (A: black dot). cmbsf: cm below surface.

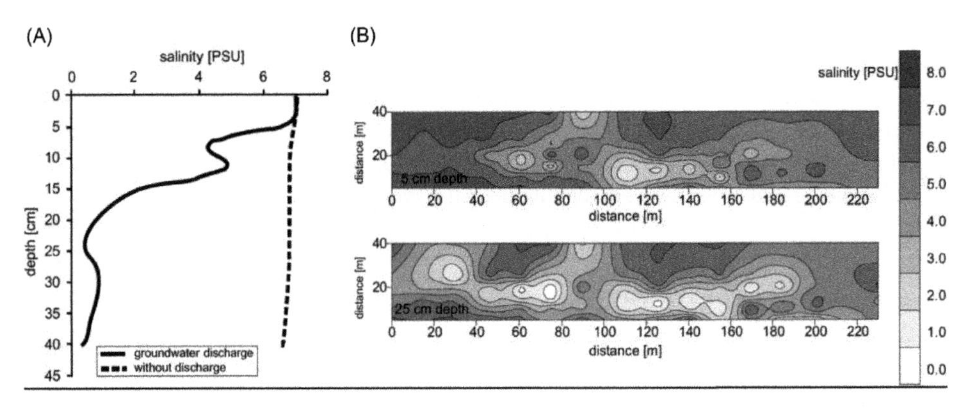

Figure 6.6 Horizontal and vertical distribution of localized seep-type SGD seepage from sands off the Hel peninsula (Figure 6.5) as visualized by the salinity (Kotwicki et al., 2014 / with permission of Elsevier). (A) typical salinity profiles for SGD impacted and (more or less) not impacted sediments. (B) Representation of a bird's-eye view of the salinity distribution in 5 cmbsf (top) and 25 cmbsf (bottom).

to bottom seawater in dissolved inorganic carbon (DIC), methane, phosphate, silica, manganese, and sulphide, but depleted in all major ions (figure 8.8; Donis et al. 2017) and ^{3}H (SGD origin essentially before the 50ies of the last century), and the water is isotopically lighter (^{2}H, ^{18}O) than reports for Holocene meteoric waters for this area (Vogler et al. 2011).

The profiles for dissolved Mn and in particular Fe indicate non-conservative processes within the benthic mixing zone, which may be associated with interactions between sulphide and sedimentary Fe and Mn oxi(hydroxi)des (figure 8.8; e.g. de Beer et al. 2005). The steep gradients in the pore waters (Figure 6.8) and the seep-type character of the areal distribution caused by intense focused upward flux of fresh groundwater (Figure 6.7) require detailed investigation on local scales. The obtained hydrochemical results cannot directly be extrapolated to a larger regional scale.

Table 6.1 Overview of methods characterizing SGD over various spatial and temporal scales.

Method	Spatial scale	Discrete vs. continuous		Aim	Application	Comment
		Spatially	Temporally			
Seepage meter	Local	Discrete	Continuous	Quantification, concentration, fluxes	Offshore	If combined with various sensors, concentration and fluxes of several SGD associated parameters can be obtained (N, P, etc.)
Pore water lance	Local	Discrete	Mostly discrete	Localization, concentration, fluxes	Onshore and offshore	Controlled by lance material and specification yields multi-element composition of pore water on temporal scales
Quasi-continuous hydrochemistry	Local	Quasi-continuous	Continuous	Localization, quantification, concentration, fluxes	Offshore (and onshore)	Requires depth information as well as clear endmember signatures for mass balance approaches (of e.g. ^{222}Rn)
Electric and electromagnetic methods	Local, regional	Continuous	Mostly discrete	Localization	On and offshore	Certain applications reflect transects rather than a complete spatially continuous image. Depends on a resistivity contrast
Travel time methods	Local	Continuous	Discrete	Localization	Offshore	
Thermal remote sensing (TIR)	Local, regional	Continuous	Mostly discrete	Localization (quantification)	Offshore	Quantification can be achieved if time-parallel in situ discharge measurements are obtained
Groundwater models	Local, regional, global	Continuous	Continuous	Concentration, fluxes, quantification	Onshore	Groundwater discharge is the output of an input function (recharge) and governed by some discrete hydrogeological information
Numerical models	Local, regional, global	Continuous	Continuous	Concentration, fluxes quantification	Onshore	Often reflects the fresh water component of SGD only, but more sophisticated models may consider recirculation
Extrapolated Ra	Local, regional, global	Continuous	Discrete	Quantification	Offshore	Although spatially coarse, it reveals all SGD water portions and provides quantitative estimates
Global water balance	Global	Continuous	Continuous	Quantification	Onshore	SGD (freshwater portion only) is deducted from input functions (rain/recharge), and known/calculated output functions, river discharge and evaporation

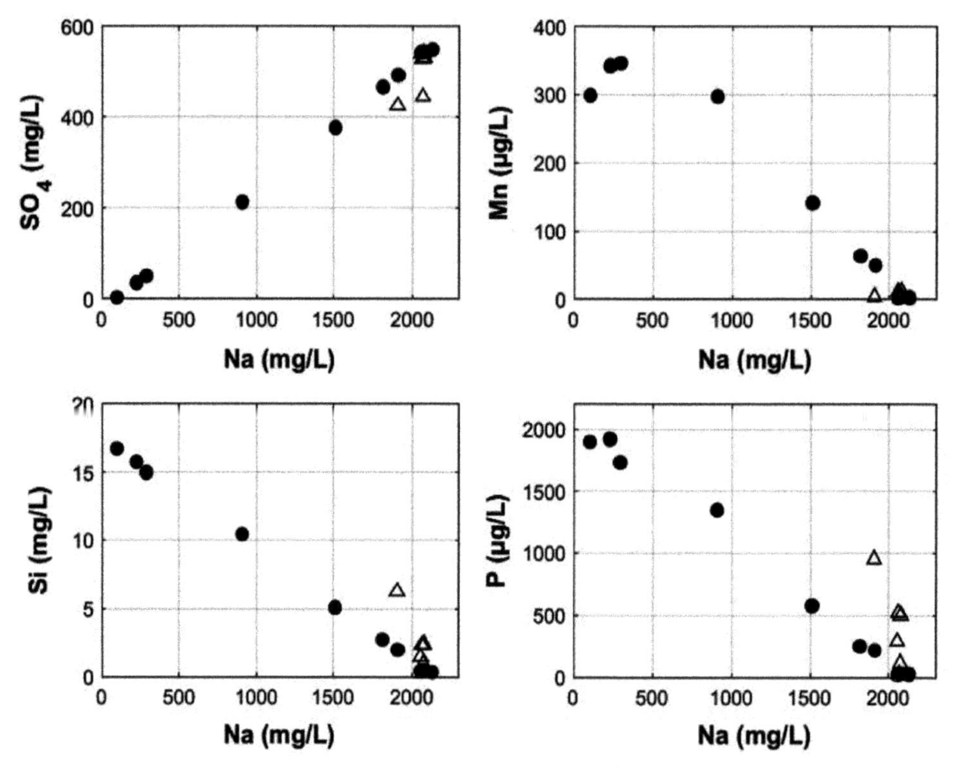

Figure 6.7 Covariation of selected concentrations of dissolved elements as a function of salinity (SO_4, Mn, and P data for the less impacted sediments (open triangles) and some of the data for the impacted site (points) are taken from Donis et al. 2017)). The Na concentrations are used as a conservative tracer for freshwater-brackish water mixing. Typical marine diagenesis is also impacting element concentrations in the top sediments of the reference site, but the influence of groundwater at the seep site is much more pronounced.

6.6 Case Study (Regional)

Interdisciplinary case studies on SGD in the context of a strong tidal regime have been conducted in the German southern North Sea, in the backbarrier sands (marine intertidal plate "Janssand") and on a barrier island (Spiekeroog, fresh and recirculated SGD) (Beck et al. 2017; Billerbeck et al. 2006; Grünenbaum et al. 2020; Moore et al. 2011; Reckhardt et al. 2017, 2015; Riedel et al. 2010; Seidel et al. 2015, 2014; Waska et al. 2019). For the backbarrier tidal flats, tide-induced pore water exchange accounts for large fluxes of CO_2, Mn, and DOC into the Wadden Sea (Al-Raei et al. 2009; Billerbeck et al. 2006; Moore et al. 2011; Santos et al. 2015).

Spiekeroog Island is part of the Eastfrisian Island chain along the North German coastline and has an approximate area of 21.3 km^2. According to Barckhausen (1969), barrier islands are formed by the interaction of wind, swell, and ocean currents. Holocene and Pleistocene sands and underlying finer-grained Pliocene sediments together form an aquifer containing the main freshwater lens below the central dune arc of Spiekeroog (Röper et al. 2014). At

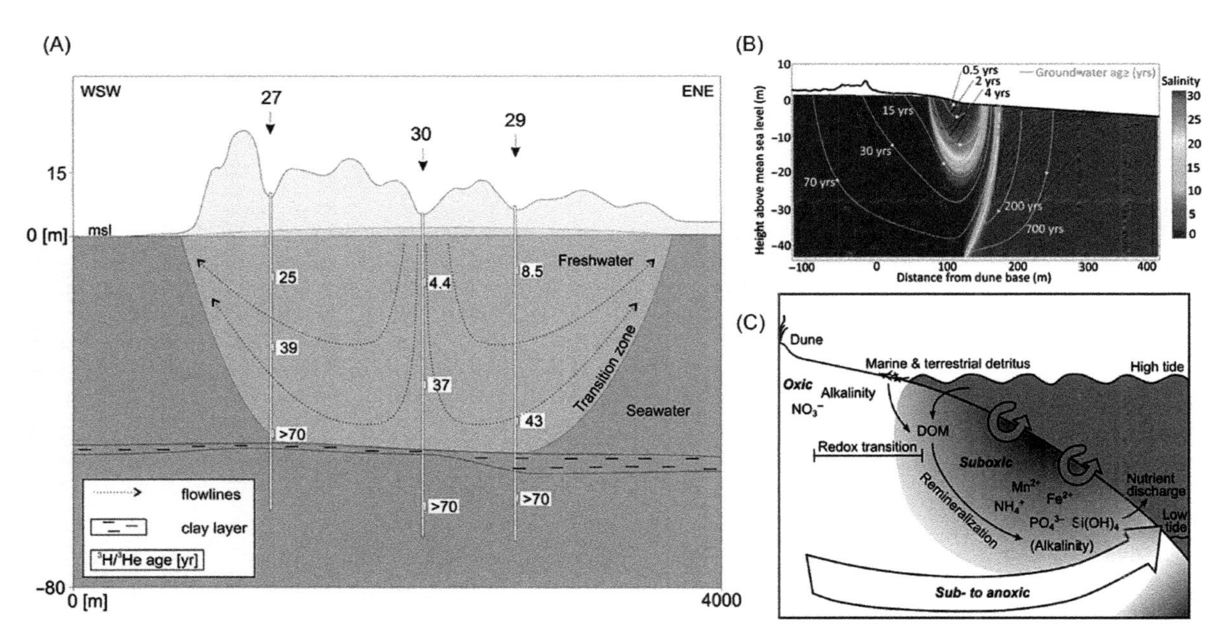

Figure 6.8 (A) Cross-section of Spiekeroog Island with increasing groundwater ages at increasing depths in three nested piezometers (27, 30, 29; Adapted from Röper et al., 2012). (B) SGD model depicting flow paths and modelled groundwater and porewater ages for South Beach, Spiekeroog (Seidel et al. 2015). (C) Conceptual diagram of geochemical pathways within the subterranean estuary of the North Beach, Spiekeroog Island (Reckhardt et al. 2015).

44–55 m below sea-level (mbsl), a continuous clay layer acts as an aquitard and restricts further extension of the freshwater lens. Groundwater recharge amounts to 300–400 mm/a and surface run-off does not exist, since most of the rainwater rapidly drains into the permeable sediments (Röper et al. 2012). The freshwater within the lens is decades old (Röper et al. 2012; Seibert et al. 2018; Figure 6.8A) and used for the island's drinking water production. Freshwater discharge locations were identified by TIR imaging in the North West (Röper et al. 2014), where fine-grained material deposited in the shelter of an offshore sand bar most likely led to the formation of preferential flow paths and local "spring"-type discharge. Evidence of diffuse freshwater discharge as well as tidally-induced saltwater circulation was obtained from porewater sampling campaigns reported in Reckhardt et al. (2015, 2017) and numerical modelling by Seidel et al. (2015), Beck et al. (2017), and Grünenbaum et al. (2020). Due to the high mean tidal range of 2.6 m, the USP is well-developed with a width of up to 100 m and a depth of up to 20 m according to preliminary numerical modelling (Seidel et al. 2015, Figure 6.8A). Modelling also suggests that porewater residence times in the USP are in the order of years, while the discharging freshwater is decades old (Figure 6.8B; Seidel et al. 2015). Depth-resolution profiles from the Spiekeroog subterranean estuary along a northern, high-exposure, and a southern, sheltered beach site indicate that tidally- and wave-driven recirculation produces relatively rapid pore water advection. This recirculation also ensures frequent inputs of marine organic matter and oxygen at the mean high water line. Due to the advective flow patterns and locally provided substrates, a lateral respiration succession is produced along a land–sea gradient, with oxygen consumption and NO_3 reduction occurring primarily in the upper part of the intertidal beach zone, followed by Fe and Mn reduction further towards the low water line (Reckhardt et al. 2015, Figure 6.8C). Although fresh groundwater only comprises a small amount of the SGD, it likely is a carrier of terrestrial new and marine-derived remineralized nutrients, as well as a source of bio-refractory terrestrial aromatic and combustion-derived DOM, to the coastal North and Wadden Sea (Reckhardt et al. 2015; Seibert et al. 2019; Seidel et al. 2015).

6.7 Case Study (Global)

While local studies of SGD have recently been published in abundance, large-scale estimates of SGD are still rare. Local studies usually report SGD in either of three units: volume per area and time (e.g. $m^3\ m^{-2}\ a^{-1}$), volume per coastline and time (e.g. $m^3\ m^{-1}\ a^{-1}$), or volume per time (e.g. $m^3\ sec^{-1}$). Area normalized SGD is often reported, e.g. by local studies using seepage meters (which have a defined area); coastline normalized SGD values are reported by regional studies extrapolating along a stretch of coast. Non-normalized discharges are reported usually for individual conduits or as total flux for a defined geographical structure, e.g. a bight or an estuary. The values of the three groups are not comparable without further knowledge. However, if the representative coastline length or area of a non-normalized value can be identified, a conversion between the units is possible, yet associated with significant spatial uncertainty.

A database extended from the literature study presented in Moosdorf et al. (2015) shows the published coastline length or area normalized values of SGD from 220 locations around the world (Figure 6.9). The database highlights the wide range of about 7 magnitudes between the published values. The 25–75 percentiles cover about 1 magnitude for area (5.8,

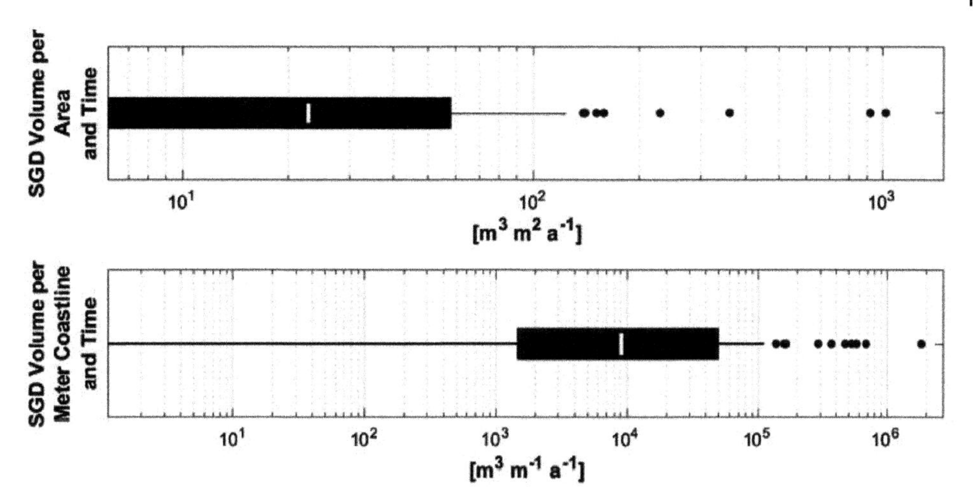

Figure 6.9 Distribution of published data reporting SGD fluxes (top) in volume per area and time and (bottom) in volume per metre coastline and time. Note the logarithmic scale in the panels. Note also, that studies with no reported SGD are not included.

59.1 $m^{-3}m^{-2}\,a^{-1}$) and length (986, 26017 $m^{-3}m^{-1}$ a) normalized values. The median reported SGD is 22.9 $m^{-3}m^{-2}\,a^{-1}$ and 4340 $m^{-3}m^{-1}\,a^{-1}$. Comparing the coastline length and area normalized values suggests that on average 190 m^2 with the reported average flux per area would be needed per metre of coastline to discharge the average reported value per metre of coastline. SGD discharging from a few 100 metres from the coast is a known phenomenon, which makes this width seem reasonable. However, assumed exponential decrease of SGD from the coastline (e.g. Taniguchi et al. 2003) would mean discharge further away from the coast.

Extrapolating the median reported SGD per metre of coastline on the length of the global coast would result in a first order estimate of global SGD. This value, however, would be highly uncertain for three reasons: (1) it is likely that the local reported values are biased towards high SGD; (2) the focus of many coastal studies is terrestrially-derived SGD, but not all report the ratio between fresh and recirculated SGD; and (3) the total length of a global coastline to use for extrapolation from local studies is unclear. The first point is important, although obvious: SGD studies are usually performed where SGD can be seen or is known. Thus, regions with very little SGD are likely underrepresented in the reported values – only very few studies report little or no SGD in their study area under the keyword "submarine groundwater discharge". These studies either do not discuss SGD at all or so little that they did not show up in the literature research leading to the body of reported local SGD values. The third point becomes obvious when recognizing that coast length is a fractal parameter, changing its value with the scale of observation (Mandelbrot 1967). The highest available resolution coastline at a global scale is 2.3 million km, by a factor of ~5 more than used in recent global studies (Moosdorf et al. 2015). However, including all uncertainties, the fresh SGD estimate based on median reported values and global coast length (based on ISCIENCES L.L.C., 2009) would be 9980 km^3/a, about a fourth of the estimated total global run-off (Fekete et al. 2002). Since the studies used here include both fresh and recirculated SGD, this estimate is reasonable but uncertain. The true global SGD estimate is highly relevant, yet still very weakly constrained.

6.8 Conclusion and Knowledge Gap

SGD is an important part of the global hydrological cycle that impacts estuarine and coastal areas. It influences a variety of aspects in these ecosystems, e.g. possible water resource availability or species distribution at local scale, as well as transport of nutrients or pollutant fluxes at a global scale. Yet, while numerous studies quantify SGD on a local scale, regional and global scale SGD amounts and their associated solute fluxes are weakly constrained and the locations of SGD are unclear for most regions worldwide.

The biogeochemical processes taking place in the fresh–salt water mixing zone are not yet well understood, despite their control of the actual hydrochemical composition of SGD entering the marine environment and thus the role of SGD in the global element cycles. As an example, in the context of metal fluxes into the ocean, the role of SGD must be better constrained, with regards to source, transport mechanism, the role of DOC-metal interactions, and microbiological activity.

One of the major challenges concerning exploration methods is the need for refined geochemical tracers as well as sampling strategies to follow the SGD continuum on all different spatial and temporal scales. This challenge especially calls for a mutual concept on SGD research from both terrestrial and marine disciplines.

Acknowledgements

MEB wishes to thank R. Endler for the allowance to present so-far unpublished results, O. Dellwig, P. Escher, and S. Vogler for field and laboratory support, and A.-K. Jenner, C. von Ahn, and V. Winde for stimulating discussions. He also wishes to thank BMBF and DFG for financial support during BONUS+ project AMBER (BMBF project No. 03F0485A) and DFG research training group BALTIC TRANSCOAST (GRK2000), respectively. This study is dedicated to the memory of Jörg-Olaf Wolff, a well respected colleague, who recently passed away. UM is grateful to the Helmholtz Association of German Research Centers funded project DESERVE. NM was funded by BMBF grant (01LN1307A SGD-NUT). HW was funded by a DFG (WA3067/2-1) and the MWK project "BIME" (ZN3184). The study contributes further to the DFG Research Group Dynadeep. This publication was furthermore supported to all authors by DFG during the KiSNet project (MA7041/6-1). The authors wish to thank J. Lewandowski for a detailed and constructive review that helped to improve the manuscript, and S. Krause for his editorial handling. This is BALTIC TRANSCOAST publication No. GRK2000/0002 and KiSNet publication No. 0001.

References

Abarca, E., Karam, H., Hemond, H.F., and Harvey, C.F. (2013). Transient groundwater dynamics in a coastal aquifer: the effects of tides, the lunar cycle, and the beach profile. *Water Resources Research* 49: 2473–2488.

Al-Raei, A.M., Bosselmann, K., Böttcher, M.E. et al. (2009). Seasonal dynamics of microbial sulfate reduction in temperate intertidal surface sediments: controls by temperature and organic matter. *Ocean Dynamics* 59: 351–370.

Bakken, T.H., Ruden, F., and Mangset, L.E. (2012). Submarine groundwater: a new concept for the supply of drinking water. *Water Resources Management* 26: 1015–1026.

Barckhausen, J. (1969). Entstehung und Entwicklung der Insel Langeoog – beiträge zur Quartärgeologie und Paläogeographie eines ostfriesischen Küstenabschnittes. *Oldenburger Jb* 68: 239–281.

Beck, M., Dellwig, O., Kolditz, K. et al. (2007). In situ pore water sampling in deep intertidal flat sediments. *Limnology & Oceanography Methods* 5: 136–144.

Beck, M., Reckhardt, A., Amelsberg, J. et al. (2017). The drivers of biogeochemistry in beach ecosystems: a cross-shore transect from the dunes to the low-water line. *Marine Chemistry* 190: 35–50.

Beusen, A.H.W., Slomp, C.P., and Bouwman, A.F. (2013). Global land-ocean linkage: direct inputs of nitrogen to coastal waters via submarine groundwater discharge. *Environmental Research Letters* 8: 034035.

Billerbeck, M., Werner, U., Polerecky, L. et al. (2006). Surficial and deep pore water circulation governs spatial and temporal scales of nutrient recycling in intertidal sand flat sediment. *Marine Ecology Progress Series* 326: 61–76.

Blöschl, G., Bierkens, M.F.P., Chambel, A. et al. (2019). Twenty-three unsolved problems in hydrology (UHP): a community perspective. *Hydrological Sciences Journal* 64: 1141–1158.

Bögli, A. (1980). *Karst Hydrology and Physical Speleology*, xiii, 284. Berlin, New York: Springer-Verlag.

Bouwman, A.F., Kram, T., and Klein, G.K. (Eds.) (2006). *Integrated modelling of global environmental change*. An overview of IMAGE 2.4 (Publication 500110002/2006). Bilthoven, N.L. Netherlands Environmental Assessment Agency.

Burnett, W.C., Aggarwal, P.K., Aureli, A. et al. (2006). Quantifying submarine groundwater discharge in the coastal zone via multiple methods. *Science of the Total Environment* 367: 498–543.

Burnett, W.C., Bokuniewicz, H., Huettel, M. et al. (2003). Groundwater and pore water inputs to the coastal zone. *Biogeochemistry* 66: 3–33.

Charette, M.A. and Sholkovitz, E.R. (2006). Trace element cycling in a subterranean estuary: part 2. Geochemistry of the pore water. *Geochimica et Cosmochimica Acta* 70: 811–826.

Church, T.M. (1996). A groundwater route for the water cycle. *Nature* 380: 579–580.

Cole, J.J., Prairie, Y.T., Caraco, N.F. et al. (2007). Plumbing the global carbon cycle: integrating inland waters into the terrestrial carbon budget. *Ecosystems* 10: 171–184.

Cyronak, T., Santos, I.R., Erler, D.V., and Eyre, B.D. (2013). Groundwater and porewater as major sources of alkalinity to a fringing coral reef lagoon (Muri Lagoon, Cook Islands). *Biogeosciences* 10: 2467–2480.

d'Elia, C.F., Webb, K.L., and Porter, J.W. (1981). Nitrate-rich groundwater inputs to discovery bay, Jamaica - a significant source of N to local coral reefs. *Bulletin of Marine Science* 31: 903–910.

Davis, S.N. (1964). Silica in streams and ground water. *American Journal of Science* 262: 870–891.

de Beer, D., Wenzhöfer, F., Ferdelman, T.G. et al. (2005). Transport and mineralization in North Sea sandy intertidal sediments, Sylt-Rømø Basin, Wadden Sea. *Limnology and Oceanography* 50: 113–127.

DeVries, T. and Primeau, F. (2011). Dynamically and observationally constrained estimates of water-mass distributions and ages in the Global Ocean. *Journal of Physical Oceanography* 41: 2381–2401.

Donis, D., Janssen, F., Liu, B. et al. (2017). Biogeochemical impact of submarine ground water discharge on coastal surface sands of the southern Baltic Sea. *Estuarine, Coastal and Shelf Research* 189: 131–142.

Duque, C., Russaniello, C.J., and Rosenberry, D.O. (2020). History and evolution of seepage meters for quantifying flow between groundwater and surface water: part 2 – marine settings and submarine groundwater discharge. *Earth-Sciences Reviews* 204: 103168.

Evans, R.J., Stewart, S.A., and Davies, R.J. (2007). Phase-reversed seabed reflections in seismic data: examples related to mud volcanoes from the South Caspian Sea. *Geo-Marine Letters* 27: 203–212.

Fekete, B.M., Vorosmarty, C.J., and Grabs, W. (2002). High-resolution fields of global runoff combining observed river discharge and simulated water balances. *Global Biogeochemical Cycles* 16: 1042.

Fleury, P., Bakalowicz, M., and de Marsily, G. (2007). Submarine springs and coastal karst aquifers: a review. *Journal of Hydrology* 339: 79–92.

Garrels, R.M. and Mackenzie, F.T. (1971). *Evolution of Sedimentary Rocks*. 1–397. New York: Norton.

Goñi, M.A. and Gardner, L.R. (2003). Seasonal dynamics in dissolved organic carbon concentrations in a coastal water-table aquifer at the forest-marsh interface. *Aquatic Geochemistry* 9: 209–232.

Gove, J.M., McManus, M., Neuheimer, A.B. et al. (2016). Near-island biological hotspots in barren ocean islands. *Nature Communications* 7: 10581. https://doi.org/10.1038/ncomms10581.

Grünenbaum, N., Greskowiak, J., Sültenfuß, J., and Massmann, G. (2020). Groundwater flow and residence times below a meso-tidal high-energy beach: a model-based analyses of salinity patterns and ^{3}H-^{3}He groundwater ages. *Journal Hydrology* 587: 124948.

Hu, C.M., Muller-Karger, F.E., and Swarzenski, P.W. (2006). Hurricanes, submarine groundwater discharge, and Florida's red tides. *Geophysical Research Letters* 33: 5.

ISCIENCES L.L.C. (2009). *Global Coastline Dataset V1*. Ann Arbor, MI.

Jankowska, H., Matciak, M., and Nowacki, J. (1994). Salinity variations as an effect of groundwater seepage through the seabed (Puck Bay, Poland). *Oceanologia* 36: 33–46.

Johannes, R.E. (1980). The ecological significance of the submarine discharge of groundwater. *Marine Ecology Progress Series* 3: 365–373.

Johannesson, K.H. and Burdige, D.J. (2007). Balancing the global oceanic neodymium budget: evaluating the role of groundwater. *Earth and Planetary Science Letters* 253: 129–142.

Johannesson, K.H., Chevis, D.A., Burdige, D.A. et al. (2011). Submarine groundwater discharge is an important net source of light and middle REEs to coastal waters of the Indian River Lagoon, Florida, USA. *Geochimica et Cosmochimica Acta* 75: 825–843.

Johnson, A.G., Glenn, C.R., Burnett, W.C. et al. (2008). Aerial infrared imaging reveals large nutrient-rich groundwater inputs to the ocean. *Geophysical Research Letters* 35: L15606 15601–15606.

Johnson, T.C., Versteeg, R.J., Day-Lewis, F.D. et al. (2015). Time-lapse electrical geophysical monitoring of amendment-based biostimulation. *Groundwater* 53: 920–932.

Jurasinski, G., Janssen, M., Voss, M. et al. (2018). Understanding the coastal ecocline: assessing sea-land-interactions at non-tidal, low-lying coasts through interdisciplinary research. *Frontiers in Marine Science* 5 (342): 1–22.

Kazemi, G.A. (2008). Editor's Message: submarine groundwater discharge studies and the absence of hydrogeologists. *Hydrogeology Journal* 16: 201–204.

Kelly, J.L., Glenn, C.R., and Lucey, P.G. (2013). High-resolution aerial infrared mapping of groundwater discharge to the coastal ocean. *Limnology and Oceanography: Methods* 11: 262–277.

Kim, T.H., Kwon, E., Kim, I. et al. (2013). Dissolved organic matter in the subterranean estuary of a volcanic island, Jeju: importance of dissolved organic nitrogen fluxes to the ocean. *Journal of Sea Research* 78: 18–24.

Kim, T.H., Waska, H., Kwon, E. et al. (2012). Production, degradation, and flux of dissolved organic matter in the subterranean estuary of a large tidal flat. *Marine Chemistry* 142–144: 1–10.

Kirkegaard, C., Sonnenborg, T.O., Auken, E., and Jørgensen, F. (2011). Salinity distribution in heterogeneous coastal aquifers mapped by airborne electromagnetics. *Vadose Zone Journal* 10: 125–135.

Knee, K.L. and Paytan, A. (2011). Submarine groundwater discharge: a source of nutrients, metals, and pollutants to the coastal ocean. In: *Treatise on Estuarine and Coastal Science*, 4 (ed. E. Wolanski and D.S. McLusky), 205–233.

Kohout, F.A. (1966). *Submarine Springs: A Neglected Phenomenon of Coastal Hydrology, Symposium on Hydrology and Water Resources Development.* 391–413. Turkey: Office of U.S. Economic Coordinator for CENTO Affairs, Central Treaty Organization, Ankera.

Korzun, V.I. (1977). *Atlas of World Water Balance: Explanatory Text.* Leningrad: UNESCO Hydrometeorological Publishing House.

Kotwicki, L., Grzelak, K., Czub, M. et al. (2014). Submarine groundwater discharge to the Baltic coastal zone: impacts on the meiofaunal community. *Journal of Marine Systems* 129: 118–126.

Kwon, E.Y., Kim, G., Primeau, F. et al. (2014). Global estimate of submarine groundwater discharge based on an observationally constrained radium isotope model. *Geophysical Research Letters* 41: 2014GL061574.

Lee, D.R. (1977). A device for measuring seepage flux in lakes and estuaries. *Limnology & Oceanography* 22: 140–147.

Lee, E., Yoon, H., Hyun, S.P. et al. (2016). Unmanned aerial vehicles (UAVs)-based thermal infrared (TIR) mapping, a novel approach to assess groundwater discharge into the coastal zone. *Limnology and Oceanography: Methods* 14: 725–735.

Lewandowski, J., Meinikmann, K., Nützmann, G., and Rosenberry, D.O. (2015). Groundwater discharge to lakes - the disregarded component in lake nutrient budgets. 2. Biogeochemical considerations. *Hydrological Processes* 29: 2922–2955.

Lin, I.T., Wang, C.-H., You, C.-F. et al. (2010). Deep submarine groundwater discharge indicated by tracers of oxygen, strontium isotopes and barium content in the Pingtung coastal zone, southern Taiwan. *Marine Chemistry* 122: 51–58.

Luijendijk, E., Gleeson, T., and Moosdorf, N. (2020). Fresh groundwater discharge insignificant for the world's oceans but important for coastal ecosystems. *Nature Communications* 11: 1260.

Macklin, P.A., Maher, D.T. & Santos, I.R. (2014). Estuarine canal estate waters: hotspots of CO_2 outgassing driven by enhanced groundwater discharge? *Marine Chemistry* 167: 82–92.

Mallast, U. and Siebert, C. (2019). Combining continuous spatial and temporal scales for SGD investigations using UAV-based thermal infrared measurements. *Hydrology & Earth System Sciences* 23: 3.

Mallast, U., Siebert, C., Wagner, B. et al. (2013). Localisation and temporal variability of groundwater discharge into the Dead Sea using thermal satellite data. *Environmental Earth Sciences* 69: 587–603. https://doi.org/10.1007/s12665-013-2371-6.

Mandelbrot, B. (1967). How long is the coast of Britain? Statistical self-similarity and fractional dimension. *Science* 156: 636–638.

Manheim, F.T. (1967). Section of geological sciences: evidence for submarine discharge of water on the Atlantic continental slope of the southern United States, and suggestions for further search. *Transactions of the New York Academy of Sciences* 29: 839–853.

Massel, S.R. (2001). Circulation of groundwater due to wave set-up on a permeable beach. *Oceanologia* 43: 279–290.

Mejías, M., Ballesteros, B.J., Antón-Pacheco, C. et al. (2012). Methodological study of submarine groundwater discharge from a karstic aquifer in the Western Mediterranean Sea. *Journal of Hydrology* 464–465: 27–40.

Michael, H.A., Mulligan, A.E., and Harvey, C.F. (2005). Seasonal oscillations in water exchange between aquifers and the coastal ocean. *Nature* 436: 1145–1148.

Miller, D.C. and Ullman, W.J. (2004). Ecological consequences of ground water discharge to delaware bay, United States. *Ground Water* 42: 959–970.

Milne, J. (1897). Sub-oceanic changes. *The Geographical Journal* 10: 129–146.

Moore, W.S. (2010). The effect of submarine groundwater discharge on the ocean. *Annual Review of Marine Science* 2: 59–88.

Moore, W.S., Beck, M., Riedel, T. et al. (2011). Radium-based pore water fluxes of silica, alkalinity, manganese, DOC, and uranium: a decade of studies in the German Wadden Sea. *Geochimica et Cosmochimica Acta* 75: 6535–6555.

Moore, W.S., Sarmiento, J.L., and Key, R.M. (2008). Submarine groundwater discharge revealed by Ra-228 distribution in the upper Atlantic Ocean. *Nature Geoscience* 1: 309–311.

Moosdorf, N., Stieglitz, T., Waska, H. et al. (2015). Submarine groundwater discharge from tropical islands: a review. *Grundwasser* 20: 53–67.

Morrow, F.J., Ingham, M.R., and McConchie, J.A. (2010). Monitoring of tidal influences on the saline interface using resistivity traversing and cross-borehole resistivity tomography. *Journal of Hydrology* 389: 96–77.

Mulligan, A.E. and Charette, M.A. (2006). Intercomparison of submarine groundwater discharge estimates from a sandy unconfined aquifer. *Journal of Hydrology* 327: 411–425.

Nelson, C.E., Donahue, M.J., Dulaiova, H. et al. (2015). Fluorescent dissolved organic matter as a multivariate biogeochemical tracer of submarine groundwater discharge in coral reef ecosystems. *Marine Chemistry* 177: 232–243.

Oberdorfer, J.A. (2003). Hydrogeologic modeling of submarine groundwater discharge: comparison to other quantitative methods. *Biogeochemistry* 66: 159–169.

Oehler, T., Bakti, H., Lubis, R.F. et al. (2019a). Nutrient dynamics in submarine groundwater discharge through a coral reef (western Lombok, Indonesia). *Limnology & Oceanography* 64: 2646–2661.

Oehler, T., Tamborski, J., Rahman, S. et al. (2019b). DSi as a tracer for submarine groundwater discharge. *Frontiers in Marine Science* 6: 563.

Paerl, H.W. (1997). Coastal eutrophication and harmful algal blooms: importance of atmospheric deposition and groundwater as "new" nitrogen and other nutrient sources. *Limnology and Oceanography* 42: 1154–1165.

Paffrath, R., Pahnke, K., Behrens, M.K. et al. (2020). Rare earth element behavior in a sandy subterranean estuary of the southern North Sea. *Frontiers in Marine Science* 7: 424.

Reckhardt, A., Beck, M., Greskowiak, J. et al. (2017). Cycling of redox-sensitive elements in a sandy subterranean estuary of the southern North Sea. *Marine Chemistry* 188: 6–17.

Reckhardt, A., Beck, M., Seidel, M. et al. (2015). Carbon, nutrient, and trace metal cycling in sandy sediments: a comparison of high-energy beaches and backbarrier tidal flats. *Estuarine, Coastal, and Shelf Science* 159: 1–14.

Riedel, T., Lettmann, K., Beck, M., and Brumsack, H.-J. (2010). Tidal variations in groundwater storage and associated discharge from an intertidal coastal aquifer. *Journal of Geophysical Research* 115: 1–10.

Robinson, C., Li, L., and Prommer, H. (2007). Tide-induced recirculation across the aquifer-ocean interface. *Water Resources Research* 43: 14.

Rocha, C., Veiga-Pires, C., Scholten, J. et al. (2016). Assessing land–ocean connectivity via submarine groundwater discharge (SGD) in the Ria Formosa Lagoon (Portugal): combining radon measurements and stable isotope hydrology. *Hydrology and Earth System Sciences* 20: 3077–3098.

Rodellas, V., Garcia-Orellana, J., Masqué, P. et al. (2015). Submarine groundwater discharge as a major source of nutrients to the Mediterranean Sea. *Proceedings of the National Academy of Sciences* 112: 3926–3930.

Röper, T., Greskowiak, J., and Massmann, G. (2014). Detecting small-scale groundwater discharge springs using handheld thermal infrared imagery. *Ground Water* 52: 936–942.

Röper, T., Kröger, K.F., Meyer, H. et al. (2012). Groundwater ages, recharge conditions and hydrochemical evolution of a barrier island freshwater lens (Spiekeroog, Northern Germany). *Journal of Hydrology* 454: 173–183.

Rosenberry, D.O., Sheibley, R.W., Cox, S.E. et al. (2013). Temporal variability of exchange between groundwater and surface water based on high-frequency direct measurements of seepage at the sediment-water interface. *Water Resources Research* 49: 2975–2986.

Russoniello, C.J., Konikow, L.F., Kroeger, K.D. et al. (2016). Hydrogeologic controls on groundwater discharge and nitrogen loads in a coastal watershed. *Journal of Hydrology* 538: 783–793.

Salt, D.E., Blayock, K., Dushnekov, N. et al. (1995). A novel strategy for the removal of toxic metals from the environment using plants. *Biotechnology* 13: 468–474.

Santos, I.R., Beck, M., Brumsack, H. et al. (2015). Porewater exchange as a driver of carbon dynamics across a terrestrial-marine transect: insights from coupled ^{222}Rn and pCO$_2$ observations in the German Wadden Sea. *Marine Chemistry* 171: 10–20.

Santos, I.R., Burnett, W.C., Dittmar, T. et al. (2009). Tidal pumping drives nutrient and dissolved organic matter dynamics in a Gulf of Mexico subterranean estuary. *Geochimica et Cosmochimica Acta* 73: 1325–1339.

Santos, I.R., Glud, R.N., Maher, D. et al. (2011). Diel coral reef acidification driven by porewater advection in permeable carbonate sands, Heron Island, Great Barrier Reef. *Geophysical Research Letters* 38: 5.

Schlüter, M., Sauter, E.J., Andersen, C.E. et al. (2004). Spatial distribution and budget for submarine groundwater discharge in Eckernförde Bay (Western Baltic Sea). *Limnology and Oceanography* 49: 157–167.

Schubert, M., Scholten, J., Schmidt, A. et al. (2014). Submarine groundwater discharge at a single spot location: evaluation of different detection approaches. *Water* 6: 584–601.

Seibert, S.L., Greskowiak, J., Prommer, H. et al. (2019). Modeling of biogeochemical processes in a barrier island freshwater lens (Spiekeroog, Germany). *Journal of Hydrology* 575: 1133–1144.

Seibert, S.L., Holt, T., Reckhardt, A. et al. (2018). Hydrochemical evolution of a freshwater lens below a barrier island (Spiekeroog, Germany): the role of carbonate mineral reactions, cation exchange and redox processes. *Applied Geochemistry* 92: 186–20.

Seidel, M., Beck, M., Greskowiak, J. et al. (2015). Benthic-pelagic coupling of nutrients and dissolved organic matter composition in an intertidal sandy beach. *Marine Chemistry* 176: 150–163.

Seidel, M., Beck, M., Riedel, T. et al. (2014). Biogeochemistry of dissolved organic matter in an anoxic tidal creek bank. *Geochimica et Cosmochimica Acta* 140: 418–434.

Siemon, B., Ibs-von Seht, M., Steuer, A. et al. (2020). Airborne electromagnetic, magnetic, and radiometric surveys at the German North Sea Coast applied to groundwater and soil investigations. *Remote Sensing* 12: 1629.

Simmons, G.M. (1992). Importance of submarine groundwater discharge (SGWD) and seawater cycling to material flux across sediment water interfaces in marine environments. *Marine Ecology Progress Series* 84: 173–184.

Slomp, C.P. and van Cappellen, P.V. (2004). Nutrient inputs to the coastal ocean through submarine groundwater discharge: controls and potential impact. *Journal Hydrology* 295: 64–86.

Steuer, A., Siemon, B., and Auken, E. (2009). A comparison of helicopter-borne electromagnetics in frequency-and time-domain at the Cuxhaven valley in Northern Germany. *Journal of Applied Geophysics* 67: 194–205.

Stieglitz, T. (2005). Submarine groundwater discharge into the near-shore zone of the Great Barrier Reef, Australia. *Marine Pollution Bulletin* 51: 51–59.

Sutter, E. and Ingham, M. (2017). Seasonal saline intrusion monitoring of a shallow coastal aquifer using time-lapse DC resistivity traversing. *Near Surface Geophysics* 55: 59–73.

Swarzenski, P.W., Burnett, W.C., Greenwood, W.J. et al. (2006). Combined time-series resistivity and geochemical tracer techniques to examine submarine groundwater discharge at Dor Beach, Israel. *Geophysical Research Letters* 33: 1–6.

Swarzenski, P.W. and Kindinger, J.L. (2003). Leaky coastal margins: examples of enhanced coastal groundwater/surface water exchange from Tampa Bay and Crescent Beach submarine spring, Florida, USA. In: *Coastal Aquifer Management and Modeling and Case Studies* (ed. A. Cheng and D. Quazar), 93–112. CRC/-Lewis Press.

Szpak, M.T., Monteys, X., O'Reilly, S. et al. (2012). Geophysical and geochemical survey of a large marine pockmark on the Malin Shelf, Ireland. *Geochemistry, Geophysics, Geosystems* 13.

Szymczycha, B., Vogler, S., and Pempkowik, J. (2012). Nutrient fluxes via submarine groundwater discharge to the Bay of Puck, southern Baltic Sea. *Science of the Total Environment* 438: 86–93.

Tamborski, J.J., Rogers, A.D., Bokuniewicz, H.J. et al. (2015). Identification and quantification of diffuse fresh submarine groundwater discharge via airborne thermal infrared remote sensing. *Remote Sensing of Environment* 171: 202–217.

Taniguchi, M., Burnett, W.C., Cable, J.E., and Turner, J.V. (2002). Investigation of submarine groundwater discharge. *Hydrological Processes* 16: 2115–2129.

Taniguchi, M., Burnett, W.C., Smith, C.F. et al (2003). Spatial and temporal distributions of submarine groundwater discharge rates obtained from various types of seepage meters at a site in the Northeastern Gulf of Mexico. *Biogeochemistry* 66: 35–53.

Viezzoli, A., Tosi, L., Teatini, P., and Silvestri, S. (2010). Surface water–groundwater exchange in transitional coastal environments by airborne electromagnetics: the Venice Lagoon example. *Geophysical Research Letters* 37: *L01402*.

Vogler, S., Dellwig, O., Escher, P. et al. (2011). A multi isotope (C,O,S,H) and trace metal study in coastal permeable sands affected by submarine groundwater discharge. *Geophysical Research Abstracts* 13: 8757.

Waska, H., Greskowiak, J., Ahrens, J. et al. (2019). Spatial and temporal patterns of pore water chemistry in the inter-tidal zone of a high energy beach. *Frontiers in Marine Science* 6: 154.

Waska, H. and Kim, G. (2011). Submarine groundwater discharge as a min nutrient source for benthic and water-column primary production in a large intertidal environment of the Yellow Sea. *Journal of Sea Research* 65: 103–113.

Winde, V., Böttcher, M.E., Escher, P. et al. (2014). Tidal and spatial variations of DI^{13}C and aquatic chemistry in a temperate tidal basin. *Journal of Marine Systems* 129: 396–404.

Xin, P., Robinson, C., Li, L. et al. (2010). Effects of wave forcing on a subterranean estuary. *Water Resources Research* 46: 17.

Xing, Q., Braga, F., Tosi, L. et al. (2016). Detection of low salinity groundwater seeping into the Eastern Laizhou Bay (China) with the aid of Landsat thermal data. *Journal of Coastal Research* 74: 149–156.

Zektzer, I.S., Ivanov, V.A., and Meskheteli, A.V. (1973). The problem of direct groundwater discharge to the seas. *Journal of Hydrology* 20: 1–36.

Zhang, Y.C., Slomp, C.P., Broers, H.P. et al. (2012). Isotopic and microbiological signatures of pyrite-driven denitrification in a sandy aquifer. *Chemical Geology* 300: 123–132.

Zhou, Y., Sawyer, A.H., David, C.H., and Famiglietti, J.S. (2019). Fresh submarine groundwater discharge to the near-global coast. *Geophysical Research Letters* 46: 5855–5863.

Zipperle, A. and Reise, K. (2005). Freshwater springs on intertidal sand flats cause a switch in dominance among polychaete worms. *Journal of Sea Research* 54: 143–150.

Section 3

7

Identifying and Quantifying Water Fluxes at Ecohydrological Interfaces

Christian Schmidt and Jan Fleckenstein

Department of Hydrogeology, Helmholtz-Center for Environmental Research – UFZ, Leipzig, Germany

7.1 Introduction

Water fluxes at ecohydrological interfaces are a key driver for biogeochemical processing and they influence habitat properties. For instance, the direction and the magnitude of water flux in the streambed control the supply of dissolved oxygen into the sediment: the downward flux of stream water supplies dissolved oxygen into the sediment while the upward flux of groundwater, which typically contains low amounts of dissolved oxygen, would result in low oxygen biogeochemical conditions. Accurately characterizing the direction and magnitude of water flow is thus an important prerequisite to adequately understand biogeochemical and ecological patterns (e.g. Gu et al. 2007; Trauth et al. 2015).

This section covers methods which are specifically suitable for the stream–groundwater interface. The point-scale approaches, however, can be applied in the general context of groundwater–surface water interactions, e.g. for lakes.

Water fluxes across the stream–streambed interface are practically never homogeneous in space. Spatial heterogeneity of water flow arises from hydromorphological features such as bedforms or meanders and from sediment heterogeneity of both the streambed and the underlying aquifer (e.g. Kalbus et al. 2009). Capturing this heterogeneity of water flow potentially helps to explain differences in biogeochemical conditions (Fox et al. 2016; Gilmore et al. 2016). Point-scale methods, when applied at multiple locations, provide insight into the spatial pattern of local water flow. Similarly at the reach scale, multiple reaches can be compared with each other (Covino et al. 2011; Patil et al. 2013). Besides spatial patterns, water fluxes at ecohydrological interfaces also vary over time, which also can have a strong influence on the biogeochemical condition. Higher stream discharges can cause transitions between gaining and losing flow conditions in the stream–groundwater connection such that under gaining conditions low dissolved oxygen (DO), dissolved organic carbon (DOC), and nutrient-depleted water is supplied from the groundwater (GW) while under losing conditions streamwater supplies DO and nutrient-rich water to the subsurface (e.g. Trauth et al. 2015).

Ecohydrological Interfaces, First Edition. Edited by Stefan Krause, David M. Hannah, and Nancy B. Grimm.

New automated data collection techniques, in situ sensors with data loggers and new sensing techniques enhance our ability to measure with high frequency over long periods (e.g. Rode et al. 2016). In this chapter we will highlight already existing as well as some emerging techniques to observe water fluxes at the stream–groundwater interfaces over time.

7.2 Point Methods

Point methods are used to detect water fluxes across the stream–groundwater interface at one point or spot. Point methods comprise direct observation of water fluxes with seepage meters, methods based on Darcy's law, as well as natural and artificial tracer techniques. These methods typically rely on the measurement of only a few variables (e.g. hydraulic head in the groundwater, stream stage) and parameters (e.g. hydraulic conductivity of the streambed). The characterization of spatial differences in water fluxes is realized by applying point methods at multiple locations.

7.2.1 Direct Quantification of Water Fluxes across the Sediment–Water Interface

The only technique to directly obtain a quantitative measure of water flux across sediment–water interfaces is via a so-called seepage meter. Seepage meters, as introduced by Lee (1977), are typically constructed from steel or plastic cylinders or drums, which are open at the bottom and closed at the top. A collection bag, which is filled half-way with water, is connected to the main chamber of the seepage meter via a hose. The seepage meter is pushed into the sediments at the water–sediment interface and fluxes into or out of the water-filled chamber are recorded via changes in the water volume in the collection bag. The recorded water volume divided by the cross-sectional area of the seepage meter directly yields the seepage flux across the water–sediment interface in units of L/T. This type of seepage meter has been applied in lakes, reservoirs, estuaries, and streams (e.g. Lee and Cherry 1978; Neumann et al. 2013; Rosenberry and Pitlick 2009; Rosenberry et al. 2012; Woessner and Sullivan 1984). In flowing water, however, seepage meters may disturb the flow field, affecting the ambient hydraulic gradients (Murdoch and Kelly 2003; Rosenberry 2008). To mitigate these effects, the collection bag can be covered with a rigid shelter (Libelo and MacIntyre 1994). Low profile seepage meters which minimally protrude into the water column reduce the disturbance of the flow field (Rosenberry 2008). Flexible, so-called streambed seepage blankets (Gilmore et al. 2016) consist of flexible rubber and are attached to the streambed by metal flanges having a very low profile of about 3 cm.

The use of a collection bag can be avoided by directly measuring the flow rate in the chamber of the seepage meter using high-precision flow meters (e.g. Rosenberry and Morin 2004). When connected to a data logger these instruments are suited for automated measurements of water fluxes over time. Methods to measure the flow rate in seepage meters comprise heat pulse techniques (Taniguchi and Fukuo 1993), ultrasonic flow measurements (Paulsen et al. 2001), dye-dilution (Sholkovitz et al. 2003), and electromagnetic flow sensors (Rosenberry and Morin 2004; Rosenberry et al. 2013).

Combining seepage meters with measurements of vertical head differences or vertical hydraulic gradients is a way to determine how hydraulic heads are influenced by the type of bag (Murdoch and Kelly 2003). With known flow rates and vertical hydraulic gradients the vertical hydraulic conductivity of the sediments can be estimated (Rosenberry and Pitlick 2009, Rosenberry et al. 2012). Since the cross-sectional area of typical seepage meters does not exceed 1 m^2, they can be considered point-scale measurements. Seepage meters provide information on the vertical direction of flow (upward or downward) and also the net water flux over the measuring period. Seepage meters can be used for monitoring water fluxes over long timescales providing that the associated flow sensors or collection bags are accessible for maintenance, i.e. the water level being sufficiently low to reach the instrument.

7.2.2 Methods Based on Darcy's Law

In most cases, streambeds and lake sediments are water saturated. Water flow in saturated sediments can be described by Darcy's law, which relates the hydraulic gradient to the hydraulic conductivity of the sediment. The two variables required to apply the Darcy equation and to solve for the water flux are the hydraulic gradient and the hydraulic conductivity.

7.2.2.1 Vertical Hydraulic Gradients

The basic tool for applying Darcy's law at water–sediment interfaces are piezometers to measure the hydraulic heads. In a typical application, piezometers are driven (often manually) into the submerged streambed or lake sediments. These piezometers typically have a relatively short screened interval of only a few centimetres. The vertical hydraulic gradient (VHG) can be obtained from the difference between the water level in the piezometer and the stream water level outside the piezometer (Δh) divided by the distance between the centre of the piezometer screen and the water sediment interface (Δl). The VHG is then the ratio of the head difference Δh and the distance Δl (VHG $= \Delta h/\Delta l$). A higher water level in the piezometer than in the surface water indicates an upward flux of water and vice versa, a higher water level in the surface water than in the piezometer indicates downward flux. The real flow field is typically three-dimensional but the water flux across the groundwater–surface water interface can often be sufficiently characterized by the one-dimensional, vertical hydraulic gradient. From the VHG one can directly infer the direction of interfacial water flux. The direction of the VHG can vary spatially as a result of the combination of local and regional effects. Temporal variation can occur due to longer term (seasonal) variations in groundwater storage and surface water levels (e.g. stream discharge) or due to short-term changes in the stream stage during single discharge events. Piezometers to observe hydraulic heads can be easily installed (Baxter et al. 2003) and measurements of vertical hydraulics' gradients are straightforward and can be easily automated with pressure transducers and data loggers (e.g. Schmidt et al. 2011). In situations with changing hydraulic conditions (Käser et al. 2009) or predominantly horizontal flow (Rosenberry et al. 2012; Ward et al. 2012) the vertical hydraulic gradient can be difficult to interpret as an indicator of the direction of flow.

7.2.2.2 Hydraulic Conductivity

In order to apply Darcy's law and to estimate water fluxes rather than the direction of flux alone, an estimate of the hydraulic conductivity of the streambed material is required. Hydraulic conductivity can vary over orders of magnitudes for different sediments. Hydraulic conductivities for coarse gravel can be in the range of 10^{-2}–10^{-3} m/s while clayey material has hydraulic conductivities down to 10^{-9} m/s. Using hydraulic tests the hydraulic conductivity can be determined using the same piezometers as for the VHG measurements. Commonly used hydraulic tests are so-called slug tests or variable head tests (e.g. Baxter et al. 2003; Conant 2004; Landon et al. 2001). The concept of a slug test is to cause a quasi-instantaneous change in water level in the piezometer by either adding or removing a "slug" of water to or from the piezometer and then to record the recovery of the water level to its prior equilibrium level over time. The rate of water level change is a function of the hydraulic conductivity of the surrounding material. As an alternative to slug tests, constant head tests can be performed. As the name implies, during these tests an artificially elevated constant head is established in the piezometer by adding water and the volumetric flux of water required to maintain this constant head is measured. Slug tests and constant head tests provide snapshots of hydraulic conductivity in time. Processess such as bioclogging or sediment deposition and erosion can alter hydraulic properties of the sediment (e.g. Newcomer et al. 2016). Independent estimates of water fluxes such as from natural tracers or seepage meters in conjunction with observations of VHGs over time can be used to evaluate the temporal variability of hydraulic conductivity (e.g. Hatch et al. 2010; Mutiti and Levy 2010). In the light of varying hydraulic conductivities, care should also be taken if the magnitude and direction of water fluxes over time are interpreted from VHG. Increased VHG can also be caused by reduced hydraulic conductivity and is not necessarily an indicator of increased water flux. Disagreement between seepage meter measurements and Darcian flux calculations have been reported (e.g. Rosenberry et al. 2012) and were attributed to sources of error in the application of seepage meters (Murdoch and Kelly 2003), differences in measurement scales (point vs. ~$0.25 m^2$) (Kennedy et al. 2009, 2010), and the distinction between total flux and vertical flux at locations with a strong horizontal flow component (Rosenberry et al. 2012). However, general spatial patterns of upward and downward flow have also been found to agree well (Kennedy et al. 2010).

As an alternative to hydraulic testing of the sediment, a grain size analysis can be performed. The determined grain size distribution of a sediment sample can, in turn, be used to obtain an estimate of hydraulic conductivity based on empirical relationships between the hydraulic conductivity and specific metrics of the grain size distribution such as the geometric mean, median, effective diameter, etc. (see Shepherd 1989 for an overview). In general, the more direct estimation of hydraulic conductivities from hydraulic tests should be preferred over grain size based estimates (Eggleston and Rojstaczer 2001). However, grain size analysis provides some information on the textural characteristics of the subsurface material, and the hydraulic conductivity values can be used as a first estimate to guide the design of further hydraulic tests.

7.2.3 Point-scale Tracer Techniques

Tracer methods, using both natural or artificial tracers, can be applied at the point scale to gain information on the local direction and magnitude of water fluxes. The most widely applied natural tracer to estimate water fluxes at the groundwater–surface water interface is heat (e.g. Constantz, 2008). The underlying idea of using heat as a natural tracer is to observe water and sediment temperatures and to invert the observed space-time pattern of temperatures for water flow based on the coupled water and heat flow equations, which account for conductive as well as convective-advective transport of heat. In this book a whole chapter is dedicated to heat tracing methods. For details on the methods the reader is referred to Chapter 8.

7.2.3.1 Other Natural Tracers

In theory, any signal that is transported by flowing water across the water–sediment interface can be used as a natural tracer. For example, natural variations in electrical conductivity (EC) have been applied to trace water flows in gravel bars (Schmidt et al. 2012; Vieweg et al. 2016) and stream–groundwater exchange (Diem et al. 2013; Vogt et al. 2010). In other applications the travel times of water to drinking water production wells from river bank filtration were estimated (Cirpka et al. 2007). EC can be interpreted as a surrogate for solute concentration. Time series of EC in surface waters often exhibit diurnal variations due to photosynthesis- and respiration-induced variations of bicarbonate concentrations (Hayashi et al. 2012; Vogt et al. 2010). These variations can be traced along subsurface flow paths. Similar to temperature, EC can be measured with automated sensors, but specifically developed measuring equipment, which is available for heat tracing, has yet to evolve for EC. The advantage of using EC is that its signal is transported over longer distances than heat, because the heat signal is strongly attenuated as it is not only transported with the flowing water through the sediment pore spaces but also by conduction into the solid matrix. In contrast, EC (and other conservative solute tracers) are exclusively transported with the flowing water and variations in solute concentrations therefore propagate over longer distances and timescales than temperature fluctuations.

To estimate water fluxes or water velocities from natural tracer signals one has to make assumptions about the flow path length. The common conceptual model when using heat as a tracer to quantify groundwater–surface water exchange is that the exchange flow is predominantly vertical and the depth of the temperature sensor in the streambed is roughly equal to the average flow path length. In settings with strong horizontal flow components, estimation of the flow path lengths can be non-trivial. Then it might be reasonable to interpret tracer signals in terms of travel times by estimating the time lag between the signal in the surface water and its occurrence in the sediment.

7.2.3.2 Artificial Tracers

Heat can also be used as an artificial tracer where instead of analysing natural temperature variations an artificial heat pulse or signal is analysed. An elaborate example of such a method for point-scale applications is the so-called heat pulse sensor (Lewandowski et al. 2011). Here, a defined heat signal, in the particular case a heat pulse, is generated and its propagation through the sediment is detected by temperature sensors surrounding the point heat source. In principle, a similar set-up could be used with solute tracers. If these small-scale tracing techniques are applied at multiple locations it is possible to derive a 3D picture of the flow field, such as in Angermann et al. (2012), where the characterization of a complex 3D flow field was demonstrated using the heat pulse sensor in a small sandbed stream. In contrast to natural tracers, artificial tracer experiments cannot simply be used for long-term monitoring of water fluxes. Although it would be possible theoretically and technically to automatically operate and control a defined heat source, it has not been done yet in the field.

In laboratory experiments, imaging of dye tracers can reveal flow patterns across the sediment–water interface. In experiments where a coupled stream–streambed system is represented by a sediment-filled flume with glass walls, imaging of dye tracer fronts has been used to quantify flow rates and to delineate the geometry of subsurface flow paths (Fox et al. 2016; Kessler et al. 2012; Salehin et al. 2004). The dye tracing experiments have been conducted by injecting a dye tracer into the circulating flume water and by subsequently monitoring the penetration of the dye into the sediment using image sequences. This can provide information on local flow velocities and directions as induced by bedforms and sediment heterogeneity (Fox et al. 2016). Moreover, a joint analysis of the local dye patterns and the resulting breakthrough curves of the dye tracer at the system outlet can help to reveal the underlying processes that generate the integral signal, which is typically observed at larger scales when conducting stream tracer tests in the field.

Electrical resistivity measurements in the streambed for observing salt tracers injected into the stream can be used to observe spatial and temporal patterns of solute transport in the streambed (Ward et al. 2010).

7.3 Reach-scale Methods

Reach-scale integral methods can be used when the spatial scale of interest is so large that multiple point measurements become infeasible. Identifying reach-scale water fluxes across the sediment–water interface helps one to better understand larger scale solute patterns (Mallard et al. 2014), to estimate hydrologic mass balances (Covino et al. 2011), and they can be useful to assess whole stream ecological processes in the light of hydrological influences (Hornbach et al. 2015). The reach scale can be defined to comprise stream segments that are typically between some hundreds of metres to a few kilometres in length.

Different to most point-scale methods, which can explicitly account for exchange flows either going upwards or downwards across the interface, the situation becomes more complex when assessing exchange fluxes at the reach scale. The observed water flux across the interface (as, for example, determined from the mass balance obtained from differential gauging) in fact only represents the net flux, but the fraction of stream water that was actually exchanged across the sediment–water interface can be much higher because of multiple additional gains and losses that occur along the observed reach. Similarly hyporheic exchange, where a certain amount of stream water enters the sediment and subsequently returns to the stream, does not change the reach-scale mass balance, but water may have been exchanged across the sediment–water interface several times. In the following section, methods to determine both the net water balance, as well as gains and losses across the sediment–water interface over stream reaches are introduced.

7.3.1 Reach-scale Mass Balance

7.3.1.1 Differential Gauging

Stream discharge measurements at successive stations along a stream reach can be used to estimate the net gaining and losing fluxes between the observation locations. However, this method can only be applied if the difference in discharge is sufficiently large, beyond the error of the discharge measurements itself. This method is thus suitable for longer reaches.

Longer reaches, however, may receive surface water from tributaries, which then also need to be accurately gauged. Stream discharge can be measured by various methods. At stationary gauges the stream stage is typically measured at a well-defined stable cross-section and the discharge is in turn obtained from a well-defined stage-discharge relationship, the so-called rating curve. This rating curve has to first be established and needs to be maintained and possibly adjusted based on regular manual flow measurements (e.g. using area-velocity methods, e.g. Herschy 1999). Such stations typically provide automated records often combined with data telemetry infrastructure. For non-permanent discharge observations, hand-held velocity meters, typically acoustic, magnetic, or older propeller-type meters are used to measure velocities at multiple points across a cross-section of the stream channel representing increments of the total channel cross-sectional area. The volumetric flow through the cross-section, i.e. total stream discharge, is simply obtained by multiplying the subsection areas with the measured average flow velocities for the respective subsections and then summing up all subsectional flows. This method is often called the area-velocity method (Herschy 1999) or simply velocity gauging. To obtain a continuous stream discharge record over time without a stationary gauge requires repeated manual flow measurements (e.g. using area-velocity methods) over a range of discharges and concurrent water level measurements. Based on this data a unique rating curve for this cross-section can be established, which allows the subsequent estimation of discharge on the basis of continuous stage measurements using pressure transducers or other automated water level recorders.

7.3.1.2 Dilution Gauging

A different approach to measure stream discharge is to apply solute tracer injections into the stream. Discharge is estimated based on the dilution of a solute tracer with a specified volume of water, which is given by the stream discharge integrated over the time period of the tracer experiment (Kilpatrick and Cobb 1985). In this method, a known quantity of solute is added to the stream water at the upstream cross-section of the stream reach of interest, and concentrations of the solute are measured at one or more points downstream over time. Discharge is then calculated from the tracer mass balance over the measurement reach according to:

$$M_0 = \int_0^t C(t)Q(t)\mathrm{d}t \tag{7.1}$$

with M_0 being the tracer mass added at the upstream point, $C(t)$ being the measured concentration over time at the downstream point, and $Q(t)$ the discharge over time at the downstream point. For a constant discharge Q in time this simplifies to:

$$Q = \frac{M_0}{\int_0^t C(t)\mathrm{d}t} \tag{7.2}$$

Where Q is the steady-state discharge in the reach. Net change of Q along a stream reach gaining or losing fluxes at the reach scale can then be determined by subtracting the discharges from consecutive reaches.

$$\Delta Q = Q_i - Q_{i-1} \tag{7.3}$$

With differential gauging the estimation of the net inflow or outflow of a stream reach is possible, but not the estimation of the actual gains or losses across the streambed over the reach.

7.3.2 Estimating Gaining and Losing Fluxes at the Reach Scale

The previous section illustrated how mass balances for stream reaches can be obtained. For solute concentrations and solute processing, the amount of gaining and losing fluxes along a reach might be more important than the net flux alone. Here we illustrate how gaining and losing fluxes can be quantified at the reach scale.

7.3.2.1 Dilution Gauging for Quantifying Gaining and Losing Fluxes

In the conventional dilution gauging to estimate stream discharge, experiments are typically performed at short reaches to minimize the loss of tracer because Equation 7.2 is strictly valid only for a full recovery of the injected tracer mass. For this reason the reach lengths for conventional dilution gauging are kept short but need to be long enough to ensure complete mixing (Covino et al. 2011). In contrast, the loss of stream tracer mass, which can be detected over longer stream reaches can be used to quantify water loss from the stream into the sediment. The losing fraction (Q_{L_i}) of the total discharge can be estimated from the tracer recovery rate and the upstream discharge (Q_{i-1}):

$$Q_{L_i} = \frac{M_0}{Q_i \int_0^t C(t)dt} Q_{i-1} \tag{7.4}$$

The water gain along a reach can be calculated as the sum of the net change of Q and the amount of water loss Q_{L_i}.

$$Q_{Gi} = \Delta Q + Q_{Li} \tag{7.5}$$

Of course, the dilution gauging approach can be combined with velocity gauging (Harvey and Wagner 2000) to estimate stream discharge, however, the loss of water can only be determined by a loss of tracer mass.

7.3.2.2 Reach-scale Natural Tracers

7.3.2.2.1 Solute and Gas Tracers

In the previous paragraph it was illustrated how a combination of differential gauging and solute injections can be used to quantify gaining and losing fluxes at the reach scale. However, the gaining component is quantified only indirectly. Thus, any direct information on gaining water fluxes will help to reduce the uncertainty of the estimated flow components. Natural tracers can be used to improve the quantification of groundwater gains to streams. By adding information obtained from natural tracers by means of tracer mass balances, the sensitivity of the differential gauging method can be improved. A high contrast of the natural tracer concentration between the surface water and groundwater is a prerequisite to apply this type of method (McCallum et al. 2012). Natural tracers that have been used in this context include chloride, calcium, magnesium, and sulphate (Cook et al.

2003; Genereux et al. 1993). A tracer particularly suitable for quantifying groundwater inflows to surface water is the noble gas radon. Radon is produced in groundwater as part of the radioactive decay of the uranium-series isotopes. Radon is added to the surface water by groundwater discharge and by hyporheic exchange fluxes (Cook et al. 2006). The half life time of radon is 3.8 days. In surface water radon outgasses to the atmosphere (Schubert et al. 2012). Both the short half life time and the outgassing maintain a high contrast between radon activities in ground and surface water, which makes radon a particularly sensitive tracer for the detection of groundwater discharge zones to streams. High radon activities indicated groundwater inflows in relative short upstream distances. Conversely, lowered radon activities in near surface groundwater indicate surface water infiltration (Hoehn and Von Gunten, 1989).

7.3.2.3 Reach-scale Heat Tracing

Heat tracing methods are primarily used at the point scale. However, fibre-optic distributed temperature sensors (FO-DTS) have enabled temperature measurements along extended fibre-optic cables with ~1 m or even lower spatial resolution. In this chapter we briefly highlight the capabilities of FO-DTS to detect groundwater surface water exchange fluxes at the reach scale. FO-DTS deployed at the streambed surface have been used to detect temperature anomalies which qualitatively indicate groundwater discharge areas, providing that both the groundwater fluxes and the temperature contrast between the groundwater and the surface water are sufficiently high. In such applications the temperature pattern from the FO-DTS have been combined with independent data, e.g. from streambed piezometers to estimate VHGs (Krause et al. 2012); with the analysis of sediment temperature or with thermal infrared imagery (Hare et al. 2015). Analysing spatially distributed FO-DTS data for local temperature anomalies to search for focused groundwater discharge can be regarded as a multi-point approach where point-scale methods and interpretations can be performed at many points in space facilitated by new measuring techniques.

Heat is also used to directly extract reach-scale information. Inverse models of the heat budgets have been used to extract hydrologic components of the heat budget such as groundwater discharge (Loheide and Gorelick 2006) or transient storage (Westhoff et al. 2007, 2011). However, given the large variety of processes that influence surface water temperatures, the isolation of single factors remains challenging (Loheide and Gorelick 2006; Sinokrot and Stefan 1993).

7.3.2.4 Reach-scale Hyporheic Exchange

We have explained how net gains and losses as well as the amount of gaining and losing flow can be determined by applying discharge measurements, stream tracers, and natural tracers. However, these gauging techniques cannot be used to quantify hyporheic exchange flows, because they do not affect the mass balance of water and conservative solutes in the stream.

The interpretation of stream tracer tests is the most common approach to derive information on hyporheic exchange at the reach sale. Here it is not the tracer mass or the tracer recovery, but the shape of the tracer breakthrough curve that is informative about the exchange process. Temporary or so-called "transient" storage of tracer in the hyporheic zone (or other storage zones like instream eddies or pockets and zones of stagnant water) results in an extended tailing of the breakthrough curve. From the stream tracer information alone it is not readily possible to separate hyporheic exchange from other storage processes, such as surface transient storage in channel side cavities (Jackson et al. 2014). Information from the tracer tests can then be extracted by fitting a tracer transport model, typically a 1D advective-dispersive model with an additional transient storage term to the data (Bencala 1983), which then yields a set of optimized model parameters that are related to, for example, water fluxes between residence time in the storage zone, or size of the storage zone. A key challenge is to separate the individual processes such as hyporheic storage and surface transient storage, which is practically impossible from the tracer breakthrough curve alone. Additional data such as in situ tracer concentration in the sediment can help to constrain parameter estimates (e.g. Harvey and Wagner 2000).

Alternatively to fitting a model to the observed data, a set of metrics can be derived directly from the tracer breakthrough curve, which may provide a more unbiased estimate of transient storage and allows a more robust comparison between different reaches without the need for an a priori defined conceptual model (Ward et al. 2013).

7.3.2.5 Modelling Interfacial Water Flow at the Reach Scale

Numerical models of coupled groundwater–surface water flow are another tool that can be used to quantify and investigate groundwater–surface water exchange fluxes and processes. The reach scale, typically between some hundreds of metres to a few kilometres, is sufficiently small that computationally demanding numerical models can be applied to estimate water fluxes accross the groundwater–surface water interface. A key challenge in the parameterization of such models is an adequate definition of the hydraulic properties of the interface, namely the hydraulic conductivities of the streambed. Often, streambed hydraulic conductivities are estimated based on a calibration of the model to observed fluxes, but this approach can be error prone depending on the quality of the calibration data.

Numerical flow models can either be used to assess field conditions at a specific field site (e.g. Brookfield et al. 2009; Fleckenstein et al. 2006; Wroblicky et al. 1998), or they can be used to evaluate generic processes and the mechanistic functioning of coupled stream–groundwater systems (e.g. Frei et al. 2010), or to test the sensitivity of a target variable (e.g. groundwater exchange with a stream), to changes in boundary conditions, system characteristics, or input parameters (Frei et al. 2009; Kalbus et al. 2009). For example, Kalbus et al. (2009) tested the influence of different parameterizations of streambed hydraulic properties on the magnitude and spatial patterns of groundwater discharge to a stream, and Frei et al. (2009) systematically evaluated the effects of different spatial arrangements of alluvial hydrofacies on stream water losses to an alluvial aquifer with deep water table.

7.3.3 Beyond the Reach Scale: Characterizing Interfacial Flux at the Watershed Scale

Reach-scale gains and losses elucidate the local hydrologic interconnection between streams and groundwater and the associated influence on stream solute concentrations. However, the solute concentrations at an observation point or a short reach are also affected by processes that take place along the entire stream network. Hence, information on gaining and losing fluxes as well as hyporheic exchange in the entire catchment, are useful for an improved understanding of catchment-scale solute dynamics.

An approach to extend from the reach scale to catchment scale is to relate data on the reach scale gaining and losing fluxes with catchment properties (see Mallard et al. 2014). They empirically described a positive relationship between the net change of Q along a reach and the contributing catchment area (the size of the flow contributing hillslope) to that reach. Moreover, they found a power law relationship where the fraction of losing fluxes decreases with increasing stream discharge (stream size). With these relationships, gains and losses can be estimated for the entire stream network with reach-scale resolution.

At catchment scale, the different temporal and spatial origins of streamflow components can be disentangled by using natural isotopic and geochemical tracers. The prerequisite is that the concentration contrast between the stream water and the subsurface water is sufficiently large. A combination of various tracers and hydrologic data likely yields the most reliable results.

Simplified, yet physically-based, spatially explicit models can be applied at large scales to quantify interfacial water fluxes. One example is the quantification of lateral, meander-driven hyporheic exchange (Kiel and Bayani Cardenas 2014) and vertical bedform driven hyporheic exchange (Gomez-Velez et al. 2015). These studies incorporated available data on stream discharge, alluvium permeability, and stream morphology. For small to meso-scale catchments also complex partial-differential-equation (PDE)-based numerical models of groundwater–surface water exchange have been applied (e.g. Huntington and Niswonger 2012; Jones et al. 2008; Yang et al. 2018). However, the computational demands of those types of models clearly inhibit their general application to larger, regional scales, unless extensive simplifying assumptions are made in model set-up and parameterization (Lemieux et al. 2008).

References

Angermann, L., Krause, S., and Lewandowski, J. (2012). Application of heat pulse injections for investigating shallow hyporheic flow in a lowland river. *Water Resources Research* 48.

Baxter, C., Hauer, F.R., and Woessner, W.W. (2003). Measuring groundwater-stream water exchange: new techniques for installing minipiezometers and estimating hydraulic conductivity. *Transactions of the American Fisheries Society* 132: 493–502.

Bencala, K.E. (1983). Simulation of solute transport in a mountain pool-and-riffle stream with a kinetic mass transfer model for sorption. *Water Resources Research* 19: 732–738.

Brookfield, A.E., Sudicky, E.A., Park, Y.-J., and Conant Jr., B. (2009). Thermal transport modelling in a fully integrated surface/subsurface framework. *Hydrological Processes* 23: 2150–2164. https://doi.org/10.1002/hyp.7282.

Cirpka, O.A., Fienen, M.N., Hofer, M. et al. (2007). Analyzing bank filtration by deconvoluting time series of electric conductivity. *Ground Water* 45: 318–328.

Conant, B. (2004). Delineating and quantifying ground water discharge zones using streambed temperatures. *Ground Water* 42: 243–257.

Constantz, J. (2008). Heat as a tracer to determine streambed water exchanges. *Water Resources Research* 44: 20.

Cook, P.G., Favreau, G., Dighton, J.C., and Tickell, S. (2003). Determining natural groundwater influx to a tropical river using radon, chlorofluorocarbons and ionic environmental tracers. *Journal of Hydrology* 277: 74–88.

Cook, P.G., Lamontagne, S., Berhane, D., and Clark, J.F. (2006). Quantifying groundwater discharge to Cockburn River, southeastern Australia, using dissolved gas tracers 222Rn and SF6. *Water Resources Research* 42: W10411.

Covino, T., McGlynn, B., and Mallard, J. (2011). Stream-groundwater exchange and hydrologic turnover at the network scale. *Water Resources Research* 47: W12521.

Diem, S., Cirpka, O.A., and Schirmer, M. (2013). Modeling the dynamics of oxygen consumption upon riverbank filtration by a stochastic-convective approach. *Journal of Hydrology* 505: 352–363.

Eggleston, J. and Rojstaczer, S. (2001). The value of grain-size hydraulic conductivity estimates: comparison with high resolution in-situ field hydraulic conductivity. *Geophysical Research Letters* 28: 4255–4258.

Fleckenstein, J.H., Niswonger, R.G., and Fogg, G.E. (2006). River-aquifer interactions, geologic heterogeneity, and low flow management. *Ground Water* 44 (6): 837–852. https://doi.org/10.1111/j.1745-6584.2006.00190.x.

Fox, A., Laube, G., Schmidt, C. et al. (2016). The effect of losing and gaining flow conditions on hyporheic exchange in heterogeneous streambeds. *Water Resources Research* 52: 7460–7477.

Frei, S., Fleckenstein, J.H., Kollet, S.J., and Maxwell, R.M. (2009). Patterns and dynamics of river-aquifer exchange with variably-saturated flow using a fully-coupled model. *Journal of Hydrology* 375: 383–393.

Frei, S., Lischeid, G., and Fleckenstein, J.H. (2010). Effects of micro-topography on surface-subsurface exchange and runoff generation in a virtual riparian wetland - A modeling study. *Advances in Water Resources* 33: 1388–1401.

Genereux, D.P., Hemond, H.F., and Mulholland, P.J. (1993). Use of radon-222 and calcium as tracers in a three-end-member mixing model for streamflow generation on the West Fork of Walker Branch Watershed. *Journal of Hydrology* 142: 167–211.

Gilmore, T.E., Genereux, D.P., Solomon, D.K. et al. (2016). Quantifying the fate of agricultural nitrogen in an unconfined aquifer: stream-based observations at three measurement scales. *Water Resources Research* 52: 1961–1983.

Gomez-Velez, J.D., Harvey, J.W., Cardenas, M.B., and Kiel, B. (2015). Denitrification in the Mississippi River network controlled by flow through river bedforms. *Nature Geoscience* 8: 941–945.

Gu, C., Hornberger, G.M., Mills, A.L. et al. (2007). Nitrate reduction in streambed sediments: effects of flow and biogeochemical kinetics. *Water Resources Research* 43: W12413.

Hare, D.K., Briggs, M.A., Rosenberry, D.O. et al. (2015). A comparison of thermal infrared to fiber-optic distributed temperature sensing for evaluation of groundwater discharge to surface water. *Journal of Hydrology* 530: 153–166. https://doi.org/10.1016/j.jhydrol.2015.09.059.

Harvey, J.W. and Wagner, B.J. (2000). *Quantifying Hydrologic Interactions between Streams and Their Subsurface Hyporheic Zones*. Streams and Ground Waters. 4–41. San Diego, San Diego: Academic Press.

Hatch, C.E., Fisher, A.T., Ruehl, C.R., and Stemler, G. (2010). Spatial and temporal variations in streambed hydraulic conductivity quantified with time-series thermal methods. *Journal of Hydrology* 389: 276–288.

Hayashi, M., Vogt, T., Maechler, L., and Schirmer, M. (2012). Diurnal fluctuations of electrical conductivity in a pre-alpine river: effects of photosynthesis and groundwater exchange. *Journal of Hydrology* 450: 93–104.

Herschy, R.W. (1999). *Hydrometry: Principles and Practice*. John Wiley & Sons Ltd.

Hoehn, E. and Von Gunten, H.R. (1989). Radon in groundwater: a tool to assess infiltration from surface waters to aquifers. *Water Resources Research* 25: 1795–1803.

Hornbach, D.J., Beckel, R., Hustad, E.N. et al. (2015). The influence of riparian vegetation and season on stream metabolism of Valley Creek, Minnesota. *Journal of Freshwater Ecology* 30: 569–588.

Huntington, J.L. and Niswonger, R.G. (2012). Role of surface-water and groundwater interactions on projected summertime streamflow in snow dominated regions: an integrated modeling approach. *Water Resources Research* 48: W11524.

Jackson, T.R., Apte, S.V., and Haggerty, R. (2014). Effect of multiple lateral cavities on stream solute transport under non-Fickian conditions and at the Fickian asymptote. *Journal of Hydrology* 519: 1707–1722.

Jones, J.P., Sudicky, E.A., and McLaren, R.G. (2008). Application of a fully-integrated surface-subsurface flow model at the watershed-scale: a case study. *Water Resources Research* 44: W03407.

Kalbus, E., Schmidt, C., Molson, J.W. et al. (2009). Influence of aquifer and streambed heterogeneity on the distribution of groundwater discharge. *Hydrology and Earth System Sciences* 13: 69–77.

Käser, D.H., Binley, A., Heathwaite, A.L., and Krause, S. (2009). Spatio-temporal variations of hyporheic flow in a riffle-step-pool sequence. *Hydrological Processes* 23: 2138–2149.

Kennedy, C.D., Genereux, D.P., Corbett, D.R., and Mitasova, H. (2009). Spatial and temporal dynamics of coupled groundwater and nitrogen fluxes through a streambed in an agricultural watershed. *Water Resources Research* 45: W09401.

Kennedy, C.D., Murdoch, L.C., Genereux, D.P. et al. (2010). Comparison of Darcian flux calculations and seepage meter measurements in a sandy streambed in North Carolina, United States. *Water Resources Research* 46: W09501.

Kessler, A.J., Glud, R.N., Cardenas, M.B. et al. (2012). Quantifying denitrification in rippled permeable sands through combined flume experiments and modeling. *Limnology and Oceanography* 57: 1217–1232.

Kiel, B.A. and Bayani Cardenas, M. (2014). Lateral hyporheic exchange throughout the Mississippi River network. *Nature Geoscience* 7: 413–417.

Kilpatrick, F.A. and Cobb, E.D. (1985). *Measurement of Discharge Using Tracers*. Department of the Interior, US Geological Survey.

Krause, S., Blume, T., and Cassidy, N.J. (2012). Investigating patterns and controls of groundwater up-welling in a lowland river by combining fibre-optic distributed temperature sensing with observations of vertical hydraulic gradients. *Hydrology and Earth System Sciences* 16: 1775–1792.

Landon, M.K., Rus, D.L., and Harvey, F.E. (2001). Comparison of instream methods for measuring hydraulic conductivity in sandy streambeds. *Ground Water* 39: 870–885.

Lee, D. (1977). Device for measuring seepage flux in lakes and estuaries. *Limnology and Oceanography* 22: 140–147.

Lee, D. and Cherry, J.A. (1978). A field exercise on groundwater flow using seepage meters and mini-piezometers. *Journal of Geological Education* 27: 6–10.

Lemieux, J.-M., Sudicky, E.A., Peltier, W.R., and Tarasov, L. (2008). Simulating the impact of glaciations on continental groundwater flow systems: 1. Relevant processes and model formulation. *Journal of Geophysical Research-Earth Surface* 113: F03017.

Lewandowski, J., Angermann, L., Nuetzmann, G., and Fleckenstein, J.H. (2011). A heat pulse technique for the determination of small-scale flow directions and flow velocities in the streambed of sand-bed streams. *Hydrological Processes* 25: 3244–3255.

Libelo, E.L. and MacIntyre, W.G. (1994). Effects of surface-water movement on seepage-meter measurements of flow through the sediment-water interface. *Applied Hydrogeology* 2: 49–54.

Loheide, S.P. and Gorelick, S.M. (2006). Quantifying stream-aquifer interactions through the analysis of remotely sensed thermographic profiles and in situ temperature histories. *Environmental Science & Technology* 40: 3336–3341.

Mallard, J., McGlynn, B., and Covino, T. (2014). Lateral inflows, stream-groundwater exchange, and network geometry influence stream water composition. *Water Resources Research* 50: 4603–4623.

McCallum, J.L., Cook, P.G., Berhane, D. et al. (2012). Quantifying groundwater flows to streams using differential flow gaugings and water chemistry. *Journal of Hydrology* 416–417: 118–132.

Murdoch, L.C. and Kelly, S.E. (2003). Factors affecting the performance of conventional seepage meters. *Water Resources Research* 39.

Mutiti, S. and Levy, J. (2010). Using temperature modeling to investigate the temporal variability of riverbed hydraulic conductivity during storm events. *Journal of Hydrology* 388: 321–334.

Neumann, C., Beer, J., Blodau, C. et al. (2013). Spatial patterns of groundwater-lake exchange – implications for acid neutralization processes in an acid mine lake. *Hydrological Processes* 27: 3240–3253.

Newcomer, M.E., Hubbard, S.S., Fleckenstein, J.H. et al. (2016). Simulating bioclogging effects on dynamic riverbed permeability and infiltration. *Water Resources Research* 52: 2883–2900.

Patil, S., Covino, T.P., Packman, A.I. et al. (2013). Intrastream variability in solute transport: hydrologic and geomorphic controls on solute retention. *Journal of Geophysical Research-Earth Surface* 118: 413–422.

Paulsen, R.J., Smith, C.F., O'Rourke, D., and Wong, T.F. (2001). Development and evaluation of an ultrasonic ground water seepage meter. *Ground Water* 39: 904–911.

Rode, M., Halbedel née Angelstein, S., Anis, M.R. et al. (2016). Continuous in-stream assimilatory nitrate uptake from high-frequency sensor measurements. *Environmental Science & Technology* 50: 5685–5694.

Rosenberry, D.O. (2008). A seepage meter designed for use in flowing water. *Journal of Hydrology* 359: 118–130.

Rosenberry, D.O., Klos, P.Z., and Neal, A. (2012). In situ quantification of spatial and temporal variability of hyporheic exchange in static and mobile gravel-bed rivers. *Hydrological Processes* 26 (4): 604–612. https://doi.org/10.1002/hyp.8154.

Rosenberry, D.O. and Morin, R.H. (2004). Use of an electromagnetic seepage meter to investigate temporal variability in lake seepage. *Ground Water* 42: 68–77.

Rosenberry, D.O. and Pitlick, J. (2009). Effects of sediment transport and seepage direction on hydraulic properties at the sediment-water interface of hyporheic settings. *Journal of Hydrology* 373: 377–391.

Rosenberry, D.O., Sheibley, R.W., Cox, S.E. et al. (2013). Temporal variability of exchange between groundwater and surface water based on high- frequency direct measurements of seepage at the sediment- water interface. *Water Resources Research* 49: 2975–2986.

Salehin, M., Packman, A., and Zaramella, M. (2004). Hyporheic exchange with gravel beds: basic hydrodynamic interactions and bedform-induced advective flows. *Journal of Hydraulic Engineering-ASCE* 130: 647–656.

Schmidt, C., Martienssen, M., and Kalbus, E. (2011). Influence of water flux and redox conditions on chlorobenzene concentrations in a contaminated streambed. *Hydrological Processes* 25: 234–245.

Schmidt, C., Musolff, A., Trauth, N. et al. (2012). Transient analysis of fluctuations of electrical conductivity as tracer in the stream bed. *Hydrology and Earth System Sciences* 16: 3689–3697.

Schubert, M., Paschke, A., Lieberman, E., and Burnett, W.C. (2012). Air–water partitioning of 222Rn and its dependence on water temperature and salinity. *Environmental Science & Technology* 46: 3905–3911.

Shepherd, R.G. (1989). Correlations of permeability and grain size. *Ground Water* 27: 633–638.

Sholkovitz, E., Herbold, C., and Charette, M. (2003). An automated dye-dilution based seepage meter for the time-series measurement of submarine groundwater discharge. *Limnology and Oceanography-Methods* 1: 16–28.

Sinokrot, B. and Stefan, H. (1993). Stream temperature dynamics - measurements and modeling. *Water Resources Research* 29: 2299–2312.

Taniguchi, M. and Fukuo, Y. (1993). Continuous measurements of groundwater seepage using an automatic seepage meter. *Ground Water* 31: 675–679.

Trauth, N., Schmidt, C., Vieweg, M. et al. (2015). Hydraulic controls of in-stream gravel bar hyporheic exchange and reactions. *Water Resources Research* 51: 2243–2263.

Vieweg, M., Kurz, M.J., Trauth, N. et al. (2016). Estimating time-variable aerobic respiration in the streambed by combining electrical conductivity and dissolved oxygen time series. *Journal of Geophysical Research: Biogeosciences* 121: 2016JG003345.

Vogt, T., Hoehn, E., Schneider, P. et al (2010). Fluctuations of electrical conductivity as a natural tracer for bank filtration in a losing stream. *Advances in Water Resources* 33: 1296–1308.

Ward, A.S., Fitzgerald, M., Gooseff, M.N. et al. (2012). Hydrologic and geomorphic controls on hyporheic exchange during base flow recession in a headwater mountain stream. *Water Resources Research* 48.

Ward, A.S., Gooseff, M.N., and Singha, K. (2010). Imaging hyporheic zone solute transport using electrical resistivity. *Hydrological Processes* 24: 948–953.

Ward, A.S., Payn, R.A., Gooseff, M.N. et al. (2013). Variations in surface water-ground water interactions along a headwater mountain stream: comparisons between transient storage and water balance analyses. *Water Resources Research* 49: 3359–3374.

Westhoff, M.C., Gooseff, M.N., Bogaard, T.A., and Savenije, H.H.G. (2011). Quantifying hyporheic exchange at high spatial resolution using natural temperature variations along a first-order stream. *Water Resources Research* 47.

Westhoff, M.C., Savenije, H.H.G., Luxemburg, W.M.J. et al. (2007). A distributed stream temperature model using high resolution temperature observations. *Hydrology and Earth System Sciences* 11: 1469–1480.

Woessner, W. and Sullivan, K. (1984). Results of seepage meter and mini-piezometer study, Lake Mead, Nevada. *Ground Water* 22: 561–568.

Wroblicky, G.J., Campana, M.E., Valett, H.M., and Dahm, C.N. (1998). Seasonal variation in surface-subsurface water exchange and lateral hyporheic area of two stream-aquifer systems. *Water Resources Research* 34: 317–328.

Yang, J., Heidbüchel, I., Musolff, A. (2018). Exploring the dynamics of transit times and subsurface mixing in a small agricultural catchment. *Water Resources Research* 54: 2317–2335. https://doi.org/10.1002/2017WR021896.

8

Heat as a Hydrological Tracer

Christian Schmidt[1], *Jörg Lewandowski*[2], *J. N. Galloway*[3], *Athena Chalari*[4], *Francesco Ciocca*[4], *Stefan Krause*[5], *Laurant Pfister*[6], and *M. Antonelli*[6,7]

[1] *Department of Hydogeology, Helmholtz Centre for Environmental Hydrology Research, Germany*
[2] *Lebniz Institute of Freshwater Ecology and Inland Fisheries, Berlin, Germany*
[3] *Department Ecohydrology, Leibniz-Institute of Freshwater Ecology and Inland Fisheries, Berlin, Germany*
[4] *Silixa, Watford, England, UK*
[5] *School of Geography, Earth and Environmental Sciences, University of Birmingham, Edgbaston, Birmingham, B15 2TT, UK*
[6] *Luxembourg Institute of Science and Technology, Department Environmental Research and Innovation, Luxembourg*
[7] *Wageningen University, Department of Environmental Sciences, Hydrology and Quantitative Water Management Group, The Netherlands*

8.1 Heat Transfer Mechanisms at Ecohydrological Interfaces (Advection, Conduction, and Radiation)

Ecohydrological interfaces comprise liquid–solid interfaces such as the stream–sediment interface, gas–solid interfaces such as the soil–atmosphere interface, liquid–liquid interfaces such as a thermocline in a lake, or gas–liquid interfaces such as the lake–atmosphere interface.

The interface at which heat as a tracer is mostly applied is the sediment–water interface at streams, lakes, and seas. The applicability of heat as a natural tracer is based on the pattern that different heat transfer mechanisms leave on the temperatures of the system of interest. In general, four basic heat transfer mechanisms can occur: conduction, advection, radiation, and latent heat transfer: (1) Heat conduction is energy transfer by microscopic diffusion. (2) Heat transfer by flowing water is often interchangeably defined as convection or advection. In this chapter we will use the term advection because heat at the sediment–water interface is transported by a large-scale flow field. Sometimes convection is specified as forced and free convection. In this light, advection is equal to forced convection. Free convection refers to fluid and heat flow due to temperature-induced density differences. In this chapter the focus is on heat as a tracer for water fluxes and will thus mainly elucidate how to estimate the effects that flowing water has on heat transport and the resulting temperature pattern. (3) Radiative heat transfer is the main mechanism how land and surface water gain solar energy but also lose heat to the atmosphere and cosmos. (4) Latent heat transfer is caused by water evaporating at some point and condensing at another point of

Ecohydrological Interfaces, First Edition. Edited by Stefan Krause, David M. Hannah, and Nancy B. Grimm.

the system and is thus only relevant at interfaces where evaporation and condensation occurs, i.e. it occurs at atmosphere–land interfaces and atmosphere–water interfaces.

Surface water bodies mainly exchange heat with the atmosphere and the subsurface (Webb et al. 2008). Heat exchange with the atmosphere occurs mainly via radiation and latent heat fluxes (Webb and Zhang 1997); however, the actual quantities vary between different climates. Heat transfer with the subsurface between the surface water and the subsurface is driven by heat advection with flowing water and by heat conduction along thermal gradients.

8.2 Difference between Heat as a Tracer and Solute Tracers

It is important to note the underlying difference between heat transport and solute transport in porous media. Both transport mechanisms can be reduced to the same mathematical equation assuming Fickian transport (deMarsily 1986). Fluid flow in saturated porous media is the product of hydraulic conductivity and hydraulic gradient (Darcy's law), which is mathematically the same as heat flow by conduction which is the product of thermal conductivity and temperature gradient (Fourier's law). The difference between solute transport and heat transport arises from the fact that heat is transported through the entire matrix of solids and fluids whereas the transport of solutes is restricted to flow paths through interconnected pores. As a consequence the hydraulic conductivity, which quantifies the resistance of a porous material to water flow, varies by orders of magnitudes depending on the material texture. In contrast, thermal conductivity, which quantifies the resistance of the material to heat flow, varies only a little between different sediments and depends mainly on the volumetric proportion of solids and fluids (porosity). Since heat is transported through the solid fluid matrix, heat signals measured as temperatures decay faster than solute signals, resulting in shorter distances over which the heat signal can be tracked compared to solutes. Hence, solute tracers are a more direct approach for tracing water fluxes than heat.

8.3 Heat as a Hydrological Tracer

Already early in the twentieth century it was recognized that subsurface temperature pattern can be utilized to detect groundwater fluxes (Slichter 1905). The underlying idea of using heat as a tracer for water flow is based on the joint transport of water and heat. Ideally, heat transport by flowing water (advective heat flow) can be inferred from temperature measurements, which can be inverted to derive the flow velocity of the water. The movement of water in the subsurface may significantly alter the temperature fields that we would expect if advection was absent and heat conduction was the only transport process. The actual shape of the temperature distribution is a result of the direction and magnitude of water flow.

An essential prerequisite for the estimation procedure is a temperature difference across the domain of interest, i.e. a temperature difference between stream water and the streambed sediment. Natural temperature differences arise from diurnal or seasonal temperature forcing at the surface. For example, in winter, colder air temperatures result in a cooling of the shallow subsurface. Conversely, warmer summer temperatures yield to subsurface heating. With increasing depth, these annual oscillations are damped until the temperatures remain virtually constant. The penetration depth of the temperature oscillations is a function of their phase length. Long-term surface temperature oscillations will leave fingerprints in greater depths than, for example, diurnal oscillations, which are usually not present below a depth of half a metre. On the other end of the timescale, the surface temperature history of past decades to millennia may be reconstructed from temperature-depth profiles from deep boreholes (e.g. Majorowicz et al. 2006). A thermal oscillation of 1000 years, for instance, can be detected down to a depth of 400 m (Pollack and Huang 2000).

The evaluation of natural temperature patterns is limited to sufficient temperature differences across the domain of interest. Heat might also be used as an artificial tracer where heat transport originating from an active heat source is tracked. Active heat tracing overcomes two key limitations of using natural temperatures differences: (1) insufficient temperature differences across the domain of interest; (2) the propagation of the heat signal can be tracked in three dimensions which is typically not possible for natural signals. A heat pulse tool specifically developed for a groundwater–surface water interface was developed by Lewandowski et al. (2011), which allows for the 3D inversion of water flux magnitudes and directions (Section 8.5.2).

8.4 Qualitative Analysis based on Heat as a Tracer

Temperature differences can be used to identify patterns of groundwater exfiltration over large areas. Thermal infrared imaging (TIR) reveals the temperature distribution of surfaces and allows the determination of the skin temperature of water bodies but not temperatures in a water body itself (Hare et al. 2015). Thus, TIR is useful for very shallow (sections of) water bodies where the entire water column shows a temperature increase due to exfiltration of groundwater (e.g. springs along the shoreline of a river). TIR is also a useful tool for the detection of groundwater exfiltration if strong up-welling of the exfiltrated groundwater in the water column transfers the temperature signal to the water surface (e.g. submarine groundwater discharge in coastal areas where discharging fresh groundwater is lighter than saline ocean water; Johnson et al. 2008; Peterson et al. 2009). TIR images can be taken by handheld cameras, by unmanned aerial vehicles (drones), by aircrafts, and satellites. The main advantage of TIR is that large areas can be easily measured with a high two-dimensional resolution. The main disadvantage of TIR is that only the skin temperature of water bodies is measured, i.e. the temperature at the water–atmosphere interface.

Another technique for pattern identification is distributed temperature sensing (DTS). Measurements are conducted along a line sensor; in contrast to TIR measurements they are not two-dimensional. The main advantage of TIR compared to DTS is that temperature patterns can be measured at the immediate sediment–water interface (Hare et al. 2015). The main disadvantages of DTS are the expensive equipment and the enormous effort required for installation. More details about DTS are provided in Section 8.6.3. Both techniques, TIR and DTS, are used to guide further sampling campaigns and conduct point measurements either at typical, representative sites or at sites with the highest exfiltration rates. A few studies even tried to derive calibrated transfer functions and calculate exfiltration rates along the DTS sensor (Blume et al. 2013).

8.5 Quantitative Analysis based on Heat as a Tracer

The transport of heat in a saturated porous medium taking into account advection and conduction was derived by Stallman (1963). When heat is used as natural tracer the heat-advection-conduction equation is typically applied in its one-dimensional form:

$$\frac{K_{fs}}{\rho c}\frac{\partial^2 T}{\partial z^2} - qz\frac{\rho fcf}{\rho c}\frac{\partial T}{\partial z} = \frac{\partial T}{\partial t} \tag{8.1}$$

where T [°C] is the measured temperature at depth, z is depth (positive downward) [m]; t is time [s]; qz is the vertical Darcy velocity [ms^{-1}] (positive upward); ρc is the volumetric heat capacity of the solid–fluid system, which can be written as $\rho c = n\rho fcf + (1-n)\rho_s c_s$ where ρfcf is the volumetric heat capacity of the fluid, $\rho_s c_s$ is the volumetric heat capacity of the solids [$Js^{-1}m^{-3}\,°C^{-1}$], and n is the porosity. K_{fs} is the thermal conductivity of the saturated sediment [$Js^{-1}m^{-1}\,°C^{-1}$]. The magnitude and direction of qz is a result of Darcian flow which is controlled by the hydraulic conductivity as the product of the hydraulic conductivity and the total head gradient. In the case of a qz of zero, Equation 8.1 reduces to the Fourier equation of heat conduction.

For the one-dimensional form of the heat-advection-conduction equation, several simple analytical solutions exist for different boundary conditions. The selection of an appropriate solution depends on the specific field conditions and on the aims of the measuring exercise. In this section, two basic examples on quantifying water flow at the stream–groundwater interface will be discussed. The first example considers the so-called mapping or point in time temperature measurements (Section 8.5.1). The second is the analysis of temperature time series and the estimation of vertical water fluxes from the propagation of temperature variations at the surface into the sediment (Section 8.5.2)

8.5.1 Steady-state Point in Time Approach

Here, a temperature sensor is temporally inserted into the sediment and point in time temperature observations are evaluated to estimate the vertical water flux. For a quantitative analysis, this method requires that at least the boundary temperatures in the surface water and the groundwater are known and that sediment temperatures are measured in at least one defined depth. However, a simultaneous observation of sediment temperature at several depths would improve the robustness of the results. Equation 8.1 can be solved analytically to estimate qz (e.g. Bredehoeft and Papadopolus 1965):

$$\frac{T(z)-T_0}{T_L-T_0} = \frac{\exp\left(\dfrac{-qz\rho fcf}{K_{fs}}z\right)-1}{\exp\left(\dfrac{-qz\rho fcf}{K_{fs}}L\right)-1} \tag{8.2}$$

where L [m] is the vertical extent of the domain and $T = T_0$ for $z = 0$, and T_L for $z = L$ are the boundary conditions. The underlying assumption of this approach is that the observed temperatures are at steady state. This assumption is practically never met in field conditions, but particularly for high upward water fluxes from the sediment to the surface and a sufficient temperature difference between surface water and groundwater this approach yields acceptable results. However, steady-state point in time temperature mapping should be regarded as an initial assessment tool to detect the spatial pattern of water fluxes across the groundwater–surface water interface. The advantage of such application is that it easily provides a snapshot in time at tens or hundreds of locations (e.g. Anibas et al. 2011; Schmidt et al. 2006).

8.5.2 Transient Analysis of Temperature Measurements

The most common analytical representation of transient heat transport is a one-dimensional domain with a sinusoidal temperature variation as upper boundary condition and a constant temperature at the lower boundary. The original solution was developed by Stallman (1965) which describes the depth-time distribution of temperatures depending on the temperature fluctuation at the surface and the direction and magnitude of the vertical water flux. This pioneering solution has not been extensively applied until an important evolution of Stallman's original work. Hatch et al. (2006) and Keery et al. (2007) separated the analytical solution into expressions for the amplitude damping and the phase shift of the surface temperature variation as a function of water flux.

This "trick" provides several practical advantages. The separation of phase shift and amplitude provides two independent equations for estimating the water flux. The method only requires the spacing between the sensors and not the absolute depths. Thus, it becomes independent from erosion or scour at the sediment surface. The method works reliably with a single complete sinusoidal oscillation. This makes it quasi transient. The method can be used to detect changes of water fluxes on timescales longer than the underlying signal wavelength, i.e. when the diurnal signal is evaluated and the temperature record captures many days, the changes of water fluxes can be estimated with a diurnal resolution. Amplitudes and phase shifts can be relatively easily extracted from the recorded signal, but it requires preprocessing to extract a single frequency sine wave (usually the diurnal signal) from the recorded signal. The processing and analysis of the temperature time series have been implemented in relatively comprehensive codes such as Ex-Stream (Swanson and Cardenas 2011) or code VFLUX (Gordon et al. 2012).

The typical method of signal extraction is spectral filtering for the frequencies of interest, or a method called Dynamic Harmonic Regression (DHR) (Young et al. 1999), which can also handle non-stationary signals. Although this type of evaluation of temperature time series has been extensively applied, it has some disadvantages. It only uses a single frequency component of the signal and disregards the remaining information of the observed time series. The evaluation procedure can only handle a single pair of observations. Incorporating data from more than two depths would yield more robust estimates of water flow. To overcome these limitations there have been attempts to incorporate the entire frequency spectrum of the recorded signal. Worman et al. (2012) computed the thermal diffusivity vertical water flux from the scaled spectrum of signal frequencies. Vandersteen et al. (2015) provided a method estimating water fluxes from the frequency response function (FRF) and also extended the analysis of temperature time series to boundary conditions that do not rely on a semi-infinite half space (temperature gradient approaches 0 as z goes infinite). This so-called local polynomial method conceptualizes the heat transport problem as a linear time invariant (LTI) system and fits an analytical FRF to the estimated FRF.

8.5.3 Numercial Models for Heat as a Tracer

Analytical methods can be relatively easy to implement but are limited to certain, particularly upper boundary conditions. Unfortunately, temperature time series often exhibit irregular patterns arising from rainstorms and associated flood events, snow-melt, etc. Under such conditions the application of a numerical heat transport model might be favourable. There are several numerical codes available that can simulate coupled water flow and heat transport in forward mode. Examples of such codes are: 2D Code VS2DH (Healy and Ronan 1996) or the 3D code HEATFLOW (Molson and Frind 2005). In these models heat transport is simulated in analogy to solute transport. Here, the Darcy equation is explicitly solved and thus they are easily applicable or extendable to variably saturated conditions. One-dimensional numerical codes specifically developed for simulating and inverting vertical temperature profiles are relatively sparse. Lapham (1989) was the first to develop a numerical routine to solve Equation (8.1) but the estimation of water fluxes was based on manual calibration. Voytek et al. (2014) developed the program 1DTempPro, which is based on VS2DH. In its original version it allowed only a manual calibration (trial and error optimization). A recent update has introduced an automated calibration procedure (Koch et al. 2016), which makes the code a really handy tool for the estimation of water fluxes from observed temperature profiles. Heat as a tracer and inverse numerical procedures such as implemented in 1DTempPro can also be used to estimate temporal variations of hydraulic conductivities due to clogging and scouring, e.g. Cox et al. (2007).

8.6 Measuring Techniques for Heat Tracing at Ecohydrological Interfaces

As mentioned at the beginning of the chapter, automated temperature sensors are relatively cheap and temperatures can be easily measured with high precision and accuracy. Although acquiring temperature data in the field is straightforward, the measuring set-up to use heat as a tracer requires a few basic considerations. The typical measuring set-up aims to collect vertical temperature profiles between the surface water and the sediment. The absolute minimum data requirements are temperature observations at the surface (the sediment–water interface) and at a defined depth in the sediment. However, a more robust interpretation of water fluxes will be obtained by the evaluation of temperature measurements at multiple depths. This would allow the analysis of different sensor pairs with the analytical methods or a joint inversion of the entire vertical profile by a numerical routine.

8.6.1 Point Temperature Sensors

There are a large number of self-contained temperature loggers commercially available starting from ~€20. The simplest versions are single channel loggers. These can be combined to arrays by placing them at defined depths in some rod or by direct installation into the sediment. The benefit of this solution is that the sensors are independent of each other and a failure of one does not affect the others. The drawback is that all sensors have to be read out individually to access the data. Moreover, the sensors have to be removed from the installation site for data reading or battery change (which is not readily possible in all models). This makes this set-up particularly useful for installations for deployment durations where internal data storage and battery capacities (e.g. weeks to a few months) are well beyond the duration of observations and no intermediate data analysis is required. Examples of sensors are the really cheap ibuttons (http://www.ibuttonlink.com/collections/temperature-logging-ibuttons) and the more precise Hobo tidbits (http://www.onsetcomp.com/products/data-loggers/utbi-001). There are also multichannel devices consisting of a lance with several temperature sensors and a data logger that is usually placed at the stream bank. In such a set-up accessing the data and maintaining the batteries is easily possible. A failure of the logger or a damaged cable connection will cause a total loss of data or at least a total stop of data acquisition. When the measuring locations are difficult to access or immediate access to the data is required, such devices can be connected to a remote data transmitter (Schmidt et al. 2014). One example of such a device with high precision is from the company UIT (http://www.uit-gmbh.de/files/download/file/0/1/0/3098.pdf).

Independent of the specific sensor set-up, it should be ensured that there is a good thermal contact between the surrounding medium (i.e. water or sediment) and the temperature sensor and a minimized heat transport along the probe. Thus, plastic or wooden probes should be preferred over metal probes. The sensor at the sediment–water interface or the probe rode protruding into the surface water can be subject to heating by solar radiation and should thus have a light colour. Placing the temperature sensor into a screened piezometer to measure depth-dependent temperatures is absolutely prohibitive because of heat conduction and convection within the water column inside the piezometer (Munz et al. 2011). In order to keep the obstruction of the natural flow field as small as possible, the diameter of the individual sensors or probe should be kept as small as possible which, however, has to be balanced with respect to mechanical stability.

8.6.2 Active Heat Tracing Sensors

Lewandowski et al. (2011) developed a heat pulse sensor (HPS) utilizing heat as an artificial tracer of hyporheic flow in three dimensions yielding flow direction and flow velocity (Lewandowski 2014). Others also used heat pulse injection techniques to measure groundwater flow velocity (Alden and Munster 1997; Ballard 1996; Ballard et al. 1996; Melville et al. 1985), soil water flux (Kamai et al. 2010; Kawanishi 1983; Ren et al. 2000; Yang and Jones 2009), and two-dimensional hyporheic flow (Greswell 2005; Greswell et al. 2008, 2009).

The HPS is basically a rod with a heating source placed inside, for example, 10 cm sediment depth and four to six temperature measurement probes placed on a virtual cylinder of 3.5 cm diameter around the probes with the heating source (Figure 8.1). Each temperature probe consists of four temperature sensors at a distance of 3 cm. A heat pulse of 1 minute duration is emitted by the heating source. The heat pulse travels by conduction and advection to the temperature sensors and breakthrough curves are recorded. A typical measurement lasts up to 1 hour. The dimensions of the HPS were chosen in accordance to the resolution of the biogeochemical sampling methods such as gel probes, dialysis samplers, and multi-level samplers (Lewandowski 2014).

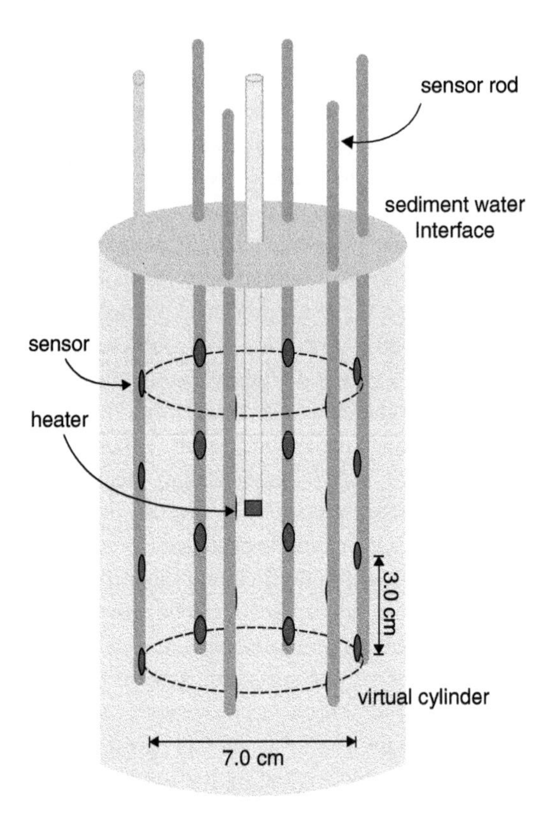

Figure 8.1 Sketch of a heat pulse sensor with the alignment of six temperature probes with four temperature sensors each on a virtual cylinder around the heating rod in its centre. (Figure from Adapted from Angermann et al. 2012 JoH).

Lewandowski et al. (2011) presented a first data analysis protocol for flow direction and flow velocity. Angermann et al. (2012a) developed an improved 3D data analysis routine for a more accurate determination of flow direction and flow velocity. To keep the routine robust and simple enough for a fast analysis of the data obtained from the HPS it relies on an analytical solution of the heat transport problem that was adapted to the specific requirements of a flexible 3D analysis (Angermann et al. 2012a).

The HPS can be used to determine the dominant flow direction and the Darcy (effective) flow velocity. Temperature differences induced by the heating element are compared to the initial temperature and are measured at all sensor points. The sensor with the largest increase in temperature is used as a first estimate for flow direction. As the sensors are not all equidistant, the dispersion of the heat cloud must be accounted for. This is done by using a modified version of the convection-dispersion equation (CDE) (Angermann et al. 2012a).

$$\Delta T(d,r,t) = \frac{Q_i}{8\pi \cdot \frac{D_t}{R} \cdot t \cdot \rho_B C_B \cdot \sqrt{\pi \cdot \frac{D_l}{R} \cdot t}} \cdot \exp - \left\{ \frac{R \cdot \left(d - \frac{t \cdot \nu_w}{R} \right)}{4\mathrm{D}_l \cdot t} - \frac{R \cdot r^2}{4\mathrm{D}_t \cdot t} \right\} \tag{8.3}$$

Where Q_i (kJ) is the initial thermal input, R is the retardation factor (–); $\rho_B C_B$ the volumetric heat capacity (kJ m^{-3} °C^{-1}); D_l and D_t are the longitudinal and transversal dispersion coefficients, respectively (m^2 s^{-1}), v is the thermal energy per volume (kJ m^{-3}); t is the time (s); r and d are the spatial coordinates of the respective sensors (m); v_w is the effective flow velocity (m s^{-1}).

The heater is used as the origin of a cylindrical coordinate system and simulated breakthrough curves are fitted to sensor data for flow direction with parameters vw, Dl, Dt, and Qi (Angermann et al. 2012a). By using the Root Mean Squared Error (RMSE) as an objective function for the fitting algorithm, a measure for the goodness-of-fit will be returned in addition to flow direction and effective flow velocity (m s^{-1}).

The HPS is most suited to coarse sand. The HPS cannot be applied in coarse gravel streams as the coarse sediment prevents the correct positioning of the device. On the other hand, the HPS is not practicable in sediments with very low hydraulic conductivities since resulting flow velocities are extremely small. At very low flow velocities in the order of 10^{-5} m s^{-1}, heat conduction becomes more important than advection and with that the heat pulse sensor is not applicable anymore. Accurate placement of the heating rod and temperature probes is essential, especially for low subsurface flow velocities. A misalignment of the rods will invalidate assumptions regarding the positional set-up of the HPS needed for the analysis of the collected data. Deployment of the HPS in high surface flow velocities is problematic as turbulence created by the HPS apparatus can lead to streambed scouring. Floating material in the water column (e.g. wood, macrophytes) can also cause problems as these debris can become entangled in the HPS apparatus.

In addition to laboratory tests of the HPS (Angermann et al. 2012a; Lewandowski et al. 2011), field applicability of the overall methodology was demonstrated by an application of device and data analysis routine to different field sites, i.e. the River Schlaube in Germany and the River Tern in the UK (Angermann et al. 2012a, 2012b Figure 8.2). For the River

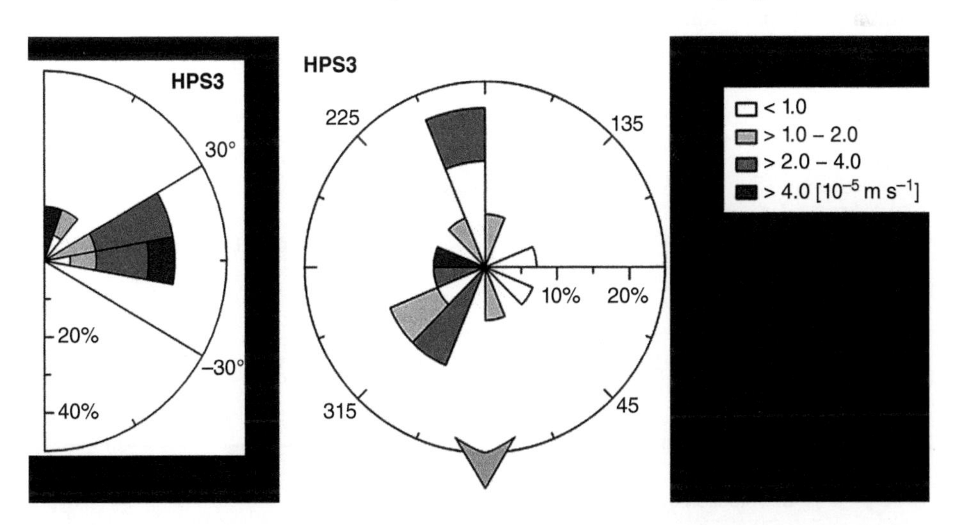

Figure 8.2 Example of a plot summarizing the results from heat pulse measurements at one site of the River Tern, UK (for details see Angermann et al. 2012a, WRR). (a) Relative frequencies (percentage of the whole amount of measurements at one site [%], radial axis) of absolute hyporheic flow velocities (grey scale) and vertical angles (in 20° clusters). (b) Relative frequencies of horizontal hyporheic flow components, i.e. the horizontal projections of flow vectors. The grey scale indicates horizontal velocity and the angles show the horizontal deviation (in 20° clusters) from the main surface flow direction. (Figure from Angermann et al. 2012a, WRR).

Schlaube, Germany (52°05′22.2″N 14°29′00.6″E), a stream with a sandy streambed, the median and maximum flow velocity in the hyporheic zone were determined with the HPS as 0.9×10^{-4} and 2.1×10^{-4} m s^{-1}, respectively. Horizontal flow components were found to be spatially very heterogeneous (Angermann et al. 2012a). At the River Tern, a lowland sandstone river, the HPS was used to determine shallow hyporheic flow at three locations representative for high versus low connectivity between stream and aquifer. Shallow hyporheic flow patterns were spatially heterogeneous. Surface water infiltration and horizontal flow coincided with inhibited groundwater up-welling, whereas locations with high streambed connectivity were characterized by increased up-welling (Angermann et al. 2012b).

8.6.3 Sensing of Space–Time Temperature Pattern with Fibre-optic Distributed Temperature Sensing

Temperature measurements carried out with conventional sensors are typically restricted to a small number of locations where temperature is measured over time. In contrast, spatial patterns of temperatures are frequently obtained by roaming surveys that map point temperatures at different locations consecutively, but not simultaneously. This can be a disadvantage in particular if the observed spatial patterns are not stationary and may change during the monitoring. Fibre-optic distributed temperature sensors (FO-DTS) have the potential to overcome the limitations of point measurements by providing high resolution (spatial and temporal) simultaneous measurements at (ten)thousands of locations and spatial scales of several kilometres.

The basic principle of FO-DTS consists in sending a series of laser pulses from an interrogator unit into a fibre-optic cable. Interactions between the laser pulse and the constituents of the fibre-optic glass lead to the backscatter of a tiny portion of the light towards the interrogator. The inelastic Raman backscatter is temperature dependent (mainly in its Anti-Stokes component, Figure 8.3) therefore, measuring backscattering intensity and travel

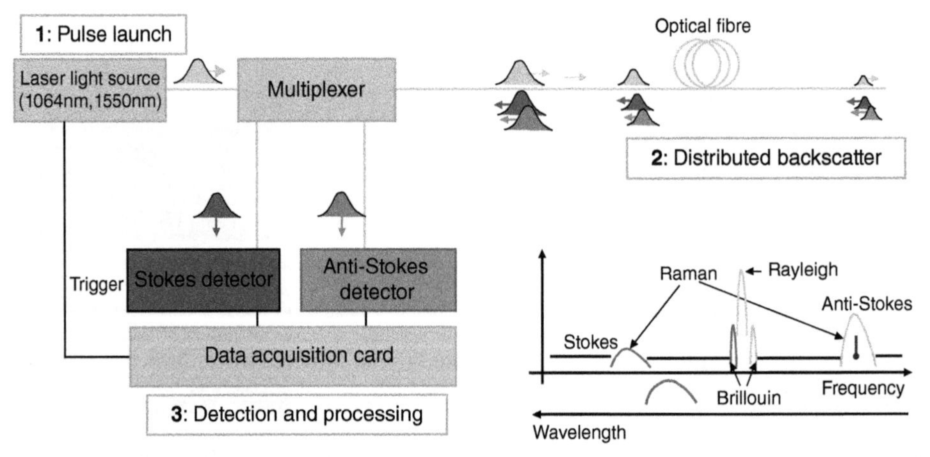

Figure 8.3 Working principle of a FO-DTS interrogator and spectrum of backscattered light inside a fibre-optic cable (bottom right). A typical FO-DTS laser source emits at 1064 nm (distances of up to 10 kilometres) and 1550 nm (above 10 kilometres). Light is backscattered along the fibre-optic cable because of Rayleigh, Brillouin, and Raman effects. Temperature information at each spatial sample along the cable is calculated from the ratio of Raman Anti-Stokes (more temperature sensitive) and Stokes (less sensitive) intensities.

time of the pulse, the interrogator can determine the temperature along the fibre-optic cable (Tyler et al. 2009; J. Selker et al. 2006). Spatial sampling resolution can nowadays be as fine as 12.5 cm, resulting in spatial resolution better than 30 cm. The sensed length of a fibre-optic cable can reach several kilometres (John S. Selker et al. 2006; J. Selker et al. 2006), at measurement intervals down to one second (Sayde et al. 2015; Thomas et al. 2012). However, as the temperature resolution of a FO-DTS is directly proportional to the amount of backscattered light available to the interrogator to perform a measurement, longer acquisition time, coarser spatial sampling, and/or shorter fibre-optic cable are often beneficial to achieve an optimal temperature resolution.

FO-DTS calibration (or configuration) also affects the temperature resolution and, in particular, the temperature accuracy. Analysis of the distributed temperature T as a function of fiber-optic length z from the intensities of Raman anti-Stokes (I_{AS}) and Stokes (I_S) backscatters requires one to determine three parameters: the differential attenuation $\Delta\alpha$, an offset factor K, and a scaling factor γ (Equation 8.4, see also Hausner et al. 2011; van de Giesen et al. 2012).

$$T(z) = \frac{\gamma}{K - \Delta\alpha * z - \ln\left(\frac{I_S(z)}{I_{AS}(z)}\right)} \tag{8.4}$$

$\Delta\alpha$ compensates the natural light losses with distance along the fibre-optic cable. It is expressed in [1/m] or [dB/km]. An incorrectly configured value for $\Delta\alpha$ leads to FO-DTS measurement diverging with distance from the real temperatures. For this reason, the differential attenuation is accounted for as a slope temperature factor, or drift.

K instead is a non-dimensional parameter that acts as an offset factor, constant in space along the entire length of the optical fibre. As K accounts for operating temperature changes of the FO-DTS interrogator, it is generally re-calculated at each temporal measurement. If this is not conducted properly, or even not done at all, measured temperatures will likely present time-varying offsets from real temperatures.

FO-DTS manufacturers set the value of the scaling factor γ [K] during the production of the instrument. γ depends on the properties of the laser used and is not supposed to vary in any conditions. Fluctuations of the scaling factor, indeed, would result in the stretching or compression of the temperature scale, respectively, leading to larger or smaller measurements of real temperature steps.

FO-DTS calibration can be performed either internally by the instrument, or by means of external reference temperatures such as water baths equipped with highly accurate temperature sensors and sections of the applied fibre-optic cable coiled inside. Number and position of the external reference temperatures depend on the number of parameters to be calibrated and on the cable deployment (Figure 8.4). Different deployment strategies applied in FO-DTS surveys require different calibration approaches. A cable deployed in single-ended configuration requires one reference per parameter; for a completely external calibration, two references are required next to each other and to the FO-DTS instrument, at different temperatures (ideally at the lower and upper boundary of the monitored

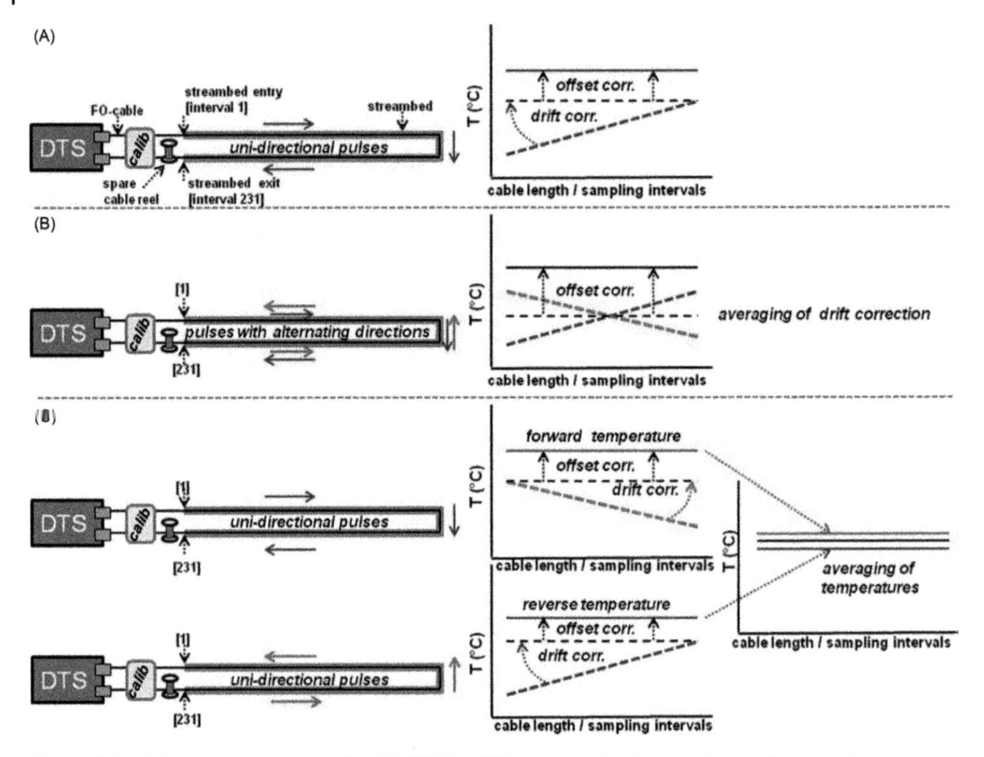

Figure 8.4 Measurement set-up for FO-DTS in different monitoring modes and respective calibration strategies: (a) single-ended monitoring with unidirectional measurements and associated correction of signal drift and offset, (b) double-ended monitoring with bidirectional measurements and associated averaging of bidirectional traces and correction of signal offsets, and (c) two-way single-ended averaging mode with alternating bidirectional single-ended tracing and associated correction of signal drift and offset before averaging the corrected temperature traces. Dashed red and blue lines represent uncorrected traces, and solid lines represent the corrected temperatures (from Adapted from (Krause & Blume 2013).

temperature fluctuations); a third bath should be placed instead at the far end of the fibre-optic cable (Hausner et al. 2011; van de Giesen et al. 2012).

Most cable deployments are conducted in a so-called "U-shaped" design, in which the far end of the cable is connected to the FO-DTS to generate a closed-loop (Figure 8.4), in order to optimize the calibration and reduce the number of external temperature references. "U-shaped" configurations do not necessarily require the entire optical cable to return to the FO-DTS; if the optical cable is equipped with at least two fibre optics (which is usually the case nowadays), their connection (called splicing) at the far end can be sued to create a loop of the then connected fibre ends inside the optical cable. Despite the splicing required and the doubling of the fibre-optic length to sense, the "U-shaped" configurations have proven to sensibly improve the accuracy as well as the resolution of the FO-DTS measurements. Those with a closed-loop, in particular, allow for effective calibration strategies such as double-ended (Figure 8.4b) and alternating two-way single-ended averaging (Figure v.c), respectively, which take advantage of the measurements performed in alternate directions along the same fibre-optic loop. In double-ended calibration, the raw data (Stokes and Anti-Stokes intensities from the two directions) are sample-by-sample combined. This results in: (i) better temperature resolution than the single-ended calibration towards the middle of the fibre-optic loop (i.e. the far end of the monitored area); (ii) a distributed differential attenuation profile with distance (rather than a constant value) that implicitly corrects for spatial changes induced by localized light losses (van de Giesen et al. 2012). Such effects can be, for instance, a result from the presence of fibre-optic junctions (splices) or kinks, often occurring when cables are deployed in harsh environments (e.g. river beds). In alternating two-way single-ended averaging (Krause and Blume 2013), instead the individually calibrated single-ended temperature measurements from alternate directions are averaged sample-by-sample (Figure 8.4c). This calibration strategy improves DTS resolution and accuracy more than the double-ended, in particular, if exact monitoring of location and values of small-scale temperature patterns is of great importance to the respective FO-DTS survey and analysis (Krause and Blume 2013).

FO-DTS has been successfully applied for the monitoring of temperature patterns at a wide range at ecohydrological interfaces (Figure 8.5) and indeed, has experienced substantial technological development and optimization by having to adapt to and deal with the challenges of ecohydrological interface monitoring at a variety of spatiotemporal scales.

The true appeal of FO-DTS lies in its strength of monitoring longitudinal temperature distributions in space and time, which has been used at a variety of soils (Steele-Dunne et al. 2010), streams (Briggs et al. 2012a; Krause et al. 2012; Krause and Blume 2013, 2013; Lowry et al. 2007; Munz et al. 2011; Westhoff et al. 2007), and lakes (Blume et al. 2013; Selker et al. 2006). In addition, in more recent applications, a fibre-optic cable has been tightly wrapped around vertical profiles to enhance the spatial sampling resolution which allowed one to obtain high-resolution measurements at centimetre and sub-centimetre scale along the profile axial direction (Briggs et al. 2012b; Vercauteren et al. 2011; Vogt et al. 2010).

Similar to streambed temperature mapping, the thermal patterns observed along the cable can provide a qualitative indicator of groundwater discharge. Localized discharge zones typically result in cold or warm temperature anomalies in summer or winter if groundwater and surface water temperatures are distinctively different from each other (Krause and Blume 2013) and may also cause a damping of the diurnal stream water temperature amplitude. FO-DTS applications have been conducted along reaches of 300–900 m. Extracting the effects of groundwater discharge or other hydrological factors on the measured temperatures can be challenging given the variety of factors influencing stream temperatures (Sinokrot and Stefan 1993) or temperatures of the fibre-optic cable. For instance, exposure to sunlight can cause erroneous signals, in particular, when black coatings are used (de Jong et al. 2015; Petrides et al. 2011). It is challenging to deploy the fibre-optic cable at the groundwater–surface water interface and assure smooth contact of the cable to the interface even if there are intense bed structures or obstacles on the sediment. Sometimes, the cable is buried to overcome these problems but that requires much effort and one has to ensure that the cable is always buried in the same depth to avoid signal differences due to different burying depths (Krause and Blume 2013).

FO-DTS applications at ecohydrological interfaces have the potential to overcome the space–time restrictions of point sensors. However, it has only been proven in a limited amount of test cases that the gained temperature data at a sediment–water interface, for instance, can actually be evaluated in the same rigid quantitative framework as temperature profiles for providing quantitative assessment of interface exchange fluxes. For

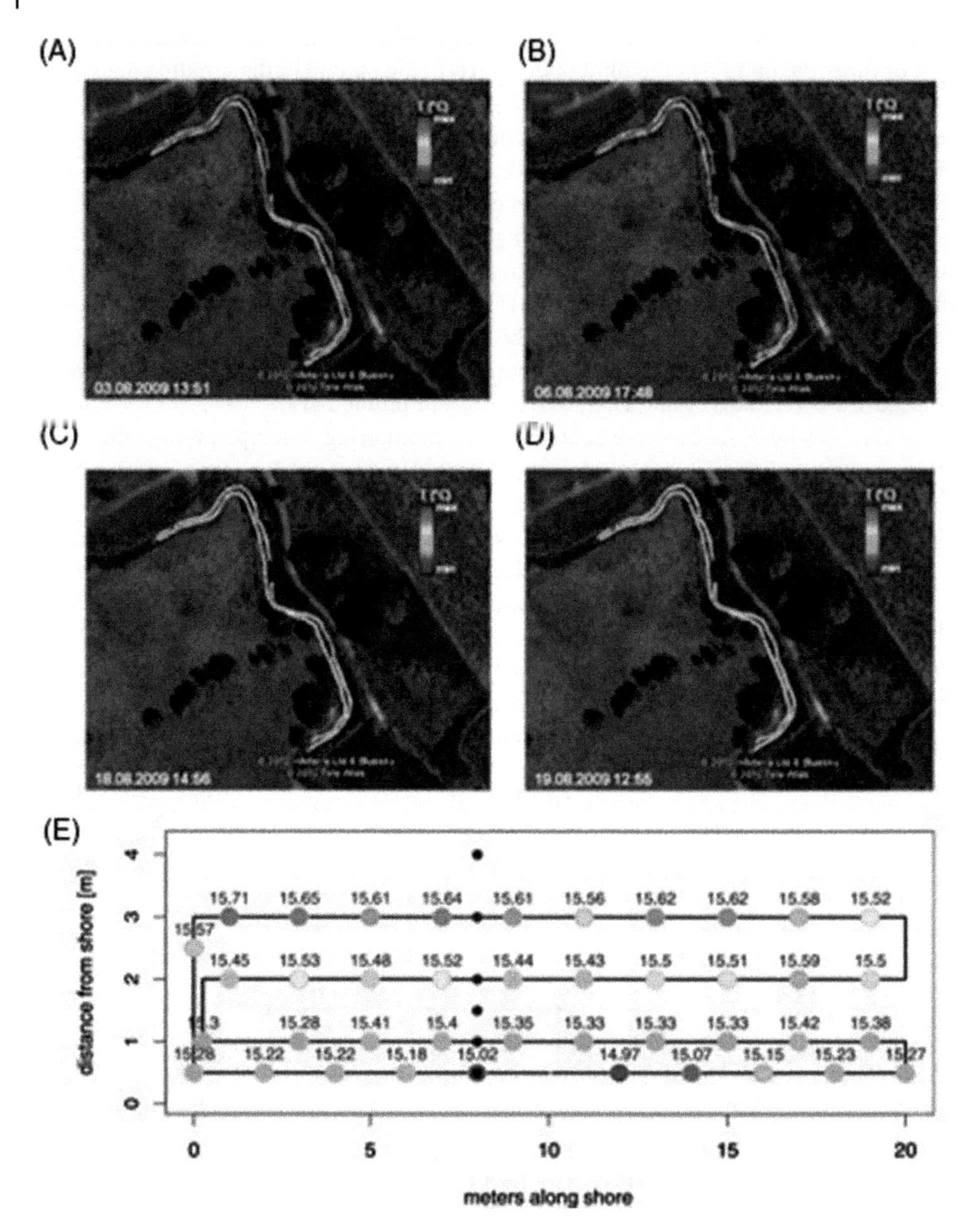

Figure 8.5 FO-DTS monitoring at groundwater–surface water interfaces with observations of groundwater up-welling into a stream (A–D) and lacustrine groundwater discharge to a lake (E) (from Krause et al., 2012 / Hydrology and Earth System Sciences / CC BY 3.0).

example, Blume et al. (2013) successfully developed a transfer function for a FO-DTS grid, monitoring patterns of lacustrine groundwater discharge to a lake but it will require additional efforts to develop a reliable upscaling technique based on FO-DTS. Vertical applications of coiled fibre-optic cables (Briggs et al. 2012b; Vogt et al. 2010) allow really high spatial and temporal resolution; however, the information gain may be justifying the technical efforts only for specific conditions, in particular, if excessive splicing of several vertical profiles in a sensor network is required. Perhaps one of the core strengths of FO-DTS application is in the identification of temperature patterns and specifically groundwater up-welling related temperature anomalies at groundwater–surface water interfaces that then guide detailed point measurements (for instance, by heat pulse sensors, piezometers, or seepage meters) at identified sites of specific interest. A way forward to provide quantitative flow assessment based on FO-DTS applications may be realized by current developments in electrical resistance heating of wires to create a linearly active heat source (Kurth et al. 2012). Actively heating FO-DTS has been applied for soil moisture sensing (Ciocca et al. 2012; Sayde et al. 2010, 2014; Striegl and Loheide 2012) or to monitor borehole flows (Coleman et al. 2015;) but has yet to be established at groundwater–surface water interfaces in streams or lakes.

8.7 Thermal Infrared Imaging

Thermal infrared imagery (TIR) refers to the part of the electromagnetic spectrum comprised between 3 and 14 μm. All objects that have a temperature above the absolute zero emit infrared radiation, which is in turn correlated to that temperature. Thermographic instruments detect and quantify variations in an object's surface radiation emission (i.e. variations of the object's surface temperature) by converting the sensed invisible infrared radiation into visible information (i.e. thermographs).

TIR has been used in hydrology for many decades (Estes 1966). The technology has been first used for its potential to investigate different hydrological processes like the detection of near-shore groundwater discharge (Banks et al. 1996; Pluhowski 1972) and the estimation of soil water content and evaporation (Estes 1966; Reginato et al. 1976; Schmugge 1978). Commissioned by agencies such as the USGS or NASA, most of these studies were based on IR images obtained from thermal infrared scanners mounted on airborne platforms.

Over the past decade, the use of TIR technology has become more widespread as thermal and digital resolutions improved and the measurement accuracy of the IR sensors was significantly increased. Simultaneously, portable IR cameras with affordable prizes have been brought to the market. In particular, the development of increasingly powerful handheld IR cameras has led to an increased number of hydrological studies employing ground-based TIR techniques.

Ground-based TIR imagery is particularly well suited for investigating hydrological processes that require non-invasive in situ observations at fine spatial scale and high temporal resolution (in comparison to airborne observations). To date, several studies have focused on the detection of groundwater exfiltration along the stream, as well as its quantitative estimation (Chen et al. 2009; Schuetz and Weiler 2011), groundwater exfiltration at the seepage face (Deitchman and Loheide 2009), and from macropores and peat pipes (Briggs et al. 2016). Pfister et al. (2010) applied ground-based TIR imagery for monitoring riparian area saturation dynamics and detecting hillslope-riparian-stream connectivity in a small forested catchment in Luxembourg (where airborne surveys were technically not feasible). Portable TIR cameras brought new momentum to the mapping of exact locations of water inflow and water mixing areas (Figure 8.6).

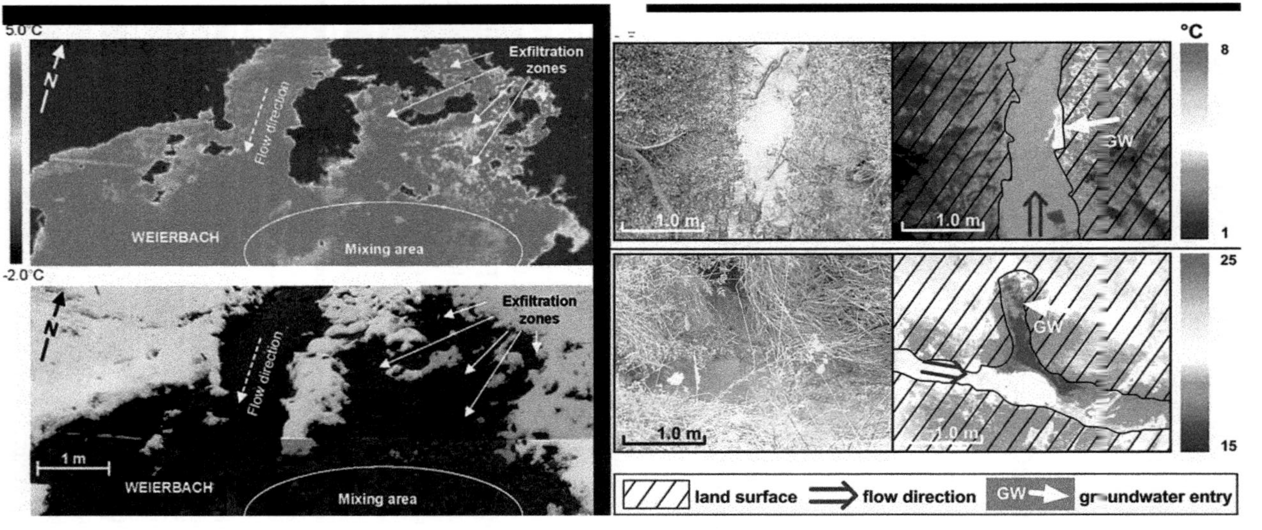

Figure 8.6 Groundwater exfiltration and water mixing revealed via ground-based IR observations. Exfiltrating groundwater exhibits notably different temperatures compared to stream water Reproduced from Pfister et al., 2010 / John Wiley & Sons; Schuetz and Weiler, 2011 / John Wiley & Sons

The combined used of ground-based and airborne TIR observations bears considerable potential, as demonstrated by Cardenas et al. (2011) in their investigation of river-hydrothermal water mixing in the Yellowstone National Park (USA). In that specific context, the complementarity of these two sensing approaches opened new vistas on non-isothermal mixing processes, as well as on in-stream temperature distributions (e.g. between thermal plumes and the bulk flow) at different spatial resolutions (from metres to centimetres). Along similar lines, Cardenas et al. (2014) documented how ground-based TIR imagery can provide key information and data required for calculating an energy balance model of a river. Moffett and Gorelick (2012) used high-resolution TIR images (obtained from a camera mounted on a tower) of a cordgrass and pickleweed canopy located in an intertidal salt marsh in the southern San Francisco Bay to estimate spatially distributed evapotranspiration. This stands as a significant improvement in comparison to the often prevailing – and physically unrealistic – assumption of uniform stomatal resistance values on a canopy exhibiting large temperature contrasts.

As demonstrated by the examples cited above, the potential for ground-based TIR imagery to generate highly valuable datasets in hydrological processes' investigations is now widely recognized. Despite this undeniable potential, almost all studies based on TIR imagery have also reported limitations of that technology. A first constraint consists in the requirement to fix various parameter settings prior to image acquisition: emissivity "ε" of an object (i.e. a measure of how much radiation an object's surface emits compared to that emitted from a blackbody at the same temperature), atmospheric temperature, relative humidity of the air, distance between the object and the instrument, reflected ambient temperature (FLIR 2010) (Figure 8.7).

The emissivity of the targeted study object (in our case, water) is influenced by the observation angle between the object and the camera, when very different from the nadir (i.e. the direction pointing directly below the observer at a particular location). Other factors that may influence the measurements are the roughness of the surface of the water (caused

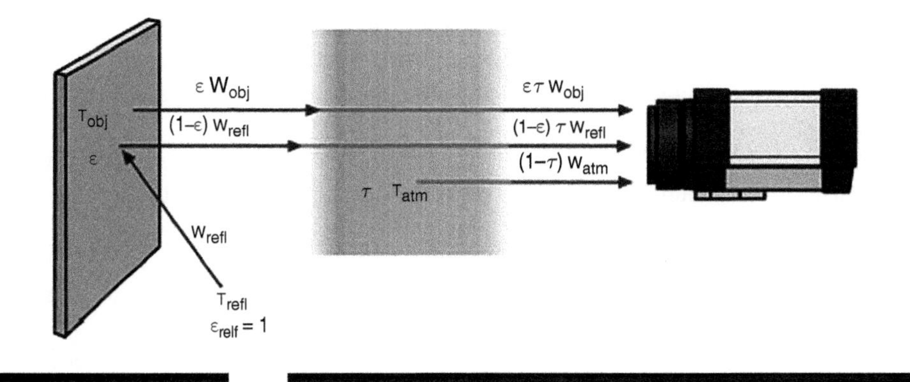

Figure 8.7 The total radiation detected by an IR camera is defined by the formula: $W_{tot} = \varepsilon\tau W_{obj} + (1 - \varepsilon)\tau W_{refl} + (1 - \tau)W_{atm}$, where ε = emissivity, W_{obj} = object's radiation, W_{refl} = reflected radiation, W_{atm} = atmospheric radiation, τ = atmospheric transmittance. T_{obj}, T_{refl}, and T_{atm} indicate, respectively, the temperature of the object, the reflected ambient temperature, and the atmospheric temperature (image modified after FLIR (2010). FLIR B series, FLIR T series User's manual).

by waves, for example), turbidity of the water, or the presence of riffle and foam (Cardenas et al. 2008, 2011; Handcock et al. 2006). Handcock et al. (2006) reported other causes that may also affect the measurements, the thermal reflection of the surrounding objects (translating in patterns of different temperatures detected by the IR camera) and the role of shading and solar irradiance in creating cooler and warmer areas, respectively.

In specific cases, when only small surface temperature differences prevail between different objects of interest (e.g. during seasonal transition), the discretization of objects via thermal IR measurements becomes highly uncertain. Moreover, since the TIR technique is sensitive only to the upper 0.1 mm of the water surface, it is not suited for obtaining information on vertical thermal profiles of the water column.

Additional difficulties may also arise during the data analysis process. Oblique images may turn out to be difficult to rectify and compare with other spatial features. Cardenas et al. (2014) designed a method based on a digital surface model to assign spatial structures to the distorted images. Another source of uncertainty is the assignation of the right temperature information to pixels containing mixed features (i.e. mixed river–river bank pixels) (Cardenas et al. 2014; Torgersen et al. 2001). However, this kind of issue is mainly related to airborne TIR surveys, because of the lower spatial resolution compared to ground-based TIR images. Despite these limitations, ground-based TIR imagery represents a very powerful tool in the hydrological investigations toolbox and its full potential has not been entirely unleashed as yet.

References

Alden, A.S. and Munster, C.L. (1997). Field test of the in situ permeable ground water flow sensor. *Ground Water Monitoring and Remediation* 17: 81–88.

Angermann, L., Krause, S., and Lewandowski, J. (2012a). Application of heat pulse injections for investigating shallow hyporheic flow in a lowland river. *Water Resources Research* 48.

Angermann, L., Lewandowski, J., Fleckenstein, J.H., and Nuetzmann, G. (2012b). A 3D analysis algorithm to improve interpretation of heat pulse sensor results for the determination of small-scale flow directions and velocities in the hyporheic zone. *Journal of Hydrology* 475: 1–11.

Anibas, C., Buis, K., Verhoeven, R. et al. (2011). A simple thermal mapping method for seasonal spatial patterns of groundwater-surface water interaction. *Journal of Hydrology* 397: 93–104.

Ballard, S. (1996). The in situ permeable flow sensor: a ground-water flow velocity meter. *Ground Water* 34: 231–240.

Ballard, S., Barker, G.T., and Nichols, R.L. (1996). A test of the in situ permeable flow sensor at Savannah River, SC. *Ground Water* 34: 389–396.

Banks, W.S.L., Paylor, R.L., and Hughes, W.B. (1996). Using thermal-infrared imagery to delineate ground-water discharge. *Groundwater* 34: 434–443. https://doi.org/10.1111/j.1745-6584.1996.tb02024.x.

Blume, T., Krause, S., Meinikmann, K., and Lewandowski, J. (2013). Upscaling lacustrine groundwater discharge rates by fiber-optic distributed temperature sensing. *Water Resources Research* 49: 7929–7944.

Bredehoeft, J.D. and Papadopolus, I.S. (1965). Rates of vertical groundwater movement estimated from the earth's thermal profile. *Water Resources Research* 1: 325–328.

Briggs, M.A., Hare, D.K., Boutt, D.F. et al. (2016). Thermal infrared video details multiscale groundwater discharge to surface water through macropores and peat pipes. *Hydrological Processes* 30: 2510–2511. https://doi.org/10.1002/hyp.10722.

Briggs, M.A., Lautz, L.K., and McKenzie, J.M. (2012a). A comparison of fibre-optic distributed temperature sensing to traditional methods of evaluating groundwater inflow to streams. *Hydrological Processes* 26: 1277–1290.

Briggs, M.A., Lautz, L.K., McKenzie, J.M. et al. (2012b). Using high-resolution distributed temperature sensing to quantify spatial and temporal variability in vertical hyporheic flux. *Water Resources Research* 48.

Cardenas, M.B., Doering, M., Rivas, D.S. et al. (2014). Analysis of the temperature dynamics of a proglacial river using time-lapse thermal imaging and energy balance modeling. *Journal of Hydrology* 519: 1963–1973. http://doi.org/10.1016/j.jhydrol.2014.09.079.

Cardenas, M.B., Harvey, J.W., Packman, A.I., and Scott, D.T. (2008). Ground-based thermography of fluvial systems at low and high discharge reveals complex thermal heterogeneity driven by flow variation and bioroughness. *Hydrological Process Interfaces with Observations* 22: 980–986.

Cardenas, M.B., Neale, C.M.U., Jaworowski, C., and Heasler, H. (2011). High-resolution mapping of river-hydrothermal water mixing: Yellowstone National Park. *International Journal of Remote Sensing* 32: 2765–2777.

Chen, X.H., Song, J., Cheng, C. et al. (2009). A new method for mapping variability in vertical seepage flux in streambeds. *Hydrogeology Journal* 17: 519–525. https://doi.org/10.1007/s10040-008-0384-0.

Ciocca, F., Lunati, I., Van de Giesen, N., and Parlange, M.B. (2012). Heated optical fiber for distributed soil-moisture measurements: a lysimeter experiment. *Vadose Zone Journal* 11. https://doi.org/10.2136/vzj2011.0199.

Coleman, T.I., Parker, B.L., Maldaner, C.H. and Mondanos, M.J. (2015). Groundwater flow characterization in a fractured bedrock aquifer using active DTS tests in sealed boreholes. *Journal of Hydrology* 528: 449–462.

Cox, M.H., Su, G.W., and Constantz, J. (2007). Heat, chloride, and specific conductance as ground water tracers near streams. *Ground Water* 45: 187–195.

de Jong, S.A.P., Slingerland, J.D., and van de Giesen, N.C. (2015). Fiber optic distributed temperature sensing for the determination of air temperature. *Atmospheric Measurement Techniques* 8: 335–339. https://doi.org/10.5194/amt-8-335-2015.

Deitchman, R.S. and Loheide, S.P. (2009). Ground-based thermal imaging of groundwater up-welling into a flow processes at the seepage face. *Geophysical Research Letters* 36 (June): 1–6. http://doi.org/10.1029/2009GL038103.

deMarsily, G. (1986). *Quantitative Hydrogeology*. San Diego, California: Academic Press.

Estes, J.E. (1966). Some applications of aerial infrared imagery. *Annals of the Association of American Geographers* 56: 673–682. https://doi.org/10.1111/j.1467-8306.1966.tb00584.x.

FLIR. (2010). *FLIR B Series, FLIR T Series, User's Manual*.

Gordon, R.P., Lautz, L.K., Briggs, M.A., and McKenzie, J.M. (2012). Automated calculation of vertical pore-water flux from field temperature time series using the VFLUX method and computer program. *Journal of Hydrology* 420: 142–158.

Greswell, R.B. (2005). *High-resolution in Situ Monitoring of Flow between Aquifers and Surface Waters*. 1–38. Bristol: Environment Agency.

Greswell, R.B., Riley, M.S., Alves, P.F., and Tellam, J.H. (2008). *A Heat Perturbation Flow Meter for Application in River Bed Sediments*. 1–45. Bristol: Environmental Agency.

Greswell, R.B., Riley, M.S., Alves, P.F., and Tellam, J.H. (2009). A heat perturbation flow meter for application in soft sediments. *Journal of Hydrology* 370: 73–82.

Handcock, R.N., Gillespie, A., Cherkauer, K.A. et al. (2006). Accuracy and uncertainty of thermal-infrared remote sensing of stream (A-D) and lacustrine groundwater discharge to a lake (E) (From Krause et al., 2012 and Blume et al., 2013) temperatures at multiple spatial scales. *Remote Sensing of Environment* 100: 427–440.

Hare, D.K., Briggs, M.A., Rosenberry, D.O. et al. (2015). A comparison of thermal infrared to fiber-optic distributed temperature sensing for evaluation of groundwater discharge to surface water. *Journal of Hydrology* 530: 153–166.

Hatch, C.E., Fisher, A.T., Revenaugh, J.S. et al. (2006). Quantifying surface water-groundwater interactions using time series analysis of streambed thermal records: method development. *Water Resources Research* 42: 10410.

Hausner, M.B., Suárez, F., Glander, K.E. et al. (2011). Calibrating single-ended fiber-optic raman spectra distributed temperature sensing data. *Sensors* 11: 10859–10879. https://doi.org/10.3390/s111110859.

Healy, R. and Ronan, A.D. (1996). *Documentation of the Computer Program VS2DH for Simulation of Energy Transport in Variably Saturated Porous Media - Modification of the U.S. Geological Survey's Computer Program VS2DT*. Reston, Virginia: USGS.

Johnson, A.G., Glenn, C.R., Burnett, W.C. et al. (2008). Aerial infrared imaging reveals large nutrient-rich groundwater inputs to the ocean. *Geophysical Research Letters* 35.

Kamai, T., Tuli, A., Kluitenberg, G.J., and Hopmans, J.W. (2008). Soil water flux density measurements near 1 cm d(-1) using an improved heat pulse probe design. *Water Resources Research* 44.

Kamai, T., Tuli, A., Kluitenberg, G.J., and Hopmans, J.W. (2010). Soil water flux density measurements near 1 cm d(-1) using an improved heat pulse probe design. *Water Resources Research 46* 44 (2008): W00D14.

Kawanishi, H. (1983). A soil-water flux sensor and its use for field studies of transfer processes in surface soil. *Journal of Hydrology* 60: 357–365.

Keery, J., Binley, A., Crook, N., and Smith, J.W. (2007). Temporal and spatial variability of groundwater-surface water fluxes: development and application of an analytical method using temperature time series. *Journal of Hydrology* 336: 1–16.

Koch, F.W., Voytek, E.B., Day-Lewis, F.D. et al. (2016). 1DTempPro V2: New features for inferring groundwater/surface-water exchange. *Groundwater* 54 (3): 434–439. http://doi.org/10.1111/gwat.12369.

Krause, S. and Blume, T. (2013). Impact of seasonal variability and monitoring mode on the adequacy of fiber-optic distributed temperature sensing at aquifer-river interfaces. *Water Resources Research* 49 (5): 2408–2423. http://doi.org/10.1002/wrcr.20232.

Krause, S., Blume, T., and Cassidy, N.J. (2012). Investigating patterns and controls of groundwater up-welling in a lowland river by combining fibre-optic distributed temperature sensing with observations of vertical hydraulic gradients. *Hydrology and Earth System Sciences* 16: 1775–1792.

Kurth, A.-M., Dawes, N., Selker, J., and Schirmer, M. (2012). Autonomous distributed temperature sensing for long-term heated applications in remote areas. *Geoscientific Instrumentation, Methods and Data Systems Discuss* 2: 855–873. https://doi.org/10.5194/gid-2-855-2012.

Lapham, W. (1989). *Use of Temperature Profiles beneath Streams to Determine Rates of Vertical Ground-Water and Vertical Hydraulic Conductivity*. USGS.

Lewandowski, J. (2014). *Coupling of Hydrodynamic and Biogeochemical Processes at Aquatic Interfaces*. Berlin, Berlin: Humboldt University.

Lewandowski, J., Angermann, L., Nuetzmann, G., and Fleckenstein, J.H. (2011). A heat pulse technique for the determination of small-scale flow directions and flow velocities in the streambed of sand-bed streams. *Hydrological Processes* 25: 3244–3255.

Lowry, C.S., Walker, J.F., Hunt, R.J., and Anderson, M.P. (2007). Identifying spatial variability of groundwater discharge in a wetland stream using a distributed temperature sensor. *Water Resources Research* 43: 1–9. https://doi.org/10.1029/2007WR006145.

Majorowicz, J., Grasby, S.E., Ferguson, G. et al. (2006). Paleoclimatic reconstructions in western Canada from borehole temperature logs: surface air temperature forcing and groundwater flow. *Climate of the Past* 2: 1–10.

Melville, J.G., Molz, F.J., and Guven, O. (1985). Laboratory investigation and analysis of a groundwater flowmeter. *Ground Water* 23: 486–495.

Moffett, K.B. and Gorelick, S.M. (2012). A method to calculate heterogeneous evapotranspiration using submeter thermal infrared imagery coupled to a stomatal resistance submodel. *Water Resources Research* 48: W01545. https://doi.org/10.1029/2011WR010407.

Molson, J.W. and Frind, E.O. (2005). *HEATFLOW: A 3D Ground-. Water Flow and Thermal Energy Transport Model, Version 4.0*. Waterloo, Ontario, Canada: University of Waterloo.

Munz, M., Krause, S., Tecklenburg, C., and Binley, A. (2011). Reducing monitoring gaps at the aquifer-river interface by modelling groundwater-surface water exchange flow patterns. *Hydrological Processes* 25: 3547–3562. https://doi.org/10.1002/hyp.8080.

Peterson, R.N., Burnett, W.C., Glenn, C.R., and Johnson, A.G. (2009). Quantification of point-source groundwater discharges to the ocean from the shoreline of the Big Island, Hawaii. *Limnology and Oceanography* 54: 890–904.

Petrides, A.C., Huff, J., Arik, A. et al. (2011). Shade estimation over streams using distributed temperature sensing. *Water Resources Research* 47. https://doi.org/10.1029/2010WR009482.

Pfister, L., McDonnell, J.J., Hissler, C., and Hoffmann, L. (2010). Ground-based thermal imagery as a simple, practical tool for mapping saturated area connectivity and dynamics. *Hydrological Processes* 24 (May): 3123–3132. http://doi.org/10.1002/hyp.7840.

Pluhowski, E.J. (1972). *Hydrologic Interpretations Based on Infrared Imagery of Long Island*. New York. Washington: U.S. Govt. Print. Off.

Pollack, H.N. and Huang, S.P. (2000). Climate reconstruction from subsurface temperatures. *Annual Review of Earth and Planetary Sciences* 28: 339–365.

Reginato, R.J., Idso, S.B., Vedder, J.F. et al. (1976). Soil water content and evaporation determined by thermal parameters obtained from ground-based and remote measurements. *Journal of Geophysical Research* 81 (9): 1617–1620. https://doi.org/10.1029/JC081i009p01617.

Ren, T., Kluitenberg, G.J., and Horton, R. (2000). Determining soil water flux and pore water velocity by a heat pulse technique. *Soil Science Society of America Journal* 64: 552–560.

Sayde, C., Buelga, J.B., Rodriguez-Sinobas, L. et al. (2014). Mapping variability of soil water content and flux across 1-1000 m scales using the actively heated fiber optic method. *Water Resources Research* 50: 7302–7317. https://doi.org/10.1002/2013WR014983.

Sayde, C., Gregory, C., Gil-Rodriguez, M. et al. (2010). Feasibility of soil moisture monitoring with heated fiber optics. *Water Resources Research* 46. https://doi.org/10.1029/2009WR007846.

Sayde, C., Thomas, C.K., Wagner, J., and Selker, J. (2015). High-resolution wind speed measurements using actively heated fiber optics. *Lett* 42 (10): 064–10,073. https://doi.org/10.1002/2015GL066729.

Schmidt, C., Bayer-Raich, M., and Schirmer, M. (2006). Characterization of spatial heterogeneity of groundwater-stream water interactions using multiple depth streambed temperature measurements at the reach scale. *Hydrology and Earth System Sciences* 10: 849–859.

Schmidt, C., Buettner, O., Musolff, A., and Fleckenstein, J.H. (2014). A method for automated, daily, temperature-based vertical streambed water-fluxes. *Fundamental and Applied Limnology* 184: 173–181.

Schmugge, T. (1978). Remote sensing of surface soil moisture. *Journal of Applied Meteorology* 17: 1549–1557.

Schuetz, T. and Weiler, M. (2011). Quantification of localized groundwater inflow into streams using ground-based infrared thermography. *Geophysical Research Letters* 38 (3): 1–5. http://doi.org/10.1029/2010GL046198.

Selker, J., van de Giesen, N., Westhoff, M. et al. (2006). Fiber optics opens window on stream dynamics. *Geophysical Research Letters* 33: L24401. https://doi.org/10.1029/2006GL027979.

Selker, J.S., Thévenaz, L., Huwald, H. et al. (2006). Distributed fiber-optic temperature sensing for hydrologic systems. *Water Resources Research* 42. https://doi.org/10.1029/2006WR005326.

Sinokrot, B.A. and Stefan, H.G. (1993). Stream temperature dynamics: measurements and modeling. *Water Resources Research* 29: 2299–2312. https://doi.org/10.1029/93WR00540.

Slichter, C.S. (1905). *Field Measurements of the Rate of Movement of Underground Waters.* Washington, DC: USGS.

Stallman, R.W. (1963). *Computation of Groundwater Velocity from Temperature Data. Methods of Collecting and Interpreting Groundwater Data.* 36–46. Washington, DC: USGS.

Stallman, R.W. (1965). Steady one-dimensional fluid flow in a semi-infinite porous medium with sinusoidal surface temperature. *Journal of Geophysical Research* 70: 2821–2827.

Steele-Dunne, S.C., Rutten, M.M., Krzeminska, D.M. et al. (2010). Feasibility of soil moisture estimation using passive distributed temperature sensing. *Water Resources Research* 46. https://doi.org/10.1029/2009WR008272.

Striegl, A.M. and Loheide, S.P. (2012). Heated distributed temperature sensing for field scale soil moisture monitoring. *Ground Water* 50: 340–347. https://doi.org/10.1111/j.1745-6584.2012.00928.x.

Swanson, T.E. and Cardenas, M.B. (2011). Ex-Stream: a MATLAB program for calculating fluid flux through sediment-water interfaces based on steady and transient temperature profiles. *Computers & Geosciences* 37: 1664–1669.

Thomas, C.K., Kennedy, A.M., Selker, J.S. et al. (2012). High-resolution fibre-optic temperature sensing: a new tool to study the two-dimensional structure of atmospheric surface-layer flow. *Boundary-Layer Meteorol* 142: 177–192. https://doi.org/10.1007/s10546-011-9672-7.

Torgersen, C.E., Faux, R.N., McIntosh, B.A. et al. (2001). Airborne thermal remote sensing for water temperature assessment in rivers and streams. *Remote Sensing of Environment* 76: 386–398.

van de Giesen, N., Steele-Dunne, S.C., Jansen, J. et al. (2012). Double-ended calibration of fiber-optic raman spectra distributed temperature sensing data. *Sensors (Switzerland)* 12: 5471–5485. https://doi.org/10.3390/s120505471.

Vandersteen, G., Schneidewind, U., Anibas, C. et al. (2015). Determining groundwater-surface water exchange from temperature-time series: combining a local polynomial method with a maximum likelihood estimator. *Water Resources Research* 51: 922–939.

Vercauteren, N., Huwald, H., Bou-Zeid, E. et al. (2011). Evolution of superficial lake water temperature profile under diurnal radiative forcing. *Water Resources Research* 47: 1–10. https://doi.org/10.1029/2011WR010529.

Vogt, T., Schneider, P., Hahn-Woernle, L., and Cirpka, O.A. (2010). Estimation of seepage rates in a losing stream by means of fiber-optic high-resolution vertical temperature profiling. *Journal of Hydrology* 380: 154–164. https://doi.org/10.1016/j.jhydrol.2009.10.033.

Voytek, E.B., Drenkelfuss, A., Day-Lewis, F.D. et al. (2014). 1DTempPro: analyzing temperature profiles for groundwater/surface-water exchange. *Groundwater* 52: 298–302.

Webb, B.W., Hannah, D.M., Moore, R.D. et al. (2008). Recent advances in stream and river temperature research. *Hydrological Processes* 22: 902–918.

Webb, B.W. and Zhang, Y. (1997). Spatial and seasonal variability in the components of the river heat budget. *Hydrological Processes* 11: 79–101.

Westhoff, M.C., Savenije, H.H.G., Luxemburg, W.M.J. et al. (2007). A distributed stream temperature model using high resolution temperature observations. *Hydrology and Earth System Sciences* 11: 1469–1480.

Worman, A., Riml, J., Schmadel, N. et al. (2012). Spectral scaling of heat fluxes in streambed sediments. *Geophysical Research Letters* 39: L23402.

Yang, C.B. and Jones, S.B. (2009). INV-WATFLX, a code for simultaneous estimation of soil properties and planar vector water flux from fully or partly functioning needles of a penta-needle heat-pulse probe. *Computers & Geosciences* 35: 2250–2258.

Young, P.C., Pedregal, D.J., and Tych, W. (1999). Dynamic harmonic regression. *Journal of Forecasting* 18: 369–394.

9

Sampling at Groundwater–Surface Water Interfaces

Jörg Lewandowski[1,2,], Jonas Schaper[3], Michael Rivett[4], and Stefan Krause[5]*

[1] *Leibniz Institute of Freshwater Ecology and Inland Fisheries, Berlin, Germany*
[2] *Geography Department, Humbold University Berlin, Germany*
[3] *Center for Applied Geoscience, Eberhard Karls University of Tübingen, Tübingen, Germany*
[4] *GroundH2O Plus Ltd., Quinton, Birmingham, UK*
[5] *School of Geography, Earth and Environmental Sciences, University of Birmingham, Edgbaston, Birmingham, UK*
[*] *Corresponding author*

9.1 Introduction

Analysis and assessment of the functioning of ecohydrological interfaces requires the ability to extract samples of water or sediment to characterize the interface properties and transport therein. Sampling of interface sediment requires extraction of material via coring methods and subsequent processing of the core ex situ and are therefore often referred to as ex situ methods. Water samples can be obtained from extracted sediment material but more commonly via in situ methods, which are less invasive and keep sediment structures more or less intact. In situ water sampling can be performed actively, i.e. by pumping or passively via diffusion-based and adsorption-based processes. Sample extraction from interfaces between groundwater on the one side and rivers, lakes, or marine environments on the other side can thus be broadly categorized as:

i) ex situ sampling methods/coring methods
ii) active in situ sampling methods
iii) passive in situ sampling methods

The principles of different sampling methods, their potential applications, advantages and disadvantages are discussed.

9.2 Ex situ Sampling Methods/Coring Methods

There are several different types and designs of corers (Lewandowski 2002) that are used to collect more or less undisturbed sediment cores at groundwater–surface water interfaces. Corers typically comprise some form of hollow tube that is inserted into the sediment either by the weight of the corer, manually by the researcher, or mechanically by a percussion

Ecohydrological Interfaces, First Edition. Edited by Stefan Krause, David M. Hannah, and Nancy B. Grimm.

hammer. The corer is then retrieved removing a column of sediment. Depending on the sediment type and coring depth this can also require much effort and special devices (e.g. winch) to support the retrieval of the core. Once the sediment core has been collected it is sliced into layers of the desired thickness. Core-slicing can be done on-site or later on in the laboratory. Pore water from the core slices is typically extracted by centrifugation or pressure filtration.

Advantages:
Core extraction methods are often the method of choice when other water sampling techniques become impractical due to sampling depth or sediment properties. For instance, in peatlands and marshes, hydraulic conductivities are often too low to allow the excessive use of active sampling methods. In such settings, coring techniques can be advantageous, especially when samples need to be obtained in depths that are not readily accessible for passive sampling methods such as peepers. The primary advantage of core extraction methods and ex situ analysis compared to in situ sampling methods is that both water, i.e. liquid phase, and sediment, i.e. solid phase, samples can be obtained simultaneously. This permits detailed side-by-side characterization of water pore quality and sediment properties that may control transport. For instance, sedimentary organic carbon amount and type controlling dissolved plume transport and attenuation (e.g. chlorinated volatile organic compounds) at the interface (Rivett et al. 2019).

Disadvantages:
The major disadvantage of coring methods is that they are, compared to other sampling methods, typically the most destructive. It is not only impossible to obtain time series of multiple samples from exactly the same site, but the sampling process also has a relatively high chance to alter the sample with respect to its chemical composition and physical configuration. Counter arguments are that the coring footprint is quite small, surrounding sediments of similar nature to those cored will quickly re-collapse in and effectively backfill the hole left (with possible exception of a punctured plastic clay horizon), and a fuller coring programme could be left as a final task after completion of the in situ monitoring programme.

Because sediment at the groundwater–surface water interface is often characterized by steep redox gradients, care must be taken to prevent atmospheric oxygen to enter the core and oxidize redox sensitive solutes and sediment material thus altering the chemical composition of the pore water. Oxidation of redox sensitive core materials can be minimized by storing and transporting cores in plastic bags that are purged with nitrogen gas directly after sampling. Core processing should then be conducted under oxygen free conditions, e.g. in a glove box in the laboratory. Temperature changes and different pressure might also contribute to alterations of the samples (Adam, 1991; Enell and Löfgren 1988). Howes et al. (1985) show that coring, compared to passive sampling, increases dissolved organic carbon (DOC) concentrations in pore water samples collected in marshes and attribute this finding to root cutting.

9.2.1 Freeze Coring

The standard coring methods suffer from a variety of shortcomings, of which some can be overcome by using a freeze core sampling technique, i.e. rapid in situ freezing of sediment (Murphy and Herkelrath 1996). There is a large variety of freeze corer designs from very simple metal tubes filled with dry ice to high-tech samplers using electric pumps, hydraulics, and liquid nitrogen (Renberg and Hansson 2010).

Advantages:
If a freeze coring (liquid CO_2 or nitrogen) technique is used, disturbances to the physical configuration of the saturated sediment and alterations to the chemical (and microbiological) composition of the sediment are reduced compared to standard coring techniques (Bianchin et al. 2015). Retrieval of larger grain sizes (cobbles, etc.) allows improved visualization of the subsurface structure (Freitas et al. 2015). Also, high (*c.* 100%) core recovery enables depth-confident and detailed profile characterization in the laboratory of physical, chemical, and microbiological parameters (Strasser et al. 2015) including biomolecular analysis – DNA profiles (Johnson et al. 2013).

Disadvantages:
Freeze coring has disadvantages of more onerous health and safety requirements and set-ups are typically developed by bespoke research rather than commercially available. There should be an awareness that the freezing process might alter the layering of the sediment due to crystal formation. Concentrations of dissolved gases might change as well as some chemical compounds. Also, the sediment structure and features such as hydraulic conductivity might change in consequence of the freezing process.

9.3 Active in situ Sampling Methods

"Active" means that the samples are obtained by low pressure (suction) with peristaltic pumps or syringes, by submersible pumps, or by gas-driven water displacement. Based on application demands (sampling volume, sampling duration, sampling frequency, sampling depth, spatial sampling resolution), parameters of interest, and site-specific conditions (substrate composition, water depth, site accessibility, surface flow velocity), active in situ samplers differ in device material and design (single to multi-depth samplers), deployment method, means of sample extraction (syringe extraction, peristaltic, or submersible pumps), sampling volume, and representative sampling area with depth specific versus depth-integrated sampling (Figure 9.1). Depending on the desired sampling volume, active in situ samplers may be conveniently used to extract volumes ranging from millilitres (Conant et al. 2004; Kalbus et al. 2006; Rivett et al. 2008) – for instance, for the analysis of water quality parameters – to several litres of water, predominantly for the analysis of invertebrate communities at ecohydrological interfaces (Bou and Rouch 1967; Hunt and Stanley 2000; Palmer and Strayer 1996).

Spatial dimensions of suction samplers range from single well piezometers with diameters from <1 mm to >50 mm (Blume et al. 2013; Conant et al. 2004), multi-level mini-piezometers of small diameters of very few millimetres (Freitas et al. 2015; Krause et al. 2013; Rivett et al. 2008; Weatherill et al. 2014), and mini-point samplers (Duff et al. 1998; Posselt et al. 2018; Sanders and Trimmer 2006), which allow single or multi-level sampling with radial or vertical sampling designs (Figure 9.1). Depth and resolution of active in situ samplers vary from mini-point samplers targeting small scales (Duff et al. 1998; Sanders and Trimmer 2006), usually for water quality analysis, to integral measurements covering decimetres to metres (Figure 9.1).

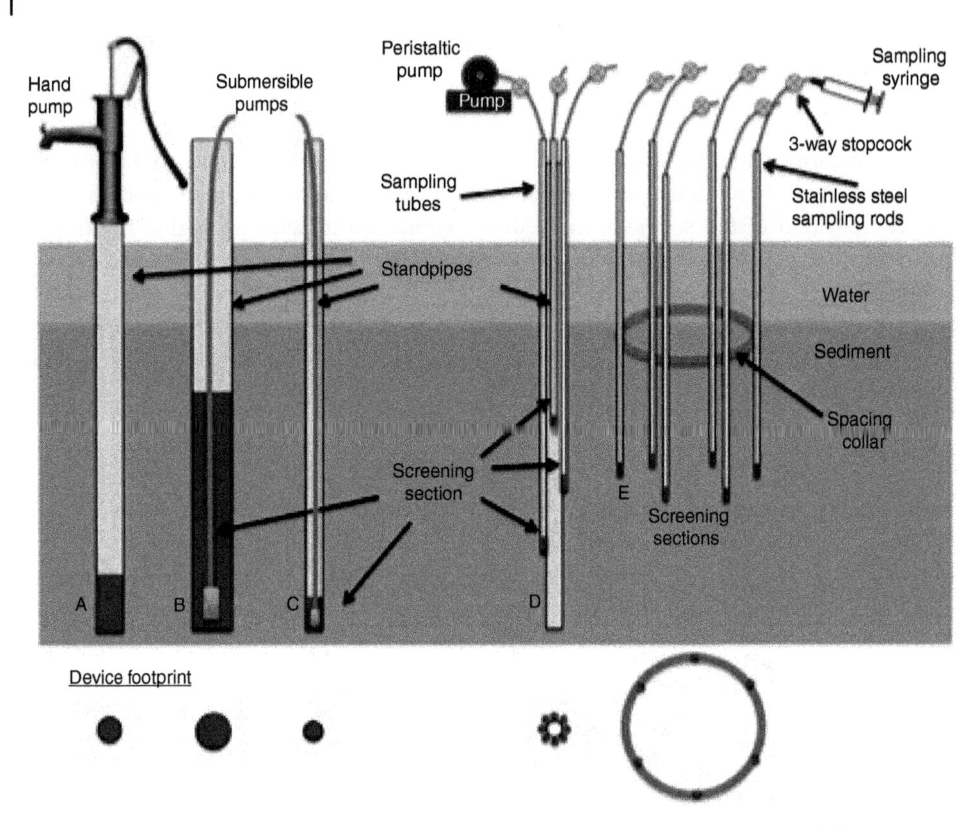

Figure 9.1 Selection of design, footprint, extraction mechanisms, and strategies of different suction-based pore water sampling techniques including Bou Rouch samplers (A), single well piezometers with integrated (B), and depth-specific (C), screening sections, multi-level samplers (D), and mini-point samplers with similar depths (E) (adapted from Gomez-Velez et al. 2016).

In addition to sampling depths, resolution and extraction volume, sampling protocols for different suction samplers can vary substantially between different types of applications and site-specific conditions, with variability in pore water pumping rates, either simultaneous or sequential pumping of water from different depths or locations (Hunt and Stanley 2000). For applications in aquatic ecology, for instance, the sampling of invertebrates requires an immediate extraction of pore water to avoid specimen's actively moving from the capture zone (Hunt and Stanley 2000; Palmer and Strayer 1996). In contrast, pore water sampling for biogeochemical applications rather requires low pumping rates, i.e. "low flow sampling", to ensure continuous flow and minimized risks of (contaminated) sediment particle mobilization and outgassing losses of dissolved gases or volatile organic compounds (Puls and Barcelona 1996; Rivett et al. 2008; Weatherill et al. 2014).

In the following, we provide an overview of different active pore water sampling devices used at the interface between groundwater and surface water, including (i) piezometers for depth-integrated and depth-specific sampling sections and Bou Rouch pumps for pore water invertebrate and suspended sediment sampling, (ii) multi-level samplers, (iii) mini-point pore water sampling, and (iv) ultrafiltration devices, and discuss their capabilities and limitations for different applications.

9.3.1 Piezometers

Piezometers (*aka*. groundwater observation wells) usually comprise steel or PVC tubes driven into the sediment furnished with a short "well screen" – a slotted (or drill-holed) section wrapped with a geotextile. The screen permits water entry from bed sediments into the piezometer while keeping the sediment out of the piezometer. It enables the collection of water level (head) data and water quality samples. In softer organic or sandy sediments piezometers can often be pressed directly into the desired depths. In gravel or clay sediments installation requires hammering at substantial force or even drilling pre-installation. The diameter of piezometers varies from <10 to >50 mm diameter. Lengths of screening sections vary from 10 (for depth-specific sampling) to several 100 of mm (for depth-integrated sampling); shorter screens are more typical. Whilst single piezometers may be used, often several may be installed at a locality completed at different depths to effectively give a multi-level profile of water quality or head gradient with depth.

Piezometers may enable relatively large volumes of water to be removed, however, good practice should keep volumes as low as practically possible to fulfil purging and sampling requirements. Before sampling (usually with a peristaltic, submersible, or hand-pump), the water standing in the piezometer and in the nearest surroundings has to be purged in order to ensure that representative pore water samples can be taken. In contrast to other active sampling methods, the water sample is not actively sucked out of the pore space but taken from the water re-filling the piezometer according to the hydrostatic gradient caused by the purging. Depending on the substrate's hydraulic conductivity, this re-filling can occur over several hour-long durations or very quickly.

Extraction of 5–10 times the piezometer volume has often been practiced in order to ensure "old" water in the piezometer and its near surroundings is removed. However, this may be overkill and low flow sampling coupled with low purge volumes (1–3 piezometer volumes) and stabilization of "wellhead parameters" (pH, EC, etc.) widely used in conventional groundwater monitoring well sampling (McMillian et al. 2018; Puls and Barcelona 1996) may often be adequate and more appropriate. This is especially the case where the piezometer screen is likely to be well flushed by natural flows through the bed sediments. Overall, the purging and sampling protocol should be considered, tested, and consistently used to ensure reproducibility as far as possible.

Bou Rouch pumps represent a specific case of piezometer sampling. They are predominantly deployed by aquatic ecologists for sampling invertebrates from the hyporheic or benthic sediment–water interfaces. The steel-based piezometer tubes of Bou Rouch pumps have usually 25 mm diameter and ~150 mm long screening sections. Because of the large demand of quickly extracted large volumes of water (to avoid specimens actively escaping from the capture zone), volumes of 0.5–2.5 L are usually extracted by hand pumps within less than 1 minute (Figure 9.1).

Advantages:

The relatively simple design and ease of installation make it possible to install and sample from clusters of large numbers of permanently installed piezometers or even use them for temporary installations or roaming surveys (e.g. Burk and Cook 2015; Meinikmann et al. 2015; Pöschke et al. 2015). In addition to water quality samples, water level measurements are often possible allowing (when compared to the stream level) the vertical hydraulic gradient and hence flow direction across the bed sediment and identification of influent or effluent stream condition.

Disadvantages:
Due to the relatively large impact of the piezometer installation, hours to days of "settling time" can be required to ensure that representative samples can be extracted. The extraction of large water volumes, in particular, by depth-integrated piezometers or the Bou Rouch method, raises questions of the exact representative sampling volume and bears the risk of accidental surface water breakthrough into the sampling zone (Figure 9.2). Such provides strong rationale for adopting low volume purging and sampling protocols.

9.3.2 Multi-level Samplers

Interfaces are often characterized by significant concentration gradients across the interface (Krause et al. 2017). Therefore, there is often the necessity to sample pore water at ecohydrological interfaces in different depths. While in some applications such sampling has been facilitated by nested designs of adjacent piezometers at different depths (Kaeser et al. 2009), the more frequently used sampling method to extract pore water

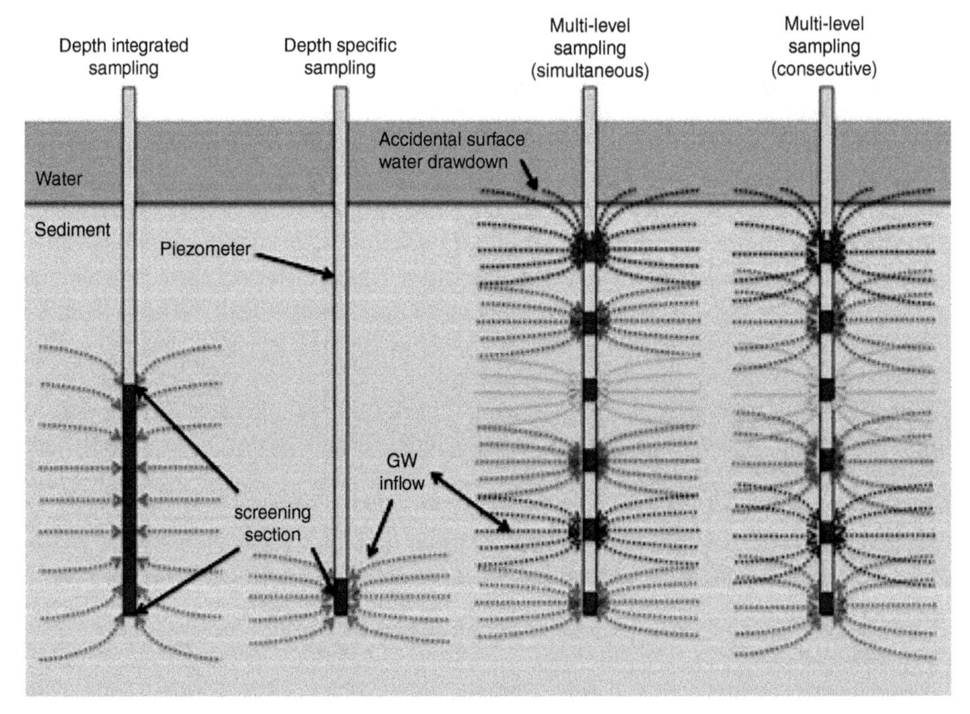

Figure 9.2 Cross-sectional geometries of pore water capture zones for depth-integrated (A) and depth-specific piezometer sampling (B), and simultaneous (C) and consecutive (D) sampling from multi-level mini-piezometers, indicating potential deflation of the vertical extent of capture zones in simultaneous sampling and risking capture zone overlaps under consecutive sampling regimes (adapted from Gomez-Velez et al. 2016).

samples from different depths with high depth resolution is based on small diameter "multi-level sampler" (MLS) (or mini-piezometer) (Freitas et al. 2015; Rivett et al. 2008). Depth and sampling intervals of MLSs can be varied based on the respective application demands. Sampling at aquifer–river interfaces usually focuses on pore water from the top 100 cm of lake or streambed sediments and hence a MLS would have sampling ports distributed over that bed's thickness.

An MLS usually comprises a ca. 1.5 mm (or greater) internal diameter PTFE or L/HDPE sampling tubes of different lengths arranged around a central rod or HDPE/PVC tube of larger diameter (8–12 mm). An MLS is installed as a bundle using temporary drive pipes subsequently removed allowing sediment re-collapse sealing as described in Rivett et al. (2008). Tubing material choice is typically based upon the analytes sampled, aiming to minimize negative bias in sampled concentrations. Wider diameter tubing (ca. >8 mm i.d.) offers the potential for both water level (head) and quality sampling.

Pore water extraction can be realized by using peristaltic pumps or by applying suction with syringes and three-way stopcocks. The pore water sampling from different depths can be carried out either simultaneously, potentially deflating the vertical extent of the capture zone and artificially enhancing its horizontal foot print (Figure 9.2), or consecutively, posing the risk of overlaps and smearing of sampling areas (Figure 9.2). For the interpretation of sampling results and uncertainties of the latter case, the direction of the consecutive sampling (bottom to top or top to bottom) can have substantial impact on the type of pore water mixing, potential overlaps, and surface water draw down (Figure 9.2). Minimizing the total volumes of water abstracted during sampling and low flow rates pumped will obviously help minimize such influences and is considered good practice. In that water is completely removed from the MLS tubing, the removal of a single tubing volume of standing water present should provide adequate purging. In narrow diameter tubing this volume is trivially small.

Advantages:
Most MLS are dimensioned to allow extraction of small discrete volumes of pore water, allowing precise interpretation of investigated parameters along the vertical profile. As suction is applied directly to the sampling tube, samples can be extracted at controlled flow rates and with minimized risk of contamination by air or by gas exchange. Required purge volumes are usually very small. The horizontal footprint and spatial scale of disturbance of the multi-level mini-piezometers is small as compared to piezometers because of the small volume of the device. Due to their rather small diameter and flexibility, MLS tubing left above the streambed usually bends with the stream flow causing less visual impairment than sturdy piezometers of greater diameter but also increasing their resilience to storm flow events. They are typically less obtrusive, become hidden by stream vegetation, and hence are less prone to vandalism.

Disadvantages:
With narrow diameter MLS tubings, usually only the central PVC or HDPE tube can be dipmetered for measuring piezometric heads, restricting observation of this information to the deepest location of the piezometer in the investigated sediment. With a usual dimension of 8–12 mm diameter, the central tubes of multi-level mini-piezometers is usually also too small to host any conventional pressure transducers for continuous monitoring of piezometric

heads. Individual tubes require resilient labelling identification systems to avoid mis-identification of tubing depths; multiple identification types are recommended – labels, different length tubing above the bed surface, fixed tube ordering around support pipe circumference.

9.3.3 Mini-point Samplers

In addition to bundles of mini-piezometers, mini-point samplers (Duff et al. 1998; Sanders and Trimmer 2006) have been used for discrete, depth-specific sampling of extremely small pore water volumes, in particular, from the top 10–30 cm of streambed or lake sediments. In many applications of mini-point samplers, extraction points for different depths are not aligned in a direct vertical profile but located along a circular footprint (Figure 9.1). Such design reduces the risk of interactions between samples of different depths such as potential overlaps or changes in the pore water flow field (Figure 9.2). In other applications, however, mini-point samplers were purposefully installed in a vertical configuration to increase the chance of obtaining water samples along a specific flow path (Posselt et al. 2018; Schaper et al. 2019).

The diameter of the circular alignment of the stainless steel rods used for sampling with mini-point samplers (Figure 9.1) has to be chosen carefully in order to minimize the impact of horizontal variability in lakebed or streambed sediment properties. Exact spacing of individual sampling probes is usually achieved by using a collar that guides and stabilizes the probes. Most mini-point samplers follow the design by Duff et al. (1998), with a fixed diameter of 10 cm and six 1/16 inch extraction rods spaced evenly along this perimeter, resulting in 5 cm distances between the probes (Figure 9.1) (Harvey and Fuller 1998; Knapp et al. 2017). Pore water is usually extracted by applying suction manually with syringes, automatically with syringe pumps, or with peristaltic pumps. For the latter, low flow rates are chosen so that the influence on the ambient pore water flow field is minimized (typically ≤1 ml/min). The sampling volume that can safely be extracted is a function of spatial and temporal sampling resolution, effective sediment porosity, and pore water flow velocities. Dead volume in the mini-points can be reduced by inserting PEEK tubing into the stainless-steel rods, which can be connected to sampling syringes via Swagelok® fittings, and NPT adapters (Posselt et al. 2018; Schaper et al. 2019).

Advantages:
The relatively easy deployment of the rather short sampling segments qualifies mini-point samplers for roaming sampling surveys where probes are removed after sample extraction. Sample processing is significantly simplified by the pre-filtration occurring within layers of glass-wool installed at the tip of the sampling rods in many mini-point sampler designs (Sanders and Trimmer 2006). Due to low dead volumes and their small size, mini-point samples can furthermore be used to collect pore water time series at a relatively high spatial (i.e. cm to dm scale) and temporal resolution. The system can furthermore be made air-tight by inserting PEEK tubing into the mini-point samplers as described previously (Posselt et al. 2018).

Disadvantages:
The installation of mini-point samplers is restricted by the sediment properties, with accurate spacing or even deployment being impossible in gravel or clay-rich sediments. Even if successfully installed at shallow depths of finer sediments, the assumption that sampling

from the radially arranged single probes is unaffected by any small-scale variability of sediment properties has yet to be demonstrated.

9.3.4 Ultrafiltration Samplers

Hollow fibres with diameters of typically 0.1–2 mm can be used to collect pore water samples by applying suction. In principle, hollow fibres are made from the same materials as dialysis membranes discussed in more detail in the following Section 9.4. The main difference of ultrafiltration to dialysis is that no equilibration is used but that suction forces the pore water through the membrane into the sampler, i.e. the sample is already filtrated when using this device. To collect a sufficient amount of sample in an adequate time the length of ultrafiltration fibres is usually 3–10 cm. Thus, the sampler integrates over some distance. To reach a high spatial resolution the sampler is usually placed horizontally at pore water–surface water interfaces. In natural systems thc horizontal placement of such samplers results in significant disturbance of the system and thus, ultrafiltration samplers are seldom used in field studies. However, in lab studies they are quite useful and often applied. They are inserted horizontally through holes in the walls both in column and flume studies. The most common brand name for such devices is the Rhizon sampler (Eijkelkamp).

Advantages:
The risk of shortcuts is minimized due to the horizontal placement of the devices in the system. High spatial and medium temporal resolutions can be achieved with ultrafiltration devices. The sampler has a small dead volume which reduces the required amount of water withdrawal. The device is relatively cheap and sampling can be easily conducted by syringes.

Disadvantages:
The sampler integrates over some centimetres (usually horizontally). In situ applications are rather complicated, i.e. Rhizon samplers are merely a lab device. The system must be prepared for later insertion of the sampler by drilling holes in the column or flume walls which is usually done before setting up an experiment.

9.4 Passive in situ Sampling Methods

Passive samplers are devices that contain a collecting medium, which either equilibrates itself via diffusion with the surrounding media or accumulates analytes via sorption over a given time period. In contrast to active sampling methods (Section 9.3), passive samplers can thus provide time-weighted average concentrations over their deployment period.

Passive samplers based on diffusive equilibrium are typically deployed for the time it takes for equilibrium to be reached between the collecting medium and the surrounding water phase, which can, depending on the size of the collecting chambers, take several days to weeks. After the sampler has been retrieved from the sampling site, time-weighted average concentrations of target compounds can directly be measured in the collective medium. Passive samplers that work based on diffusive equilibrium include dialysis samplers (also called peepers) and DET (diffusional equilibration in thin films) samplers.

Sorption-based passive samplers, often also referred to as dynamic or kinetic samplers, differ from diffusive equilibration based passive samplers mainly in that target compounds accumulate in or on the collecting medium, i.e. the binding phase, via sorption processes. Time-weighted average concentrations in the surrounding water are calculated from analyte diffusive fluxes into the collecting medium. Typically, analyte diffusive fluxes are calculated by relating accumulated mass of the analyte to be related to sampling time, diffusive properties of the target compound, and device-specific parameters such as the thickness of protective membranes or binding phases. Sorption-based passive samplings include diffusive gradients in thin films (DGT) samplers for nutrients and metals and related devices specifically developed for organic compounds such as polar organic chemical integrative samplers (POCISs), Chemcatcher® samplers, and DGT samplers for organic compounds (o-DGT).

The main advantage of passive sampling methods is that they are a relatively cost-effective tool to provide time-weighted average concentrations under minimal supervision and maintenance during the sampling period. DGT samplers provide a time-integrated water quality sample over the time period of their deployment in the riverbed. The main disadvantage of peepers, DET, and DGT samplers is the loss of temporal resolution. Another advantage of kinetic samplers (DGT) over diffusive equilibrium samplers is the accumulation of the compound of interest in the sampler, which is most relevant for compounds that occur in concentrations that are low compared to analytical detection thresholds. The main disadvantage of adsorption-based passive samplers is that both the dependence of diffusive fluxes' environmental conditions and the diffusion coefficients or target analytes may be unknown a priori. Deployment times may typically vary from minutes to months depending on the technique, the set-up, and the design of the sampler.

9.4.1 Dialysis Samplers

The first dialysis samplers (also called peepers) were introduced in 1976 in the international literature by Hesslein (1976) and Mayer (1976). The basic principle is the diffusion-driven equilibration across a dialysis membrane between distilled water in the sampler and the pore water surrounding the device (Adams 1991; Hesslein 1976). The pore size of the membrane is chosen such that ions can cross the membrane while larger molecules and particles are excluded from the sample (Hesslein 1976). Carignan (1984b) suggested using microbially inert membranes made of polysulfone (e.g. Gelman HT-450, HT-Tuffryn 200®).

In the past, a large number of different designs and variations were developed to meet the needs of the different applications (Adams 1991). Quite often peepers are made of transparent plastics such as Perspex since it is possible to visually check the content in the chambers for gas bubbles and precipitates. Figure 9.3 shows a common peeper design similar to the one introduced by Hesslein (1976) (Lewandowski 2002). Since 1976, the design remained pretty much the same and nowadays they can be considered a standard method in limnology. The peeper chambers are milled into a 2 cm thick base plate of Perspex which is covered by a dialysis membrane and a 0.6 cm thick cover plate for fixing the membrane. Openings in the cover plate above the chambers allow a free exchange between distilled water in the peeper chambers and pore water. The cover plate is fixed by plastic screws onto the base plate. Usually two rows of chambers are used; the distance of the chambers is quite often 1 cm and the volume of the chambers approximately 5 ml. Depending on the

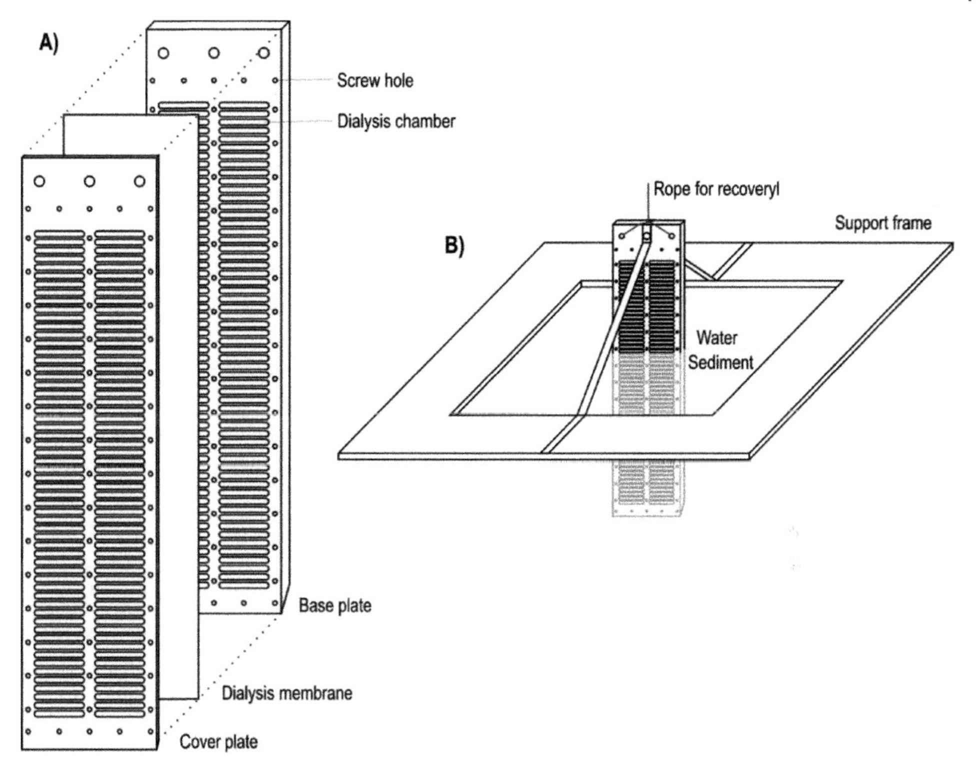

Figure 9.3 (A) Common design of one-dimensional pore water sampler according to the peeper design suggested by Hesslein (1976). (B) Installation of a one-dimensional pore water sampler attached to a rope in a lake from a boat using a heavy support frame that assures the correct positioning at the sediment–water interface (Adapted from Lewandowski, 2002).

number of parameters that should be analysed in the pore water, the small sample volume might be a challenge for analytics.

One-dimensional peepers can be used at different interfaces. The major restriction is that there must be water, pore water, groundwater, or other water in front of the peeper chambers. Originally, peepers were developed in marine sciences for the sediment–water interface in the sea or coastal areas (Hesslein 1976; Mayer 1976). Limnologists copied the method, used it first at the sediment–water interfaces of lakes and later on at the sediment–water interfaces of streams (Schaper et al. 2018; Schroeder et al. 2020). Peepers have also been deployed in fens, wetlands, estuaries, streambed sediments, and lab experiments. In shallow environments or lab settings peepers can be exposed quite easily by wading. Inserting the peeper is quite easy if the sediment is soft enough. In deeper zones of lakes and oceans divers might be used for the deployment of the peepers. However, usually different types of stands that assure partial penetration of the sediment by the peeper are used in deep zones to assure a correct positioning of the peepers by simply lowering them attached to a rope from a boat.

On the one hand, the exposition duration should be short so that there is only little time for biofilm development on the membrane. On the other hand, the duration must be long enough so that a dynamic equilibrium can establish between both sides of the membrane

(Brandl and Hanselmann 1991) and also so that initial disturbances of the redox zonation of the sediment due to the insertion of the sampler has recovered. The required time depends on temperature, geometry of the sampler's chambers, the porosity and thickness of the membrane, the porosity of the sediment, and the chemical composition of the pore water (Brandl and Hanselmann 1991; Enell and Löfgren 1988). According to Adams (1991), durations between 6 days and 1 month are commonly used. Enell and Löfgren (1988) state that exposition durations of 2–10 days are sufficient for equilibrium. Carignan (1984a) suggests 15 days at temperatures of 20–25 °C and 20 days at 4–6 °C. Hesslein (1976) used exposition durations of 1 week. Some authors already consider that in addition to simple equilibration also some time is necessary to reduce the impact of the peeper deployment on the pore water composition. For example, the redox zonation in the sediment might be slightly disturbed by relocating sediment and importing some oxygen with the sampler. Furthermore, pore water directly in front of the peeper is diluted by the equilibration with distilled water of the sampler. Some additional exposition duration assures that original conditions can re-establish (Harper et al. 1997; Hesslein 1976).

After recovery of the peeper its membrane is punctured by a syringe or pipette, the sample is withdrawn, transferred into sample containers, acidified if applicable, and cooled until analysis (Adams 1991; Hesslein 1976). The time span between recovery of the peeper and conservation of the samples should be as short as possible to avoid changes of the sample composition, for example, due to oxygen import or changes of temperature. Some authors use helium or nitrogen atmospheres in glove boxes (Adams 1991; Hesslein 1976), others determine the rate of change and calculate original concentrations based on the time span between recovery and sample conservation (Brandl and Hanselmann 1991), others try to finish the sampling of all peeper chambers within 5 minutes (Adams 1991), others used Saran®-films to cover the sampler and avoid oxygen import (Brandl and Hanselmann 1991). We recommend cleaning the peeper immediately after recovery by immersing it into the lake water or stream water and drying it afterwards completely with paper tissues. That is important because water on the membrane will seriously alter sample composition: Water in front of the membrane has a different composition than the water inside the chambers and there will be a diffusive equilibration. Furthermore, when puncturing the membrane, the water standing on the membrane might mix with the water in the chamber. Additionally, oxygen transport across the membrane is much faster with water on both sides of the membrane instead of air on one side and water on the other side.

Two-dimensional (2D) peepers are based on Hesslein's (1976) one-dimensional (1D) pore water sampler (Lewandowski et al. 2002). Instead of a column of chambers in a 1D arrangement (Figure 9.3) there are circular bore holes in a 2D arrangement (Figure 9.4A). The challenge of the 2D peeper is the large number of samples (e.g. 2280 in the standard design). This is not only a problem because of the workload coupled to such a large number of samples but also because fast processing of the samples after recovery of the sampler is required to avoid redox and temperature changes that will seriously alter sample composition. The basic idea of the 2D peeper was to design the distance of its chambers (9 mm) analogously to microtiter plates. With that trick it is possible to use multichannel pipettes for transferring 8 (or 12) samples at a time in one pipetting step from the sampler into microtiter plates (Lewandowski et al. 2002). Additionally, the analytics had to be miniaturized

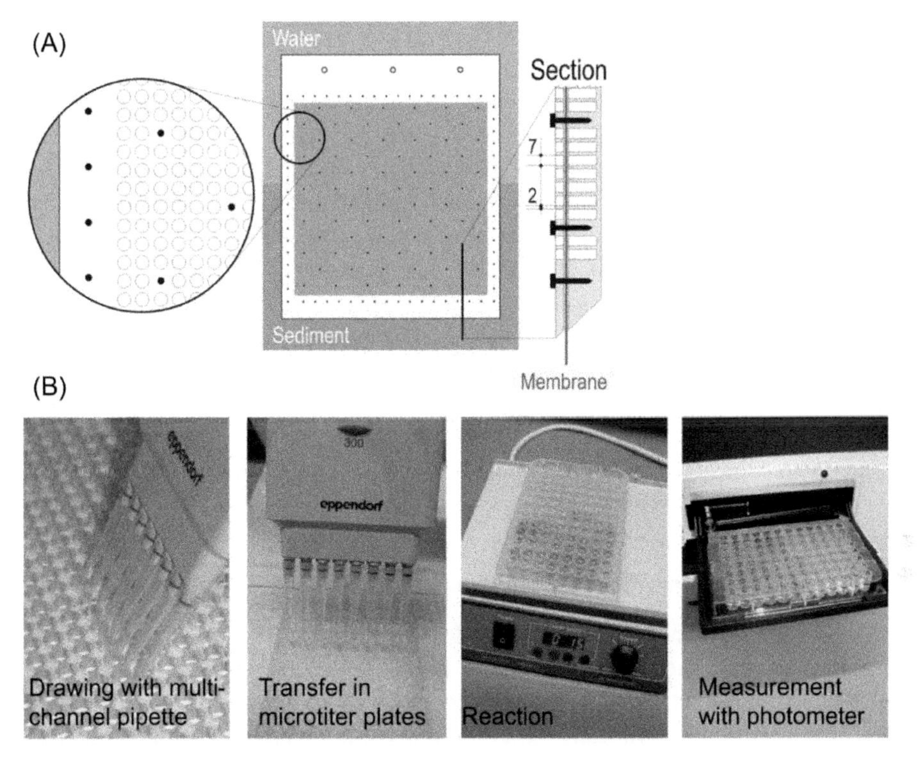

Figure 9.4 Two-dimensional pore water sampler according to Adapted from Lewandowski, 2002 (A) design of sampler, (B) analytical procedure.

so that the analytics can be conducted in microtiter plates and that the small volume in the peeper chambers (600 ml) is sufficient for the analytics. There are miniaturized and validated methods for phosphate, ammonium, iron, and sulphate (Laskov et al. 2007). The sampling volume is sufficient for the analytics of three of the four aforementioned ions (Lewandowski et al. 2007). 2D peepers have been used in lakes (Lewandowski et al. 2002, 2005), in mesocosm studies to investigate the impacts of macrozoobenthos (Lewandowski and Hupfer 2005; Lewandowski et al. 2007) and macrophytes (Laskov et al. 2007) on biogeochemical patterns in the sediment, and in streams to study turnover along transport paths in the hyporheic zone (Lewandowski and Nützmann 2010).

Advantages:
The main advantage of dialysis samplers is that sampling of pore water is conducted in situ. Sample oxidation and temperature changes are reduced compared to the collection of cores and later separation of pore water. Furthermore, peepers allow a large spatial resolution (Hesslein 1976), time consumption and required equipment are moderate, and reproducibility is quite high (Carignan 1984a). According to Brandl and Hanselmann (1991), peepers are the best method for pore water sampling when focusing on microbiological and geochemical methods. The main advantage of 2D peepers compared to 1D peepers is the fast processing of the samples and the 2D resolution.

Disadvantages:
Carignan (1984a) reports several potential causes of errors and sampling artefacts: e.g. membrane type, material for peeper construction, chemical composition of initial water in the chambers, design of the sampler, and insufficient exposition duration. He states that most researchers forget to show the suitability of their combinations of design and sampling technique for their respective study system. Also, it should be taken into account that peepers do not allow a snapshot of pore water profiles. Bandl and Hanselmann (1991) state for that reason that peepers should only be used in systems that are in a steady state so that there are no changes of pore water compositions during the measurement. In case there is a change of the pore water composition, the liquids in the peeper chambers will equilibrate into the direction of that new pore water composition. The peeper reflects generally more the pore water composition at the end of the exposition period than at its beginning. Early concentration features might be still visible as "shadows". The main disadvantage of 2D peepers compared to 1D peepers is the limited sample volume and the restricted number of parameters for which miniaturized methods are available.

9.4.2 Diffusional Equilibration in Thin Film Gel Samplers (DET Technique)

A special form of dialysis samplers are gel samplers (Davison et al. 1991; Kankanamge et al. 2020; Krom et al. 1994; Reeburgh and Erickson 1982). In such gel samplers the concentration gradients of dissolved ions are not collected in chambers initially filled with distilled water but are a snapshot in a (usually 0.1–1 mm) thin continuous water-containing gel, for example, made of polyacrylamide (Krom et al. 1994; Reeburgh and Erickson 1982). In front of the gel there is a dialysis membrane or filter as in a regular peeper. Before installation of the gel sampler it is stored analogously to a regular peeper in a container filled with distilled water. To deoxygenate the sampler, nitrogen is bubbled through the container's water (Krom et al. 1994). In general, there are two different gel sampler types and techniques: DET (diffusional equilibration in thin films) (Davison et al. 1991) and DGT (diffusive gradients in thin films) (Davison and Zhang 1994). When using the DET technique the sampler is exposed in situ until there is equilibrium between the pore water and the water in the gel (Davison et al. 1991). In a DGT gel sampler there is a binding agent placed behind the gel that binds the species of interest via sorption processes. This is irreversible under in situ conditions. (Davison and Zhang 1994). Thus, the latter technique is a sorption-based sampler and not described in the present chapter but in Section 9.4.3.

After recovery of the gel sampler its gel is sliced (resolution up to 1 mm), the ions of interest are re-dissolved and measured analytically. Basically, every solute can be measured as long as it is smaller than the pores of the dialysis membrane. Alternatively, the unsliced gel is measured with a beam technique measuring the entire gel with a high spatial resolution (Harper et al. 1997). When applying the DET technique the ions are usually immobilized after recovery of the gel sampler before applying the beam technique. For example, Davison et al. (1994) suggests immersing the recovered gel sampler in NaOH to oxidize and immobilize Fe^{2+} and Mn^{2+} to three-valent, precipitated hydroxides. However, there is also a technique for phosphate determination without previous fixation but with a fast processing and scanning of the distribution pattern within less than 25 minutes. After recovery the gel is immediately removed from the sampler and placed on a transparency sheet. A second gel layer previously soaked for 2 hours with the reagents of the standard molybdate ascorbic acid method is placed on the first gel layer and finally a transparence sheet is placed on top. After approximately 20 minutes of colour development of the layers, it is scanned with a conventional office flatbed scanner, converted into monochrome channels, and calibrated by gel pieces exposed to defined concentrations (Pages et al. 2011). Robertson et al. (2008, 2009) developed and applied a similar technique based on a flatbed scanner for Fe(II) determination. In contrast to these imaging techniques, the direct analytics of the ions requires that the samples are first re-dissolved in a defined volume of elution medium. Mobile ions are re-dissolved by simple equilibration with the elution medium.

Advantages:
An advantage of the gel sampler is the short exposition time, which depends on gel thickness and technique (DET or DGT) between some minutes (Davison et al. 1994) and a few hours (Krom et al. 1994). Furthermore, gel samplers allow a very high spatial resolution up to 1 mm (slicing techniques) or up to 0.1 mm (beam techniques). That is important since all features smaller or equal to the sampler resolution are lost during the investigations (Harper et al. 1997). With a resolution of 1 cm as is usual for conventional peepers, processes at the immediate sediment–water interface and flux estimates can be incorrect (Fones et al. 1998; Urban et al. 1997).

Disadvantages:
A disadvantage of the method is the vertical diffusion in the gel that does not occur in peepers with chambers. This vertical diffusion in the gel results during exposition and after recovery of the sampler to a smearing of concentration gradients (Harper et al. 1997; Krom et al. 1994). Thus, after recovery, gel samplers should be sliced or fixed as fast as possible. Davison et al. (1994) recommend 20 seconds. Davison et al. (1994) and Harper et al. (1997) suggest using chambers filled with gel to reduce the problem of diffusion in the gel layer but simultaneously use the advantage of a beam technique.

9.4.3 Sorption-based Passive Samplers

Diffusive gradient in thin films (DGT) samplers (similar to DET samplers) consist typically of a protective membrane and a relatively thick (~1 mm) gel layer followed by a binding phase. In contrast to the DET technique, for the DGT technique the optimum exposure time is much more important: if the exposition time is too short, concentrations are below the limit of quantification. If the exposition time is too long, all binding sites are blocked and further uptake is prohibited (Reeburgh and Erickson 1982). As in photography, the optimum picture is between over- and underexposure.

DGT is a robust in situ passive sampling technique for the measurement of labile concentrations, species, and distribution of various solutes in soil, sediment, and waters (Li et al. 2019). DGT gel samplers were developed for major ions and charged metal species. For instance, gel samplers (DET and DGT) were used for sulphide (Reeburgh and Erickson 1982), chloride, nitrate, sulphate, ammonium (Krom et al. 1994), iron, manganese (Davison et al. 1994), zinc, arsenic (Davison et al. 1997), calcium, magnesium, sodium, and potassium (Fones et al. 1998). As a binding agent in the DGT gel sampler, Davison and Zhang (1994) used a resin forming chelate complexes with metals (Chelex 100); Reeburgh and

Erickson (1982) used lead acetate for sulphide. After recovery of the gel sampler its gel is either sliced or the unsliced gel is measured with a beam technique. Thus, when applying the slicing technique, the ions that are fixed in the gel have to be re-dissolved by applying some kind of strong solvent, for example, a strong acid (Harper et al. 1997; Krom et al. 1994). Alternatively, the entire gel is measured with a large spatial resolution (up to 100 μm), e.g. X-ray analysis (Harper et al. 1997). Davison et al. (1994) used proton-induced X-ray emission (PIXE) for the measurement of the gel. PIXE is only applicable for elements with larger atomic numbers than sodium (Davison et al. 1994). For lighter elements, Rutherford backscattering can be used (Davison et al. 1994). An image-delivering semi-quantitative method for sulphide was developed by Reeburgh and Erickson (1982). Lead acetate in the gel reacts with sulphide to become black lead sulphide. Depending on the sulphide concentration, the black colour is more or less intensive. Robertson et al. (2008, 2009) developed and applied an alternative DGT technique for sulphide and combined it with a simultaneous Fe(II) determination based on DET.

A variety of sorption-based passive samplers have been developed to sample both hydrophobic and relatively polar organic compounds. The o-DGT sampler was specifically designed to sample organic chemicals. Similar to the original DGT sampler, o-DGT samplers consist of a protective membrane followed by a relatively thick (~1 mm) agarose gel layer and an agarose-based binding phase, which contains sorbent materials (Chen et al. 2012).

In contrast, polar organic chemical integrative samplers (POCIS) (Alvarez et al. 2004) and Chemcatcher® are made from disks of sorbent material, which varies depending on the polarity of the compound of interest. POCIS and Chemcatcher® samplers have primarily been used in surface water, waste water treatment plants, and in groundwater monitoring wells (Berho et al. 2013). In contrast to the DGT technique they provide – at least in their classic design – no spatial resolution, i.e. only one value per POCIS or Chemcatcher®. Mechelke et al. (2019) developed a sampler based on o-DGT and Chemcatcher® to assess concentration gradients across the surface water–groundwater interface in an urban river.

Advantages:
At low concentrations there is an enrichment of the ion of interest that allows a reliable analysis. In contrast to the DET technique it is not necessary to slice the gel immediately after sampling. There is no further diffusion that smears the concentration features. Compared to other adsorption-based passive samplers, DGT methods are typically less sensitive to environmental conditions such as hydrodynamic flow rates.

Disadvantages:
The DGT technique is only applicable for ions for which a binding agent has been developed and for each ion or group of ions a different type of sampler is required with a binding agent specifically developed for the ion of interest. The DGT technique is only applicable when there is a fast supply of the ion of interest from the matrix (Fones et al. 1998). In many systems this might cause interpretation problems and artefacts: due to the binding characteristics of the binding layer there is a strong uptake of the ion of interest. As a consequence, the concentrations of the ion in front of the gel sampler are drastically decreased. Equilibrium reactions might result in a mobilization of the ion of interest from the sediment matrix as a consequence of the decreased concentrations. However, even equilibrium reactions have a thermodynamic component and it takes some time to release the ion. At the same time, the

permanent uptake of the binding agent keeps the concentrations in front of the sampler low. Due to the low concentrations there is a strong concentration gradient in front of the sampler. Diffusive and advective processes result in a transport of the ion of interest to the surface of the gel sampler. Thus, the concentrations measured by the DGT technique are usually a complex overlap of the original concentrations, of the equilibrium processes close to the sampler, and of the diffusive and advective transport processes.

9.5 Outlook

The development of the sampling methods discussed in this chapter is primarily driven by the motivation to capture the steep biogeochemical, biological, and physical gradients encountered at groundwater–surface water interfaces. Although many sampling devices have been miniaturized, challenges remain to sample very steep gradients on the sub-cm scale. For instance, in hyporheic sediments oxygen can be depleted within several centimetres or even millimetres. To date, the only methods available to assess sub-mm scale gradients are in situ sensors.

Whether a coring method, or active or passive sampling method should be chosen, not only depends on the parameter of interest, but also on the study design and overall study goal. If measured concentration depth profiles are used to estimate turnover or removal rates via a reactive transport model, the temporal resolution of the sampling method should match the temporal resolution at which water flow velocities are estimated. Active sampling methods can be combined more readily than passive methods with methods that yield flux information at high temporal resolution (i.e. hours), such as heat pulse sensors (Angermann et al. 2012; Banks et al. 2018). On the contrary, if water fluxes are assessed using methods that provide flux values over longer timescales, for instance, methods utilizing daily temperature fluctuations, sampling of time-weighted average concentrations may be sufficient. Active sampling methods with high temporal resolution may capture both chemical tracers and target analytes thus providing information on concentration gradients and transport parameters at the same location within the groundwater–surface water interface. When passive sampling devices are used, however, additional instrumentation is required, which may not capture the transport characteristics at the sampling location. Recent developments, such as the sediment bed passive flux meter (Kunz et al. 2017; Layton et al. 2017; Ottosen et al. 2020) that allows the simultaneous assessment of both water and contaminant flux, seek to overcome this issue, but further testing under field conditions is required.

References

Adams, D.D. (1991). Sediment pore water sampling. In: *CRC Handbook of Techniques for Aquatic Sediments Sampling* (ed. A. Mudroch and S.D. Mac Knight), 171–202. Boca Raton: CRC Press.

Alvarez, D.A., Petty, J.D., Huckins, J.N. et al. (2004). Development of a passive, *in situ*, integrative sampler for hydrophilic organic contaminants in aquatic environments. *Environ. Toxicol. Chem.: Internat. J.* 23: 1640–1648.

Angermann, L., Krause, S., and Lewandowski, J. (2012). Application of heat pulse injections for investigating shallow hyporheic flow in a lowland river. *Water Resourc. Res.* 48: W00P02.

Banks, E.W., Shanafield, M., Noorduijn, S. et al. (2018). Active heat pulse sensing of 3-D-flow fields in streambeds. *Hydrol. Earth Sys. Sci.* 22: 1917–1929.

Berho, C., Togola, A., Coureau, C. et al. (2013). Applicability of polar organic compound integrative samplers for monitoring pesticides in groundwater. *Environ. Sci. Pollut. Res.* 20: 5220–5228.

Bianchin, M., Smith, L., and Beckie, R. (2015). Freeze shoe sampler for the collection of hyporheic zone sediments and porewater. *Groundwater* 53: 328–334.

Bou, C. and Rouch, R. (1967). Un nouveau champ de recherches sur la faune aquatique souterraine. Comptes Rendus de l'Académie des Sciences de Paris, *Sciences de la Vie*. 265: 369–370.

Blume, T., Krause, S., Meinikmann, K., and Lewandowski, J. (2013). Upscaling lacustrine groundwater discharge rates by fiber-optic distributed temperature sensing. *Water Resour Res.*. 49: 7929–7944.

Brandl, H. and Hanselmann, K.W. (1991). Evaluation and application of dialysis porewater samplers for microbiological studies at sediment-water interfaces. *Aquat. Sci.* 53: 55–73.

Burk, L. and Cook, P.G. (2015). A simple and affordable system for installing shallow drive point piezometers. *Groundwater Monit.Remediat.* 35: 101–104.

Carignan, R. (1984a). Interstitial water sampling by dialysis: methodological notes. *Limnol. Oceanogr.* 29: 667–670.

Carignan, R. (1984b). Sediment geochemistry in a eutrophic lake colonized by the submersed macrophyte *Myriophyllum spicatum*. *Verh. Internat. Verein. Limnol.* 22: 355–370.

Chen, C.E., Zhang, H., and Jones, K.C. (2012). A novel passive water sampler for *in situ* sampling of antibiotics. *J. Environ. Monit.* 14: 1523–1530.

Conant, B. Jr., Cherry, J.A., and Gillham, R.W. (2004). A PCE groundwater plume discharging to a river: influence of the streambed and near-river zone on contaminant distributions. *J. Contam. Hydrology* 73: 249–279.

Davison, W., Fones, G.R., and Grime, G.W. (1997). Dissolved metals in surface sediment and a microbial mat at 100-μm resolution. *Nature* 387: 885–887.

Davison, W., Grime, G.W., Morgan, J.A.W., and Clarke, K. (1991). Distribution of dissolved iron in sediment pore waters at submillimetre resolution. *Nature* 352: 323–325.

Davison, W. and Zhang, H. (1994). *In situ* speciation measurements of trace components in natural waters using thin-film gels. *Nature* 367: 546–548.

Davison, W., Zhang, H., and Grime, G.W. (1994). Performance of gel probes used for measuring the chemistry of pore waters. *Environ. Sci. Technol.* 28: 1623–1632.

Duff, J.H., Murphy, F., Fuller, C.C., Triska, F.J., Harvey, J.W., and Jackman, A.P. (1998). A mini drivepoint sampler for measuring pore water solute concentrations in the hyporheic zone of sand-bottom streams. *Limnol. Oceanogr.* 43: 1378–1383.

Enell, M. and Löfgren, S. (1988). Phosphorus in interstitial water: methods and dynamics. *Hydrobiologia* 170: 103–132.

Fones, G.R., Davison, W., and Grime, G.W. (1998). Development of constrained DET for measurements of dissolved iron in surface sediments at sub-mm resolution. *Sci. Total Envir.* 221: 127–137.

Freitas, J.G., Rivett, M.O., Roche, R.S. et al. (2015). Heterogeneous hyporheic zone dechlorination of a TCE groundwater plume discharging to an urban river reach. *Sci. Tot. Environ.* 505: 236–252.

Harper, M.P., Davison, W., and Tych, W. (1997). Temporal, spatial and resolution constraints for *in situ* sampling devices using diffusional equilibration: dialysis and DET. *Environ. Sci. Technol.* 31: 3110–3119.

Harvey, J.W. and Fuller, C.C. (1998). Effect of enhanced manganese oxidation in the hyporheic zone on basin-scale geochemical mass balance. *Water Resource Res.* 34: 623–636.

Hesslein, R.H. (1976). An *in situ* sampler for close interval pore water studies. *Limnol. Oceanogr.* 22: 912–914.

Howes, B.L., Dacey, J.W.H., and Wakeham, S.G. (1985). Effects of sampling technique on measurements of porewater constituents in salt-marsh sediments. *Limnol. Oceonogr.* 30: 221–227.

Hunt, G.W. and Stanley, E.H. (2000). An evaluation of alternative procedures using the Bou-Rouch method for sampling hyporheic invertebrates. *Can. J. Fish. Aquat. Sci.* 57: 1545–1550.

Johnson, R.L, Brow, C.N., Johnson, R.O., and Simon, H.M. (2013). Cryogenic core collection and preservation of subsurface samples for biomolecular analysis. *Groundwater Monit. Remediat.* 33: 38–43.

Kaeser, D.H., Binley, A., Heathwaite, A.L., and Krause, S. (2009). Spatio-temporal variations of hyporheic flow in a riffle-step-pool sequence. *Hydrol. Process.* 23: 2138–2149.

Kalbus, E., Reinstorf, F., and Schirmer, M. (2006). Measuring methods for groundwater - surface water interactions: a review. *Hydrol. Earth Syst. Sci.* 10: 873–887.

Kankanamge, N.R., Bennett, W.W., Teasdale, P.R. et al. (2020). A new colorimetric DET technique for determining mm-resolution sulfide porewater distributions and allowing improved interpretation of iron(II) co-distributions. *Chemosphere* 244: 125388.

Knapp, J.L., González-Pinzón, R., Drummond, J.D. et al. (2017). Tracer-based characterization of hyporheic exchange and benthic biolayers in streams. *Water Resour. Res.* 53: 1575–1594.

Krause, S., Lewandowski, J., Grimm, N.B. et al. (2017). Ecohydrological interfaces as hotspots of ecosystem processes. *Water Resour. Res.* 53: 6359–6376.

Krause, S., Tecklenburg, C., Munz, M., and Naden, E. (2013). Streambed nitrogen cycling beyond the hyporheic zone: Flow controls on horizontal patterns and depth distribution of nitrate and dissolved oxygen in the upwelling groundwater of a lowland river. *J. Geophys. Res. Biogeosci.* 118: 54–67.

Krom, M.D., Davison, P., Zhang, H., and Davison, W. (1994). High-resolution pore-water sampling with a gel sampler. *Limnol. Oceanogr.* 39: 1967–1972.

Kunz, J.V., Annable, M.D., Cho, J. et al. (2017). Quantifying nutrient fluxes with a new hyporheic passive flux meter (HPFM). *Biogeosciences* 14: 631–649.

Laskov, C., Herzog, C., Lewandowski, J., and Hupfer, M. (2007). Miniaturized photometrical methods for the rapid analysis of phosphate, ammonium, ferrous iron, and sulfate in pore water of freshwater sediments. *Limnol. Oceanogr. Methods* 4: 63–71.

Layton, L., Klammler, H., Hatfield, K. et al. (2017). Development of a passive sensor for measuring vertical cumulative water and solute mass fluxes in lake sediments and streambeds. *Adv. Water Resour.* 105: 1–12.

Lewandowski, J. (2002). *Untersuchungen zum Einfluss seeinterner Verfahren auf die Phosphor-Diagenese in Sedimenten (Investigations of the influence of in-lake measures on the phosphorus diagenesis in sediments, in German with abstract and supporting information in English). Dissertation*, Humboldt University Berlin, 144.

Lewandowski, J. and Hupfer, M. (2005). Effect of macrozoobenthos on two-dimensional small-scale heterogeneity of pore water phosphorus concentrations in lake sediments: a laboratory study. *Limnol. Oceanogr.* 50: 1106–1118.

Lewandowski, J., Laskov, C., and Hupfer, M. (2007). The relationship between *Chironomus plumosus* burrows and the spatial distribution of pore-water phosphate, iron and ammonium in lake sediments. *Freshwater Biol.* 52: 331–343.

Lewandowski, J. and Nützmann, G. (2010). Nutrient retention and release in a floodplain's aquifer and in the hyporheic zone of a lowland river. *Ecol. Eng.* 36: 1156–1166.

Lewandowski, J., Rüter, K., and Hupfer, M. (2002). Two-dimensional small-scale variability of pore water phosphate in freshwater lakes: results from a novel dialysis sampler. *Environ. Sci. Technol.* 36: 2039–2047.

Lewandowski, J., Schadach, M., and Hupfer, M. (2005). Impact of macrozoobenthos on two-dimensional small-scale heterogeneity of pore water phosphorus concentrations: *in situ* study in Lake Arendsee (Germany). *Hydrobiologia* 549: 43–55.

Li, C., Ding, S.M., Yang, L.Y. et al. (2019). Diffusive gradients in thin films: devices, materials and applications. *Environ. Chem. Lett.* 17: 801–831.

Martens, C.S. and Val Klump, J. (1980). Biogeochemical cycling in an organic-rich coastal marine basin: i. Methane sediment-water exchange process. *Geochim. Cosmochim. Acta* 44: 471–490.

Mayer, L.M. (1976). Chemical water sampling in lakes and sediments with dialysis bags. *Limnol. Oceanogr.* 21: 909–912.

McMillan, L.A., Rivett, M.O., Wealthall, G.P. et al. (2018). Monitoring well utility in a heterogeneous DNAPL source zone area: insights from proximal multilevel sampler wells and sampling capture-zone modelling. *J. Contam. Hydrol.* 210: 15–30.

Mechelke, J., Vermeirssen, E.L., and Hollender, J. (2019). Passive sampling of organic contaminants across the water-sediment interface of an urban stream. *Water Res.* 165: 114966.

Meinikmann Hupfer, M., and Lewandowski, J. (2015). Phosphorus in groundwater discharge – a potential source for lake eutrophication. *J. Hydrol.* 524: 214–226.

Murphy, F. and Herkelrath, W.N. (1996). A sample-freezing drive shoe for a wire line piston core sampler. *Groundwater Monit. Remediat.* 16: 86–90.

Ottosen, C.B., Rønde, V., McKnight, U.S. et al. (2020). Natural attenuation of a chlorinated ethene plume discharging to a stream: integrated assessment of hydrogeological, chemical and microbial interactions. *Water Res.* 186: 116332.

Pages, A., Teasdale, P.R., Robertson, D. et al. (2011). Representative measurement of two-dimensional reactive phosphate distributions and co-distributed iron(II) and sulfide in seagrass sediment porewaters. *Chemosphere* 85: 1256–1261.

Pöschke, F., Lewandowski, J., and Nützmann, G. (2015). Impacts of alluvial structures on small-scale nutrient heterogeneities in near-surface groundwater. *Ecohydrology* 8: 682–694.

Posselt, M., Jaeger, A., Schaper, J.L. et al. (2018). Determination of polar organic micropollutants in surface and pore water by high-resolution sampling-direct injection-ultra high performance liquid chromatography-tandem mass spectrometry. *Environ. Sci. Process. Imp.* 20: 1716–1727.

Puls, R.W. and Barcelona, M.J. (1996). *Low-flow (minimal drawdown) groundwater sampling procedures*. EPA/540/S-95/504.

Reeburgh, W.S. and Erickson, R.E. (1982). A "dipstick" sampler for rapid continuous chemical profiles in sediments. *Limnol. Oceanogr.* 27: 556–559.

Renberg, I. and Hansson, H. (2010). Freeze corer No. 3 for lake sediments. *J. Paleolimnol.* 44: 731–736.

Rivett, M.O., Ellis, R., Greswell, R.B., et al. (2008). Cost-effective mini drive-point piezometers and multilevel samplers for monitoring the hyporheic zone. *Quart. J. Eng. Geol. Hydrogeology* 41: 49–60.

Rivett, M.O., Roche, R.S., Tellam, J.H., and Herbert, A.W. (2019). Increased organic contaminant residence times in the urban riverbed due to the presence of highly sorbing sediments of the Anthropocene. *J. Hydrol X* 3: 100023.

Robertson, D., Teasdale, P.R., and Welsh, D.T. (2008). A novel gel-based technique for the high resolution, two-dimensional determination of iron (II) and sulfide in sediment. *Limnol. Oceanogr. Meth.* 6: 502–512.

Robertson, D., Welsh, D.T., and Teasdale, P.R. (2009). Investigating biogenic heterogeneity in coastal sediments with two-dimensional measurements of iron(II) and sulfide. *Environ. Chem.* 6: 60–69.

Sanders, I.A. and Trimmer, M. (2006). In situ application of the 15NO3− isotope pairing technique to measure denitrification in sediments at the surface water-groundwater interface. *Limnol. Oceanogr. Methods.* 4: 142–152.

Schaper, J.L., Posselt, M., Bouchez, C. et al. (2019). Fate of trace organic compounds in the hyporheic zone: influence of retardation, the benthic biolayer, and organic carbon. *Environ. Sci. Technol.* 53: 4224–4234.

Schaper, J.L., Seher, W., Nützmann, G. et al. (2018). The fate of polar trace organic compounds in the hyporheic zone. *Water Res.* 140: 158–166.

Schroeder, H., Duester, L., Fabricius, A.-L. et al. (2020). Sediment water (interface) mobility of metal(loid)s and nutrients underundisturbed conditions and during resuspension. *J. Hazardous Mater.* 394: 122543.

Strasser, D., Lensing, H.J., Nuber, T. et al. (2015). Improved geohydraulic characterization of river bed sediments based on freeze-core sampling – development and evaluation of a new measurement approach. *J. Hydrol.* 527: 133–141.

Urban, N.R., Dinkel, C., and Wehrli, B. (1997). Solute transfer across the sediment surface of a eutrophic lake: i. Porewater profiles from dialysis samplers. *Aquat. Sci.* 59: 1–25.

Weatherill J., Krause S., Voyce K. et al. (2014). Nested monitoring approaches to delineate groundwater trichloroethene discharge to a UK lowland stream at multiple spatial scales. *J. Contam. Hydrol.* 158: 38–54. http://dx.doi.org/10.1016/j.jconhyd.2013.12.001.

10

Automated Sensing Methods for Dissolved Organic Matter and Inorganic Nutrient Monitoring in Freshwater Systems

Phillip J. Blaen[1,2,3], Kieran Khamis[1], Charlotte E.M. Lloyd[4], Chris Bradley[1], David Hannah[1], and Stefan Krause[1,2]

[1] *School of Geography, Earth and Environmental Sciences, University of Birmingham, Edgbaston, Birmingham, B15 2TT, UK*
[2] *Birmingham Institute of Forest Research (BIFoR), University of Birmingham, Edgbaston, Birmingham, B15 2TT, UK*
[3] *Yorkshire Water, Halifax Road, Bradford, BD6 2SZ UK*
[4] *Organic Geochemistry Unit, Bristol Biogeochemistry Research Centre, School of Chemistry, University of Bristol, Cantocks Close, Bristol, BS8 1TS, UK*

10.1 Introduction and Historical Context

Rivers act as key conveyors of nutrients and organic matter through catchments from land surfaces to the ocean, and thus fluvial networks constitute important linkages at the interfaces of terrestrial, aquatic, marine, and atmospheric systems. Hydrological transport processes, coupled with in-stream chemical and biological transformations, are a fundamental component of global biogeochemical cycles and it is thus important to understand process dynamics at finer reach and watershed-scales (Battin et al. 2008; Ensign and Doyle 2006; Hood et al. 2015). In many landscapes, anthropogenic activities such as farming and urban development contribute significantly to elevated riverine nutrient loads if not managed correctly, which can substantially impact aquatic communities and ecosystem functioning (Smith and Schindler 2009). Eutrophication is a particular concern and is associated with algal blooms that lower dissolved oxygen concentrations, alter pH, and increase turbidity levels, which in turn are deleterious to river habitats and aquatic biodiversity (Camargo et al. 2005; Friberg et al. 2009). Elevated riverine nutrient levels also pose threats to human health and well-being, either directly or indirectly. For example, high nitrate concentrations in drinking water are associated with diabetes and cancer, and concentrations >50 mg L^{-1} in drinking water can lead to methemoglobinemia, especially in infants (Ward et al. 2005). Indirect threats include reductions in key aquatic ecosystem services, such as freshwater provision and maintenance of suitable habitats for fisheries (Bennett et al. 2009; MEA 2005).

The environmental consequences associated with excessive nutrient concentrations provide a clear rationale for understanding how, where, and when nutrients are transported and transformed through fluvial networks. Catchment managers can use data supplied by nutrient monitoring programmes to identify the effect of natural environmental variability (e.g. seasonal droughts, annual biomass production, and senescence) or anthropogenic

Ecohydrological Interfaces, First Edition. Edited by Stefan Krause, David M. Hannah and Nancy B. Grimm.

impacts (e.g. point-source discharges and land-use change) on river water quality. This information can also help to determine the quantity and quality of water required to sustain aquatic ecosystems (i.e. environmental flows) and identify the most suitable times or places for water abstraction (Bartram and Rees 1999; Palmer and Bernhardt 2006). Over time, the accumulated data from nutrient monitoring programmes can provide insights into temporal variability in ecosystem behaviour (Burt et al. 2010) and the processes that underpin these patterns (Rode et al. 2016). For example, long-term monitoring of river nitrate concentrations in the UK and USA reveal substantial increases over the last 50 years, corresponding with similar increases in the application of inorganic nitrogen fertilizer to farmland over the same period (Burt et al. 2010; McIsaac and Libra 2003). Such long-term datasets are invaluable for quantifying the benefits of mitigation measures, underpinning future decisions relating to catchment management (Bowes et al. 2009) and understanding the environmental drivers of nutrient loads in order to predict future trends (Whitehead et al. 2009).

Over the last 50 years, monitoring of river nutrient concentrations has developed from occasional sampling of local rivers (Casey and Clarke 1979) to the establishment of national-scale standardized sampling programmes, such as the UK Harmonised Monitoring Scheme and General Quality Assessment Scheme (Simpson 1980). More recently, multinational schemes, such as the European Union Water Framework Directive or WFD, 2000/60/EC, have been established to form holistic assessment protocols across several different member states to improve the effectiveness of transboundary water resource management (Hering et al. 2010). Traditionally, river water sampling was undertaken manually using discrete grab samples, followed by laboratory analysis to determine the water quality parameters of interest. As such, the spatial and temporal resolution of nutrient monitoring systems and observational networks were relatively low due to cost restraints associated with both fieldwork and laboratory analysis (Bende-Michl and Hairsine 2010). However, studies now suggest that river nutrient concentrations can change rapidly (i.e. minutes to hours) in response to variability in climatic and environmental factors, particularly when interacting with anthropogenic management practices and engineered structures (Bieroza and Heathwaite 2015; Halliday et al. 2015; Khamis et al. 2020; Wade et al. 2012). As such, these changes are unlikely to be captured fully by the monitoring frequency employed by standard water quality sampling approaches. Inadequate sampling frequencies can result in missing important pulses of nutrient export from river catchments associated with short-term storm events, and also in underestimating annual nutrient loads by up to an order of magnitude (Bowes et al. 2009; Lloyd et al. 2016; Mellander et al. 2012).

The use of in situ sensors in river networks for quantification of key macronutrient concentrations (i.e. C, N, P), species (e.g. NO_3^-, NO_2^-, NH_4^+) and composition (e.g. dissolved organic matter, DOM); spectral slope and fluorescence spectra, has increased in recent years (Blaen et al. 2016; Khamis et al. 2018; Ruhala and Zarnetske 2017). This trend has been driven by technological improvements and increased recognition of the challenges and limitation of manual sampling approaches for monitoring nutrient concentrations (e.g. uncertainty in flux calculations). In situ sensors can generate data at sub-minute frequencies and can therefore enable paired flow and nutrient measurements at the same resolution. Additional advantages of in situ nutrient monitoring include the capacity to: (i) better understand the drivers and controls on catchment nutrient dynamics (Blaen et al. 2017; Bowes et al. 2009), (ii) capture non-linear behaviour and tipping points in lotic ecosystems (Wade et al. 2012), (iii) improve legislative monitoring strategies (Halliday et al. 2015), and (iv) to provide early-warning indicators of potentially hazardous bacteria or waste by-products (Bridgeman et al. 2015). However, equipment for in situ nutrient monitoring is often expensive to purchase and maintain, and presents its own unique set of challenges that must be overcome to produce robust and meaningful data. In this chapter, we firstly review the principles of measurement associated with most common contemporary in situ nutrient sensing techniques. We then describe the challenges faced by in situ nutrient monitoring sampling programmes and outline how these can be overcome. Subsequently, we demonstrate how previous applications of in situ monitoring have improved our understanding of ecosystem dynamics at the interfaces of lotic systems and their associated catchments. Finally, we conclude by discussing future directions for this rapidly-developing field of research.

10.2 Technological Principles of Techniques for in situ Nutrient Monitoring

Recent advances in electrochemical detection, colorimetry, and optical methods using absorbance or fluorescence offer increasing opportunities for automated in situ determinations of macronutrient concentrations, species, and compositions. The following section outlines the principles of these measurement techniques.

10.2.1 Electrochemical Detection

The application of electrochemical detection using ion-selective electrodes (ISEs) to determine nutrient concentrations has been well established in laboratory environments. More recently, miniaturization of the technology has enabled in situ quantification of ionic activity, including nitrate, ammonium, and pH (Le Goff et al. 2002; Merks 1975). A doped membrane (made typically of glass or polymers), specific to a particular ion of interest, is placed between a sensing electrode and a reference electrode. The principle of ISE measurement is based on the voltage potential between the electrodes induced by the activity of the dissolved ion of interest in solution (Figure 10.1a). The main advantages of ISEs are their ease of use, and their relatively low upfront purchase cost relative to other sensor types, which means they can be deployed in arrays to improve monitoring accuracy (Belikova et al. 2019). Moreover, ISEs do not require consumable reagents and provide relatively fast measurements, and can thus be used to monitor trends in ionic activity at sub-minute resolution if required. The total measurement range is larger than many other sensor techniques (see below) and can typically span at least three orders of magnitude (Hach 2016; YSI 2016), thus making them suitable for aquatic environments where system dynamics can change rapidly. ISEs are not affected significantly by water colour or turbidity, and therefore results are accurate and stable even in sediment-rich waters in which alternative (e.g. optical) measurement techniques may require post-processing corrections. However, data resolution tends to be relatively low compared with other techniques (errors are typically up to 10%) and instruments are also often subject to significant drift and interference from non-target ions (De Marco et al. 2007), which can be especially problematic in heavily-polluted systems. Consequently, field instruments often require regular recalibration and cleaning (usually weekly but in some cases daily), although recent developments in sensor technology have overcome some of these issues. For example, some nitrate ISEs are now capable of providing stable, drift-free measurements for up to four months in the field (Le Goff et al. 2002), albeit over a reduced concentration range.

10.2.2 Colorimetry

Colorimetric measurement is based on mixing samples with appropriate reagents to trigger a chemical reaction that results in a coloured solution, the intensity of which is proportional to the concentration of the analyte of interest. This solution is then measured using photometry and absorbance converted into concentration using a predefined calibration curve (Figure 10.1b). Colorimetric methods have been utilized to measure nitrate, nitrite, ammonium, phosphate (usually as total reactive phosphorus, TRP), and total phosphorus (TP; with an extra digestion step) using either submersible microfluidic technology or bank-side analysers. In comparison to ISEs, colorimetric techniques usually provide greater precision and accuracy. However, their measurement range is generally limited to only 1–2 orders of magnitude (Patton and Kryskalla 2011). Most instruments require calibration, which can often be achieved in situ and at regular programmed

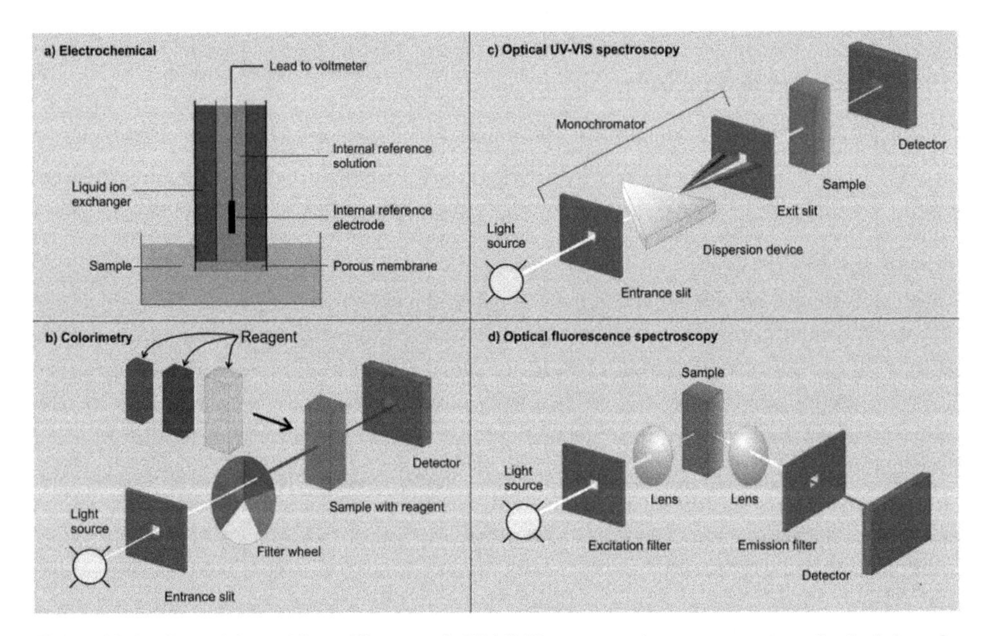

Figure 10.1 (from Adapted from Blaen et al., 2016) Diagrammatic representation of principles of in situ nutrient monitoring techniques for (a) electrochemical, (b) colorimetry, (c) optical UV-VIS spectroscopy, and (d) optical fluorescence spectroscopy.

intervals. Drawbacks to colorimetric methods include high initial set-up costs and high running costs associated with consumption of reagents. Moreover, safe disposal facilities are required for waste reagents because some chemicals are toxic to the environment. The volume of waste produced depends on the technology used; a typical lab-on-a-chip microfluidic approach for measuring nitrate will produce 200 ml of waste per day (Beaton et al. 2012), while a bank-side analyser measuring TP will produce 2400 ml per day (Hach Lange 2016). Given the temperature dependency of most colorimetric reactions, post-correction of measurements is necessary if the water temperature undergoes significant change during deployment. Furthermore, measurements are sensitive to the background colour of the water being tested, which can present problems where waters are highly turbid. In-line filtration can mitigate this problem but at the expense of increased costs and maintenance. Temporal sampling resolutions can be as low as 30 min but are limited by the time required for the reagent colour to develop fully. Given the nature of the analysis, colorimetric methods can have relatively high power requirements because some reactions need to be completed at high temperature and high pressure to accelerate reaction times. However, new sensors are being developed with lower power requirements that can be run feasibly from batteries over longer time periods than earlier generations of instruments.

10.2.3 Optical Ultraviolet–Visible (UV-VIS) Spectroscopy

Optical UV-VIS sensors use ultraviolet spectroscopy to measure the absorption of a sample at a defined wavelength, that is then processed to determine a range of parameters including dissolved organic carbon (DOC) and nitrate (Figure 10.1c). A key advantage of this technology is that no reagents are consumed during the measurement process. The optical signal absorbance spectra is converted to the parameter of interest using a variety of algorithms (Lepot et al. 2016). Therefore, unlike ISEs or colorimetric methods, the technique constitutes a proxy measurement as it does not quantify the parameter of interest directly. The resulting data are usually highly precise and accurate, although this can vary from parameter to parameter (Li et al. 2016; Van Den Broeke et al. 2006), and measurements can be acquired at up to sub-minute resolution if required. The collected data stream can also include additional spectral data from which useful information can be derived. For example, spectral slope can provide insights into molecular weight (Spencer et al. 2007). Instrument costs for optical sensors are typically high. Moreover, regular cleaning is necessary to limit fouling of the optics as wiper mechanisms and air-blast kits are often unable to inhibit biofouling in some systems (Jones et al. 2014). As with colorimetry, optical interferences can arise from in situ applications, such river debris, water discolouration, or high sediment loads. All instruments require initial calibration but, compared with other sensor technologies, calibrations can remain stable for extended periods. For example, NO_3-N calibrations can remain stable for up to two years (Huebsch et al. 2015). Historically, such instruments have been associated with high power usage, but the recent introduction of LED light sources, which are low in both cost and power consumption, has made optical sensors appropriate for a wider range of applications than was possible previously (e.g. Blaen et al. 2017).

10.2.4 Optical Fluorescence Spectroscopy

The recent development of in situ fluorescence spectroscopy has been driven by advances in technology (e.g. UV LED light sources) coupled with decreases in sensor costs (Khamis et al. 2018; Ruhala and Zarnetske 2017). In situ fluorometers generally target a single excitation-emission pair, and measure the light emitted from a sample at a given wavelength when the sample is excited by light of a known wavelength (Figure 10.1d). Thus, in situ sensors have a relatively narrow wavelength range compared with benchtop equivalents. However, these wavelength pairs can be selected to target particular compound classes based on their fluorescent properties. Many commercially-available sensors are designed to target relatively low molecular weight compounds such as proteineous material and hydrocarbons. However, sensors are also capable of detecting larger, more complex, macromolecular materials that comprise the coloured DOM (CDOM) fraction. Using specific wavelengths, additional parameters, such as chlorophyll *a* and dissolved oxygen (fluorescence lifetime measurement), can be quantified. The temporal resolution of fluorescent probes can be very high and offers the possibility of sub-minute resolution data. Fluorescence sensors can be expensive to install, although these costs are usually less than those for colorimetric sensors. As with UV-VIS spectroscopy instruments, a robust field calibration is required for all sensors, however on-going maintenance costs are minimal and calibrations are stable for ~6 months. Until recently, fluorescence sensors were designed primarily for marine research (Daniel et al. 1995) and earlier generations of instruments were robust to the point of being over-engineered for river systems. However, the advent of LED light sources (Conmy et al. 2014), and a growing market for non-marine monitoring applications, has led to newer generations of lower-cost instruments designed specifically for use in freshwater environments. Fluorescence sensors suffer from similar issues as UV-VIS sensors in that they are subject to interference from turbidity, bubbles, temperature solar radiation, and fouling (Downing et al. 2012; Khamis et al. 2015; Saraceno et al. 2017; Watras et al. 2011). In addition to the fluorescence signal, it is also important that field instruments monitor absorbance at the relevant wavelengths to enable accurate post-processing of data, because inner-filter effects caused by reabsorption of emitted photons by other molecules in the sample can affect the data (Lakowicz 2013). Installation locations should also be chosen carefully to ensure the appropriate measurement angles as UV solar radiation can cause interference with measurements, though this can be mitigated by using a shade cap (e.g. Watras et al. 2011).

10.2.5 Capabilities and Limitations of Automated Techniques for in situ Nutrient Monitoring

A comparison of the in situ nutrient monitoring techniques described in the preceding text is provided in Table 10.1 which summarizes the available measurement parameters, range, accuracy, and key maintenance requirements of each technique. It is important to note that the measurement accuracy of most field instruments is almost always lower than would be achievable using laboratory equipment (but see Bieroza and Heathwaite 2016). Regular maintenance of all instruments is necessary to prevent further decreases in accuracy; as a general rule, wherever battery operated or mechanical equipment is employed, maintenance frequency is proportional to the sampling frequency of the instrument.

Table 10.1 Summary of commonly-used measurement methods for in situ nutrient monitoring and the associated accuracy, range, and maintenance requirements for each method (Adapted from Blaen et al., 2016).

Measurement method	Measurement principle	Parameters	Measurement range[a]	Accuracy[a]	Maintenance	References
Absorbance	Spectrophotometric method to measure the amount of light of a specific wavelength absorbed by a sample	Nitrate/ nitrite	0.1–500 mg L^{-1} NO_3-N	Up to 10% of reading	• Regular calibration (3–6 months) • Regular cleaning due to fouling • Check power supply • Check light source • Site specific calibration	Huebsch et al. (2015)
		TOC/DOC	1–150 mg L^{-1} TOC 0.5–75 mg L^{-1}- DOC	<10% of reading		Waterloo et al. (2006), Jeong et al. (2012), Lee et al. (2015)
		Phosphate	0–0.3 mg L^{-1} PO_4-P	2% of reading		Subchem (2016)
Fluorescence	Spectrophotometric method to measure the emission of light at specific wavelengths after excitation with a specific wavelength of light	Tryptophan-like	0–20000 ppb	2% of reading	• Regular calibration (3–6 months) • Regular cleaning due to fouling • Check power supply • Check light source • Site specific calibration	Khamis et al. (2015)
		CDOM	0–500 μg L^{-1}	± 0.09 μg L^{-1}		Chen (1999)
		Chlorophyll a	0–125 μg L^{-1}	± 0.02 μg L^{-1}		Del Castillo et al. (1999)
Colorimetric	Wet chemical technique, detection via photometry	Nitrate/ nitrite	0.3–50 mg L^{-1}	<3% of reading	• Regular calibration (once a week or once a day if automated) • Regular cleaning due to fouling • Check power supply • Replacement of reagents • Disposal of reagents	Greenway et al. (1999), Petsul et al. (2001)
		Phosphate	1–10 mg L^{-1}	<5% of reading		Doku and Haswell (1999), Cleary et al. (2008)
		Total phosphorus	0.1–5 mg L^{-1} PO_4-P	2% of reading		Jordan et al. (2007); Wade et al. (2012), Lloyd et al. (2016)

(Continued)

Table 10.1 (Continued)

Measurement method	Measurement principle	Parameters	Measurement range[a]	Accuracy[a]	Maintenance	References
Electrochemistry–Ion Selective Electrodes	Direct potentiometry between a sensing electrode and a reference electrode	Nitrate	0–422 mg L^{-1} NO_3-N	Up to 10% or 2 mg L^{-1}, whichever is greatest, LOD 0.25 mg L^{-1}	• Calibration at least once a day due to drift • Regular cleaning due to fouling • May need reconditioning	Adsett and Zoerb (1991), Le Goff et al. (2002)
		Ammonium	0–9000 mg L^{-1} NH_4-N	Up to 10% or 2 mg L^{-1}, whichever is greatest		Merks (1975), Toda et al. (2011)
		pH	0–14 units	±0.2 units		Adamchuk et al. (1999), Collings et al. (2003)

a) Measurement accuracy compiled from commercially available sensors. Note that accuracy under field conditions may be less than under laboratory test conditions.

10.3 General Technological Challenges of in situ Nutrient Monitoring Techniques

In situ nutrient monitoring systems pose a number of common barriers that must be addressed to produce robust, high-quality datasets. The following section discusses these technological challenges and outlines the potential options available to overcome them.

10.3.1 Power Supply

Given the remote locations of many in situ nutrient monitoring systems, the most common source of instrument power is from batteries, fuel cells, or solar panels. In some cases (e.g. for bankside analysers) generators or mains electricity may be required to meet the additional power demands associated with pumping water to the analyser, or the heating required for certain digestion steps (Wade et al. 2012). Power requirements vary between sensors and their associated dataloggers and telemetry systems. Consequently, an accurate estimation of power requirements should be incorporated into the design process of each monitoring system. When using batteries or fuel cells, regular power checks should be made to minimize instrument downtime, particularly during winter, because battery efficiency decreases at lower temperatures. Solar panels may also operate sub-optimally in winter when low solar altitudes and more frequent cloud cover reduce incident radiation levels. Solar panel efficiency can also decrease in summer in forested areas due to the shading effects of vegetation. Therefore, careful positioning of solar panels is necessary for optimal efficiency. Using panels in combination with a battery and charging regulator is recommended to buffer the effects of diurnal and seasonal variability in electricity generation.

10.3.2 Fouling

Almost all in situ sensors are subject to fouling. When submerged in the water column, it is common for inorganic sediments, organic biofilms, and other coarse debris to accumulate around the sensor. For bankside analysers, it is also common for biofilms to develop in the pump tubing. Technologies to reduce the incidence of sensor fouling include automatic sensor wipers, compressed air pumps to clear accumulated particles and biofilms, and anti-fouling coatings (Conmy et al. 2014; Kotamäki et al. 2009). Despite these, however, regular manual cleaning of sensors is almost always required to ensure reliable data measurements because automated equipment is often unable to remove larger pieces of debris. Post-processing of data following cleaning is also necessary to correct for temporal drift patterns caused by the gradual build-up of sediments and biofilms over time (Jones et al. 2014).

10.3.3 External Environmental Factors

Many in situ nutrient monitoring sensors are influenced to some extent by external environmental factors. These can include pH, temperature, electrical conductivity, and turbidity (Downing et al. 2012; Khamis et al. 2015). If these environmental variables are not accounted for by the sensor, they should be monitored and recorded separately to enable post-processing corrections of the data at a later stage. In addition, regular water samples should be acquired and run on laboratory instruments to cross-check in situ sensor calibrations and provide an independent measure of data accuracy.

Taking these issues into account when designing a sensor monitoring system will help to reduce sensor down-time and improve the accuracy and reliability of data outputs. Nonetheless, choosing the most appropriate monitoring system requires careful consideration of several factors that are both application- and site-specific (Bende-Michl and Hairsine 2010), such as site security and accessibility. Moreover, mechanical failures are always possible, particularly during storms or spring thaw events where the risk of damage increases. As such, regular sensor checks should be performed to confirm the monitoring system is performing to standard. Here, telemetry systems can be particularly helpful because they enable remote monitoring of in situ sensors and can be programmed to alert the user if sensor performance or power supplies are compromised (Glasgow et al. 2004). A summary of the key factors that should be considered when designing an in situ nutrient monitoring system is presented in Table 10.2.

Table 10.2 Key factors to consider when designing an in situ nutrient monitoring system.

Topic	Consideration	Potential option
Location	Is the site secure from humans and animals?	House equipment in locked cabinet
		Ensure cables are buried or enclosed
	Is the site accessible?	Arrange access rights with landowner prior to installation
		Ensure route to site is clear and suitable for carrying field equipment
Power	Is mains power available?	Install new mains electricity cable to site (n.b. usually extremely expensive)
	Will batteries need to be replaced frequently?	Use high-capacity batteries or arrange multiple batteries in parallel
		Reduce sampling frequency to decrease power use
	Are light levels sufficient for solar panel operation?	Calculate power requirements and conduct light survey prior to installation
Fouling	How frequently will cleaning take place?	Use anti-fouling measures to reduce cleaning frequency requirements
		Employ telemetry system to monitor system remotely and alert user when cleaning is required

(Continued)

Table 10.2 (Continued)

Topic	Consideration	Potential option
Environmental	Is the site likely to flood?	Ensure equipment is either capable of withstanding submersion for extended periods of time, or install further up river bank to reduce flood risk
	What is the likely minimum and maximum temperature at the site?	Check operating temperature range of equipment exceeds likely environmental temperature range
		Insulate equipment to prevent frost damage in winter
	Are large pieces of debris common at the site?	Ensure equipment is robust enough to withstand debris impacts
Communications	Where will data be stored?	Local data storage (n.b. most cost-effective option)
		Install telemetry system to transfer data to remote server
		Use both local and remote storage to decrease risk of data loss
Other	Will monitoring equipment be integrated with other equipment in the river catchment?	Ensure sampling frequency matches or exceeds existing equipment in sampling network
		Position monitoring site in location that maximizes information provided by new sensor and reduces data redundancy with existing equipment

10.4 Applications of Automated in situ Sensing Methods for Dissolved Organic Matter and Inorganic Nutrient Monitoring

The increasing availability of automated in situ nutrient monitoring systems in recent years has enabled insights into hydrological and biogeochemical processes that would be difficult, if not impossible, to quantify using traditional manual water sampling and laboratory analysis techniques. The following section highlights some examples of where the application of in situ nutrient monitoring has improved our hydroecological understanding at the interfaces of lotic systems and their associated catchments.

10.4.1 Estimation of Nutrient Loads

Traditionally, the estimation of nutrient fluxes and export loads from river catchments has been achieved through discrete spot samples, acquired either manually or using autosamplers (e.g. Johnes 2007). However, in contrast to relatively stable lake systems (Pobel et al. 2011), lotic systems are often highly dynamic and require higher frequency measurements for accurate characterization than can be achieved using spot samples. Several studies have emphasized the serious errors in load estimation that can arise from inadequate sampling resolutions (Cassidy and Jordan 2011; Johnes 2007; Lloyd et al. 2016). In this respect, automated in situ nutrient samplers can offer continuous data at higher resolution than could otherwise be achieved. However, it is important to note that sampling frequency will vary according to the research question. For example, Bieroza et al. (2014) discuss how in situ monitoring at sub-hourly resolution is necessary to provide information on nutrient export during extreme flows, but suggest that weekly sampling may be sufficient to characterize inter-annual changes in N and P loads. Similarly, Cassidy and Jordan (2011) highlight the need to consider individual catchment responses to rainfall events (e.g. flashy vs. muted flow behaviour) when designing an automated sampling programme to prevent either inadequate characterization of short-term event dynamics (temporal sampling resolution too low) or excessive data redundancy (temporal sampling resolution too high). As an example, Cassidy and Jordan (2011) present data for a relatively flashy catchment and suggest that 20 min resolution monitoring is required to characterize P export loads with confidence during storm events. In contrast, the temporal stability of nutrient dynamics under baseflow conditions may require less intensive sampling to capture inter-annual changes. However, Bieroza and Heathwaite (2016) suggest in situ monitoring can capture diurnal patterns that would otherwise be missed by automated grab sampling due to uncertainties introduced by post-sampling storage transformations.

10.4.2 Physical Controls on Nutrient Export

Continuous datasets provided by in situ nutrient sensors allow the identification of nutrient sources within river catchments and can clarify how these are mobilized under different environmental conditions. For example, Jeong et al. (2012) use UV-VIS technology to demonstrate how rainfall intensity acts as a key control on the type of C (i.e. dissolved vs. particulate) transported through networks during storm events. In addition, data generated from in situ nutrient N and P sensors have been used to explore linkages between meteorological conditions (event and antecedent conditions) and in-channel nutrient fluxes (Outram et al. 2014), and also to highlight how nutrient catchment source areas can vary between storm events depending upon prevailing hydro-climatological conditions (Blaen et al. 2017). Such research has implications for predictions of riverine nutrient export under future climate scenarios and developing appropriate water resource management and adaptation strategies, particularly in the face of anticipated changes in the magnitude and frequency of extreme meteorological and hydrological events (Kendon et al. 2014).

10.4.3 Hydroecological Process Understanding

In situ sensing technologies can improve our understanding of hydroecological processes over diurnal to seasonal temporal scales. Over short timescales, distinct diel cycles of organic C have been observed in both rivers and lakes that are believed to be linked primarily to biological activity, and possibly also to photolysis (Spencer et al. 2007; Watras et al. 2015; Worrall et al. 2015). In addition, Worrall et al. (2015) employed hourly UV-VIS data to suggest that in-stream processing removed almost one-third of the incoming DOC from a 1700 km^2 river catchment in western England. Over longer timescales, Voss et al. (2015) presented fluorescence data for a river draining a snowmelt dominated river basin in the Canadian Rockies that shows clear seasonal changes in DOM optical properties, with increased DOC concentrations and a shift towards higher aromaticity (molecular weight) during the spring freshet period. Wilson et al. (2013) also use in situ fluorescence DOM measurements to characterize and track long-term changes in DOC export from a forested stream catchment. They suggest that variation in forest productivity can increase the pool of leachable soil carbon during summer months, which enters the stream at the soil–water interface, and maximum DOC export during late summer and autumn months. Gilbert et al. (2013) deployed in situ sensors in an estuarine environment in the Pacific Northwest and found high rates of inorganic N removal that they attributed to biological uptake or denitrification in freshwater tidal flat areas. Finally, recent studies have utilized paired deployments of in situ sensors to quantify temporal variability in reach-scale nutrient processing dynamics (Kraus et al. 2017; Kunz et al. 2017).

10.4.4 Management and Legislation

When compared to traditional discrete sampling and laboratory analysis methods, the additional information provided by in situ nutrient monitoring can inform sampling regimes and assist managers of river catchments to ensure that water quality meets legislative requirements. For example, Skeffington et al. (2015) investigated variability in P, temperature, and dissolved oxygen in an urban stream in southern England and demonstrated that WFD chemical status may be incorrectly assigned if only daily sampling regimes are used. Similarly, Bowes et al. (2009) employed in situ monitoring data to separate P loads between those from sewage treatment works (STW) and those from diffuse and in-channel sources, to confirm that the STW was not discharging P in excess of its consent limits.

There is considerable potential to deploy in situ nutrient monitoring systems more widely to monitor compliance with environmental quality standards. In water bodies affected by sewage effluent, the fluorescence of aquatic organic matter emitted at 350 nm when excited at 280 nm has been correlated with microbial water quality. Baker et al. (2015) examined the relationship between the fluorescence at this wavelength and *Escherichia coli*, finding a log correlation of 0.74 for catchments in both the West Midlands (UK) and Durban (South Africa). While further developments in this area require catchment-specific calibrations, they have the potential to yield real-time water quality data and reduce the extent to which catchment managers may rely upon costly and time-consuming laboratory analyses of water samples.

10.5 Future Directions for in situ Nutrient Sensing Applications

Advances in technology, coupled with lower development costs, have led to a significant uptake in the use of in situ nutrient sensors in river environments in recent years. This final section discusses some potential future directions for this rapidly-developing field.

10.5.1 Adaptive Monitoring Techniques

In situ nutrient sensors enable significantly higher temporal monitoring resolutions than otherwise achievable using traditional discrete sampling methods. However, these high sampling frequencies pose their own unique set of obstacles, including increased power and reagent consumption, high data storage requirements, and increased physical wear and maintenance needs. In addition, many aquatic ecosystems display pulsed behaviour (Junk et al. 1989; Raymond et al. 2016; Tockner et al. 2000), in which extended periods of low biogeochemical variability, often during low flows, are interspersed with shorter intense biogeochemically "active" periods. This presents a challenge for in situ nutrient sensors operating at a fixed sampling resolution because sampling at high resolutions will result in increased costs and substantial data redundancy, while sampling at lower resolutions may be too coarse to characterize short-term event dynamics. To resolve this, Blaen et al. (2016) argue that adaptive monitoring approaches are required for investigating dynamic river ecosystems that exhibit strong non-linear behaviour (cf. Krause et al. 2015). Adaptive monitoring approaches are structured around a range of sampling frequencies that are triggered by defined trigger variables. These may include the nutrient of interest (e.g. N or P), a proxy variable (e.g. river level), or a combination of multiple variables (Figure 10.2). The design and set-up of an adaptive monitoring system is strongly system-specific, because inter-catchment differences in land use, geology, and vegetation will affect the lag time between trigger and response. Adaptive monitoring approaches offer the potential to characterize in-stream nutrient dynamics during key hydrological events and thereby contribute greatly to our understanding of the catchment processes that underpin these patterns (Blaen et al. 2016).

10.5.2 Improvements in Sensor Design

Recent increases in the availability of low-cost and open-source in situ nutrient sensors have potential to increase the density of sensor networks across river catchments (Mao et al. 2019). For example, in situ fluorescence sensors for chlorophyll *a* are available at a unit cost (in 2013) of US $150 (Leeuw et al. 2013). Low-cost open-source nitrate sensors that utilize the Nitrate Reductase Nitrate – Nitrogen Analysis Method, controlled by an Arduino microcontroller, are available for less than 15% of the cost of a similar commercial proprietary system (Wittbrodt et al. 2015). When coupled with telemetry systems, such sensors can form semi-autonomous networks capable of monitoring catchment biogeochemistry and hydrology at similar spatial and temporal resolutions (Watras et al. 2015). Miniaturization of sensors, combined with lower power requirements, also offers the possibility of improving monitoring coverage in remote areas that would be logistically unfeasible to access in person on a regular basis. However, the lack of reliable and low-cost automated cleaning systems to stop the fouling of sensor nodes within distributed monitoring networks remains a key challenge that urgently needs to be addressed.

10.5.3 Integration with Other Disciplines and Across Environmental Interfaces

Datasets generated from in situ nutrient monitoring technology provide greatly improved estimates and understanding of nutrient cycling through river catchments from their

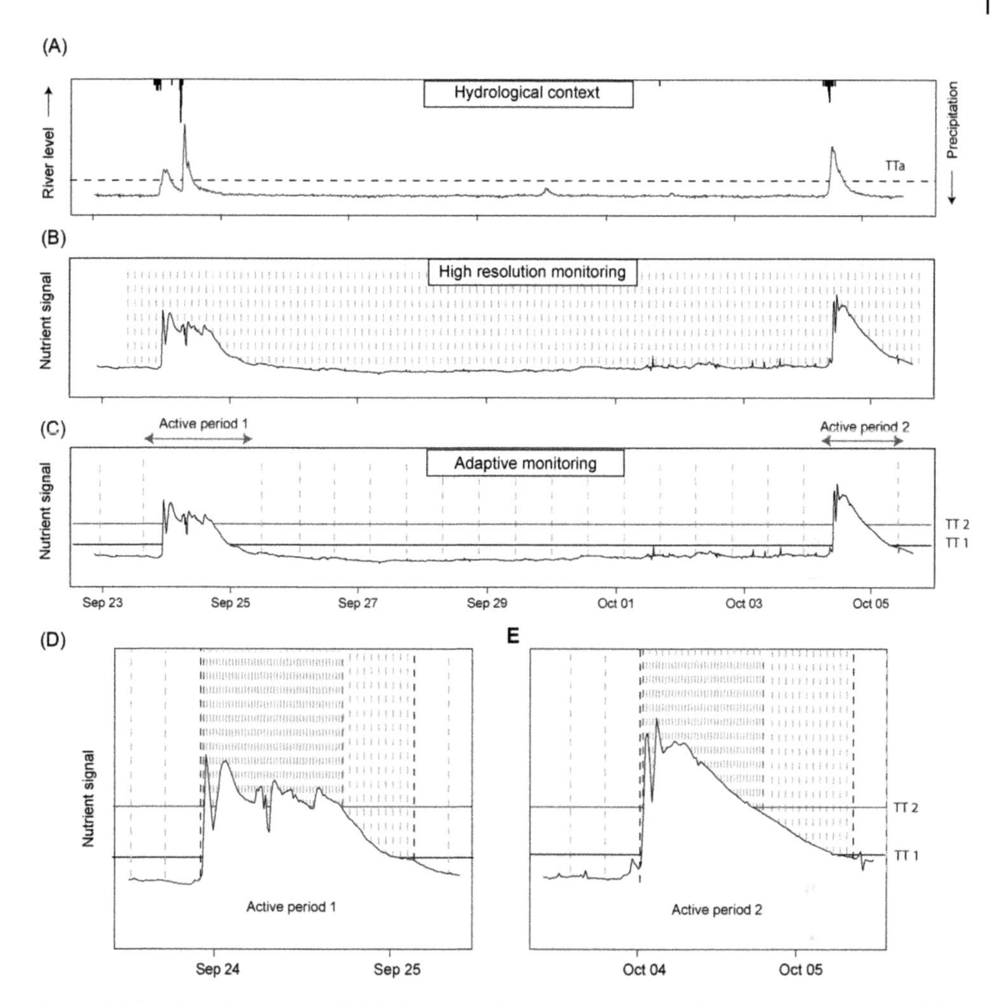

Figure 10.2 (from Blaen et al. 2016): Conceptual representation of adaptive monitoring principles based on Krause et al. (2015). Vertical dashed lines represent sample points, TT1 is trigger threshold 1 (black line), TT2 is trigger threshold 2 (red line), and TTa is an additional trigger threshold (horizontal dashed line). (a) Hydrological context (precipitation and river level) and TTa, (b) "traditional" high resolution monitoring of a variable (nutrient) of interest, (c) adaptive monitoring approach with TT1 and TT2 highlighted, (d and e) events of interest or "active periods" highlighting the transition between the three monitoring frequency states based on threshold exceedance under rapidly changing environmental conditions. Note that for a change from baseline monitoring frequency to higher resolution monitoring states TT1 and TTa must be exceeded to avoid inappropriate triggering.

headwaters to the freshwater–marine transitional zone and beyond (Blaen et al. 2016; Tyler et al. 2016). However, interpreting these data patterns in the context of physical catchment processes requires cross-disciplinary working across a wide range of research fields, including the atmospheric, soil, hydrological, and biogeochemical sciences. In addition, while the abundance of data generated by in situ sensor networks may dramatically increase our understanding of environmental complexity, the size of these datasets represents a major statistical and computational challenge for accurate analysis and meaningful interpretation. In this respect, "big data" management approaches are likely to be required in the near future to exploit the scientific benefits of information provided by such sensor networks and capture the full extent of environmental variability and system behaviour (Crawford et al. 2015; Krause et al. 2015). If these challenges can be met, then such monitoring systems offer huge potential to improve our understanding of nutrient dynamics within freshwater environments.

References

Adamchuk, V., Morgan, M.T., and Ess, D.R. (1999). An automated sampling system for measuring soil pH. *Transactions of the ASAE. American Society of Agricultural Engineers* 42 (4): 885.

Adsett, J.F. and Zoerb, G.C. (1991). Automated field monitoring of nitrate levels. In *Automated Agriculture for the 21st Century*. 326–335. ASAE Publication No. 1191. St. Joseph, MI., USA: American Society of Agricultural Engineers.

Baker, A., Cumberland, S.A., Bradley, C. et al. (2015). To what extent can portable fluorescence spectroscopy be used in the real-time assessment of microbial water quality? *The Science of the Total Environment* 532: 14–19.

Bartram, J. and Rees, G. (1999). *Monitoring Bathing Waters: A Practical Guide to the Design and Implementation of Assessments and Monitoring Programmes*. CRC Press.

Battin, T.J., Kaplan, L.A., Findlay, S. et al. (2008). Biophysical controls on organic carbon fluxes in fluvial networks. *Nature Geoscience* 1 (2): 95–100.

Beaton, A.D., Cardwell, C.L., Thomas, R.S. et al. (2012). Lab-on-chip measurement of nitrate and nitrite for in situ analysis of natural waters. *Environmental Science & Technology* 46 (17): 9548–9556.

Belikova, V., Panchuk, V., Legin, E. et al. (2019). Continuous monitoring of water quality at aeration plant with potentiometric sensor array. *Sensors and Actuators. B, Chemical* 282: 854–860.

Bende-Michl, U. and Hairsine, P.B. (2010). A systematic approach to choosing an automated nutrient analyser for river monitoring. *Journal of Environmental Monitoring: JEM* 12 (1): 127–134.

Bennett, E.M., Peterson, G.D., and Gordon, L.J. (2009). Understanding relationships among multiple ecosystem services. *Ecology Letters* 12 (12): 1394–1404.

Bieroza, M.Z. and Heathwaite, A.L. (2015). Seasonal variation in phosphorus concentration–discharge hysteresis inferred from high-frequency in situ monitoring. *Journal of Hydrology* 524: 333–347. https://doi.org/10.1016/j.jhydrol.2015.02.036.

Bieroza, M.Z. and Heathwaite, A.L. (2016). Unravelling organic matter and nutrient biogeochemistry in groundwater-fed rivers under baseflow conditions: uncertainty in in situ high-frequency analysis. *The Science of the Total Environment* 572: 1520–1533.

Bieroza, M.Z., Heathwaite, A.L., Mullinger, N.J. et al. (2014). Understanding nutrient biogeochemistry in agricultural catchments: the challenge of appropriate monitoring frequencies. *Environmental Sciences: Processes and Impacts* 16 (7): 1676–1691.

Blaen, P., Khamis, K., Lloyd, C. et al. (2016). Real-time monitoring of nutrients and dissolved organic matter in rivers: adaptive monitoring strategies, technological challenges and future directions. *The Science of the Total Environment* 569–570: 647–660.

Blaen, P., Khamis, K., Lloyd, C. et al. (2017). High-frequency monitoring of catchment nutrient exports reveals highly variable storm event responses and dynamic source zone activation. *Journal of Geophysical Research: Biogeosciences*. https://doi.org/10.1002/2017JG003904.

Bowes, M.J., Smith, J.T., and Neal, C. (2009). The value of high-resolution nutrient monitoring: a case study of the River Frome, Dorset, UK. *Journal of Hydrology* 378 (1): 82–96.

Bridgeman, J., Baker, A., Brown, D. et al. (2015). Portable LED fluorescence instrumentation for the rapid assessment of potable water quality. *The Science of the Total Environment* 524–525: 338–346.

Burt, T.P., Howden, N.J.K., Worrall, F. et al. (2010). Long-term monitoring of river water nitrate: how much data do we need? *Journal of Environmental Monitoring: JEM* 12 (1): 71–79.

Camargo, J.A., Alonso, A., and Salamanca, A. (2005). Nitrate toxicity to aquatic animals: a review with new data for freshwater invertebrates. *Chemosphere* 58 (9): 1255–1267.

Casey, H. and Clarke, R.T. (1979). Statistical analysis of nitrate concentrations from the River Frome (Dorset) for the period 1965-76. *Freshwater Biology* 9 (2): 91–97.

Cassidy, R. and Jordan, P. (2011). Limitations of instantaneous water quality sampling in surface-water catchments: comparison with near-continuous phosphorus time-series data. *Journal of Hydrology* 405 (1): 182–193.

Chen, R.F. (1999). In situ fluorescence measurements in coastal waters. *Organic Geochemistry* 30 (6): 397–409.

Cleary, J., Slater, C., McGraw, C., and Diamond, D. (2008). An autonomous microfluidic sensor for phosphate: on-site analysis of treated wastewater. *IEEE Sensors Journal* 8 (5): 508–515.

Collins, K., Christy, C., Lund, E. et al. (2003). Developing an automated soil pH mapping system. *ASAE Paper No. MC03-205*. St. Joseph, Michigan.

Conmy, R.N., Del Castillo, C.E., Downing, B.D. et al. (2014). Experimental design and quality assurance: in situ fluorescence instrumentation. In: *Aquatic Organic Matter Fluorescence* (ed. R.G. Noble, J. Lead, A. Baker et al.), 190–230. Cambridge University Press.

Crawford, J.T., Loken, L.C., Casson, N.J. et al. (2015). High-speed limnology: using advanced sensors to investigate spatial variability in biogeochemistry and hydrology. *Environmental Science & Technology* 49 (1): 442–450.

Daniel, A., Birot, D., Blain, S. et al. (1995). A submersible flow-injection analyser for the in-situ determination of nitrite and nitrate in coastal waters. *Marine Chemistry* 51 (1): 67–77. https://doi.org/10.1016/0304-4203(95)00052-s.

De Marco, R., Clarke, G., and Pejcic, B. (2007). Ion-selective electrode potentiometry in environmental analysis. *Electroanalysis: an International Journal Devoted to Fundamental and Practical Aspects of Electroanalysis* 19 (19–20): 1987–2001.

Del Castillo, C.E., Coble, P.G., Morell, J.M. et al. (1999). Analysis of the optical properties of the Orinoco River plume by absorption and fluorescence spectroscopy. *Marine Chemistry* 66 (1): 35–51.

Doku, G.N. and Haswell, S.J. (1999). Further studies into the development of a micro-FIA (μFIA) system based on electroosmotic flow for the determination of phosphate as orthophosphate. *Analytica Chimica Acta* 382 (1): 1–13.

Downing, B.D., Pellerin, B.A., Bergamaschi, B.A. et al. (2012). Seeing the light: the effects of particles, dissolved materials, and temperature on in situ measurements of DOM fluorescence in rivers and streams. *Limnology and Oceanography: Methods* 10 (10): 767–775.

Ensign, S.H. and Doyle, M.W. (2006). Nutrient spiraling in streams and river networks. *Journal of Geophysical Research: Biogeosciences* 111 (G4).

Environment Agency. (2016). *Historic river quality (GQA) 1990 to 2009*. Retrieved from http://apps.environment-agency.gov.uk/wiyby/37811.aspx (accessed 19 July 2016).

Friberg, N., Skriver, J., Larsen, S.E. et al. (2009). Stream macroinvertebrate occurrence along gradients in organic pollution and eutrophication. *Freshwater Biology* 55 (7): 1405–1419. https://doi.org/10.1111/j.1365-2427.2008.02164.x.

Gilbert, M., Needoba, J., Koch, C. et al. (2013). Nutrient loading and transformations in the Columbia River estuary determined by high-resolution in situ sensors. *Estuaries and Coasts* 36 (4): 708–727. https://doi.org/10.1007/s12237-013-9597-0.

Glasgow, H.B., Burkholder, J.M., Reed, R.E. et al. (2004). Real-time remote monitoring of water quality: a review of current applications, and advancements in sensor, telemetry, and computing technologies. *Journal of Experimental Marine Biology and Ecology* 300 (1–2): 409–448. https://doi.org/10.1016/j.jembe.2004.02.022.

Greenway, G.M., Haswell, S.J., and Petsul, P.H. (1999). Characterisation of a micro-total analytical system for the determination of nitrite with spectrophotometric detection. *Analytica Chimica Acta* 387 (1): 1–10.

Hach (2016). *Water quality meters & probes; electrochemistry selection guide*. Retrieved from http://www.hach.com/asset-get.download.jsa?id=7664029560 (accessed 19 July 2016).

Hach Lange (2016). *Phosphax sigma phosphate analyser*. Available at: https://uk.hach.com/phosphate-analysers/phosphax-sigma-phosphate-analyser/family?productCategoryId=25033233213 (accessed 15 June 2023).

Halliday, S.J., Skeffington, R.A., Wade, A.J. et al. (2015). High-frequency water quality monitoring in an urban catchment: hydrochemical dynamics, primary production and implications for the water framework directive. *Hydrological Processes* 29 (15): 3388–3407. https://doi.org/10.1002/hyp.10453.

Hering, D., Borja, A., Carstensen, J. et al. (2010). The European water framework directive at the age of 10: a critical review of the achievements with recommendations for the future. *The Science of the Total Environment* 408 (19): 4007–4019.

Hood, E., Battin, T.J., Fellman, J. et al. (2015). Storage and release of organic carbon from glaciers and ice sheets. *Nature Geoscience* 8 (2): 91–96.

Huebsch, M., Grimmeisen, F., Zemann, M. et al. (2015). *Field experiences using UV/VIS sensors for high-resolution monitoring of nitrate in groundwater*. Available at: http://t-stor.teagasc.ie/handle/11019/796 (accessed 25 June 2016).

Jeong, J.-J., Bartsch, S., Fleckenstein, J.H. et al. (2012). Differential storm responses of dissolved and particulate organic carbon in a mountainous headwater stream, investigated by high-frequency, in situ optical measurements. *Journal of Geophysical Research: Biogeosciences* 117 (G3). https://doi.org/10.1029/2012jg001999.

Johnes, P.J. (2007). Uncertainties in annual riverine phosphorus load estimation: impact of load estimation methodology, sampling frequency, baseflow index and catchment population density. *Journal of Hydrology* 332 (1–2): 241–258. https://doi.org/10.1016/j.jhydrol.2006.07.006.

Jones, T.D., Chappell, N.A., and Tych, W. (2014). First dynamic model of dissolved organic carbon derived directly from high-frequency observations through contiguous storms. *Environmental Science & Technology* 48 (22): 13289–13297.

Jordan, P., Arnscheidt, A., McGrogan, H. et al. (2007). Characterising phosphorus transfers in rural catchments using a continuous bank-side analyser. *Hydrology and Earth System Sciences* 11 (1): 372–381.

Junk, W.J., Bayley, P.B., and Sparks, R.E. et al. (1989). The flood pulse concept in river-floodplain systems. *Canadian Special Publication of Fisheries and Aquatic Sciences/ Publication Speciale Canadienne Des Sciences Halieutiques Et Aquatiques* 106 (1): 110–127.

Kendon, E.J., Roberts, N.M., Fowler, H.J. et al. (2014). Heavier summer downpours with climate change revealed by weather forecast resolution model. *Nature Climate Change* 4 (7): 570–576. https://doi.org/10.1038/nclimate2258.

Khamis, K., Bradley, C., and Hannah, D.M. (2018). Understanding dissolved organic matter dynamics in urban catchments: insights from in situ fluorescence sensor technology. *Wiley Interdisciplinary Reviews: Water* 5 (1): e1259. https://doi.org/10.1002/wat2.1259.

Khamis, K., Bradley, C., and Hannah, D.M. (2020). High frequency fluorescence monitoring reveals new insights into organic matter dynamics of an urban river, Birmingham, UK. *The Science of the Total Environment* 710: 135668.

Khamis, K., Sorensen, J.P.R., Bradley, C. et al. (2015). In situ tryptophan-like fluorometers: assessing turbidity and temperature effects for freshwater applications. *Environmental Science. Processes & Impacts* 17 (4): 740–752.

Kotamäki, N., Thessler, S., Koskiaho, J. et al. (2009). Wireless in-situ sensor network for agriculture and water monitoring on a river basin scale in southern finland: evaluation from a data user's perspective. *Sensors* 9 (4): 2862–2883.

Kraus, T.E.C., O'Donnell, K., Downing, B.D. et al. (2017). Using paired in situ high frequency nitrate measurements to better understand controls on nitrate concentrations and estimate nitrification rates in a wastewater-impacted river. *Water Resources Research* 53 (10): 8423–8442.

Krause, S., Lewandowski, J., Dahm, C.N. et al. (2015). Frontiers in real-time ecohydrology - a paradigm shift in understanding complex environmental systems. *Ecohydrology* 8 (4): 529–537 https://doi.org/10.1002/eco.1646.

Kunz, J.V., Hensley, R., Brase, L. et al. (2017). High frequency measurements of reach scale nitrogen uptake in a fourth order river with contrasting hydromorphology and variable water chemistry (Weiße Elster, Germany). *Water Resources Research* 53 (1): 328–343.

Lakowicz, J.R. (2013). *Principles of Fluorescence Spectroscopy*. Springer Science & Business Media.

Lee, E.-J., Yoo, G.-Y., Jeong, Y. et al. (2015). Comparison of UV-VIS and FDOM sensors for in situ monitoring of stream DOC concentrations. *Biogeosciences* 12 (10): 3109–3118.

Le Goff, T., Braven, J., Ebdon, L. et al. (2002). An accurate and stable nitrate-selective electrode for the in situ determination of nitrate in agricultural drainage waters. *The Analyst* 127 (4): 507–511.

Leeuw, T., Boss, E.S., and Wright, D.L. (2013). In situ measurements of phytoplankton fluorescence using low cost electronics. *Sensors* 13 (6): 7872–7883.

Lepot, M., Torres, A., Hofer, T. et al. (2016). Calibration of UV/Vis spectrophotometers: a review and comparison of different methods to estimate TSS and total and dissolved COD concentrations in sewers, WWTPs and rivers. *Water Research* 101: 519–534.

Li, W-T, Jin, J., Li, Q. et al. (2016). Developing LED UV fluorescence sensors for online monitoring DOM and predicting DBPs formation potential during water treatment. *Water Research* 93: 1–9.

Lloyd, C.E.M., Freer, J.E., Johnes, P.J. et al. (2016). Discharge and nutrient uncertainty: implications for nutrient flux estimation in small streams. *Hydrological Processes* 30 (1): 135–152. https://doi.org/10.1002/hyp.10574.

Mao, F., Khamis, K., Krause, S. et al. (2019). Low-cost environmental sensor networks: recent advances and future directions. *Frontiers in Earth Science* 7: https://doi.org/10.3389/feart.2019.00221.

McIsaac, G.F. and Libra, R.D. (2003). Revisiting nitrate concentrations in the Des Moines River. *Journal of Environmental Quality* 32 (6): 2280–2289.

MEA (2005). *Ecosystems and Human Well-being*. Washington, DC: Island press.

Mellander, P.-E., Melland, A.R., Jordan, P. et al. (2012). Quantifying nutrient transfer pathways in agricultural catchments using high temporal resolution data. *Environmental Science & Policy* 24: 44–57.

Merks, A.G.A. (1975). Determination of ammonia in sea water with an ion-selective electrode. *Netherlands Journal of Sea Research* 9 (3): 371–375.

Outram, F.N., Lloyd, C.E.M., Jonczyk, J. et al. (2014). High-frequency monitoring of nitrogen and phosphorus response in three rural catchments to the end of the 2011–2012 drought in England. *Hydrology and Earth System Sciences* 18 (9): 3429–3448. https://doi.org/10.5194/hess-18-3429-2014.

Palmer, M.A. and Bernhardt, E.S. (2006). Hydroecology and river restoration: ripe for research and synthesis. *Water Resources Research* 42 (3): 1127.

Patton, C.J. and Kryskalla, J.R. (2011). Colorimetric determination of nitrate plus nitrite in water by enzymatic reduction, automated discrete analyzer methods. *US Geological Survey Techniques and Methods* 34. Available at: https://pubs.usgs.gov/tm/05b08.

Petsul, P.H., Greenway, G.M., and Haswell, S.J. (2001). The development of an on-chip micro-flow injection analysis of nitrate with a cadmium reductor. *Analytica Chimica Acta* 428 (2): 155–161.

Pobel, D., Robin, J., and Humbert, J.-F. (2011). Influence of sampling strategies on the monitoring of cyanobacteria in shallow lakes: lessons from a case study in France. *Water Research* 45 (3): 1005–1014.

Raymond, P.A., Saiers, J.E., and Sobczak, W.V. (2016). Hydrological and biogeochemical controls on watershed dissolved organic matter transport: pulse-shunt concept. *Ecology* 97 (1): 5–16.

Rode, M., Wade, A.J., Cohen, M.J. et al. (2016). Sensors in the stream: the high-frequency wave of the present. *Environmental Science & Technology* 50 (19): 10297–10307.

Ruhala, S.S. and Zarnetske, J.P. (2017). Using in-situ optical sensors to study dissolved organic carbon dynamics of streams and watersheds: a review. *The Science of the Total Environment* 575: 713–723.

Saraceno, J.F., Shanley, J.B., Downing, B.D. et al. (2017). Clearing the waters: evaluating the need for site-specific field fluorescence corrections based on turbidity measurements. *Limnology and Oceanography: Methods* 15 (4): 408–416. https://doi.org/10.1002/lom3.10175.

Simpson, E.A. (1980). The harmonization of the monitoring of the quality of rivers in the United Kingdom/Contrôle harmonisé de la qualité des rivières au Royaume Uni. *Hydrological Sciences Journal* 25 (1): 13–23.

Skeffington, R.A., Halliday, S.J., Wade, A.J. et al. (2015). Using high-frequency water quality data to assess sampling strategies for the EU water framework directive. *Hydrology and Earth System Sciences* 19 (5): 2491–2504. https://doi.org/10.5194/hess-19-2491-2015.

Smith, V.H. and Schindler, D.W. (2009). Eutrophication science: where do we go from here? *Trends in Ecology & Evolution* 24 (4): 201–207.

Spencer, R.G.M., Pellerin, B.A., Bergamaschi, B.A. et al. (2007). Diurnal variability in riverine dissolved organic matter composition determined by in situ optical measurement in the San Joaquin River (California, USA). *Hydrological Processes* 21 (23): 3181–3189.

Subchem (2016). *Autonomous profiling nutrient analyzer specifications*. Retrieved from http://subchem.com/Autonomous%20Profiling%20Nutrient%20Analyzer%20(APNA).pdf (accessed 25 June 2016).

Tockner, K., Malard, F., and Ward, J.V. (2000). An extension of the flood pulse concept. *Hydrological Processes* 14 (16–17): 2861–2883.

Toda, K., Yawata, Y., Setoyama, E. et al. (2011). Continuous monitoring of ammonia removal activity and observation of morphology of microbial complexes in a microdevice. *Applied and Environmental Microbiology* 77 (12): 4253–4255.

Tyler, A.N., Hunter, P.D., Spyrakos, E. et al. (2016). Developments in Earth observation for the assessment and monitoring of inland, transitional, coastal and shelf-sea waters. *The Science of the Total Environment* 572: 1307–1321.

Van Den Broeke, J., Brandt, A., Weingartner, A. et al. (2006). Monitoring of organic micro contaminants in drinking water using a submersible UV/Vis spectrophotometer. In: *Security of Water Supply Systems: from Source to Tap* (ed. J. Pollert and B. Dedus), 19–29. Netherlands: Springer.

Voss, B.M., Peucker-Ehrenbrink, B., Eglinton, T.I. et al. (2015). Seasonal hydrology drives rapid shifts in the flux and composition of dissolved and particulate organic carbon and major and trace ions in the Fraser River, Canada. *Biogeosciences* 12 (19): 5597–5618.

Wade, A.J., Palmer-Felgate, E.J., Halliday, S.J. et al. (2012). Hydrochemical processes in lowland rivers: insights from in situ, high-resolution monitoring. *Hydrology and Earth System Sciences* 16 (11): 4323–4342.

Ward, M.H., DeKok, T.M., Levallois, P. et al. (2005). Workgroup report: drinking-water nitrate and health—recent findings and research needs. *Environmental Health Perspectives* 113 (11): 1607–1614.

Waterloo, M.J., Oliveira, S.M., Drucker, D.P. et al. (2006). Export of organic carbon in run-off from an Amazonian rainforest blackwater catchment. *Hydrological Processes: an International Journal* 20 (12): 2581–2597.

Watras, C.J., Hanson, P.C., Stacy, T.L. et al. (2011). A temperature compensation method for CDOM fluorescence sensors in freshwater. *Limnology and Oceanography, Methods / ASLO* 9 (7): 296–301.

Watras, C.J., Morrison, K.A., Crawford, J.T. et al. (2015). Diel cycles in the fluorescence of dissolved organic matter in dystrophic Wisconsin seepage lakes: implications for carbon turnover: Diel CDOM fluorescence cycles. *Limnology and Oceanography* 60 (2): 482–496.

Whitehead, P.G., Wilby, R.L., Battarbee, R.W. et al. (2009). A review of the potential impacts of climate change on surface water quality. *Hydrological Sciences Journal* 54 (1): 101–123.

Wilson, H.F., Saiers, J.E., Raymond, P.A., and Sobczak, W.V. (2013). Hydrologic drivers and seasonality of dissolved organic carbon concentration, nitrogen content, bioavailability, and export in a forested New England stream. *Ecosystems* 16 (4): 604–616.

Wittbrodt, B.T., Squires, D.A., Walbeck, J. et al. (2015). Open-source photometric system for enzymatic nitrate quantification. *PloS One* 10 (8): e0134989.

Worrall, F., Howden, N.J.K., and Burt, T.P. (2015). Understanding the diurnal cycle in fluvial dissolved organic carbon–The interplay of in-stream residence time, day length and organic matter turnover. *Journal of Hydrology* 523: 830–838.

YSI (2016). *YSI ion selective electrodes*. Retrieved from https://www.ysi.com/File%20Library/Documents/Specification%20Sheets/YSI-ISE-Electrodes-W74-1014-spec-sheet.pdf (accessed 19 July 2016).

11

Tracing Hydrological Connectivity with Aerial Diatoms

L. Pfister[1], J. Klaus[1], C.E. Wetzel[1], M. Antonelli[1,2], and N. Martínez-Carreras[1]

[1] Luxembourg Institute of Science and Technology, Department Environmental Research and Innovation, Catchment and Eco-hydrology research group, 41 rue du Brill, L-4422 Belvaux, Luxembourg.
[2] Wageningen University, Department of Environmental Sciences, Hydrology and Quantitative Water Management Group, Droevendaalsesteeg 3a, Building 100, 6708PB Wageningen, The Netherlands.

11.1 Introduction

The first mainstream theories (e.g. the role of infiltration excess – or Hortonian – overland flow Horton (1933); the partially contributing area concept (Dunne and Black 1970)) that had emerged from research studies carried out in experimental catchments had a major shortcoming, in that they clearly underestimated the enormous spatial and temporal variability of the processes taking place within the hydrological cycle (Beven 2000; McDonnell and Woods 2004).

Monitoring environmental processes are of intricate complexity and remain measurement-limited to a large extent. New insights into water sources, flow paths, and transit times came with the work of Pinder and Jones (1969) and Sklash and Farvolden (1979) that introduced chemical and isotope tracers for hydrograph separation and ultimately changed our knowledge on time- and geographic sources of water. Even though these techniques have been a major subject of investigation in experimental hydrology, the thrill that had originally surrounded their application potential has partly faded away by repetitive application to temperate-forest catchments (Burns 2002).

Diatoms have the strong advantage of being a ubiquitous and a widely used group of organisms in the study of environmental issues (Smol and Stoermer 2010): they are already consolidated as a main tool in palaeo and modern studies spanning several timescales (from the Cenozoic to a few hundred years) (Boeff et al. 2016; Davies and Kemp 2016) to study river and lake dynamics, nutrient enrichment (Juggins et al. 2016), climatic events (Pajunen et al. 2016; Smol and Douglas 2007), water levels of lakes (Leira et al. 2015) and oceans (Palmer and Abbott 1986), reconstruction of disastrous flooding (Matsumoto et al. 2016), tsunamis (Hayward et al. 2004, Hemphill-Haley 1996), heavy metal contamination (Pandey et al. 2014), alongside their already well-established and routine use as bioindicators of water quality of lakes and rivers in Europe (European Committee for Standardization 2003).

Ecohydrological Interfaces, First Edition. Edited by Stefan Krause, David M. Hannah, and Nancy B. Grimm.

In this context, Pfister et al. (2009) have attempted to overcome some of the limitations of conventional tracers used for determining the spatial origin of water, by exploring the potential for aerial diatoms (unicellular algae) to trace the onset/cessation of rapid surface and subsurface flowpaths in the hillslope-riparian zone-stream continuum of two small experimental catchments in Luxembourg. This chapter introduces the concept of using aerial diatoms for hydrological process research. It summarizes current research, presents the potential and limitations for the use of aerial diatoms as tracers of hydrological connectivity, and proposes prospective ideas for further research into aerial diatom tracing.

11.2 Aerial Diatoms and What Makes Them an Interesting Tracer in Hydrology

Although the vast majority of diatom species are known from aquatic environments, several species belonging to a large number of genera can be found in terrestrial habitats such as soils and humid rocks (Ettl and Gärtner 2014). Soils are the habitats of terrestrial algae (Petersen 1935). This category can be then divided into three sub-categories, based on their appearance (growth) on: (1) bare soil (aeroterrestrial), (2) soil that periodically dries out (euterrestrial), and (3) on soil that is always moist (hydroterrestrial). Petersen also differentiated the species into surface-soil algae (epiterranean) from those occurring below the surface (subterranean), where a large number of species have been reported (Falasco et al. 2014; Ivarsson et al. 2013). Hereafter, "aerial diatom" communities are defined as living exposed to the air, outside of lentic and lotic environments, following the definition of Johansen (2010). This definition stipulates that most species that are found in baseflow drift samples are not strictly terrestrial.

Species classification is based mainly on the ecological requirement values provided by Van Dam et al. (1994). This classification relies on soil moisture content and is based on the knowledge of over half a century of diatom research by many authors. Here "aerial diatoms" are categorized with values 4 and 5 in Van Dam's classification. These two categories correspond to diatoms "mainly occurring on wet and moist or temporarily dry places", as well as "nearly exclusively occurring outside water bodies", respectively (Figure 11.1). The work of Van Dam et al. (1994) is a more recent compilation of ecological information for diatoms gathered in a single publication.

Early work by Pfister et al. (2009) in the schistous and forested Weierbach catchment (0.45 km^2), as well as in the sandstone dominated and forested Huewelerbach catchment (2.8 km^2), both located in Luxembourg, explored the potential for aerial diatoms to monitor the onset and cessation of rapid surface and/or subsurface water flow paths in the hillslope-riparian zone-stream continuum. They investigated the flushing of aerial diatoms to the creek during two rainfall run-off events. The experimental protocol consisted in sampling water from the two creeks before and during the flood hydrograph. For each sample, the relative proportions of aquatic and aerial diatoms were determined via countings and identification of individual diatom frustules from microscopic slides. Samples taken during the rising and falling limbs of the storm hydrographs clearly showed the occurrence of aerial diatoms flushing to the creek.

Building on these first experiments, additional aerial diatom tracing experiments were carried out. In the Weierbach catchment, Klaus et al. (2015) took 203 samples in the HRS

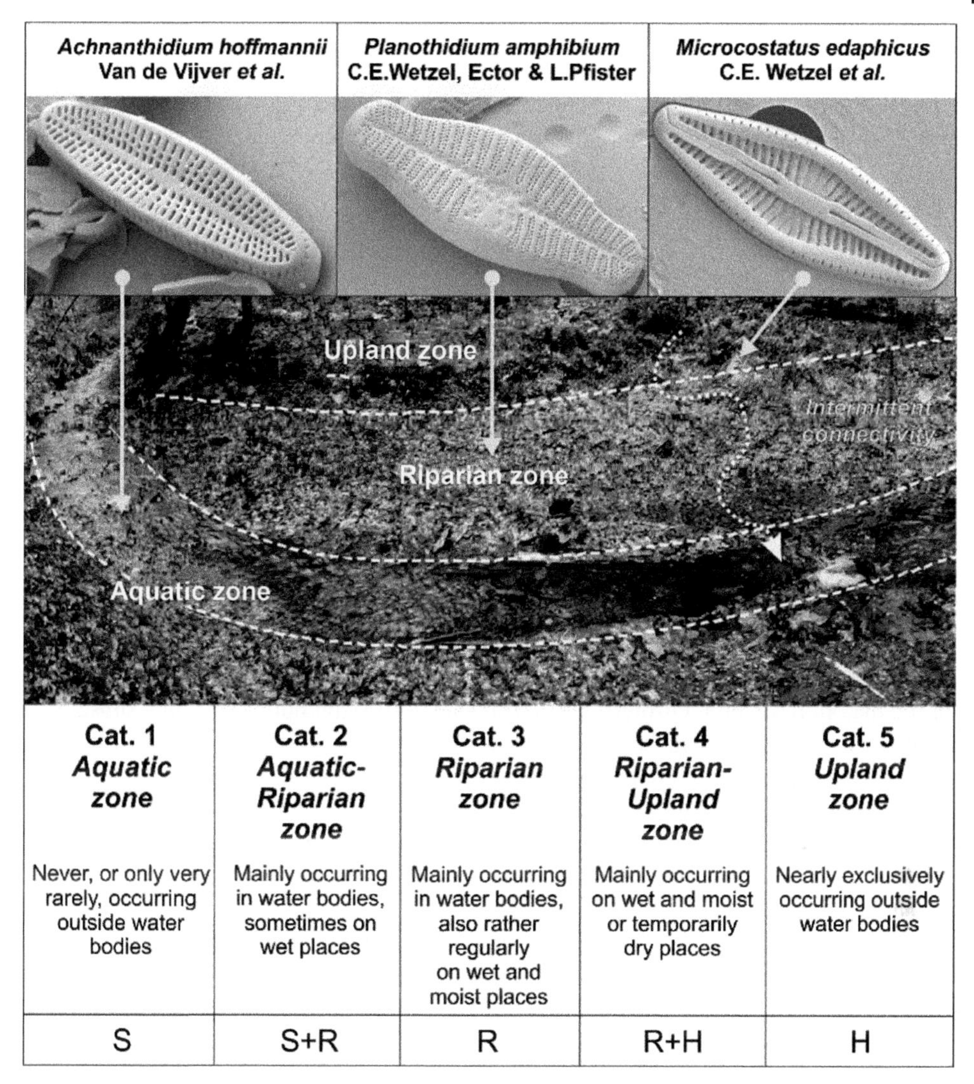

Figure 11.1 Conceptual classification of diatom occurrence in relation to moisture content (categories 1 to 5) based on Van Dam et al. (1994). Hypothesized-related hydrological functional units (aquatic to upland zones) based on Pfister et al. (2009), with exemplified diatom species. (Reproduced from Klaus et al., 2015 / John Wiley & Sons.)

continuum and identified 213 different species. After ordination of these samples, they were able to show distinct differences between each landscape unit's centroids (hillslope, riparian zone, stream), thereby documenting for each unit proper diatom communities. In the same catchment, Martínez-Carreras et al. (2015) tested the hypothesis that different diatom species assemblages map onto individual moisture domains, as a prerequisite for recording the onset/cessation of hydrological connectivity within the hillslope-riparian zone-stream system. Their investigations span over four complete meteorological seasons. They first characterized diatom assemblages in various habitats within the Weierbach catchment collecting samples from rock surfaces (epilithon), stream sediment surfaces (epipelon), overland flow on hillslopes, stream water, and various other substrates, such as litter, bryophytes, vegetation, and soils. Martínez-Carreras et al. (2015) showed that (i) the relative abundance of aerial diatoms increased in stream water during all monitored storm events – thereby confirming the potential for aerial diatoms to quantify riparian zone-stream connectivity, (ii) riparian zones appeared to constitute the largest aerial diatom reservoir, and (iii) aerial diatoms were rather scarce in habitats located further upslope.

11.3 Linking Aerial Diatom Reservoir Dynamics and Physiographic Catchment Characteristics

As a follow-up to the proof-of-concept work exploring the potential for aerial diatoms as hydrological tracers, investigations next focused on specific aspects related to diatom reservoir dynamics on the one hand and the potential for aerial diatoms to percolate through the soil matrix. Both aspects are key for further application of aerial diatoms in hydrological processes research.

Coles et al. (2016) have carried out artificial water sprinkling experiments in the Weierbach catchment on various habitats along the hillslope-riparian zone-stream continuum in order to assess a potential depletion of aerial diatom reservoirs and their subsequent recovery. They found evidence for a quite substantial depletion effect in the case of successive (artificial) rainfall events – however, they never observed a complete exhaustion of the diatom reservoirs. Furthermore, Tauro et al. (2015) did not observe any substantial percolation of aerial diatom frustules through soil columns (Ø = 7.2 cm, length = 6 cm) in controlled laboratory conditions. On the contrary, laboratory work with larger soil columns and columns filled with glass beads of different sizes (to simulate different soil grain sizes) suggests that diatom percolation may well occur under certain circumstances.

Relying on the ubiquitous character of aerial diatoms, Pfister et al. (2015) tested their potential for tracing hydrological connectivity in three contrasted physiographic settings, located in Luxembourg, Slovakia, and Oregon (USA). Their experiments confirmed the existence of aerial diatom reservoirs (of variable sizes) and their intermittent connectedness to the river network in all investigated catchments. However, strictly aerophytic communities exhibited contrasting behaviour between the three catchments. They would increase with the rising limbs of storm hydrographs in Oregon (see also Wetzel et al. 2014) and Luxembourg, while their abundance did not change in the mountainous catchment located in Slovakia. Further research will be needed to better understand aerial diatom assemblages in the landscape, as well as their temporal dynamics.

11.4 Aerial Diatoms as a Hydrological Tracer for Mesoscale Connectivity

While structural connectivity describes how different landscape elements connect to each other, functional connectivity describes the way in which interactions between these elements take place (Wainwright et al. 2011). Hydrological connectivity is part of functional connectivity and describes the ability of energy, matter, and organisms to transfer between and within compartments of the hydrological cycle (Pringle 2003). With this concept in mind, aerial diatoms can be an optimal tracer of hydrological connectivity within the hydrological cycle. Martínez-Carreras et al. (2015) showed that the relative importance of aerial diatoms species systematically increased during run-off events in a headwater catchment, which suggests connectivity between the soil surface of near-stream areas and the stream (see Section 11.2 of this chapter). At larger scale, Rimet et al. (2004) showed that the species distribution of diatoms in streams is controlled by bedrock geology and associated water quality within the individual streams and corresponding catchments. Based on this prior knowledge, Klaus et al. (2015) hypothesized that (aerial) diatoms could serve as tracers of hydrological connectivity beyond headwater scale in catchments of different bedrock geology and land use.

To test their hypothesis, Klaus et al. (2015) sampled one rainfall run-off event in the Attert River basin in Luxembourg. Automatic samplers were installed in six sub-catchments (0.45–43.97 km^2) and the main catchment outlet (247 km^2). While a total of 421 different diatom species were detected in the 149 stream samples taken during the rainfall run-off event, generally less than 25% were aerial species in all investigated catchments. An ordination of samples and locations with a non-metric multidimensional scaling (NMDS) approach (Minchin 1987) revealed that the samples from the individual catchments were generally distinguishable, forming distinct point clouds. This indicates that physiographic differences result in different species compositions.

During the sampled event, the aerial diatom compositions showed pronounced dynamics in all catchments. The relative abundance of aerial diatoms increased with each of the four main pulses of the monitored precipitation event. In most catchments the documented occurrence of aerial diatom frustules and relative abundances were highest as a response to the most intense precipitation sequences and the subsequent highest discharge values. In two of the studied catchments, where near-stream saturated areas can play an important role on run-off generation, the highest abundances were not directly related to the highest discharge peaks. During the last occurring pulse of precipitation, the increase of stream discharge coincided with a decrease in aerial diatom abundance (Klaus et al. 2015). This could indicate a depletion of the aerial diatom reservoir over time, as documented by Coles et al. (2016). Overall, aerial diatom tracing clearly indicated increased connectivity from various habitats in the hillslope-riparian-stream continuum during the observed event over various sized catchments.

Besides the documented dynamics during the observed precipitation events, aerial diatoms also showed a clear scale effect. A difference in the respective share of aerial diatoms compared to that of the total species abundance was obvious in the various sized catchments. The median relative abundance of the aerial diatoms (based on all samples for one catchment) was significantly decreasing with increasing catchment size, following a power law. This result indicates that fewer aerial diatoms can be found with increasing catchment size and that the application of aerial diatoms as a tracer of hydrological connectivity may become limited at the upper mesoscale.

Furthermore, Klaus et al. (2015) classified all occurring diatom species based on the catchment bedrock geologies of the streams where they were found. They used this information to calculate the contribution of every geological setting to the diatom populations found at the main catchment outlet (247 km^2). While this is not a quantitative mixing analysis for separating the stream hydrograph in different source areas (Hooper et al. 1990; Klaus and McDonnell 2013), it is a qualitative indication of where water comes from, eventually indicating connectivity of sub-catchments to the main outlet. This analysis identified clearly how the contributions of different areas varied over the observed event, yet most of the species that were observed during the discharge peaks were non-unique in terms of their geological allocation.

11.5 The Future of Aerial Diatoms as Ecohydrological Tracers

Aerial diatom assemblages are largely controlled by local characteristics of physiographic settings as demonstrated by Klaus et al. (2015); and they are confirmed to be flushed from their terrestrial habitats into streams during rainfall events either in natural conditions (e.g. Klaus et al. 2015; Martínez-Carreras et al. 2015) or in controlled experiments (Coles et al. 2016).

In general, aerial diatom communities have lower species richness when compared to freshwater assemblages and are thought to be "simpler" than diatom communities that are proper to aquatic habitats (Johansen 2010). However, this is not always the case for catchments that have high soil moisture content during most of the year. Antonelli et al. (2017) reported more than 250 species from soil habitats of the Attert basin (Luxembourg) and the list of species includes several poorly known taxa (Stanek-Tarkowska et al. 2016; Wetzel et al. 2013). Further studies of aerial communities may help to further understanding of catchment hydrological functioning, involving diatom fluxes (and consequently, biogenic silica) from soils to freshwater ecosystems and how the species are structured in a complex landscape context. In this context, a better characterization of aerial diatoms' ecology and habitat preferences would ease the use of aerial diatoms as hydrological tracers by facilitating the necessary step of spatially characterizing the aerial communities in the area of interest. Specific species were found to be more or less abundant, depending on prevailing land-use types, such as cultivated soils (Antonelli et al. 2017; Stanek-Tarkowska and Noga 2012), or forested areas (Coles et al. 2016; Wetzel et al. 2013).

Diatoms are used as tracers with high detectability and moderate reactivity (Abbott et al. 2016). However, a limitation to their more widespread use remains – their high specificity. Consequently, sample analysis continues to require substantial taxonomic expertise. Further quantitative polymerase chain reaction techniques to automate diatom identification and quantification (Hamilton et al. 2015) could improve the application of this approach with time- and cost-saving high-throughput analysis.

Ultimately, widening their application potential towards and beyond hydrological processes research is about to open new vistas for environmental monitoring. As an extension to the exploratory work in the Attert basin in Luxembourg, we may consider using aerial diatom communities as a surrogate of soil quality since some common parameters like soil pH, total C, and N soil content have an influence on the development of these communities. Ultimately, this type of aerial diatom assemblage could potentially indicate distinct anthropogenic disturbances. As pointed out by Johansen (2010), aerial diatoms also may have potential as clean water indicators. In a world where it becomes increasingly difficult to find unpolluted, oligotrophic lentic and lotic waters (Kociolek and Stoermer 2009), aerial habitats may provide additional diatom taxa that could be used as indicators of clean water.

References

Abbott, B.W., Baranov, V., Mendoza-Lera, C. et al. (2016). Using multi-tracer inference to move beyond single-catchment ecohydrology. *Earth-Science Reviews* 160: 19–42.

Antonelli, M., Wetzel, C.E., Ector, L. et al. (2017). On the potential for terrestrial diatom communities and diatom indices to identify anthropic disturbance in soils. *Ecological Indicators* 75: 73–81. https://doi.org/10.1016/j.ecolind.2016.12.003.

Beven, K. (2000). On the future of distributed modelling in hydrology. *Hydrological Processes* 14: 3183–3184.

Boeff, K.A., Strock, K.E., and Saros, J.E. (2016). Evaluating planktonic diatom response to climate change across three lakes with differing morphometry. *Journal of Paleolimnology* 56: 33–47.

Burns, D.A. (2002). Stormflow-hydrograph separation based on isotopes: the thrill is gone — what's next?. *Hydrological Processes* 16: 1515–1517.

Burt, T.P. and McDonnell, J.J. (2015). Whither field hydrology? The need for discovery science and outrageous hydrological hypotheses. *Water Resources Research* 51: 5919–5928, https://doi.org/10.1002/2014WR016839.

Coles, A.E., Wetzel, C.E., Martínez-Carreras, N. et al. (2016). Diatom as a tracer of hydrological connectivity: are they supply limited? *Ecohydrology* 9: 631–645, https://doi.org/10.1002/eco.1662.

Davies, A. and Kemp, A.E. (2016). Late Cretaceous seasonal palaeoclimatology and diatom palaeoecology from laminated sediments. *Cretaceous Research* 65: 82–111.

Dunne, T. and Black, R.D. (1970). Partial area contributions to storm runoff in a small New England watershed. *Water Resources Research* 6: 1296–1311.

Ettl, H. and Gärtner, G. (2014). *Syllabus der Boden-, Luft- und Flechtenalgen*, 2e. Berlin Heidelberg: Springer.

European Committee for Standardization. (2003). *Water quality - Guidance standard for the routine sampling and pretratment of benthic diatoms from rivers.* EN 13946:2003. Brussels: Comité Européen de Normalisation.

Falasco, E., Ector, L., Isaia, M. et al. (2014). Diatom flora in subterranean ecosystems: a review. *International Journal of Speleology* 43: 231–251.

Hamilton, P.B., Lefebvre, K.E., and Bull, R.D. (2015). Single cell PCR amplification of diatoms using fresh and preserved samples. *Frontiers in Microbiology* 6: https://doi.org/10.3389/fmicb.2015.01084.

Hayward, B.W., Cochran, U., Southall, K. et al. (2004). Micropalaeontological evidence for the Holocene earthquake history of the eastern Bay of Plenty, New Zealand, and a new index for determining the land elevation record. *Quaternary Science Reviews* 23: 1651–1667.

Hemphill-Haley, E. (1996). Diatoms as an aid in identifying late-Holocene tsunami deposits. *The Holocene* 6: 439–448.

Hooper, R.P., Christophersen, N., and Peters, N.E. (1990). Modelling streamwater chemistry as a mixture of soilwater end-members — an application to the Panola Mountain catchment, Georgia, U.S.A. *Journal of Hydrology* 116: 321–343.

Horton, R.E. (1933). The role of infiltration in the hydrologic cycle. *Transactions, American Geophysical Union* 14: 446–460.

Ivarsson, L.N., Ivarsson, M., Lundberg, J. et al. (2013). Epilithic and aerophilic diatoms in the artificial environment of Kungsträdgården metro station, Stockholm, Sweden. *International Journal of Speleology* 42: 289–297.

Johansen, J.R. (2010). *Diatoms of Aerial Habitats. In the Diatoms: Applications for the Environmental and Earth Sciences*, 2e (ed. J.P. Smol and E.F. Stoermer), 465–472. Cambridge: Cambridge University Press.

Juggins, S., Kelly, M., Allott, T. et al. (2016). A water framework directive-compatible metric for assessing acidification in UK and Irish rivers using diatoms. *Science of the Total Environment* 568: 671–678.

Klaus, J. and McDonnell, J.J. (2013). Hydrograph separation using stable isotopes: review and evaluation. *Journal of Hydrology* 505: 47–64.

Klaus, J., Wetzel, C.E., Martínez-Carreras, N. et al. (2015). A tracer to bridge the scales: on the value of diatoms for tracing fast flow path connectivity from headwaters to meso-scale catchments. *Hydrological Processes* 29: 5275–5289. https://doi.org/10.1002/hyp.10628.

Kociolek, J.P. and Stoermer, E.J. (2009). Oligotrophy: the forgotten end of an ecological spectrum. *Acta Botanica Croatica* 68: 465–472.

Leira, M., Filippi, M.L., and Cantonati, M. (2015). Diatom community response to extreme water-level fluctuations in two Alpine lakes: a core case study. *Journal of Paleolimnology* 53: 289–307.

Martínez-Carreras, N., Wetzel, C.E., Frentress, J. et al. (2015). Hydrological connectivity inferred from diatom transport through the riparian-stream system. *Hydrology and Earth System Sciences* 19: 3133–3151. https://doi.org/10.5194/hess-19-3133-2015.

Matsumoto, D., Sawai, Y., Yamada, M. et al. (2016). Erosion and sedimentation during the September 2015 flooding of the Kinu River, central Japan. *Nature - Scientific Reports* 6: 34168. https:/doi.org/10.1038/srep34168.

McDonnell, J.J. and Woods, R. (2004). On the need for catchment classification. *Journal of Hydrology* 299: 2–3.

Minchin, P.R. (1987). An evaluation of the relative robustness of techniques for ecological ordination. *Vegetatio* 69: 89–107.

Pajunen, V., Luoto, M., and Soininen, J. (2016). Stream diatom assemblages as predictors of climate. *Freshwater Biology* 61: 876–886.

Palmer, A.J.M. and Abbott, W.H. (1986): Diatoms as indicators of sea-level change. In: *Sea-level Research. A Manual for the Collection and Evaluation of Data* (ed. O. van de Plassche), 457–487. The Netherlands: Springer.

Pandey, L.K., Kumar, D., Yadav, A. et al. (2014). Morphological abnormalities in periphytic diatoms as a tool for biomonitoring of heavy metal pollution in a river. *Ecological Indicators* 36: 272–279.

Petersen, J.B. (1935). Studies on the biology and taxonomy of soil algae. *Dansk Botanisk Arkiv* 8 (9): 1–183.

Pfister, L., McDonnell, J.J., Wrede, S. et al. (2009). The rivers are alive: on the potential for diatoms as a tracer of water source and hydrological connectivity. *Hydrological Processes* 23 (19): 2841–2845.

Pfister, L., Wetzel, C.E., Martínez-Carreras, N. et al. (2015). Examination of aerial diatom flushing across watersheds in Luxembourg, Oregon and Slovakia for tracing episodic hydrological connectivity. *Journal of Hydrology and Hydromechanics* 63 (3): 235–245.

Pinder, G.F. and Jones, J.F. (1969). Determination of the ground-water component of peak discharge from the chemistry of total runoff. *Water Resources Research* 5: 438–445.

Pringle, C. (2003). What is hydrologic connectivity and why is it ecologically important? *Hydrological Processes* 17: 2685–2689.

Rimet, F., Ector, L., Cauchie, H.M., and Hoffmann, L. (2004). Regional distribution of diatom assemblages in the headwater streams of Luxembourg. *Hydrobiologia* 520 (1–3): 105–117.

Sklash, M.G. and Farvolden, R.N. (1979). The role of groundwater in storm runoff. *Journal of Hydrology* 43: 45–65.

Smol, J.P. and Douglas, M.S. (2007). From controversy to consensus: making the case for recent climate change in the Arctic using lake sediments. *Frontiers in Ecology and the Environment* 5: 466–474.

Smol, J.P. and Stoermer, E.F. (eds.) (2010). *The Diatoms: Applications for the Environmental and Earth Sciences*. Cambridge University Press.

Stanek-Tarkowska, J. and Noga, T. (2012). Diversity of diatoms (Bacillariophyceae) in the soil under traditional tillage and reduced tillage. *Inz Ekol* 30: 287–296.

Stanek-Tarkowska, J., Wetzel, C.E., Noga, T., and Ector, L. (2016). Study of the type material of Navicula egregia Hustedt and descriptions of two new aerial Microcostatus (Bacillariophyta) species from Central Europe. *Phytotaxa* 280 (2): 163–172.

Tauro, F., Martínez-Carreras, N., Barnich, F. et al. (2015). Diatom percolation through soils - a proof of concept laboratory experiment. *Ecohydrology* 9: 753–764.

Van Dam, H.M., Mertens, A., and Sinkeldam, J. (1994). A coded checklist and ecological indicator values of freshwater diatoms from the Netherlands. *Netherlands Journal of Aquatic Ecology* 28: 117–133.

Wainwright, J., Turnbull, L., Ibrahim, T.G. et al. (2011). Linking environmental régimes, space and time: interpretations of structural and functional connectivity. *Geomorphology* 126: 387–404.

Wetzel, C.E., Martínez-Carreras, N., Hlúbiková, D. et al. (2013). New combinations and type analysis of Chamaepinnularia species (Bacillariophyceae) from aerial habitats. *Cryptogamie, Algologie* 34: 149–168.

Wetzel, C.E., Van de Vijver, B., Kopalová, K. et al. (2014). Type analysis of the South American diatom Achnanthes haynaldii (Bacillariophyta) and description of planothidium amphibium sp. nov., from aerial and aquatic environments in Oregon (USA). *Plant Ecology and Evolution* 147: 439–454.

12

Measurement of Metabolic Rates at the Sediment–Water Interface Using Experimental Ecosystems

Alba Argerich[1] *and Janine Rüegg*[2]

[1] *School of Natural Resources, University of Missouri, Columbia, Missouri 65211, USA.*
[2] *Stream Biofilm and Ecosystem Research Laboratory, École Polytechnique Fédéral de Lausanne, 1015 Lausanne, Switzerland. Current address: Interdisciplinary Center for Mountain Research, University of Lausanne, Lausanne, Switzerland.*

12.1 Introduction

Ecohydrological interfaces have been recognized to be hotspots of biodiversity (Boulton et al. 1998; Gilbert et al. 1990), and more recently, also hotspots of metabolic activity (e.g. Krause et al. 2017; Lautz and Fanelli 2008; McClain et al. 2003), as these transition zones often present distinct physicochemical conditions and house more diverse biological communities than their adjacent ecological systems. Because physiochemical conditions and biotic diversity at ecohydrological interfaces have the potential to impact the surrounding environment, there is a growing interest in quantifying and understanding the driving factors of metabolic transformations in these transition zones (Krause et al. 2017; Lewandowski et al. 2019; Stegen et al. 2018).

The study of metabolic rates in interfaces is complex because of the inaccessibility of the system and because any attempt at observation usually leads to an impact on the system under study, i.e. the observer effect. In order to overcome these difficulties, scientists use artificial human-made ecosystems which can be generally classified depending on the level of simplification in relation to its natural counterpart: from relatively simple methods, often small scale, which we will term microcosms (e.g. dark bottles, sediment columns), to small-scale mesocosms (e.g. open bottom or recirculating chambers), to larger-scale mesocosms such as artificial ponds and streams. Here we will focus on the use of microcosms and small-scale mesocosms to study metabolic rates at the sediment–water interface. We will provide a short historical overview of the development of artificial ecosystems, before presenting the methodologies which are more adequate to study metabolic rates at different spatial scales, ranging from laboratory set-ups to small-scale mesocosms designed to be run in the field. We will focus on freshwater interfaces and discuss the methodologies that are more suitable for the study of lentic (i.e. standing water) and lotic (i.e. flowing) sediment–water interfaces. Finally, we will end by presenting special considerations such as minimizing the observer effect, potential method improvements, and new research avenues.

Ecohydrological Interfaces, First Edition. Edited by Stefan Krause, David M. Hannah, and Nancy B. Grimm.

12.2 Development of the Experimental Ecosystem Approach

12.2.1 Some Landmark Developments

The first attempts to reproduce complex systems in the laboratory by using micro-/mesocosm experiments started in the 1700s. Initially, these experiments were limited to the use of columns to study the movement of water through soil (De la Hire 1703) and saturated sediments (Darcy 1856), and thus provided the quantitative basis for modern hydrology (Ritzi and Bobeck 2008; Simmons 2008). The first studies of metabolic rates from freshwater epilithic communities using experimental ecosystems did not happen until approximately 70 years ago (McIntire and Phinney 1965). A few precursor studies focused on the use of different in situ methods for the study of respiration and photosynthetic rates using epilithic algal cultures (Ryther 1956) and created the basis for the actual recirculating flow chambers by constructing a small flowing water microcosm using a glass condenser tube (Odum and Hoskin 1957).

These first attempts at producing and using experimental ecosystems were followed by their use as replicable units that could be manipulated during long-term periods. For instance, a series of five-gallon carboys containing not only algae but also invertebrates and young fish were used to measure the effects of different combinations of nutrient additions on ecosystem respiration and photosynthetic rates (McConnell 1962). The duration of these experiments lasted between five and seven months, depending on the treatment, and evidenced the problems related to the representativeness of the system, an issue McConnell pointed out: "...microcosms may differ from natural habitats in having unusual concentrations of metabolic gases, low rates of bacterial activity, and accumulations of biotic substances which may be inhibitory to primary and secondary production".

The study of metabolic rates using experimental streams, defined as relatively small constructed channels with controlled flows (Warren and Davis 1971), started in the second half of the twentieth century and mostly related to the understanding of the self-cleaning capacity of the streams (Wuhrmann 1954). Experimental streams were considered a better representation of the natural environment but, at the same time, presented inherent difficulties related to the control of environmental factors. These channels were open to the atmosphere, making the mass balance of gases difficult, and contained sediments, which affected the distribution of solutes.

The second half of the twentieth century also marked the development of radiotracers, a growing awareness of human impacts on the environment, and a focus on freshwater ecosystems by systems ecologists (Larned 2010). These new approaches expanded the research of metabolic activity at the sediment–water interface marking the relevance of the role of periphyton (i.e. a combination of algae, bacteria, and fungi growing on submerged surfaces and often referred to as a biofilm; Larned 2010) on nutrient cycling. Experiments with radioactive isotopes were developed to understand the transfer of matter and energy in ecosystems. For example, phosphorus ^{32}P was used in aquatic microcosms to study the dynamics of phosphorus uptake and transformation in aquatic communities (Whittaker 1961). Whittaker's technique was later improved by Patten and Witkamp (1967), which demonstrated that it was possible to add progressive complexity to the microcosms by adding one new labelled part of the ecosystem at a time (e.g. ^{134}Cessium-labelled oak leaves, mineral soil, microflora, millipedes, and aqueous leachate). Finally, the last decades have been characterized by technological advancements related to the development of data loggers and sensors able to fit mesocosms without excessively disrupting the medium.

12.2.2 Selection and Use of Experimental Ecosystems to Measure Metabolic Rates

Experimental ecosystems, either micro- or mesocosms, try to replicate natural processes under a controlled environment. The definition of micro- and mesocosms varies greatly in the literature. In this chapter, we will use these terms to refer to experimental ecosystems that span a spatial scale of tens of square centimetres to tens of square metres. Examples of experimental ecosystems are sediment columns (Figure 12.1a), stream recirculation chambers (Figure 12.1b), parafluvial tanks (Figure 12.1c), or artificial channels (Figure 12.1d). One of the advantages of using experimental ecosystems is that they can be easily altered and replicated, allowing the study of metabolic processes under different scenarios with sufficient statistical powcr.

The measurement of metabolic rates across the sediment–water interface often requires the use of complementary approaches, i.e. a mixture of field observations and the use of experimental ecosystems. This is because in many cases, ecosystems are too large to be encompassed by a single spatial-scale of measurement; even large-scale methods such as the whole-stream approach to measure metabolism (Marzolf et al. 1994; Odum 1956) or the measurement of gas exchange in lakes and oceans (Wanninkhof 1992) are constrained to a spatial scale of hundreds of square metres. Larger-scale manipulations (i.e. whole lakes, whole watersheds), while very informative, are invasive (and thus rarely permitted), and lack replication.

The estimation of metabolic rates based on field observations often integrates an average heterogeneity in ecological processes that might be important for the understanding of the functionality of different parts of the ecosystem. Experimental ecosystems are commonly

Figure 12.1 Images of different mesocosms: (a) sediment column, (b) flow chambers, (c) parafluvial ponds, and (d) artificial stream channels at the River Urban Lab The River Urban Lab / https://urbanriverlab.com/ last accessed 28 April, 2023

used to identify parts of the ecosystem that could present antagonistic responses to changes in the environment. Only with both types of information, from large and small spatial scales, will we achieve a mechanistic understanding of ecosystem processes. Experimental ecosystems represent an intermediate scale of measurement (larger than individual organisms but smaller than entire ecosystems). They allow the study of heterogeneity within an ecosystem (e.g. compare metabolic rates at near-shore lake sediments vs. rates at sediments from the lake's deepest area). Moreover, they are designed to isolate specific factors of interest (e.g. isolating the effects of light from changes in temperature on aquatic primary production, parameters that are difficult to tease apart in the natural environment).

The methods for measuring metabolic rates at the sediment–water interface are surprisingly similar for lotic and lentic ecosystems. The key difference resides in the need to simulate the effect of flowing water when estimating metabolic rates in lotic ecosystems. This can be achieved by using flow-through sediment columns (Figure 12.1a) instead of bottles, recirculation chambers (Figure 12.1b) instead of the open-bottom chambers used in lentic environments, or using the whole-stream reach in situ approach instead of the whole-lake approach when dealing with larger spatial scales of measurement.

These different methods can be used to acquire estimates of metabolic activity such as aquatic primary production and aerobic respiration. These two metrics have become a common tool to assess river health (Fellows et al. 2006; Mulholland et al. 2005; Young et al. 2008) since they integrate the functioning of the entire aquatic community. Traditionally, aquatic respiration and primary production have been estimated by examining changes in dissolved oxygen over time. However, dissolved oxygen in flowing waters changes not only because of primary production (which produces oxygen as a by-product when fixing inorganic carbon) or aerobic respiration (which consumes oxygen when oxidizing organic matter) but because of gas exchange with the atmosphere (reaeration). This last component is one of the largest sources of uncertainty when measuring stream metabolism at a reach scale (Aristegi et al. 2009). Methodologies to measure metabolic rates at a small scale circumvent this problem by using contained, gas-tight, chambers. By avoiding gas exchange, enclosed mesocosms allow for a relatively simple calculation of production and respiration as the gain and loss terms in dissolved oxygen concentration, respectively. Rates based on dissolved oxygen can then be converted to carbon units using photosynthesis and respiration coefficients obtained experimentally (e.g. 1.89 and 0.85, respectively, for stream communities; Bott et al. 1978) to compare to the generally accepted units for metabolic rates in other ecosystems (e.g. terrestrial primary production).

12.3 Measuring Metabolic Rates of Benthic Sediments in Lotic Environments

12.3.1 Column Experiments

Column experiments have been widely used to study solute transport and transformation in sediments, to predict the fate and mobility of pollutants in the subsurface. The interface between water and sediments usually presents a steep gradient of physicochemical conditions and high abundance and activity of microorganisms, which generally leads to the enhancement of transformation processes. However, accessing this ecotone without disrupting it is difficult, so in situ measurements of processes that take place in this interface are rare. A common approach is the use of column experiments. Although the reproduction of natural flow patterns within the columns is difficult to achieve, and repacked columns are not able to reproduce the heterogeneity in texture and materials of natural sediments, columns are considered a suitable, simplified method to measure transformation rates and microbial activity in sediments (Kessler et al. 2019).

Columns have been used to estimate dissolved inorganic carbon percolation fluxes in the unsaturated zone (Thaysen et al. 2014), to measure microbial respiration (Hendry et al. 2001), the kinetics of microbial communities (Muñoz-Leoz et al. 2011), as well as to understand the fate and transformation of contaminants (e.g. waste water irrigation Richter et al. 2015, mercury mobilization, and transformation Poulin et al. 2016), to name just a few research areas where column studies have improved our understanding of sediment–water interface properties and functionality.

Columns are usually made of inert materials to prevent interaction with solutes. Materials used to build columns include acrylic glass (which allows visual control of the sediments, especially useful when using dye tracers), stainless steel, PVC, aluminium, and polyethylene (Banzhaf and Hebig 2016). In addition to column material, the overall size of the columns and the length-to-diameter ratio are also important column design factors since they will determine the reproducibility of the experiment and how easily the results can be modelled, and eventually, scaled to natural systems. Regarding size, columns vary from a few centimetres to metres in both diameter and length, depending on the research question (Lewis and Sjöstrom 2010). It is recommended to maintain a length-to-diameter ratio of 1:4 to avoid sidewall effects (i.e. water slowing down at the sides of the column), which might be an issue in unsaturated columns and to a lesser extent for saturated columns (Bergström 2000; Lewis and Sjöstrom 2010).

Column experiments can be classified according to the level of saturation of the environment that they are trying to reproduce (unsaturated vs. saturated columns) and/or according to the method of construction (i.e. packed columns vs. monoliths) (Lewis and Sjöstrom 2010). Unsaturated columns are used in the study of soil respiration but may also represent areas of the hyporheic zone (i.e. zone between the stream–sediment interface and groundwater) where stream sediment is no longer fully saturated. Saturated columns are used to study the behaviour of solutes and water in the saturated zone of different ecosystems (e.g. lakes, rivers, ponds, floodplains). The difference in construction is related to the way sediments are handled. Packed columns are built from sieved sediments collected from the field, with the goal to create a homogeneous medium, with a similar bulk density as the natural environment while avoiding preferential flow paths by removing heterogeneity. Despite not representing the heterogeneity of natural sediments, packed columns are used frequently because they are highly reproducible and they can be used to study different treatments with replication. In contrast to packed columns, monoliths are intact pieces of sediment, collected without altering their structure, and therefore a closer representation of reality. For an extensive review of column construction of both packed sediments and monoliths see Lewis and Sjöstrom (2010).

12.3.2 Flow Chambers

The sealed chamber approach has been used in various forms to determine metabolic rates and other sediment properties at small spatial scales. In this approach, benthic substrates are removed from the stream and placed in a closed-system chamber, which contains a small propeller capable of reproducing streamflow (e.g. Bott et al. 1997). Chambers with sediments can then be transported to the laboratory or placed at the stream side and used as field chambers.

Fundamental properties of chamber designs addressing water circulation in a laboratory setting have been discussed by Vogel (1994), who considered attaining laminar flow, moving water with minimal energy costs, and economical construction as the key characteristics of a good chamber design. Dodds and Brock (1998) brought the flow chamber approach to field settings and incorporated a modular construction to ensure the ability to disassemble the chambers for cleaning and trouble-shooting as the fourth component of good design.

All chamber designs have inherent limitations and present trade-offs between ease of use, accurate reproduction of natural environmental conditions, and cost of construction. Common design limitations related to the reproduction of natural environmental conditions include the inability to reproduce natural flow, the alteration of the spectral irradiance arriving at the sediments, water temperature rise, and elevated internal chamber pressure (Dodds and Brock 1998). Simpler designs may be easier to use and cheaper to build but are not able to recreate field flow conditions. Similarly, chambers that can replicate natural light conditions are preferred but they often come with the cost of increased chamber temperature as sunlight adds heat to the chambers.

A recirculating chamber (Figure 12.1b), generally has a main body that contains water and sediments, a propeller to move the water, and an (external) motor/power source to move the propeller. The main body of recirculating chambers is often made from clear acrylic plastic to allow the penetration of light and is sealable to prevent gas exchange. Within the main body is a location to place the sediments that may be recessed to prevent water from flowing through the sediment. An internal propeller moves the water through a return portion of the chamber (either to the side or below the sediment-containing portion of the chamber), then over the sediments (before which there might be a section designed to create laminar flow), and then returns to the propeller to be moved through the cycle again. The chamber requires a gas-tight sealed port to place the measurement probe and/or collect a water sample without introducing air. The motor and the power source are often external to prevent heating of the water.

The use of chambers to measure rates of respiration and primary production requires a dark/light conditions approach where changes in oxygen are measured over time (for a period long enough to detect changes in oxygen but short enough to not significantly change other sediment and water characteristics). Similarly to open-channel methods to measure stream metabolism, chamber measurements of dissolved oxygen in the dark are assumed to be the result of aerobic respiration. In contrast, measurements in the light reflect net primary production, which can be converted to gross primary production after subtracting the measured respiration. Other metabolic rates, such as nutrient uptake, can be measured by adding solutes at time zero and collecting samples through time.

A recent field chamber design that has been extensively tested and proven useful for stream-side measurements from tropical to arctic streams allows the use of 10 cm × 30 cm × 6 cm of sediment in approximately 10 L of water (Rüegg et al. 2015). The chamber is designed to accommodate baskets full of sediments that have been incubated in the stream for a determined period or with sediment transferred directly from the field. The transfer of sediments (alone or in baskets) introduces a methodological artefact that, unfortunately, cannot be avoided but, at the same time, allows the method to be applied to material other than benthic sediments such as submerged vegetation or leaf detritus. Adjustments in regard to the type of incubated material, location of the incubation, and replication can be made depending on the research question.

12.3.3 Artificial Streams

Measurements of metabolic rates in artificial streams often follow the same approach as for natural streams (i.e. open-channel method). Estimation of respiration and primary production rates in aquatic ecosystems has been traditionally based on the fact that changes in dissolved oxygen concentration across time and/or space are due to the physical exchange of oxygen between water and the atmosphere (which can result in a positive or negative oxygen gain in water), oxygen consumption (through aerobic respiration by heterotrophic bacteria, algae, and other aquatic organisms), oxygen production (by primary producers), and changes in dissolved oxygen because of groundwater accrual. Rates are obtained by analysing the behaviour of 24-hour diel dissolved oxygen curves in conjunction with irradiance measurements, water temperature, streamflow, and water volume, or wetted channel area (Demars et al. 2015). Changes in dissolved oxygen during night-time are assumed to result from respiration (once gas exchange between water and air has been accounted for), and representative to respiration occurring during daytime. Any further change in oxygen concentration during daytime is then attributed to primary production. As some artificial streams are a closed system in terms of water, a single location of measurement will be enough to measure the diel dissolved oxygen change while a flow-through artificial channel will require the measurement of dissolved oxygen at both the inlet and outlet.

The chamber approach discussed in the previous section can also be used in combination with the larger-scale mesocosm if smaller spatial-scale heterogeneity within the artificial stream is of particular interest to the researcher, and be extrapolated to the entire stream using the sediment area and an area-weighted average. Respiration and primary production rates estimated with the chambers are expected to be different from those measured using the open-channel approach: firstly, because not all habitats can be represented in chambers, and secondly, because of the error associated with the estimation of gas exchange coefficients in the open-channel method, which is avoided when using flow chambers. Indeed, uncertainty in the reaeration coefficient measurements is one of the main limitations for the study of stream ecosystem metabolic rates (Aristegi et al. 2010) using an open-channel approach. The use of chambers eliminates the need to estimate gas exchange coefficients but, on the other hand, chambers do not capture the weight of the processes occurring in the hyporheic zone (e.g. the chamber presented above only includes 6 cm of sediment while hyporheic zones can be much deeper). Chamber measurements are often thought to underestimate heterotrophic conditions and favour autotrophic conditions compared to the open-channel approach (Bott et al. 1978). Algae, which are the primary producers in aquatic ecosystems, require light and are thus found in the benthic surface layer, which is well captured by the shallow sediments in chambers. On the other hand, the bulk of respiration is mostly due to heterotrophic bacteria inhabiting the deeper sediments which are not usually included in chambers. Altogether, the use of chambers can result in an over-emphasis of photosynthesis over respiration. Also, because chambers measure metabolic activity for only a small volume of sediment, much replication is needed to capture the natural spatial and temporal variability usually encountered in heterogeneous artificial streams.

12.4 Measuring Sediment Metabolic Rates in Lentic Environments

12.4.1 Incubation Experiments without Recirculation

Incubation experiments where sediments are placed in bottles or jars and are incubated for some time is a widespread method to measure metabolic rates. They can be carried out in the lab or the field. There is no generally approved incubation container size or material, but rather both are adjusted to the specific research questions.

Sediment incubations in jars have been used to study respiration and primary production, the decomposition of different types of organic matter in anaerobic sediments (Emilson et al. 2018), and to test the amendment of sediments with different materials for the removal of contaminants (Öztürk and Kavak 2005). Respiration and primary production rates can be measured using the light/dark bottle approach. The light and dark methods allow the calculation of net primary production and respiration, respectively, as described previously in this chapter. Benchtop bottles can be small plastic or glass bottles filled with the sediment of interest, with or without an overlying water column and with or without headspace. If only a single measurement in time is required, a simple lid that seals gas-tight will suffice. After the incubation period, the lid can be removed to measure the dissolved oxygen. Metabolic rates are then determined by the change in oxygen over the incubation time, from the beginning of the incubation to the end, assuming a linear change over time. Something as simple as 50 ml centrifuge tubes can be used (e.g. Johnson et al. 2009). If continuous measurements over time are wanted or needed, a lid with a sealable port for a dissolved oxygen sensor needs to be fashioned. Such bottles will need to be large enough to accommodate the oxygen probe.

Incubation experiments in the field (i.e. in situ approach) are similar. Substrates are placed in a container, dissolved oxygen is measured at the beginning of the incubation, and the container is then sealed and placed at the same place where the sediment has been collected to run the incubation under natural light and temperature conditions. After incubation, dissolved oxygen in the water is remeasured, and water is usually replaced for a second incubation, this time under dark conditions using a light-impenetrable cover (e.g. commercial grade black trash bags). If only a single time point is measured for the entire incubation, the incubation time needs to be long enough to detect a measurable change in the gas concentration but short enough to avoid a significant change in the physicochemical characteristics of water and sediment (e.g. a depletion in nutrients). If water is not replaced between light and dark treatments, the choice of which treatment to run first (i.e. dark or light) will depend on the water's oxygen saturation level. If water is well oxygenated, the dark method is generally used first to draw down the oxygen and avoid oversaturation; in oxygen-depleted water, the light treatment is run first to increase oxygen levels and prevent anoxia during the dark incubation. The use of a modified lid with a port for a dissolved oxygen sensor allows for a continuous measurement of changes in dissolved oxygen and temperature. Thus, neither the benchtop nor the in situ bottle methods are challenging to implement. However, certain considerations are essential to make, the main one is whether a single measurement with a linear approximation of gas change through time is valid.

12.4.2 Open-bottom Chambers for in situ Measurements

Open-bottom chambers have developed from something as simple as a carboy cut in half and placed upside down on top of the sediments of interest, with valves at the top to allow the collection of measurements or sample water (Fellows et al. 2007), to free-vehicle benthic chamber instruments used in deeper waters (e.g. Jahnke and Christiansen 1989; Riebesell et al. 2013).

The principle of the open-bottom chambers is similar to the recirculating chambers (Section 12.2.2) in that dissolved oxygen is measured in the water above the sediment to determine aerobic metabolic rates of respiration and production using a light and dark method. However, there are a few critical differences. Firstly, due to the standing nature of water in lakes, the recreation of flow, critical in chambers used to estimate metabolic rates in streams, is usually omitted. Secondly, as the chambers are deployed and filled within the pond, lake, or marine sediments, they generate genuine in situ measurements. A critical factor is to deploy the chamber with as little disturbance to the sediment as possible. And thirdly, the open-bottom chamber is not a closed system, as its name implies. Thus, pore water can seep into the chamber from deeper sediments, and such chambers have actually been used to test for seepage water amounts (e.g. Taniguchi et al. 2006). In terms of gas exchange, such seepage might bring in water with a different dissolved oxygen concentration complicating the estimation of respiration and primary production rates. However, sediments of many lentic water bodies are anoxic within the first centimetres of depth, and as diffusion is slow, dissolved oxygen measured at the surface is unlikely to be affected. However, anaerobic metabolism could contribute a good portion of the metabolism of the sediment–water interface and should thus not be underestimated.

12.4.3 Artificial Pond and Lake Mesocosms

Measurements of metabolic rates in artificial ponds follow whole-lake methods, similar to how the open-channel approach to measure respiration and primary production in streams is also used in artificial streams. Whole-lake approaches usually use the diel oxygen curve to determine respiration and primary production (e.g. Staehr et al. 2010). Similarly, the assumption stands that night-time respiration is the same as daytime respiration to allow for the calculation of gross primary production (e.g. Staehr et al. 2012). One major difference between the lentic and lotic approaches is how the gas exchange coefficients are measured and the placement of the probe that measures the dissolved gas of interest. The water surface of lentic water bodies is (often) smoother than those of streams, and the current is, in many cases, negligible, allowing for the deployment of floating chambers to measure the exchange of gases at the water–air interface. In short, an inert gas (e.g. SF_6) is pumped into the water, and its accumulation measured in an open-bottom chamber that floats on the surface. The accumulation can then be converted to a gas exchange coefficient of oxygen. Secondly, the placement of probes is critical. Lentic waters often exhibit stratification, which inhibits gas exchange between different sections of the water column. So, if the objective is understanding the processes occurring at the sediment–water interface, a probe in the upper part of the water column is unlikely to provide the metabolic rates of interest. For that reason, open-bottom chambers are often used in conjunction with whole-lake approaches. Chambers can test for different sediment types with different properties, allowing for a more detailed characterization of the ecosystem under study.

12.5 Challenges and Opportunities

Experimental ecosystems provide numerous opportunities to study different physical, chemical, and biotic conditions occurring at the sediment–water interface. The possibility to manipulate environmental conditions and to study interactions among factors are manifold. However, experimental ecosystems entail some challenges that need to be addressed to ensure accurate results.

For example, artefacts due to the shape and dimensions of the experimental ecosystems (i.e. the enclosure effect) can be significant (Mine Berg et al. 1999; Petersen et al. 1997, 1999). However, many studies fail to report key details of the containers used. Any unintentional deviations in environmental factors, such as temperature, nutrient concentrations, and flow (to name just a few major ones), due to the methodology employed, may affect the results and their interpretation. Still, such deviations are generally not measured, or at least not reported, and rarely taken into account in the analysis of data. A second critical factor to consider when using mesocosms is the problem of scale (Chen et al. 2000; Gerhart and Likens 1975). Heterogeneity in natural habitat and environmental conditions (e.g. substrate, macrophyte and periphyton density, incident solar radiation) makes scaling measurements from micro- and mesocosms to the whole system challenging. Free-water techniques, on the other hand, offer the promise of integrating a signal over an entire section of a freshwater ecosystem (e.g. for whole-reach stream metabolism; Bernot et al. 2010; Mulholland et al. 2001; Uehlinger and Naegeli 1998), but at the same time do not allow for the detection of small-scale heterogeneity. Finally, experimental ecosystems often fail in replicating natural conditions in their totality. The self-contained nature of mesocosms prevents a natural exchange of water, solutes, heat, and gases with the surrounding environment. This lack of exchange creates unintended changes in water velocity, temperature, nutrients, and gas concentrations, and pressure inside the mesocosm during the incubations. Here we are going to discuss how to minimize the effects on two of the main critical variables, in our view, temperature and water velocity.

12.5.1 Temperature Effects

One of the most critical environmental factors in the measurement of metabolic rates is that of temperature. Roughly, per every 10 °C increase in temperature, metabolic rates double (Bothwell 1988). Temperature effects rarely need to be considered in column or bottle experiments conducted in the laboratory since temperature can be held stable. However, unintended changes in temperature can become a problem when measuring in situ metabolic rates using enclosed chambers in the field. Such changes are especially evident if mesocosms are placed stream-side rather than within the stream. In the case of enclosed chambers, the very nature of the mesocosm prevents the exchange of water with the surrounding environment; therefore, heat easily accumulates in the system. Unintended increases in temperature can be minimized by using a climate-controlled chamber. Control in temperature though, comes with the alteration of other environmental factors (e.g. light) and often requires transporting the sediments to the lab, increasing the time between sediment collection and the measurement of metabolic rates, which has its own set of problems.

Increases in temperature in flow chambers can be due to the exposure of a confined volume of water to light and the heat created by the motor. Motors generate heat and thus are often placed outside the chamber. Because they are rarely waterproof, chambers cannot be fully submerged to prevent warming in the surrounding air. Similarly, shading would defeat the purpose of placing the chamber to capture in situ light conditions.

There are two main options to approach this problem: minimizing temperature increases during the measurement and applying correction factors after the measurement. The first option includes submerging the chambers as far into the water body as possible or into a side-pool (e.g. a kids pool) where water is continuously exchanged with a pump. The second option is to apply a retroactive correction factor. Converting changes in dissolved oxygen to metabolic rates can be modelled, which allows the inclusion of temperature directly, and the standardization of metabolic rates to a common temperature. The post-correction method has clear advantages over the control for temperature during measurements. Still, it might be problematic if temperature differences among water bodies are a point of interest when comparing replicates (e.g. from different streams with different temperature regimes). As with most method adjustments, the research question will determine which control measures are acceptable. If temperature is critical, then perhaps the transport of sediment and water off-site (to the laboratory or climate chamber) might be the best option (always depending on the remoteness of sites). If the original properties of the sediments need to be preserved as much as possible, any in situ adjustments or later correction factors will provide the best approach.

12.5.2 Flow Reproducibility

The reproduction of natural flow patterns is a major challenge in experimental ecosystems. Column experiments, if not carefully designed, might present unintended preferential flowpaths that could lead to the estimation of metabolic rates far from those in natural environments (Franklin et al. 2019). Recirculating chambers, on the other hand, although useful to characterize benthic metabolic rates, cannot reproduce hyporheic water exchange and therefore do not recreate the natural complexity of streamflow patterns.

In columns, the use of baffles at the inlet and outlet of the column can help disperse the flow uniformly to avoid unintended preferential flowpaths. It is recommended that the baffle used is at least as thick as the column diameter and perpendicular to its longitudinal axis to create uniform flow within the column (Barry 2009). Sidewall effects (i.e. water slowing down at the sides of the column), might be an issue in unsaturated columns (and to a lesser extent for saturated columns) and can be minimized by observing a recommended diameter-to-length ratio of 1:4 (Bergström 2000; Lewis and Sjöstrom 2010).

In chambers, the creation of laminar flow (i.e. "ideal" flow) requires chambers to be long (many times the width of the chamber). Vogel (1994) suggests that the chamber flow area should be three times longer than the working area and still may not precisely reflect natural streamflow. In addition to choosing the right chamber length, laminar flow can be supported by the use of flow directors (Dodds and Brock 1998). Furthermore, natural streamflow patterns at the sediment–water interface are difficult to measure without specialized equipment. The complexity in measurement may explain why studies based on chamber measurements rarely test and describe water velocity at the sediment–water interface. Measuring velocity in the chamber water column is easier and allows for an approximate comparison with flow velocities in streams. However, even such measurements are rarely presented in publications, nor are velocity profiles across and along the chambers (however, see Rüegg et al. 2015). Most flow chambers intended to be run in the field generate slower water velocities than those found in natural conditions. Mainly, because the motors used to create flow have to run on batteries, which restrict the size of the motors and thus the range in velocities generated (e.g. Rüegg et al. 2015). Moving the mesocosms into a laboratory setting reduces the need for a mobile power source but, at the same time, modifies environmental conditions. Light conditions in the lab will surely differ from those in the field, and the physicochemical characteristics of water and sediments will be most likely changed during transportation.

12.5.3 Moving Beyond Aerobic Metabolic Rates

Mesocosms are useful to measure other processes at the sediment–water interface beyond aerobic respiration and primary production. Processes related to the use and transformation of nutrients, bacterial community dynamics, or ecotoxicology (Caquet et al. 2000), are just a few examples of the type of applications of mesocosms.

Nutrient uptake, incorporation, and remineralization and release back into the water column are critical parts of the nutrient cycle (Mulholland et al. 1985; Stream Solute 1990). Often, researchers are only interested in net uptake to understand how freshwaters process and remove excess nutrients. The methods described in Sections 12.2 and 12.3 are well suited to measure nutrient cycling as they can be manipulated with the addition of one or more nutrients without affecting the overall ecosystem. One example is the measurement of ammonium uptake rates using recirculating chambers by adding a set amount of ammonium and measuring its decline over time (O'Brien and Dodds 2008). The slope of the decline gives information about the nutrient uptake rate. The methodology is straightforward, though similar issues than those described for respiration and primary production rates about temperature and flow can apply. Any changes in the mesocosms that affect metabolic activity will likely also affect nutrient uptake. Those changes will require some knowledge about the ecosystems, especially to find the best function to describe nutrient dynamics (e.g. linear, second-order kinetics, exponential decay, Michaelis–Menten; Trentman et al. 2015). Combining the nutrient uptake method with detailed analyses of the sediments' organic and nutrient content can provide information about where nutrients might be stored (e.g. deep vs. shallow sediment). Mesocosms, being closed systems, are ideal for building mass balances and for studying the fate of nutrients.

The use of mesocosms is also critical for the study of anaerobic rates; in fact, studies for the production of methane or nitrogen gas from sediments often rely on the use of batch experiments. The quantification and characterization of nitrogen removal processes in freshwaters have been of great interest in the last decades due to the increased concentrations in nitrate observed in surface- and groundwaters. Denitrification rates (i.e. the conversion of nitrate (NO_3) to nitrogen gas (N_2)) has been traditionally measured using the acetylene inhibition technique (Sørensen 1978), which uses the properties of acetylene in blocking one of the intermediate steps in the denitrification reaction (the reduction of nitrous oxide (N_2O) to N_2). Sediments are incubated in a sealed reactor, such as a bottle, and exposed to acetylene. Then, samples are collected over time for the analysis of N_2O. Observed changes in N_2O concentrations are used to calculate the denitrification rate of the sediment. This approach originated in the study of marine sediments and has been later applied to stream and lake sediments with success (e.g. Christensen et al. 1990). However, the acetylene block method has some limitations, including the inhibition of nitrification, the inhibition of methanogenesis, and the inhibition of sulphate-respiring bacteria (Seitzinger et al. 1993). One of the alternatives to the acetylene inhibition method is the use of isotopic nitrogen tracer techniques (i.e. ^{15}N) that can be used both in situ and in mesocosms (Mulholland et al. 2004). As with all the other methods mentioned, the research question is again key to select the appropriate method.

12.6 Summary and Conclusions

There are many open questions related to the understanding and measurement of metabolic rates using artificial ecosystems. These questions range from testing the effects of "artefacts" created by the methods employed (e.g. how to correct and minimize the effect for increased water temperature in recirculating chambers, sediment handling, flow variation), to research areas opening up due to the development of new technologies (e.g. new analytical techniques for the analysis of micropollutants), as well as interdisciplinary research (e.g. holistic view of the interfaces rather than just discipline-driven questions).

Methodological-related studies related to the "artefacts" generated by the mesocosms were more difficult to find than expected. The topic of scale (Petersen et al. 2009; Spivak et al. 2011) and relevance of mesocosms (Carpenter 1996; Drake and Kramer 2012) has been extensively reviewed. However, aspects related to other limitations of experimental ecosystems have received less attention. For example, we are aware that temperature changes in chambers are a problem. Still, studies do not generally report temperature or explain any corrections done during data analysis because of unintended increases in temperature, making comparisons among studies or compilation of datasets extremely challenging.

The development of new technologies has opened a new set of possibilities for the measurement of metabolic rates in the sediment–water interface. Traditionally, measurements of metabolites have been done by chemical analysis of water samples or by using sensors with electrodes. The recent incorporation of fibre optics into sensor systems has allowed the miniaturization of devices and the measurement of small sample volumes, making them excellent choices for the measurement of metabolic rates in columns and the field. In general, fibre-optic sensors can be classified into two types depending on their mode of operation: (a) chemical sensors – that deliver real-time information on the presence of specific compounds, and (b) biosensors that make use of biological components (e.g. enzymes, microorganisms, antibodies, nucleic acids) to detect and/or quantify target compounds (Espinosa Bosch et al. 2007). For more information, see the review by Pospíšilová et al. (2015) and other chapters of this book. Accompanying the development of new sensors had been the development of a smart tracer (e.g. the resazurin-resorufin die tracer, Haggerty et al. 2008) that can capture aerobic metabolic activity at different spatial scales (patch and reach scale) and quantify the contribution of different ecosystem compartments (e.g. hyporheic zone, pools, etc.) to the overall ecosystem respiration (Argerich et al. 2011; González-Pinzón and Haggerty 2013; González-Pinzón et al. 2012). These new developments allow for methods to be potentially applied truly in situ, as well as providing faster results and thus reducing the time of mesocosm measurement and any of the artefacts that come with the use of mesocosms.

Finally, one of the most important questions that remains to be answered is the scaling of results obtained in mesocosms to the ecosystem level. Studies that have applied the same treatment at more than one spatial scale evidence that results can change with the size of the experimental unit (e.g. Cooper et al. 1998; Hendry et al. 2001) while others remain unaffected (Spivak et al. 2011). In addition to the spatial-scale-dependence of the results, the lack of reproducibility of the temporal behaviour in mesocosms imposes additional challenges to extrapolation. A time lag often exists between treatments and responses that is not always well captured by artificial ecosystems, and that can vary depending on the spatial scale (Englund and Cooper 2003).

A meta-analysis of 360 experiments using mesocosms for the study of aquatic ecosystems evidenced the need for multi-scale experiments, both for time and space, to better capture natural processes (Petersen et al. 1999). We would add that in order to understand metabolic rates at an ecosystem level, we need to use complementary methods (i.e. different types of mesocosms, field observations, modelling, etc.). Mechanistic understanding of processes in general, and at interfaces in particular, needs the integration of several techniques (and maybe different levels of complexity of mesocosms), including the study across spatial and temporal scales.

The trend of mesocosms being more and more realistic opens new opportunities for inter- and transdisciplinary research with a holistic view of the ecosystem (in contrast to traditional disciplinary studies that portion research into physical, chemical, and biological aspects), and will probably facilitate the scaling of metabolic rates measured in artificial ecosystems to the natural ecosystem.

References

Argerich, A., Haggerty, R., Marti, E. et al. (2011). Quantification of metabolically active transient storage (MATS) in two reaches with contrasting transient storage and ecosystem respiration. *Journal of Geophysical Research* 116 (G3): G03034. https://doi.org/10.1029/2010JG001379.

Aristegi, L., Igartua, O.I., and Irurtia, A.E. (2010). Metabolism of Basque streams measured with incubation chambers. *Limnetica* 29 (2): 301–310.

Aristegi, L., Izagirre, O., and Elosegi, A. (2009). Comparison of several methods to calculate reaeration in streams, and their effects on estimation of metabolism. *Hydrobiologia* 635 (1): 113–124.

Banzhaf, S. and Hebig, K.H. (2016). Use of column experiments to investigate the fate of organic micropollutants–a review. *Hydrology and Earth System Sciences* 20: 3719–3737.

Barry, D. (2009). Effect of nonuniform boundary conditions on steady flow in saturated homogeneous cylindrical soil columns. *Advances in Water Resources* 32 (4): 522–531.

Bergström, L. (2000). Leaching of agrochemicals in field lysimeters—a method to test mobility of chemicals in soil. In: *Pesticide/Soil Interactions. Some Current Research Methods* (ed. J. Cornejo, P. Jamet, and F. Lobnik), 279–285. Paris: INRA.

Bernot, M.J., Sobota, D.J., Hall, R.O., Jr et al. (2010). Inter-regional comparison of land-use effects on stream metabolism. *Freshwater Biology* 55 (9): 1874–1890.

Bothwell, M.L. (1988). Growth rate responses of lotic periphytic diatoms to experimental phosphorus enrichment: the influence of temperature and light. *Canadian Journal of Fisheries and Aquatic Sciences* 45 (2): 261–270.

Bott, T., Brock, J., Baatrup-Pedersen, A. et al. (1997). An evaluation of techniques for measuring periphyton metabolism in chambers. *Canadian Journal of Fisheries and Aquatic Sciences* 54 (3): 715–725.

Bott, T.L., Brock, J.T., Cushing, C.E. et al. (1978). A comparison of methods for measuring primary productivity and community respiration in streams. *Hydrobiologia* 60 (1): 3–12.

Boulton, A.J., Findlay, S., Marmonier, P. et al. (1998). The functional significance of the hyporheic zone in streams and rivers. *Annual Review of Ecology and Systematics* 29: 59–81.

Caquet, T., Lagadic, L., and Sheffield, S.R. (2000). Mesocosms in ecotoxicology (1): outdoor aquatic systems. In: *Reviews of Environmental Contamination and Toxicology* (ed. G.W. Ware), 1–38. New York: Springer.

Carpenter, S.R. (1996). Microcosm experiments have limited relevance for community and ecosystem ecology. *Ecology* 77 (3): 677–680.

Chen, C.C., Petersen, J.E., and Kemp, W.M. (2000). Nutrient uptake in experimental estuarine ecosystems: scaling and partitioning rates. *Marine Ecology Progress Series* 200: 103–116.

Christensen, P.B., Nielsen, L.P., Sørensen, J. et al. (1990). Denitrification in nitrate-rich streams: diurnal and seasonal variation related to benthic oxygen metabolism. *Limnology and Oceanography* 35 (3): 640–651.

Cooper, S.D., Diehl, S., Kratz, K. et al. (1998). Implications of scale for patterns and processes in stream ecology. *Australian Journal of Ecology* 23 (1): 27–40.

Darcy, H. (1856). *Les fontaines publiques de la ville de Dijon: Exposition et application*. Paris: Dalmont.

De la Hire, P. (1703). Sur l'origine des rivières. In: *Historie de l'Academie Royale des Sciences*, 1–6. Paris.

Demars, B.O., Thompson, J., and Manson, J.R. (2015). Stream metabolism and the open diel oxygen method: principles, practice, and perspectives. *Limnology and Oceanography: Methods* 13 (7): 356–374.

Dodds, W.K. and Brock, J. (1998). A portable flow chamber for in situ determination of benthic metabolism. *Freshwater Biology* 39 (1): 49–59.

Drake, J.M. and Kramer, A.M. (2012). Mechanistic analogy: how microcosms explain nature. *Theoretical Ecology* 5 (3): 433–444.

Emilson, E.J.S., Carson, M.A., Yakimovich, K.M. et al. (2018). Climate-driven shifts in sediment chemistry enhance methane production in northern lakes. *Nature Communications* 9 (1): 1801.

Englund, G. and Cooper, S.D. (2003). Scale effects and extrapolation in ecological experiments. *Advances in Ecological Research* 33: 161–213.

Espinosa Bosch, M., Ruiz Sánchez, A.J., Sánchez Rojas, F. et al. (2007). Recent development in optical fiber biosensors. *Sensors* 7 (6): 797–859.

Fellows, C., Wos, M., Pollard, P. et al. (2007). Ecosystem metabolism in a dryland river waterhole. *Marine and Freshwater Research* 58 (3): 250–262.

Fellows, C.S., Clapcott, J.E., Udy, J.W. et al. (2006). Benthic metabolism as an indicator of stream ecosystem health. *Hydrobiologia* 572 (1): 71–87.

Franklin, S., Vasilas, B., and Jin, Y. (2019). More than meets the dye: evaluating preferential flow paths as microbial hotspots. *Vadose Zone Journal* 18 (1): 190024.

Gerhart, D.Z. and Likens, G.E. (1975). Enrichment experiments for determining nutrient limitation: four methods compared 1. *Limnology and Oceanography* 20 (4): 649–653.

Gilbert, J., Dole-Olivier, M., Marmonier, P. et al. (1990). Surface-groundwater ecotones. In: *Ecology and Management of Aquaticterrestrial Ecotones* (ed. R. Naiman and H. Décamps), 199–225. Paris: Unesco.

González-Pinzón, R. and Haggerty, R. (2013). An efficient method to estimate processing rates in streams. *Water Resources Research* 49: 6096–6099. https://doi.org/10.1002/wrcr.20446.

González-Pinzón, R., Haggerty, R., and Myrold, D.D. (2012). Measuring aerobic respiration in stream ecosystems using the resazurin-resorufin system. *Journal of Geophysical Research* 117: G00N06. https://doi.org/10.1029/2012JG001965.

Haggerty, R., Argerich, A., and Martí, E. (2008). Development of a "smart" tracer for the assessment of microbiological activity and sediment-water interaction in natural waters: the resazurin-resorufin system. *Water Resources Research* 44: W00D01. https://doi.org/10.1029/2007WR006670.

Hendry, M.J., Mendoza, C.A., Kirkland, R. et al. (2001). An assessment of a mesocosm approach to the study of microbial respiration in a sandy unsaturated zone. *Ground Water* 39 (3): 391–400.

Jahnke, R. and Christiansen, M. (1989). A free-vehicle benthic chamber instrument for sea floor studies. *Deep Sea Research Part A. Oceanographic Research Papers* 36 (4): 625–637.

Johnson, L.T., Tank, J.L., and Dodds, W.K. (2009). The influence of land use on stream biofilm nutrient limitation across eight North American ecoregions. *Canadian Journal of Fisheries and Aquatic Sciences* 66 (7): 1081–1094.

Kessler, A.J., Chen, Y.-J., Waite, D.W. et al. (2019). Bacterial fermentation and respiration processes are uncoupled in anoxic permeable sediments. *Nature Microbiology* 4 (6): 1014–1023.

Krause, S., Lewandowski, J., Grimm, N.B. et al. (2017). Ecohydrological interfaces as hot spots of ecosystem processes. *Water Resources Research* 53 (8): 6359–6376.

Larned, S.T. (2010). A prospectus for periphyton: recent and future ecological research. *Journal of the North American Benthological Society* 29 (1): 182–206.

Lautz, L.K. and Fanelli, R.M. (2008). Seasonal biogeochemical hotspots in the streambed around restoration structures. *Biogeochemistry* 91 (1): 85–104.

Lewandowski, J., Arnon, S., Banks, E. et al. (2019). Is the hyporheic zone relevant beyond the scientific community? *Water* 11 (11): 2230.

Lewis, J. and Sjöstrom, J. (2010). Optimizing the experimental design of soil columns in saturated and unsaturated transport experiments. *Journal of Contaminant Hydrology* 115 (1): 1–13.

Marzolf, E.R., Mulholland, P.J., and Steinman, A.D. (1994). Improvements to the diurnal upstream-downstream dissolved oxygen change technique for determining whole-stream metabolism in small streams. *Canadian Journal of Fisheries and Aquatic Sciences* 51 (7): 1591–1599.

McClain, M.E., Boyer, E.W., Dent, C.L. et al. (2003). Biogeochemical hot spots and hot moments at the interface of terrestrial and aquatic ecosystems. *Ecosystems* 6: 301–312.

McConnell, W.J. (1962). Productivity relations in carboy microcosms. *Limnology and Oceanography* 7 (3): 335–343.

McIntire, C.D. and Phinney, H.K. (1965). Laboratory studies of periphyton production and community metabolism in lotic environments. *Ecological Monographs* 35 (3): 238–258.

Mine Berg, G., Glibert, P.M., and Chen, C.C. (1999). Dimension effects of enclosures on ecological processes in pelagic systems. *Limnology and Oceanography* 44 (5): 1331–1340.

Mulholland, P.J., Fellows, C.S., Tank, J.L. et al. (2001). Inter-biome comparison of factors controlling stream metabolism. *Freshwater Biology* 46: 1503–1517.

Mulholland, P.J., Houser, J.N., and Maloney, K.O. (2005). Stream diurnal dissolved oxygen profiles as indicators of in-stream metabolism and disturbance effects: Fort Benning as a case study. *Ecological Indicators* 5 (3): 243–252.

Mulholland, P.J., Newbold, J.D., Elwood, J.W. et al. (1985). Phosphorus spiralling in a woodland stream: seasonal variations. *Ecology* 66 (3): 1012–1023.

Mulholland, P.J., Valett, H.M., Webster, J.R. et al. (2004). Stream denitrification and total nitrate uptake rates measured using a field 15N tracer addition approach. *Limnology and Oceanography* 49 (3): 809–820.

Muñoz-Leoz, B., Antigüedad, I., Garbisu, C. et al. (2011). Nitrogen transformations and greenhouse gas emissions from a riparian wetland soil: an undisturbed soil column study. *Science of the Total Environment* 409 (4): 763–770.

O'Brien, J.M. and Dodds, W.K. (2008). Ammonium uptake and mineralization in prairie streams: chamber incubation and short-term nutrient addition experiments. *Freshwater Biology* 53 (1): 102–112.

Odum, H.T. (1956). Primary production in flowing waters. *Limnology and Oceanography* 1 (2): 102–117.

Odum, H.T. and Hoskin, C.M. (1957). Metabolism of a laboratory stream microcosm. *Publications from the Institute of Marine Science Austin, Texas* 4: 115–133.

Öztürk, N. and Kavak, D. (2005). Adsorption of boron from aqueous solutions using fly ash: batch and column studies. *Journal of Hazardous Materials* 127 (1–3): 81–88.

Patten, B.C. and Witkamp, M. (1967). Systems analysis of 134cesium kinetics in terrestrial microcosms. *Ecology* 48 (5): 813–824.

Petersen, J.E., Chen, C.C., and Kemp, W.M. (1997). Scaling aquatic primary productivity: experiments under nutrient-and light-limited conditions. *Ecology* 78 (8): 2326–2338.

Petersen, J.E., Cornwell, J.C., and Kemp, W.M. (1999). Implicit scaling in the design of experimental aquatic ecosystems. *Oikos* 85: 3–18.

Petersen, J.E., Kennedy, V.S., Dennison, W.C., and Kemp, W.M. (eds.) (2009). *Enclosed Experimental Ecosystems and Scale: Tools for Understanding and Managing Coastal Ecosystems*. Springer.

Pospíšilová, M., Kuncová, G., and Trögl, J. (2015). Fiber-optic chemical sensors and fiber-optic bio-sensors. *Sensors* 15 (10): 25208–25259.

Poulin, B.A., Aiken, G.R., Nagy, K.L. et al. (2016). Mercury transformation and release differs with depth and time in a contaminated riparian soil during simulated flooding. *Geochimica et Cosmochimica Acta* 176: 118–138.

Richter, E., Hecht, F., Schnellbacher, N. et al. (2015). Assessing the ecological long-term impact of wastewater irrigation on soil and water based on bioassays and chemical analyses. *Water Research* 84: 33–42.

Riebesell, U., Czerny, J., von Bröckel, K. et al. (2013). Technical note: a mobile sea-going mesocosm system – new opportunities for ocean change research. *Biogeosciences* 10 (3): 1835–1847.

Ritzi, R.W. and Bobeck, P. (2008). Comprehensive principles of quantitative hydrogeology established by Darcy (1856) and Dupuit (1857). *Water Resources Research* 44: W10402. https://doi.org/10.1029/2008WR007002.

Rüegg, J., Brant, J.D., Larson, D.M. et al. (2015). A portable, modular, self-contained recirculating chamber to measure benthic processes under controlled water velocity. *Freshwater Science* 34 (3): 831–844.

Ryther, J.H. (1956). The measurement of primary production. *Limnology and Oceanography* 1 (2): 72–84.

Seitzinger, S.P., Nielsen, L.P., Caffrey, J. et al. (1993). Denitrification measurements in aquatic sediments: a comparison of three methods. *Biogeochemistry* 23 (3): 147–167.

Simmons, C.T. (2008). Henry Darcy (1803–1858): immortalised by his scientific legacy. *Hydrogeology Journal* 16 (6): 1023–1038.

Sørensen, J. (1978). Denitrification rates in a marine sediment as measured by the acetylene inhibition technique. *Applied and Environmental Microbiology* 36 (1): 139–143.

Spivak, A.C., Vanni, M.J., and Mette, E.M. (2011). Moving on up: can results from simple aquatic mesocosm experiments be applied across broad spatial scales? *Freshwater Biology* 56 (2): 279–291.

Staehr, P.A., Bade, D., Van de Bogert, M.C. et al. (2010). Lake metabolism and the diel oxygen technique: state of the science. *Limnology and Oceanography: Methods* 8 (11): 628–644.

Staehr, P.A., Testa, J.M., Kemp, W.M. et al. (2012). The metabolism of aquatic ecosystems: history, applications, and future challenges. *Aquatic Sciences* 74 (1): 15–29.

Stegen, J.C., Johnson, T., Fredrickson, J.K. et al. (2018). Influences of organic carbon speciation on hyporheic corridor biogeochemistry and microbial ecology. *Nature Communications* 9 (1): 1–11.

Taniguchi, M., Burnett, W.C., Dulaiova, H. et al. (2006). Submarine groundwater discharge measured by seepage meters in Sicilian coastal waters. *Continental Shelf Research* 26 (7): 835–842.

Thaysen, E.M., Jacques, D., Jessen, S. et al. (2014). Inorganic carbon fluxes across the vadose zone of planted and unplanted soil mesocosms. *Biogeosciences* 11 (24): 7179–7192.

Trentman, M.T., Dodds, W.K., Fencl, J.S. et al. (2015). Quantifying ambient nitrogen uptake and functional relationships of uptake versus concentration in streams: a comparison of stable isotope, pulse, and plateau approaches. *Biogeochemistry* 125 (1): 65–79.

Uehlinger, U. and Naegeli, M.W. (1998). Ecosystem metabolism, disturbance, and stability in a prealpine gravel bed river. *Journal of the North American Benthological Society* 17 (2): 165–178.

Vogel, S. (1994). *Life in Moving Fluids: The Physical Biology of Flow*. Princeton University Press.

Wanninkhof, R. (1992). Relationship between wind speed and gas exchange over the ocean. *Journal of Geophysical Research: Oceans* 97 (C5): 7373–7382.

Warren, C.E. and Davis, G.E. (1971). Laboratory stream research: objectives, possibilities, and constraints. *Annual Review of Ecology and Systematics* 2: 111–144.

Whittaker, R.H. (1961). Experiments with radiophosphorus tracer in aquarium microcosms. *Ecological Monographs* 31 (2): 157–188.

Stream Solute Workshop. (1990). Concepts and methods for assessing solute dynamics in stream ecosystems. *Journal of the North American Benthological Society* 9 (2): 95–119.

Wuhrmann, K. (1954). High-rate activated sludge treatment and its relation to stream sanitation: II. Biological river tests of plant effluents. *Sewage and Industrial Wastes* 26 (2): 212–220.

Young, R.G., Matthaei, C.D., and Townsend, C.R. (2008). Organic matter breakdown and ecosystem metabolism: functional indicators for assessing river ecosystem health. *Journal of the North American Benthological Society* 27 (3): 605–625.

13

Using Diel Solute Signals to Assess Ecohydrological Processing in Lotic Systems

Marie J. Kurz[1,] and Julia L.A. Knapp[2]*

[1] Environmental Sciences Division, Oak Ridge National Laboratory, Oak Ridge, Tennessee, United States of America
[2] Department of Earth Sciences, Durham University, Durham, United Kingdom
* This submission was written by the author acting in her own independent capacity and not on behalf of UT-Battelle, LLC, or its affiliates or successors.

13.1 Introduction

Lotic systems are a prominent ecohydrological interface within which complex, coupled interactions occur between ecological, hydrological, and geochemical processes. Aquatic ecosystems both respond to and alter the transport and cycling of solutes within streams and rivers (Clarke 2002). Hydrologic conditions and the availability of nutrients required for metabolism regulate the health, productivity, distribution, and composition of the stream ecosystem. In return, the functions and structure of the ecosystem control energy turnover and elemental cycling (Julian et al. 2011), and physically engineer hydrologic controls on solute transport and cycling (Clarke 2002; Gurnell 2014). Thus, disentangling temporal signals in solute concentrations, especially at the stream reach scale, can provide insight into temporal dynamics in ecosystem functions, such as metabolism and nutrient cycling, resulting from ecohydrological process interactions.

Many of the ecological and geochemical processes controlling solute cycling in streams operate on diel (24-hour) timescales in response to the solar photocycle (Figure 13.1): increased daytime evapotranspiration of catchment vegetation and daytime melting of snow/ice mediate the relative flow paths and fluxes of water and solutes into and out of streams (Bond et al. 2002; Czikowsky and Fitzjarrald 2004; Duncan et al. 2015; Flewelling et al. 2014; Kirchner et al. 2020; Oviedo-Vargas et al. 2022; Schmadel et al. 2016). Photosynthesis and respiration by in-stream organisms alternately produce and consume oxygen and carbon dioxide, driving diel variations in dissolved oxygen (DO) concentrations, redox state, and pH (Desmet et al. 2011; Odum 1956; Simonsen and Harremoës 1978). In turn, these metabolically-driven changes can indirectly drive further diel element retention and/or release elements via a number of oxygen- and pH-dependent reactions (Nimick et al. 2011). Assimilatory uptake of nutrients by primary producers can also result in diel signals in major and trace element concentrations (e.g. Cohen et al. 2013; Heffernan and Cohen 2010; Kurz et al. 2013).

Ecohydrological Interfaces, First Edition. Edited by Stefan Krause, David M. Hannah, and Nancy B. Grimm.

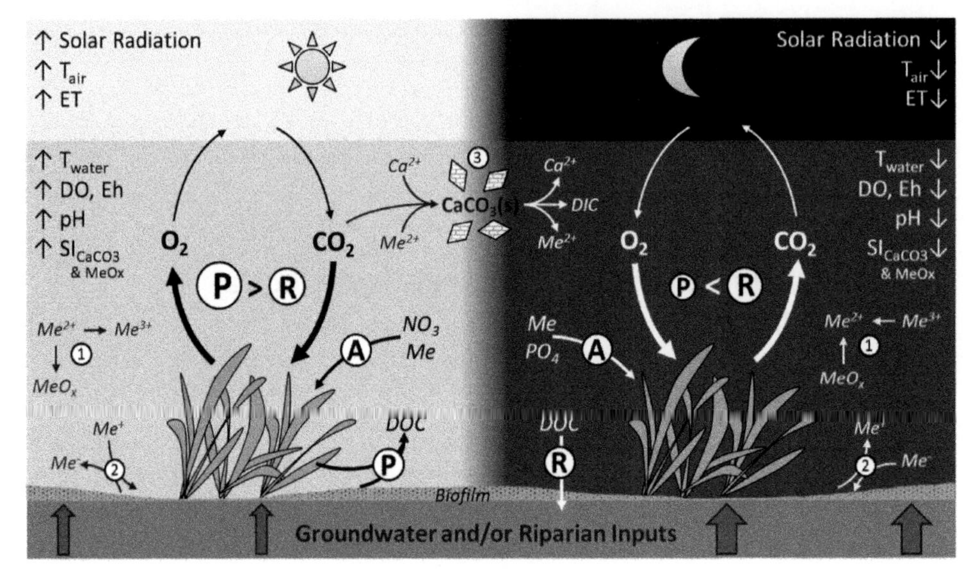

Figure 13.1 Major processes affecting element concentrations in streams at diel frequency, including processes directly driven by metabolic activity: photosynthesis (P), respiration (R), and assimilation (A); and processes indirectly driven by metabolism: speciation (1), pH-dependent sorption (2), and carbonate mineral (co)precipitation (3). T = temperature, ET = evapotranspiration, DO = dissolved oxygen, Eh = redox potential, SI = saturation index, *Me* = cationic (+) and anionic (-) trace elements, and DIC = dissolved inorganic carbon. Adapted from Nimick et al. (2011) and Kurz et al. (2013).

The magnitude, timing, and geometry of diel signals reflect the cumulative effects of the driving processes, thereby providing information on the timing and rates of these processes and, potentially, other ecological parameters such as limitation status, physiological traits, and nutrient coupling (Appling and Heffernan 2014; Cohen et al. 2013). Our ability to gain hydro-biogeochemical understanding from diel signals hinges on our ability to measure and disentangle the effects of overlapping diel and non-diel processes that can alter, and even obscure, the appearance of diel signals. The severity of these effects can vary considerably between different solutes and stream conditions, hence, diel signals being more commonly observed for certain solutes and during baseflow. Atmospheric equilibration of gaseous solute (e.g. DO and CO_2) results in the diel signals continually being "reset", whereas for non-gaseous solutes, upstream signals will overlap with the effect of downstream processing (Hensley and Cohen 2016). Determining the length of stream (and associated transport timescale) over which diel solute signals emerge, i.e. the integration length, is equally important for process interpretation. Gaseous solutes and non-gaseous solutes with short uptake lengths (e.g. soluble Fe) will be more responsive to reach and smaller-scale process interactions, while non-gaseous solutes with longer integration lengths (e.g. NO_3, organic carbon, and Cl) will be more responsive to larger-scale structural controls. Lastly, it should be noted that concentration changes in the bulk water column may not fully reflect the magnitude or timing of all processing occurring within the stream water column (Battin et al. 2003; Beck et al. 2009).

One of the earliest uses of diel solute signals to inform biogeochemical processing in streams can be traced back to the seminal study by Odum (1956), which showed that diel oxygen variations in rivers could be used to calculate ecosystem productivity. Since then, the presence of diel signals in streams and their multitude of physical and biogeochemical drivers have received increasing attention (see reviews by Gammons et al. 2015; Nimick et al. 2011). The following chapter reviews the ecological processes driving diel signals in oxygen as well as major nutrients and other trace elements; discusses how diel signals can be used to evaluate functioning and interactions within stream ecosystems; and considers potential future research directions and outstanding questions still to be addressed by or about diel solute signals and associated processes.

13.2 Diel Oxygen Signals and Whole-stream Metabolism

Ecosystem metabolism is defined as the difference between Primary Productivity (*PP*) and Ecosystem Respiration (*ER*). In this, *PP* is the synthesis of organic compounds from carbon dioxide and light, leading to the production of oxygen and biomass, thus converting solar energy to chemical energy:

$$CO_2 + H_2O + \text{Energy} \rightarrow CH_2O + O_2$$

ER, on the other hand, is the consumption of a part of this same chemical energy, resulting in the production of inorganic carbon. Even though many kinds of anaerobic respiration exist, this chapter will focus on aerobic respiration, which uses oxygen as a terminal electron acceptor and converts it back to the waste product carbon dioxide. While primary productivity takes place in shallow and well-lit parts of the stream's water column and sediment surface, a substantial part of respiration occurs in the hyporheic zones, i.e. within the streambed, where many aquatic microorganisms can be found (Bencala et al. 2011) due to the greatly decreased flow velocities.

Primary productivity can be further refined into Net Primary Productivity (*NPP*) and Gross Primary Productivity (*GPP*). *NPP* refers to the amount of energy stored in biomass, whereas *GPP* includes both stored and respired energy.

$$NPP = GPP - ER$$

The magnitude of individual metabolic processes changes over the course of a day, with the diurnal character of metabolic activity resulting mainly from the changing availability of light. *PP* is proportional to the available amount of photosynthetically active solar radiation, resulting in increased concentrations of dissolved oxygen (DO) during the day (Figure 13.2A). A second, key factor in regulating metabolic processes is temperature. It stimulates organism activity and influences energy demand by regulating cellular activity. The non-linear dependence of metabolic activity and oxygen demand on temperature follows an asymmetric bell shape with a slow increase at lower temperatures and a sharp decrease at high temperatures, and the range and the precise shape of this temperature response curve is species dependent. Even though *ER* is theoretically temperature dependent, for the sake of simplifying rate calculations it is often assumed to be constant over time (Figure 13.2B).

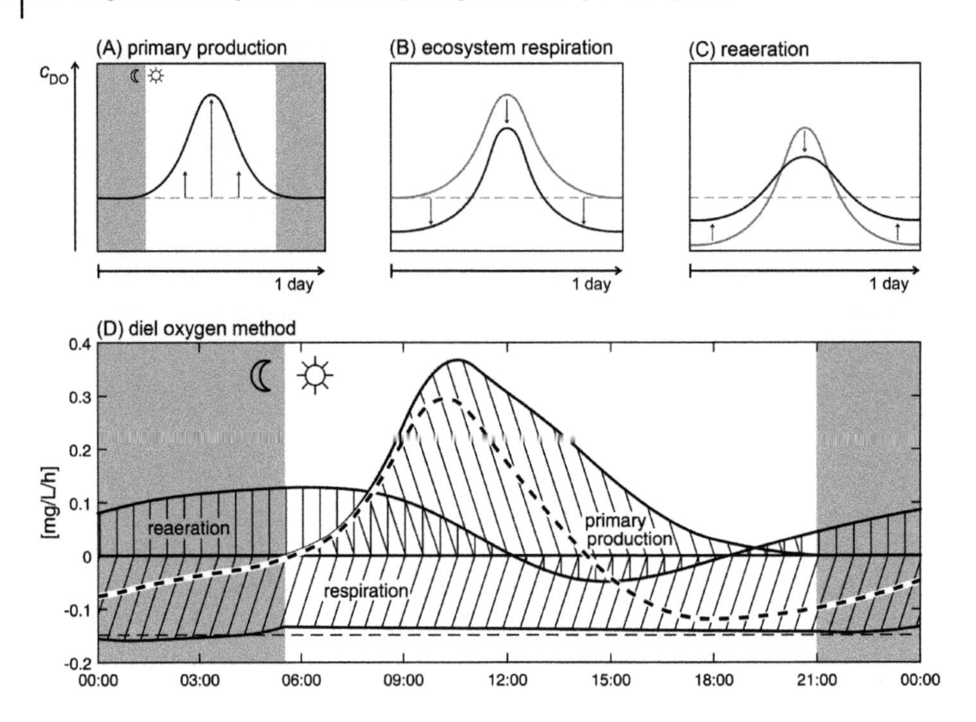

Figure 13.2 (A) Primary productivity increases the concentration of dissolved oxygen (DO) during the day. (B) Ecosystem respiration results in an assumed constant decrease of the dissolved oxygen concentration at night and day. (C) Reaeration decreases the concentration of dissolved oxygen if the water is oversaturated, and increases it in case of undersaturation through gas exchange across the air–water interface. (D) In the diurnal method, the rate of change of oxygen (*dc/dt*) is calculated over time (dashed line). Reaeration (vertically hatched area) is calculated from the reaeration rate constant (estimated either from gas tracer tests or empirical relationships and the saturation deficit). This reaeration rate is positive if the actual concentration of DO is smaller than its saturation concentration (see C), and negative in the opposite case. The respiration rate is then calculated as the remainder during the night (right slanted hatching), and interpolated linearly during the day. During the day, primary productivity is calculated as the remainder (left slanted hatching). Together, the three processes of reaeration, respiration, and primary productivity make up the rate of change of oxygen at any given time. Adapted from Knapp (2017).

13.2.1 Quantifying Metabolic Rates: The Diel Oxygen Method

13.2.1.1 Components of the Oxygen Balance

Metabolic rates can be calculated from the changes in dissolved concentrations of atmospheric gases in streamwater, because diurnal signals of dissolved gases are closely linked and typically resulting from diurnal cycles in metabolic activity. Technically, metabolic rates represent uptake rates, with respiration being the uptake of DO, while photosynthesis involves the uptake of CO_2. Because of feedback mechanisms and the interdependence of the involved processes, however, inferring metabolic rates from CO_2 concentrations is not straightforward. For this reason, changes in DO over time are typically used to quantify metabolic activity of both respiration and photosynthesis instead. This "diel oxygen method" was first presented by Odum (1956), and has been extensively applied in a variety of ecosystems (e.g. Caffrey 2003; Fernández Castro et al. 2021; Gattuso et al. 1993; Hall 1972; Staehr and Sand-Jensen 2007).

Metabolic rates are calculated by evaluating the different oxygen consuming and producing processes that affect the rate of change of oxygen, to isolate the effect of *PP*:

$$\frac{dc_{DO}}{dt} = PP - ER + k_2(c_{sat} - c_{DO})$$

In addition to primary production and respiration, this relationship includes a third process that affects concentrations of dissolved oxygen: reaeration across the air–water interface (Hall and Ulseth, 2020). Reaeration is a purely physical process, made up of the reaeration rate constant $k_2\left[T^{-1}\right]$ and the saturation deficit $(c_{sat} - c_{DO})$ defined as the difference between the actual and the saturation concentration of DO, c_{DO} and c_{sat}, respectively. Reaeration drives the actual concentration of DO towards the saturation concentration, thereby minimizing the saturation deficit (Figure 13.2C). In streams or lakes with high primary productivity, the system is likely to be over-saturated with respect to oxygen. Hence, oxygen is removed from the water phase to the atmosphere via the process of reaeration. Whereas the sign of the saturation deficit determines the direction of the gas-exchange, the rate at which the system is driven towards the saturation equilibrium is reflected in its reaeration rate coefficient k_2. Additional processes may also affect the oxygen balance in some systems, but in near-pristine streams, the three processes of primary production, respiration, and reaeration jointly drive diurnal changes in DO concentrations.

13.2.1.2 Quantifying Reaeration

While reaeration is not directly related to metabolism, it still needs to be quantified to reliably assess metabolic rates from oxygen measurements using the diel oxygen approach. In the simplest approach, one can avail of the large number of empirical relationships (e.g. O'Connor and Dobbins 1958; Tsivoglou and Wallace 1972; see Cox 2003 for an overview) linking reaeration to hydraulic parameters including discharge, in-stream velocity, stream width and bed slope. These relationships, however, are only valid under very specific conditions., and are known to underestimate reaeration especially for smaller streams (Soares et al. 2013; Young and Huryn 1999).

A more exact, but also more time-consuming way of determining reaeration rate coefficients are gas tracer tests (e.g. Wanninkhof et al. 1990). A volatile compound like propane (e.g. Knapp et al. 2015; Marzolf et al. 1994; Young and Huryn 1999), SF_6 (e.g. Ho et al., 2011; Huisman et al. 2004), or a noble gas (Benson et al. 2014; Knapp et al. 2019; Reid et al. 2007) is continuously injected into the stream. The reaeration rate coefficient can then be determined from the gas loss along the reach (i.e. following the methods detailed by Wanninkhof et al. 1990; Marzolf et al. 1994), with gas exchange coefficients of the tracer gas being converted to the one of oxygen (i.e. the reaeration rate coefficient) through a conversion factor (Rathbun et al. 1978).

Another promising approach for the determination of the reaeration rate coefficient is that of Pennington et al. (2018), who calculated reaeration rates from measurements of DO, pCO_2, pH, and temperature data. The required data can easily be recorded with sensors and the method is thus much less labour intensive than performing gas tracer injections.

Reaeration rate coefficients typically increase with the area of the air–water interface, because oxygen can escape (or enter) the river over a larger area. Additionally, reaeration rate coefficients typically increase with in-stream velocity of the water and turbulence

(Ulseth et al. 2019). Since reaeration is a purely physical process, the determined coefficient remains valid while the environmental conditions do not change. To deal with changing hydraulic conditions, Zappa et al. (2007) related the gas exchange rate to turbulent dissipation rates, whereas Beaulieu et al. (2013) used a two-parameter exponential model to interpolate the relationship between reaeration rates and stream discharge. Some methods designed for longer oxygen time series use reaeration rate coefficients scaled to a Schmidt number. This way, one reaeration rate coefficient can be used for the whole time series while still allowing for fluctuations in gas exchange related to temperature (Appling et al. 2018).

13.2.1.3 Quantifying Primary Productivity and Ecosystem Respiration

Once the reaeration rate coefficient is known, the calculation of respiration and primary production rates is relatively straightforward, following either the one-station method (Demars et al. 2015; Odum 1956), where the DO concentrations recorded at one location are considered representative for the whole stream, or the two-station method (Demars et al. 2015; Marzolf et al. 1994), where both the difference in DO between an upstream and downstream location, lagged by travel time, and the average DO concentrations are considered. Reaeration is quantified and subtracted from night-time data of the rate of change in oxygen concentrations (dc / dt), and respiration is calculated as the remainder (Figure 13.2D). Finally, *NPP* can be calculated as the unaccounted for rate of change of oxygen during the day. This method is described in detail by Marzolf et al. (1994) and illustrated in Figure 13.2D.

An additional complication to the mass balance method of calculating metabolism is the potential impact of low-DO groundwater inputs and how best to account for these (Hall and Tank 2005). Calculations of *GPP* are generally less biased by groundwater than calculations of *ER* (Hall and Tank 2005). Nonetheless, not accounting for groundwater gains can lead to falsely characterizing streams as autotrophic or heterotrophic, highlighting the importance of disentangling co-occurring physical (i.e. groundwater–surface water exchange) and ecological (i.e. metabolism) processes for an accurate understanding of carbon cycling.

13.2.2 Other Methods for the Estimation of Metabolic Rates

Several alternative approaches exist for the quantification of metabolic rates, many of which are loosely based on the diel oxygen method, including techniques relying on the paired measurements of CO_2 and DO (Hanson et al. 2003; Vachon et al. 2020; Wright and Mills 1967). Alternatively, proxies for whole-stream aerobic respiration can be determined from "smart" tracer tests with the reactive tracer resazurin (e.g. Haggerty et al. 2008, 2009; Knapp et al. 2018; Ledford et al. 2021b) or through the assessment of the diel change in $\delta^{18}O$ (Hotchkiss and Hall 2014; Tobias et al. 2007; Tromboni et al. 2022).

Metabolic rates can also be assessed directly through in vitro methods, which are performed in closed-off artificial systems instead of the natural whole stream system. These include the "bottle" method (Gaarder 1927), the ^{14}C incorporation technique (Nielsen 1951), and an approach based on ^{18}O labelled oxygen (Grande et al. 1989). These methods are relatively easy to implement, and the procedures are well established. Like all ex situ methods, however, they suffer from two major shortcomings: The in vitro conditions regarding light, turbulence, and nutrient abundance differ greatly from the ambient conditions of aquatic systems, and the results are representative for natural ecosystems only in a very limited sense. Furthermore, the upscaling of these results to whole river reaches, larger coastal systems, or lakes depends on several assumptions that have to be made which may not always be justified (Hanson et al. 2011; Staehr et al. 2010).

13.2.3 New Developments in Oxygen Measurements: Optical DO Sensors

The above calculations of metabolic rates rely on precise and accurate measurements of DO concentrations at a sufficiently high temporal resolution. While the Winkler titration method (Winkler 1914) is undoubtedly one of the most accurate techniques to measure DO concentrations (e.g. Furuya and Harada 1995), it is not practical for high-frequency, long-term monitoring. However, an increasing number of sensors for in situ long-term monitoring of DO concentrations are now available, many of which are optical sensors. The principle of these optical sensors is based on the interaction between DO and a luminescent dye. The presence of oxygen leads to the quenching of photons that are emitted by the exciting light source. It thus reduces the luminescence lifetime to an extent inversely proportional to the concentration of DO in the sample. Optical sensors are usually more accurate than electrochemical sensors and the measurements are less strongly affected by instrument drift. It is thus not surprising that optical sensors have become the new standard for field-based oxygen measurements (Almeida et al. 2014).

The development of these in situ sensors has revolutionized our ability to measure metabolic rates across space and time at near-continuous frequencies and has enabled us to ask and answer far-reaching questions. Only 10 years ago, the analysis of metabolic rates across a stream network would have involved many hours of field and laboratory work, whereas most of this work can now be accomplished by in situ sensors providing continuous oxygen measurements even from remote locations that may be difficult to access. The wealth of data has also given rise to novel data analysis techniques. Rather than using some few, but highly accurate field measurements to constrain rate coefficients, we can feed large data sets into numerical models to obtain best estimates of rate coefficients across time and space. For example, Cox et al. (2015) developed a Fourier method to estimate *GPP* based on the relationship between time-averaged *GPP* and the amplitude of the diel harmonic of an oxygen recording, whereas others estimate all or some of the components of the oxygen balance by inverse modelling (e.g. Hall and Beaulieu 2013; Riley and Dodds 2013 and others) Recently, Diamond et al. (2021) proposed several proxies for *GPP* and *ER* that can be directly derived from the diel DO range and the maximum daily DO deficit, respectively. Bayesian approaches have also proven promising, and some are freely available as R packages, e.g. "streamMetabolizer" (Appling et al. 2018) or "BASE" (Grace et al. 2015), facilitating the production of high-frequency time series of metabolic rates.

13.3 Diel Element Signals and Ecosystem Processing

Primary productivity is the synthesis of organic compounds from carbon dioxide and light, leading to the production of oxygen and biomass, whereas respiration consumes oxygen and organic carbon. Thus, metabolic activity in streams can be inferred from changes in oxygen concentrations, as outlined in the previous section, as well as changes in organic and inorganic carbon concentrations. Other major and trace elements are also required to produce biomass, based on the material requirements of metabolic processes such as RNA production and protein synthesis (Elser et al. 2003; Glass et al. 2009; Vrede et al. 2004). To maintain compositionally homeostatic biomass, organisms acquire (or "assimilate") elements from the environment in specific ratios (Frost et al. 2005; Kerkhoff et al. 2005). Diel fluctuations in element acquisition and export result from diel changes in the rates of metabolic processes, which in turn reflect light-sensitive circadian rhythms in enzyme activity and genetic expression (Ashworth et al. 2013). In aquatic ecosystems, such elemental acquisition and export from and to the water column can result in observable changes in in-stream element concentrations at diel timescales.

If element dynamics at the ecosystem level were controlled solely by assimilatory uptake, the timing, magnitude, and coupling of environmental element signals would reflect ecosystem metabolic activity and stoichiometry. In reality, multiple ecological, geochemical, and hydrologic processes complicate the relationships between ecosystem metabolism and elemental dynamics. First, autotroph stoichiometry, although rigid, is not fixed, and can respond to changes in nutrient supplies and light (Sterner et al. 1998). Uptake of nutrients in excess of contemporary metabolic demand, by ecological processes such as luxury uptake, biosorption, and bioaccumulation, can decouple assimilation from growth (Khoshmanesh et al. 2002; Krems et al. 2013). Second, nutrient limitation status and physiological traits can also alter the dynamics of coupled element uptake (Appling and Heffernan 2014). Third, multiple abiotic geochemical and hydrologic processes, often indirectly linked to metabolism, can also drive fluctuations in element concentrations and speciation with variable timing (Gammons et al. 2015; Nimick et al. 2011 and references therein). Fourth, the appearance of resulting element signals can be altered by local or upstream processes. Whereas the short integration length of oxygen and other gaseous solutes typically allows for accurate evaluation of reach-scale metabolism, the signals of non-gaseous solutes with longer integration lengths can reflect a mixture of local and upstream controls on in-stream processing, such that these non-gaseous signals may be wholly or partially disconnected from the timing and magnitude of local controls. Variable inputs, dispersion, and mixing of waters with different residence times can all shift, dampen, and obscure observable diel signals in non-gaseous solutes (e.g. Appling and Heffernan 2014; Chamberlin et al. 2021; Chittoor Viswanathan et al. 2015; Hensley and Cohen 2016; Kunz et al. 2017), although the application of a two-station approach, similar to oxygen, may help to isolate the effects of reach-scale processing (e.g. Kunz et al. 2017).

As a result of these complications, the magnitude, timing, and shape of diel element signals often vary or disappear completely between locations and times, even for major nutrients (e.g. Halliday et al. 2013; Hansen and Singh 2018; Lupon et al. 2016; Oviedo-Vargas et al. 2022; Pellerin et al. 2009; Worrall et al. 2015). While there have been successes disentangling multiple biological and abiotic drivers of observed diel signals by both inference and direct measurement of the contributing processes in times or systems with predictable flow (e.g. Cohen et al. 2013; Heffernan and Cohen 2010; Kurz et al. 2013), fewer studies have attempted to disentangle diel ecologically-driven signals from hydrologically-driven variability (but see Aubert and Breuer 2016; Chamberlin et al. 2021; Heathwaite and Bieroza 2021). The latter is particularly critical as dynamic flow and associated background variability in element concentrations is far more common than stationarity and the first step in gaining hydro-biogeochemical understanding from diel signals across a range of systems and conditions is distinguishing them in the first place.

As our understanding of the processes driving diel signals improves, so does our ability to disentangle observed signals in order to resolve the timing and magnitude of specific controlling processes. Mechanistic theories and models can help further guide our understanding of the links between the exact magnitude, timing, and shape of diel signals and their underlying ecological controls, from organismal traits to ecosystem-scale processing (e.g. Appling and Heffernan 2014; Huang et al. 2022). Thus, diel element signals, particularly when correlated to diel metabolism, have the potential to resolve questions related to the magnitude and timing of metabolic processes, coupling between metabolism and nutrient cycling at various scales, nutrient limitation, spatiotemporal changes in ecosystem functioning, etc.

13.3.1 Quantifying Diel Element Signals

The magnitude of a diel element signal reflects the rate of net whole-stream element removal from or release to the water column. Diel signals can therefore be used to quantify rates of whole-stream metabolic processes such as photosynthetic oxygen production, as described in Section 13.2, and assimilatory uptake as successfully demonstrated for nitrate by Heffernan and Cohen (2010). Removal (or release), in units of g-element/m^3/d, is evaluated as the integral between the observed concentration and either (a) the interpolation between adjacent diel maxima (or minima) as in Figure 13.3A, or (b) the preceding maximum (or minimum) concentration. The difference between flow weighted inputs, where known, and the diel maxima (or minima) can be further used to estimate removal (or release) via non-diurnal processes such as denitrification or heterotrophic assimilation.

If there is only a single diel process controlling the timing and magnitude of a diel signal (a rare assumption, but see Heffernan and Cohen 2010), the removal rate quantification can be reasonably straightforward (Figure 13.3A). When multiple diel processes simultaneously remove and/or release an element from the water column, the magnitude and timing of both processes must be taken into account in order to avoid the over- or underestimation of retention by both the diel and any non-diel processes. Figure 13.3 illustrates these overlapping effects when quantifying assimilatory uptake. If assimilation occurs *out of* phase with another diel processes (e.g. calcite co-precipitation), the observed diel maxima is reduced relative to that which would occur if only assimilation were occurring (Figure 13.3B vs. A). As a result, both the assimilatory and geochemical removal is underestimated, and any non-diurnal removal is overestimated (e.g. Cohen et al. 2013). If the two diel processes occur *in* phase, the observed diel maxima should not be affected because removal by both processes is zero at the time of the diel maxima (Figure 13.3C). In this case, not accounting for the geochemical removal would result in the overestimation of assimilatory uptake but have no effect on estimated non-diurnal removal (Figure 13.3C vs. A). Cohen et al. (2013) provided the first successful illustration of this multi-signal deconvolution for nutrients, when separating out the relative role of calcite co-precipitation and assimilation on total diel P removal (34% vs. 66%, respectively). As a result, the magnitude of P assimilation, the ecosystem C:P stoichiometry, and the lag between primary productivity and adjusted assimilatory uptake could be more accurately evaluated.

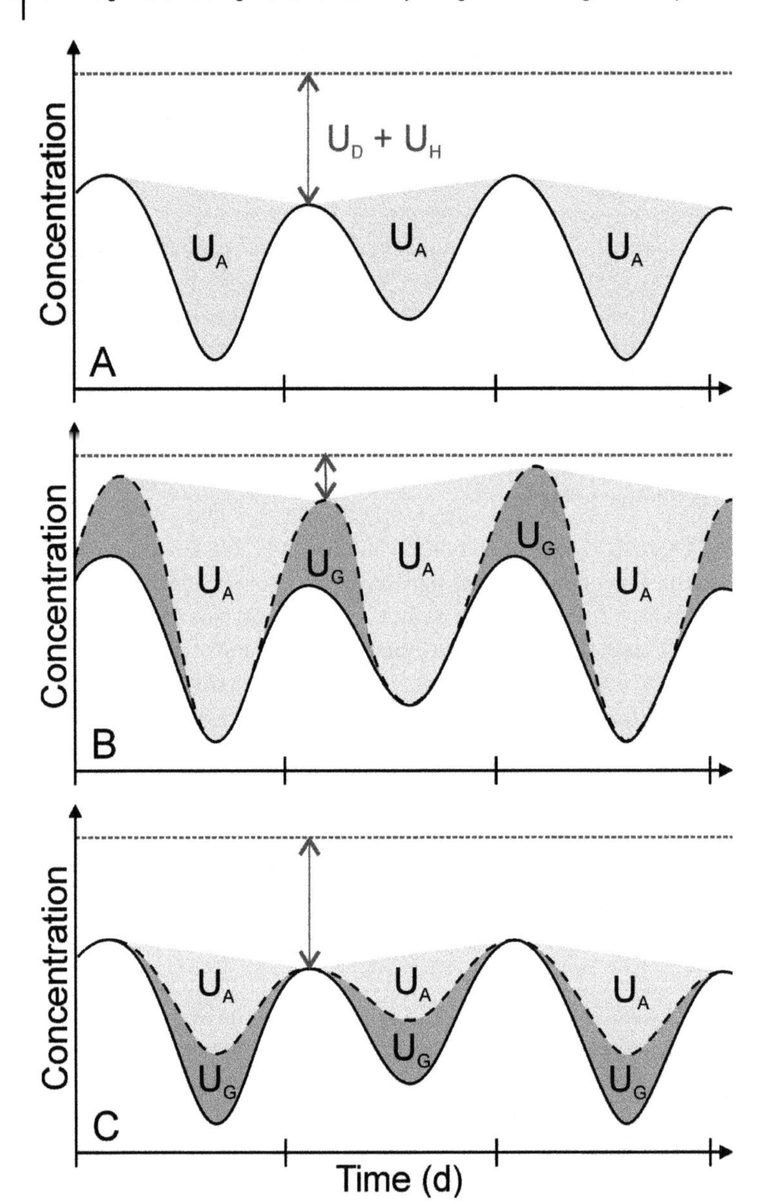

Figure 13.3 Estimating removal rates from observed diel element signals (solid black line). (A) Effect of a single uptake process, such as assimilation. Assimilatory uptake (U_A) evaluated as the integrated concentration difference between the observations and the interpolation between diel maxima (light grey area). The difference between the flow weighted inputs (dashed grey line) and the diel maxima reflects removal from non-diurnal processes ($U_D + U_H$). (B) Overlapping effect of two diel processes occurring out of phase, such as assimilatory uptake (U_A) and geochemical uptake (U_G, dark grey area). (C) Overlapping effect of two diel processes occurring in phase, where removal by both processes is zero at the time of the diel maxima. Black dashed lines represent the diel signals which would occur from assimilation alone. Adapted from Cohen et al. (2013) and Heffernan and Cohen (2010).

This diel method for estimating nutrient uptake is particularly useful in large rivers, where nutrient and isotopic tracer addition-based estimates of uptake are impractical, providing a relatively scale-independent method for estimating element retention (e.g (Kunz et al. 2017)). Unlike uptake estimates based on the stoichiometry and metabolism of primary producers, the diel method is independent of metabolism measurements and does not rely on uncertain estimates of dominant stoichiometric ratios. When combined with independent measurements of whole-stream productivity, diel-based estimates of uptake can be used to identify lags and stoichiometric variation in the relationship between primary productivity and assimilation and to evaluate ecosystem stoichiometry and fine-scale coupling between ecosystem processing (e.g. Cohen et al. 2013; Khamis et al. 2021; Lupon et al. 2016; Rode et al. 2016a).

As we expand our evaluation of diel signals to an increasing number of elements, we need effective methods to evaluate and compare the diagnostic parameters of multiple signals. Diel signals are often reported in terms of their peak or trough timing (timing being the most diagnostic indicator of the controlling process) and their per cent increase (the difference between the daily maximum and minimum, divided by the minimum). Neither metric accounts for whether, or to what degree, an observed diel signal is statistically significant, which can be important when evaluating lower resolution data, signals with a small % increase, or signals resulting from multiple, overlapping controls. To quantify key parameters of diel signals, Kurz et al. (2013) introduced an analytical model which calculates the daily mean, amplitude, phase, and statistical significance of an observed diel signal. Mean, amplitude, and phase (timing of the modelled diel maximum) are calculated by fitting a sine function with 24-hr periodicity to an observed concentration time series. Significance is evaluated by how well the observed data conforms to the sine fit, as calculated by comparing the F-statistic of the sine fit relative to a null model of the time-series fit to the daily mean. A diel signal is considered significant if the p-value of the F-test is <0.1. To visually compare multiple diel signals, the modelled phase and significance is plotted as shown in Figure 13.4, with the phase of all signals plotted relative to the timing of the solar maximum because the solar cycle is presumed to be the ultimate driver of all diel signals. This diagram makes it possible to compare the key parameters (modelled or actual) of multiple diel signals across times and ecosystems. The "significance" axis in the plot can alternately be replaced with other variables, such as the per cent increase or daily magnitude.

13.3.2 Developments in Nutrient and Other Sensors

As with dissolved oxygen, the expanding development of in situ major nutrient sensors in recent decades has revolutionized our ability to resolve continuous, fine-scale fluctuations in dissolved organic carbon, nitrate, ammonium, and phosphate concentrations over multi-day to multi-year deployments (see in-depth discussions by Bieroza et al. 2023; Burns et al. 2019; Pellerin et al. 2016; Rode et al. 2016b). These sensor-based observations have given us many insights, summarized in the following section, into the mechanistic drivers and instantaneous coupling of nutrient cycling in streams. In situ nutrient sensors, which range from optical to "lab-in-a-can" designs, are generally more expensive and maintenance intensive than DO sensors. As a result, insights into broader temporal and spatial scale diel nutrient dynamics have lagged that of oxygen-based metabolism and been largely limited

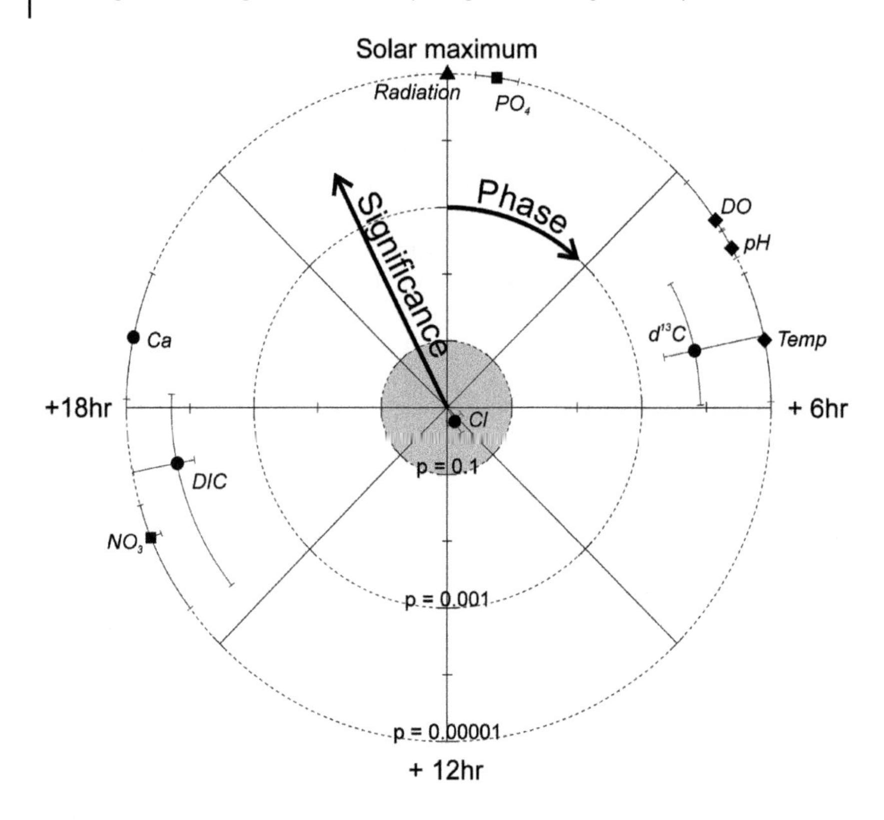

Figure 13.4 Example plot summarizing two key parameters (phase and statistical significance) of multiple diel signals. The timing of the diel maximum relative to the peak in solar radiation is plotted as the clockwise distance from + *y* axis. The distance from the origin reflects the significance of the diel signal, plotted as the log of the *p*-values derived using an F-test. Parameters with $p < 0.1$ are not considered to have statistically significant diel signals. Solutes with $p \leq 0.00001$ are plotted on the outer ring. The points represent the seasonal mean and the error bars the range in observed values. Model, figure, and example data Adapted from Kurz et al. (2013).

to cross-study syntheses. However, in the last decade, a growing number of extended duration deployments and/or distributed nutrient sensor networks have been used to resolve regimes in diel biogeochemical processes across seasons, flow conditions, and systems (e.g. Aubert and Breuer 2016; Hansen and Singh 2018; Khamis et al. 2021; Rode et al. 2016a).

In contrast to major nutrients, the observation of diel signals in major and trace elements and a wider range of nutrient species is still largely constrained by the need for time-intensive (auto)collection and analysis of discrete water samples. Corresponding system-specific studies have revealed intriguing hints into the ecological drivers of biologically active solutes over short deployments (e.g. Kurz et al. 2013), but expanding our mechanistic understanding of the fine-scale ecological drivers and emergent patterns in the coupled diel signals of a larger array of elemental species remains an open area for future research. Advances in sensors and related high temporal frequency techniques will likely continue to direct our path forward. Promising efforts include the development of a freshwater, autonomous alkalinity sensor (Shangguan et al. 2021); the use of UV-spectrophotometers aided by statistical algorithms to predict high-frequency signals in the concentrations of constituents that lack well-defined spectral peaks, including Fe, Mn, Si, and multiple phosphorus species (Birgand et al. 2016; Hammond et al. 2022; Vaughan et al. 2018); and the use of machine learning to predict completely non-sensed solutes from proxy solute time series, including major ions and trace metals from conductivity and Ca^{2+} and Al^{3+} from multi-parameter sonde data (Galella et al. 2021; Green et al. 2021; Ross et al. 2018). These latter examples further illustrate how the relative ease of producing sensor-based datasets of long duration and sub-hourly frequency, compared to discrete samples, provides a more robust basis for understanding, from providing sufficient computation data for signal prediction, and strengthening the statistical analysis of diel signals (as discussed in the previous section), to enabling more nuanced signal interpretation from nonlinearities and sub-hourly coupling in diel processes (e.g. Heffernan and Cohen 2010).

13.3.3 Controlling Ecological Processes

13.3.3.1 Carbon (C)

Diel signals have been observed in both dissolved organic and inorganic carbon (DOC and DIC, respectively) and carbon isotopes, and reflects a number of ecological processes as well as geochemical drivers. Both total DOC concentrations and the fraction of labile C compounds (those which break down quickly) typically increase during the day, in phase with DO (Chittoor Viswanathan et al. 2015; Kaplan and Bott 1982; Parker et al. 2010a, 2010b; Rutherford and Hynes 1987). These changes are associated with daytime excretion of labile DOC during photosynthesis, in excess of immediate consumption by respiration (Parker et al. 2010b) and, in the absence of photosynthesis, night-time DOC consumption by heterotrophic organisms (Bertilsson and Jones 2003; Kuserk et al. 1984). The rate of night-time consumption is a function of microbial community composition, the lability of available DOC, and temperature (Kaplan and Bott 1982; Volk et al. 1997). Diel DOC signals are not consistently present, but, where found, daytime concentrations have been observed to increase by as much as 100% relative to night-time (Nimick et al. 2011 and references therein). Diel variations in DOC quality are not always observed concurrently with DOC signals, and can occur even in the absence of diel variation in bulk DOC concentrations (Spencer et al. 2007). Photolysis, photo-oxidation, and sorption can increase or decrease the concentration of DOC (Jardim et al. 2009; McKnight and Duren 2004; Scott et al. 2003; Vesper and Smilley 2010), thus complicating the appearance of diel DOC signals, as well as altering the quality of dissolved organic matter (DOM) available to the aquatic ecosystem at diel timescale (e.g. Westhorpe et al. 2012; Wilson and Xenopoulos 2013). Cycling in the availability of DOM can in turn affect metabolic activity (Findlay et al. 2003).

The same processes that produce diel variations in bulk DOC concentrations can also produce variation in DOC isotopic composition. Photosynthesis preferentially consumes ^{12}C faster than ^{13}C (Falkowski and Raven 2013), producing biomass typically enriched in ^{12}C relative to atmospheric CO_2 (–20‰ to –30‰ vs. –7.5‰ to –8.5‰, respectively; Clark and Fritz 1997; NOAA 2010) and daytime decreases in water column $\delta^{13}C$-DOC (Parker et al. 2010b; Ziegler and Fogel 2003). This labile, isotopically light DOC is subsequently consumed preferentially by heterotrophs at night, resulting in increasing night-time $\delta^{13}C$ -DOC.

In addition to the production of organic C (biomass), photosynthesis also consumes inorganic C in the form of CO_2, driving diel variations in total DIC concentrations and the isotopic composition of that DIC. Net consumption of CO_2 by daytime photosynthesis and production of CO_2 by night-time respiration drives increases in DIC concentrations at night and decreases during the day (Parker et al. 2010b), inverse to the phase of DO and DOC. Although diel variations in inorganic C concentrations are directly linked to metabolism, DIC signals are rarely used as a direct measure of metabolic activity due to the overlapping effects of diel changes in pH and carbonate minerals saturation state on DIC concentrations (de Montety et al. 2011; Guasch et al. 1998; Hartley et al. 1996; Liu et al. 2008; Spiro and Pentecost 1991), all of which occur in phase, making them difficult to disentangle (Figure 13.3C). Hence, diel signals in DIC are often greatest (up to 400% increase) in streams draining carbonate terrains (Nimick et al. 2011 and references therein).

One way to distinguish between diel variations in DIC concentrations resulting from metabolic processing vs. mineral reactions is to calculate the slope of the relationship between DO and dissolved CO_2 concentrations. If changes in DIC concentrations are the direct result of metabolic processing, the slope of the relationship should be close to –1. Slopes steeper than –1 reflect CO_2 contributions due to carbonate precipitation (e.g. de Montety et al. 2011), whereas carbonate dissolution produces the opposite effect.

Diel variations in isotopic DIC composition can also be used to distinguish between controlling processes due to both fractionation and to the distinct isotopic composition of varying DIC sources. Opposite to the effect on δ^{13}C-DOC, night-time respiration decreases water column δ^{13}C -DIC while daytime photosynthesis increases δ^{13}C-DIC, the magnitude of which has been positively correlated to stream productivity (Parker et al. 2010a). In contrast, if carbonate dissolution and precipitation are the dominant controls on DIC cycling, the opposite isotopic signal is produced. Night-time dissolution of carbonate results in an increase in water column δ^{13}C-DIC ($\delta^{13}C–CaCO_3$ ~0‰), and daytime precipitation in a decrease in δ^{13}C-DIC of 0.5 to 1‰ under equilibrium conditions (de Montety et al. 2011 and references therein).

13.3.3.2 Nitrogen (N)

Diel signals have been extensively reported in nitrate and, to a lesser degree, ammonium, nitrite, and dissolved gaseous nitrogen species. Multiple simultaneous and often interacting processes drive diel variability in inorganic N species in streams making it particularly challenging to explicitly disentangle the resulting reach-scale diel N signals. Autotrophic assimilatory uptake is one of the dominant controls on diel nitrate and ammonium removal in streams. This results in diel signals that exhibit daytime minima when N demand is highest (Halliday et al. 2013; Heffernan and Cohen 2010; Martí et al. 2020; Nimick et al. 2011; Roberts and Mulholland 2007; Rusjan and Mikoš 2010; among many others). When autotrophic assimilation dominates N removal, daily estimates of uptake based on diel nitrate or ammonium signals are strongly correlated to daily estimates of GPP (such correlation has even been extended to hourly estimates with nitrate; Heffernan and Cohen 2010), and to estimates of uptake based on stoichiometry and GPP. Over longer timescales, the amplitude (including the lack) of diel nitrate and ammonium signals, and corresponding magnitude of assimilatory uptake, have been found to be a function of light conditions, temperature, nutrient concentrations, and, ultimately, whole-stream primary productivity (Heffernan and Cohen 2010; Kent et al. 2005; Roberts and Mulholland 2007; Rode et al. 2016a; Rusjan and Mikoš 2010). In nutrient-enriched systems especially, ammonium can be the energetically favoured species to meet autotrophic N demand, reducing daytime assimilatory nitrate removal and the amplitude of corresponding diel nitrate signals (e.g. Kunz et al. 2017).

Nitrification and denitrification can also drive diel signals in nitrate, ammonium, nitrogen, and oxygen gas, although the potential controls on these processes become increasingly complex and interacting. Diel variability in nitrification rates most commonly results from diel changes in pH, temperature, DO, and ammonium concentrations, but can be inhibited by light (Kunz et al. 2017; Martí et al. 2020; Warwick 1986; and references therein). Due to all but the latter, nitrification rates are generally greater during the day, reducing daytime ammonium concentrations, mirroring the signal from assimilatory ammonium removal, and increasing daytime nitrate concentrations, countering the signal from assimilatory nitrate removal. Diel variability in denitrification rates can likewise be driven by diel changes in pH and temperature, benthic metabolism responding to the availability of light, nitrate, labile organic carbon and oxygen to denitrifying zones, and indirectly to autotrophic control on these same variables (Heffernan and Cohen 2010; Kunz et al. 2017; Laursen and Seitzinger 2004; Nimick et al. 2011; and references therein). These multiple process drivers can produce opposing higher or lower denitrification rates during the day, making it challenging to disentangle their relative importance on observed diel N signals, although there appears to be organization in relative process importance with system light and productivity regimes. Coupling between GPP and estimates of potential assimilatory vs. dissimilatory uptake may also help to separate probable removal pathways (e.g. Heffernan and Cohen 2010).

There is potential that diel variations in isotopic N and O composition could be used to quantitatively isolate process-specific diel N signals. N isotopes have been used to track diurnally fluctuating nitrate inputs (Ohte et al. 2004) and the coupling of dual N isotopes can be an indicator of denitrification (2:1 residual enrichment in $\delta^{15}N_{NO3}$ and $\delta^{18}O_{NO3}$; Kendall et al. 2007). Despite this, only limited success has been had in using N isotopes to separate out biogeochemical N removal pathways controlling the diel behaviour of N (Cohen et al. 2012; Gammons et al. 2011; Pellerin et al. 2009).

13.3.3.3 Phosphorus (P)

After C and N, phosphorus is the next most abundant element in biomass and arguably the most important given its role as a limiting nutrient. Changes in stream P concentrations have been correlated to productivity patterns at longer timescales (Hill et al. 2001); however, diel P signals have rarely been reported in streams, even when other diel signals are present (Hatch et al. 1999; Kuwabara 1992; Parker et al. 2007; Triska et al. 1989). Unlike N uptake, which is primarily controlled by biologic processes, P cycling is complicated by multiple geochemical interactions, including redox sensitive sorption and co-precipitation (Diaz et al. 1994; House 1990; Reddy et al. 1999; Withers and Jarvie 2008). Many of these geochemical interactions can vary diurnally as the result of stream metabolism and other physiochemical factors, thereby obscuring any visible diel signal (e.g. Figure 13.3B; Cohen et al. 2013).

Phosphorus has no stable isotope with which to evaluate stream P cycling or help distinguish between controlling diel processes. In part due to these geochemical interactions, environmental availability of P is often scarce, and primary productivity is often limited by P (Elser et al. 2007). This limitation could also be a primary reason why diel P signals are uncommon in non-altered streams (Appling and Heffernan 2014).

When present, P concentrations have been observed to increase during the day and decrease at night (Cohen et al. 2013; Gammons et al. 2011; Kuwabara 1992; Vincent and Downes 1980). Cohen et al. (2013) were able to further isolate the magnitude and timing of assimilatory P uptake relative to the out-of-phase removal by calcite co-precipitation, in a reasonably homeostatic ecosystem. Notably, the timing of observed P signals lags behind that of primary productivity and assimilatory-driven diel N signals by 7–10 hours. Although in primary productivity, N and/or P cycling can be decoupled at larger scales (Hall et al. 2013), this is unlikely to be the case over short spatiotemporal scales when assimilation is the primary driver of diel signals. More plausible explanations include P luxury uptake capacity (Appling and Heffernan 2014) or that the lag in P uptake reflects night-time assimilatory uptake. Plasticity in assimilatory timing could allow night-time uptake if P concentrations were higher than due to decreased daytime geochemical removal, however, this is unlikely to be consistent across systems given the range of potential geochemical interactions. There is also evidence that the metabolic demand for P is highest at night (Ashworth et al. 2013; Cohen et al. 2013), illustrating a plausible link between diel element cycling and organismal physiology.

13.3.3.4 Trace Elements

Trace elements are key requirements for metabolic processes such as chlorophyll production (Fe and Mg; Pushnik et al. 1984), and enzyme and protein synthesis (Fe and Mo; Glass et al. 2009). However, few studies have isolated the direct effect of assimilatory uptake or other ecological processes on diel trace element signals (but see Kurz et al. 2013; Kuwabara 1992). The abiotic geochemical interactions affecting trace element cycling at diel timescales are often even more complex than for the major nutrients. Although the timing of these processes is generally known, quantifying the magnitude of their removal is complicated, making it difficult to in turn isolate assimilatory uptake, such as was done for phosphorus by Cohen et al. (2013). In addition, the focus to date has generally been on trace-metal enriched systems (e.g. Gammons et al. 2015), where biological controls on element cycling are likely to be minor in relation to geochemical drivers. Assimilatory uptake can be estimated from in situ organism stoichiometry and rates of primary productivity, assuming that uptake and biomass stoichiometry are not decoupled from metabolic demand. Using this approach, Kurz et al. (2013) estimated that autotrophic assimilation could account for a significant portion of some observed diel trace elements' signals. A better understanding of the reciprocal interactions between ecological and geochemical processes controlling the cycling and availability of metals is critical to evaluating ecosystem functioning, particularly in systems where low concentrations may create micronutrient limitation.

13.4 Looking Forward: Challenges and Open Questions

Diel signals can provide useful information to evaluate the timing and magnitude of ecohydrological processes in lotic systems, including GPP and ER (Section 13.2) and assimilatory element uptake (Section 13.3), and the coupling of these and other process interactions. Advances in in situ sensor technology have, and should continue to, greatly expand our capability to measure concentrations of a growing suite of solutes at near-continuous temporal resolution (see Bieroza et al. 2023; Kirchner et al. 2004; Pellerin et al. 2016; Rode et al. 2016b). The significance of sensor technology in transforming how we study coupled ecohydrological processes and the types of questions we can address cannot be overstated. Together with refinements in quantitative and statistical methods, the high-frequency measurements provided by in situ solute and gas sensors have greatly increased our capability to resolve fine-scale solute dynamics and calculate processes rates with reduced uncertainty compared to the previously available in situ methods. This has, in turn, provided a methodologically consistent and scale-independent way to assess ecosystem processing and coupling across multiple systems. As sensor costs have declined, there has been a proliferation of increasingly long-term time series of coupled solute concentrations and process rates from stream and river systems across the world.

Grounded by traditional sample-based data, sensor time series are enabling more systematic analysis of emergent patterns in metabolism and uptake rates across climatic regimes and land covers, nuanced evaluation of fine-scale ecological processes and theories, and interrogation of the complex effects of dynamic forcing on metabolic rates and ecosystem functioning. The advances gained from our expanded capabilities are being supported by and, in turn, driving creativity in statistical and model approaches to resolving and interpreting diel signals, and creativity in finding novel ways to use sensor data to evaluate diel processes. Beyond research, high-resolution sensor-based time series (esp. of DO) and associated tools and conceptual advances are being integrated into environmental management (e.g. Jankowski et al. 2021; Ledford et al. 2021a). Our ability to synthesize understanding across these growing data and model resources is being facilitated by invaluable efforts to conduct research that is multi-disciplinary and promotes mutual benefit and transferability of research findings (Goldman et al. 2022), including sharing and following consistent protocols; ensuring (meta)data and model codes are openly accessible using common platforms and architecture; and placing greater value on impactful, well-curated datasets and models.

The following sections envision areas for future advances in ecohydrological research using diel signals and explore associated challenges and open questions.

13.4.1 Larger-scale Emergent Regimes in Coupled Ecological Processes

The availability of oxygen time series from many different lotic systems has resulted in a shift from considering rivers as continuous gradients (Vannote et al. 1980) to analysing rivers as networks. It has been well established, both through theory and empirical findings, that *GPP* increases with stream order, due to increasing light availability caused by increasing channel width and reduced shading (Finlay 2011; Vannote et al. 1980). Thanks to the wealth of available data sets, machine learning techniques now allow one to extrapolate stream metabolism and its dependence on temperature, light, and nutrient availability to the network scale (Segatto et al. 2021). Additionally, the classification of productivity regimes based on annual patterns of *GPP* (Savoy et al. 2019) for different sections of the river network can aid in understanding the influence and range of the magnitude and timing of network productivity (Bernhardt et al. 2018; Koenig et al. 2019). The availability of a large number of continuous oxygen measurements also supported the comparison of metabolic rates across more than 200 US streams (Bernhardt et al. 2022). The study found that annual rates of *GPP* and *ER* vary substantially across river ecosystems. They also observed a much weaker seasonality in metabolic rates in river ecosystems compared to terrestrial ecosystems, with light availability and flow stability as the strongest controls on the timing and magnitude of metabolic rates in aquatic systems. Seasonality was found to be strongest in streams with high light availability and stable flows, highlighting how changes in land use and climate resulting in more frequent disturbances may directly affect metabolic rates on annual scales.

Corresponding large-scale comparison of the rates of other ecological processes across seasons, flow conditions, and systems, and the coupling of these (micro)nutrient regimes to those of stream metabolism is a promising area for future research. For example, Rode et al. (2016a) used a 2-year, high-frequency sensor time series to estimate and compare continuous rates of assimilatory nitrate uptake and metabolism in two stream reaches at daily and yearly timescales. Nitrate uptake and metabolism were strongly correlated in both reaches and variability in nitrate uptake could largely be explained by reach nitrate concentrations and light availability associated with watershed land cover and season. Data-informed process models have also shown promise in leveraging high-frequency time series of flow, nitrate, and DO to constrain estimates of the rates of multiple, interacting pathways of inorganic nitrogen uptake processes and disentangle their relative importance and coupling and drivers across seasons (Huang et al. 2022). Exploring how different ecological processes and cycles are interlinked by combining diel-based process estimates with simpler proxy metrics and/or sample-based process estimates is another promising path when high-frequency time series of multiple biologically active solutes are not available. Diamond et al. (2021) demonstrated how this approach could be used to relate dynamics in DO, metabolism, denitrification, and dissolved organic matter dynamics and scaling relationships across a regional network of headwater streams.

Further studies and modelling approaches like those highlighted previously, which take advantage of the scale-independence of the diel approach for estimating ecological process rates, should help advance growing efforts to resolve the relative controls and organization of coupled ecological processes within and across lotic networks. The complexity of the watershed structure presents a key challenge here. Because of the hierarchical organization of stream networks, large-scale watershed structure (land cover, bedrock geology, regional climate, etc.) influences hydro-biogeochemical conditions and processes at smaller scales. However, a large-scale watershed structure cannot fully explain the heterogeneity observed in ecohydrological processes (for an example, see Ward et al. 2022), including stream metabolism and (micro)nutrient uptake. Instead, reach and smaller-scale process interactions may be sufficiently independent of larger-scale controls or variable enough to obscure any coherent emergent signal. Furthermore, the effects of local heterogeneity may become increasingly integrated (e.g. Abbott et al. 2018), especially when evaluating solutes with long integration lengths. Key questions to address include how the complex interactions between multi-scale watershed structure, hydrologic forcing, etc., control emergent patterns in ecosystem processes and process coupling at different spatial scales; and how well these network scale patterns and controls align with finer-scale mechanistic process understanding synthesized from reach-scale studies and conventional experimental approaches (saturating pulse nutrient additions, isotopic tracer additions, etc.).

13.4.2 Linking Fine-scale Solute Dynamics to Fine-scale Ecological Process Understanding

Analysis of high-frequency diel signals is also proving critical to advancing finer-scale mechanistic understanding of ecological processes and ecological theory by allowing us to disentangle the timing and magnitude of individual ecological process rates, interrogate direct coupling of ecological processes at daily and sub-daily timescales, more accurately calculate ecosystem-level stoichiometry, and theoretically resolve ecosystem limitation status and physiological traits, all for an increasing range of elements. As our capacity to resolve diel signals expands, so does the need for mechanistic theories and models relating the timing, magnitude, and shapes of these signals to their causes and implications. The model by Appling and Heffernan (2014) does just this, by evaluating how parameters including uptake, excretion, growth, respiration, and mortality effect coupled nutrient uptake and export, and thus in-stream concentrations. The presence and absence of diel signals, as well as lags and changes in magnitude were all modelled. Notably, they found that temporal nutrient coupling could be an indicator of ecosystem limitation status as the input of one nutrient was coupled to the export of another only when the input nutrient is limiting. Thus, diel signals should not occur if a nutrient is limiting. Lags in element uptake relative to primary productivity could be used to evaluate luxury uptake capacity and distinguish between primary and secondary limitation vs. co-limitation.

Validating the predictions laid out in this and other theoretical models and conceptual frameworks (e.g. Schade et al. 2005) would allow us to utilize the presence and coupling of diel solute signals to answer questions relating to what controls direct (de)coupling in biologically active solute cycling; how are productivity and nutrient demand partitioned between autotrophic communities (à la Cohen et al. 2013) and, conversely, what are the relative contributions of different ecological communities or reaction pathways to "whole-stream" processes (e.g. autotrophic vs. heterotrophic contributions to ER; Hall and Beaulieu 2013); how does physiology mediate the relationship between resource input pulses and ecosystem nutrient dynamics; how are ecosystem-level stoichiometric relationships influenced by temporal scales, community composition, and light, nutrient availability, and other environmental conditions; and how does the stoichiometry of ecosystem nutrient demand and limitation scale along stream networks. Along the same vein, isolating individual process rates from solute time series should advance our ability to validate or refine long-held assumptions about the lack of diel variation in some ecological processes such as ER and non-assimilatory pathways (heterotrophic uptake, nitrification, and denitrification). Addressing these areas of research is perhaps more dependent than others on the availability of comparable high-frequency time series of multiple biologically active solutes. In situ sensors have unquestionably enabled this to date and future sensor developments will likely direct which elements we are able to evaluate going forward. Coupling analysis of diel signals with new approaches such as reactive stream tracers (e.g. Ledford et al. 2021b) may be another approach for addressing some of these questions.

13.4.3 Capturing Dynamic Forcing, States, and Systems

Lotic systems are highly dynamic and we know that the drivers and controls of ecohydrological processes are time-sensitive. Despite this, the majority of experiments and models address steady conditions. As our understanding of ecohydrological processing under steady conditions has matured, questions relating to if and how (co)variation in process dynamics responds to dynamic forcing, from short-term hydrologic events to long-term climate change, have received widespread and growing attention. Long-duration, high-frequency time series have made it possible to record not only baseline conditions, but to identify the effects of dynamic forcing on metabolic rates and ecosystem processing. For instance, Roberts et al. (2007) investigated the factors controlling the seasonal, day-to-day, episodic (i.e. storm events) and inter-annual variability in stream metabolism; O'Donnell and Hotchkiss (2022) examined the resistance and resilience of stream metabolism to isolated high flow events in a human-impacted stream; and Mulholland et al. (2009) documented the cascading effects of a late spring freeze on stream light levels, metabolism, nitrate uptake, and macroinvertebrates. Others have examined the effect of climate change on the metabolic balance of the oceans (López-Urrutia et al. 2006), natural streams (Demars et al. 2011), and how climate-change-related alterations in metabolic activity influence an ecosystems' capability for carbon sequestration (Yvon-Durocher et al. 2010).

Diel solute signals are observed most frequently during stable and low flow periods, when biogeochemical cycling dominates, and typically disappear during and following storm events or other time-varying conditions (e.g. Hensley et al. 2019; Kirchner and Neal 2013). Distinguishing if the lack of diel signals during these dynamic periods is due to concurrent shifts in controlling process dynamics or due to the signals being masked by background variability is a challenge. If it is the latter, accurately distinguishing and quantifying diel process signals from hydrologic or other sources of variability is a further challenge that inhibits our ability to resolve the effect of dynamic forcings on metabolic rates and ecosystem functioning.

Quantitative tools to disentangle diel biogeochemically-driven signals from other sources of variability in continuous time series are growing. Leveraging novel statistical tools, Aubert and Breuer (2016) successfully quantified diel nitrate signals out of a larger high-frequency dataset of nitrate and associated environmental variables. Seasonality in the magnitude and timing of the identified signals, which were not otherwise visible in the dataset, most likely reflect control by evapotranspiration rather than more common ecological processes. Chamberlin et al. (2021) developed an empirically-based signal decomposition method to filter out diel solute signals from background variability in solute concentrations induced from environmental factors and sensor noise. This method was generally successful in maintaining information needed to quantify the decomposed diel signals for ecological processes. Heathwaite and Bieroza (2021) used harmonic regression to partition solute concentration signals during storm events between hydrologic flushing and diel biogeochemical cycling, classify multiple water quality parameters based on their relative sensitivity to these respective forcings, and evaluate how storm magnitude and timing affected this relative sensitivity.

Going forward, these and related approaches will allow us to more confidently link diel (and other frequency) solute signals to different dynamic hydrologic and biogeochemical drivers and address questions including: if and how ecological processes respond to episodic or repeat disturbances such as storm pulses, droughts, water extraction, and time variable waste water treatment plant inputs; how do antecedent conditions moderate these responses; and how do different types and scales of dynamic forcing affect the resistance and resilience of metabolism and ecosystem process across scales. Increased quantification of how ecosystem processes respond to dynamic forcing will also aid the development of a predictive framework for understanding controls on these ecological process responses (O'Donnell and Hotchkiss 2022).

13.4.4 New Tools and Theories for Analysing Diel Signals and Their Information

All of the future research areas already discussed are being aided, if not driven, by new modelling and statistical approaches to resolve and interpret diel signals from solute time series. To summarize briefly: Refined statistical approaches and machine learning are opening up novel capabilities to identify and quantify diel signals from background variability (e.g. Aubert and Breuer 2016; Chamberlin et al. 2021; Halliday et al. 2013; Heathwaite and Bieroza 2021; Kirchner and Neal 2013) and to leverage sensor networks to predict solute concentrations and rates of ecological processes in stream reaches that lack such data (Green et al. 2021; Segatto et al. 2021). Data-informed inverse modelling is being used to guide coupled diel process interpretation of and extrapolation from solute time series (e.g. Appling et al. 2018; Huang et al. 2022; Segatto et al. 2021). While these modelling approaches may be less certain than the diel approach to quantifying process rates, they present an alternative option when the diel approach is not possible due to a failure in inherent assumptions (e.g. see discussions in Kunz et al. 2017) or is too labour intensive for a large time series. Developing multidimensional forward models that enable accurate parameter estimation and spatial scaling of ecological processes and account for watershed context and processes such as channel-subsurface exchange, remains more challenging. Lastly, theoretical models of diel process dynamics (e.g. Appling and Heffernan 2014) are helping to spur hypothesis generation and new ways of approaching diel signals in lotic systems. Continuing to embrace and expand such explicit, iterative integration of experimental and field observations with mechanistic and system models is essential to improving hypothesis generation, guiding experimental and monitoring designs, and, ultimately, to advancing a more predictive, transferable understanding of ecohydrological processes and process interactions.

References

Abbott, B.W., Gruau, G., Zarnetske, J.P. et al. (2018). Unexpected spatial stability of water chemistry in headwater stream networks. *Ecol. Lett.* 21: 296–308. https://doi.org/10.1111/ele.12897.

Almeida, G.H., Boëchat, I.G., and Gücker, B. (2014). Assessment of stream ecosystem health based on oxygen metabolism: which sensor to use? *Ecol. Eng.* 69: 134–138.

Appling, A.P., Hall, R.O., Jr, Yackulic, C.B., and Arroita, M. (2018). Overcoming equifinality: leveraging long time series for stream metabolism estimation. *J. Geophys. Res. Biogeosciences* 123: 624–645.

Appling, A.P. and Heffernan, J.B. (2014). Nutrient limitation and physiology mediate the fine-scale (de)coupling of biogeochemical cycles. *Am. Nat.* 184: 384–406. https://doi.org/10.1086/677282.

Ashworth, J., Coesel, S., Lee, A. et al. (2013). Genome-wide diel growth state transitions in the diatom *Thalassiosira pseudonana*. *Proc. Natl. Acad. Sci.* 110: 7518–7523. https://doi.org/10.1073/pnas.1300962110.

Aubert, A.H. and Breuer, L. (2016). New seasonal shift in in-stream diurnal nitrate cycles identified by mining high-frequency data. *PLOS ONE* 11: e0153138. https://doi.org/10.1371/journal.pone.0153138.

Battin, T.J., Kaplan, L.A., Newbold, J.D., and Hendricks, S.P. (2003). A mixing model analysis of stream solute dynamics and the contribution of a hyporheic zone to ecosystem function. *Freshw. Biol.* 48: 995–1014. https://doi.org/10.1046/j.1365-2427.2003.01062.x.

Beaulieu, J.J., Arango, C.P., Balz, D.A., and Shuster, W.D. (2013). Continuous monitoring reveals multiple controls on ecosystem metabolism in a suburban stream. *Freshw. Biol.* 58: 918–937. https://doi.org/10.1111/fwb.12097.

Beck, A.J., Janssen, F., Polerecky, L. et al. (2009). Phototrophic biofilm activity and dynamics of diurnal Cd cycling in a freshwater stream. *Environ. Sci. Technol.* 43: 7245–7251. https://doi.org/10.1021/es900069y.

Bencala, K.E., Gooseff, M.N., and Kimball, B.A. (2011). Rethinking hyporheic flow and transient storage to advance understanding of stream-catchment connections. *Water Resour. Res.* 47.

Benson, A., Zane, M., Becker, T.E. et al. (2014). Quantifying reaeration rates in alpine streams using deliberate gas tracer experiments. *Water* 6: 1013–1027.

Bernhardt, E.S., Heffernan, J.B., Grimm, N.B. et al. (2018). The metabolic regimes of flowing waters. *Limnol. Oceanogr.* 63: S99–S118. https://doi.org/10.1002/lno.10726.

Bernhardt, E.S., Savoy, P., Vlah, M.J. et al. (2022). Light and flow regimes regulate the metabolism of rivers. *Proc. Natl. Acad. Sci.* 119: e2121976119.

Bertilsson, S., and Jones, J.B. (2003). Supply of dissolved organic matter to aquatic ecosystems: autochthonous sources. In: *Aquatic Ecosystems: Interactivity of Dissolved Organic Matter* (ed. S.E.G. Findlay and R.L. Sinsabaugh), 3–24. Academic Press. https://doi.org/10.1016/B978-012256371-3/50002-0.

Bieroza, M., Acharya, S., Benisch, J. et al. (2023). Advances in catchment science, hydrochemistry, and aquatic ecology enabled by high-frequency water quality measurements. *Environ. Sci. Technol.* https://doi.org/10.1021/acs.est.2c07798.

Birgand, F., Aveni-Deforge, K., Smith, B. et al. (2016). First report of a novel multiplexer pumping system coupled to a water quality probe to collect high temporal frequency in situ water chemistry measurements at multiple sites. *Limnol. Oceanogr. Methods* 14: 767–783. https://doi.org/10.1002/lom3.10122.

Bond, B.J., Jones, J.A., Moore, G. et al. (2002). The zone of vegetation influence on baseflow revealed by diel patterns of streamflow and vegetation water use in a headwater basin. *Hydrol. Process.* 16: 1671–1677. https://doi.org/10.1002/hyp.5022.

Burns, D.A., Pellerin, B.A., Miller, M.P. et al. (2019). Monitoring the riverine pulse: applying high-frequency nitrate data to advance integrative understanding of biogeochemical and hydrological processes. *Wiley Interdiscip. Rev. Water* 6: e1348. https://doi.org/10.1002/wat2.1348.

Caffrey, J.M. (2003). Production, respiration and net ecosystem metabolism in U.S. estuaries. In: *Coastal Monitoring through Partnerships: Proceedings of the Fifth Symposium on the Environmental Monitoring and Assessment Program (EMAP) Pensacola Beach, FL, U.S.A., April 24–27, 2001* (ed. B.D. Melzian, V. Engle, M. McAlister, et al.), 207–219. Netherlands, Dordrecht: Springer. https://doi.org/10.1007/978-94-017-0299-7_19.

Chamberlin, C.A., Katul, G.G., and Heffernan, J.B. (2021). A multiscale approach to timescale analysis: isolating diel signals from solute concentration time series. *Environ. Sci. Technol.* 55: 12731–12738. https://doi.org/10.1021/acs.est.1c00498.

Chittoor Viswanathan, V., Molson, J., and Schirmer, M. (2015). Does river restoration affect diurnal and seasonal changes to surface water quality? A study along the Thur River, Switzerland. *Sci. Total Environ.* 532: 91–102. https://doi.org/10.1016/j.scitotenv.2015.05.121.

Clark, I.D. and Fritz, P. (1997). *Environmental Isotopes in Hydrogeology*. CRC press.

Clarke, S.J. (2002). Vegetation growth in rivers: influences upon sediment and nutrient dynamics. *Prog. Phys. Geogr.* 26: 159–172. https://doi.org/10.1191/0309133302pp324ra.

Cohen, M.J., Heffernan, J.B., Albertin, A., and Martin, J.B. (2012). Inference of riverine nitrogen processing from longitudinal and diel variation in dual nitrate isotopes. *J. Geophys. Res.* 117: G01021. https://doi.org/10.1029/2011JG001715.

Cohen, M.J., Kurz, M.J., Heffernan, J.B. et al. (2013). Diel phosphorus variation and the stoichiometry of ecosystem metabolism in a large spring-fed river. *Ecol. Monogr.* 83: 155–176. https://doi.org/10.1890/12-1497.1.

Cox, B.A. (2003). A review of dissolved oxygen modelling techniques for lowland rivers. *Sci. Total Environ.* 314: 303–334.

Cox, T.J., Maris, T., Soetaert, K. et al. (2015). Estimating primary production from oxygen time series: a novel approach in the frequency domain. *Limnol. Oceanogr. Methods* 13: 529–552.

Czikowsky, M.J. and Fitzjarrald, D.R. (2004). Evidence of seasonal changes in evapotranspiration in Eastern U.S. hydrological records. *J. Hydrometeorol.* 5: 974–988. https://doi.org/10.1175/1525-7541(2004)005<0974:EOSCIE>2.0.CO;2.

de Montety, V., Martin, J.B., Cohen, M.J. et al. (2011). Influence of diel biogeochemical cycles on carbonate equilibrium in a karst river. *Chem. Geol.* 283: 31–43. https://doi.org/10.1016/j.chemgeo.2010.12.025.

Demars, B.O., Russell Manson, J., Olafsson, J.S. et al. (2011). Temperature and the metabolic balance of streams. *Freshw. Biol.* 56: 1106–1121.

Demars, B.O., Thompson, J., and Manson, J.R. (2015). Stream metabolism and the open diel oxygen method: principles, practice, and perspectives. *Limnol. Oceanogr. Methods* 13: 356–374.

Desmet, N.J.S., Van Belleghem, S., Seuntjens, P. et al. (2011). Quantification of the impact of macrophytes on oxygen dynamics and nitrogen retention in a vegetated lowland river. *Phys. Chem. Earth Parts ABC* 36: 479–489. https://doi.org/10.1016/j.pce.2008.06.002.

Diamond, J.S., Bernal, S., Boukra, A. et al. (2021). Stream network variation in dissolved oxygen: metabolism proxies and biogeochemical controls. *Ecol. Indic.* 131: 108233. https://doi.org/10.1016/j.ecolind.2021.108233.

Diaz, O.A., Reddy, K.R., and Moore, P.A., Jr. (1994). Solubility of inorganic phosphorus in stream water as influenced by pH and calcium concentration. *Water Res.* 28: 1755–1763. https://doi.org/10.1016/0043-1354(94)90248-8.

Duncan, J.M., Band, L.E., Groffman, P.M., and Bernhardt, E.S. (2015). Mechanisms driving the seasonality of catchment scale nitrate export: evidence for riparian ecohydrologic controls. *Water Resour. Res.* 51: 3982–3997. https://doi.org/10.1002/2015WR016937.

Elser, J.J., Acharya, K., Kyle, M. et al. (2003). Growth rate-stoichiometry couplings in diverse biota. *Ecol. Lett.* 6: 936–943. https://doi.org/10.1046/j.1461-0248.2003.00518.x.

Elser, J.J., Bracken, M.E.S., Cleland, E.E. et al. (2007). Global analysis of nitrogen and phosphorus limitation of primary producers in freshwater, marine and terrestrial ecosystems. *Ecol. Lett.* 10: 1135–1142. https://doi.org/10.1111/j.1461-0248.2007.01113.x.

Falkowski, P.G. and Raven, J.A. (2013). *Aquatic Photosynthesis*. Princeton University Press.

Fernández Castro, B., Chmiel, H.E., Minaudo, C. et al. (2021). Primary and net ecosystem production in a large lake diagnosed from high-resolution oxygen measurements. *Water Resour. Res.* 57: e2020WR029283.

Findlay, S.E., Sinsabaugh, R.L., Sobczak, W.V., and Hoostal, M. (2003). Metabolic and structural response of hyporheic microbial communities to variations in supply of dissolved organic matter. *Limnol. Oceanogr.* 48: 1608–1617.

Finlay, J.C. (2011). Stream size and human influences on ecosystem production in river networks. *Ecosphere* 2: 1–21.

Flewelling, S.A., Hornberger, G.M., Herman, J.S. et al. (2014). Diel patterns in coastal-stream nitrate concentrations linked to evapotranspiration in the riparian zone of a low-relief, agricultural catchment. *Hydrol. Process.* 28: 2150–2158. https://doi.org/10.1002/hyp.9763.

Frost, P.C., Evans-White, M.A., Finkel, Z.V. et al. (2005). Are you what you eat? Physiological constraints on organismal stoichiometry in an elementally imbalanced world. *Oikos* 109: 18–28.

Furuya, K. and Harada, K. (1995). An automated precise Winkler titration for determining dissolved oxygen on board ship. *J. Oceanogr.* 51: 375–383.

Gaarder, T. (1927). Investigations of the production of plankton in the Oslo Fjord. Rapports et Proces-verbaux des Reunions. *Cons. Int. Pour 1. Exploration Mer* 42: 1–48.

Galella, J.G., Kaushal, S.S., Wood, K.L. et al. (2021). Sensors track mobilization of 'chemical cocktails' in streams impacted by road salts in the Chesapeake Bay watershed. *Environ. Res. Lett.* 16: 035017. https://doi.org/10.1088/1748-9326/abe48f.

Gammons, C.H., Babcock, J.N., Parker, S.R., and Poulson, S.R. (2011). Diel cycling and stable isotopes of dissolved oxygen, dissolved inorganic carbon, and nitrogenous species in a stream receiving treated municipal sewage. *Chem. Geol.* 283: 44–55. https://doi.org/10.1016/j.chemgeo.2010.07.006.

Gammons, C.H., Nimick, D.A., and Parker, S.R. (2015). Diel cycling of trace elements in streams draining mineralized areas—A review. *Appl. Geochem.* 57: 35–44. https://doi.org/10.1016/j.apgeochem.2014.05.008.

Gattuso, J.-P., Pichon, M., Delesalle, B., and Frankignoulle, M. (1993). Community metabolism and air-sea CO_2 fluxes in a coral reef ecosystem (Moorea, French Polynesia). *Mar. Ecol. Prog. Ser.* 96: 259–267.

Glass, J.B., Wolfe-Simon, F., and Anbar, A.D. (2009). Coevolution of metal availability and nitrogen assimilation in cyanobacteria and algae. *Geobiology* 7: 100–123. https://doi.org/10.1111/j.1472-4669.2009.00190.x.

Goldman, A.E., Emani, S.R., Pérez-Angel, L.C. et al. (2022). Integrated, coordinated, open, and networked (ICON) science to advance the geosciences: introduction and synthesis of a special collection of commentary articles. *Earth and Space Science* 9: e2021EA002099. https://doi.org/10.1029/2021EA002099.

Grace, M.R., Giling, D.P., Hladyz, S. et al. (2015). Fast processing of diel oxygen curves: estimating stream metabolism with BASE (BAyesian Single-station Estimation). *Limnol. Oceanogr. Methods* 13: 103–114.

Grande, K.D., Marra, J., Langdon, C. et al. (1989). Rates of respiration in the light measured in marine phytoplankton using an 18O isotope-labelling technique. *J. Exp. Mar. Biol. Ecol.* 129: 95–120.

Green, M.B., Pardo, L.H., Bailey, S.W. et al. (2021). Predicting high-frequency variation in stream solute concentrations with water quality sensors and machine learning. *Hydrol. Process.* 35: e14000. https://doi.org/10.1002/hyp.14000.

Guasch, H., Armengol, J., Martí, E., and Sabater, S. (1998). Diurnal variation in dissolved oxygen and carbon dioxide in two low-order streams. *Water Res.* 32: 1067–1074. https://doi.org/10.1016/S0043-1354(97)00330-8.

Gurnell, A. (2014). Plants as river system engineers. *Earth Surf. Process. Landf.* 39: 4–25. https://doi.org/10.1002/esp.3397.

Haggerty, R., Argerich, A., and Martı, E. (2008). Development of a "smart" tracer for the assessment of microbiological activity and sediment-water interaction in natural waters: the resazurin-resorufin system. *Water Resour Res.* 44: W00D01.

Haggerty, R., Martí, E., Argerich, A. et al. (2009). Resazurin as a "smart" tracer for quantifying metabolically active transient storage in stream ecosystems. *J. Geophys. Res. Biogeosciences* 114. https://doi.org/10.1029/2008JG000942.

Hall, C.A. (1972). Migration and metabolism in a temperate stream ecosystem. *Ecology* 53: 585–604.

Hall, R.O., Baker, M.A., Rosi-Marshall, E.J. et al. (2013). Solute-specific scaling of inorganic nitrogen and phosphorus uptake in streams. *Biogeosciences* 10: 7323–7331. https://doi.org/10.5194/bg-10-7323-2013.

Hall, R.O. and Beaulieu, J.J. (2013). Estimating autotrophic respiration in streams using daily metabolism data. *Freshw. Sci.* 32: 507–516. https://doi.org/10.1899/12-147.1.

Hall, R.O. and Tank, J.L. (2005). Correcting whole-stream estimates of metabolism for groundwater input. *Limnol. Oceanogr. Methods* 3: 222–229.

Hall, R.O., Jr, and Ulseth, A.J. (2020). Gas exchange in streams and rivers. *Wiley Interdiscip. Rev. Water* 7: e1391.

Halliday, S.J., Skeffington, R.A., Wade, A.J. et al. (2013). Upland streamwater nitrate dynamics across decadal to sub-daily timescales: a case study of Plynlimon, Wales. *Biogeosciences* 10: 8013–8038. https://doi.org/10.5194/bg-10-8013-2013.

Hammond, N., Birgand, F., Carey, C.C. et al. (2022). High-frequency sensor data capture short-term variability in Fe and Mn cycling due to hypolimnetic oxygenation and seasonal dynamics in a drinking water reservoir (preprint). *Environmental Sciences* https://doi.org/10.1002/essoar.10512927.1.

Hansen, A. and Singh, A. (2018). High-frequency sensor data reveal across-scale nitrate dynamics in response to hydrology and biogeochemistry in intensively managed agricultural basins. *J. Geophys. Res. Biogeosciences* 123: 2168–2182. https://doi.org/10.1029/2017JG004310.

Hanson, P.C., Bade, D.L., Carpenter, S.R., and Kratz, T.K. (2003). Lake metabolism: relationships with dissolved organic carbon and phosphorus. *Limnol. Oceanogr.* 48: 1112–1119.

Hanson, P.C., Hamilton, D.P., Stanley, E.H. et al. (2011). Fate of allochthonous dissolved organic carbon in lakes: a quantitative approach. *PLoS One* 6: e21884.

Hartley, A.M., House, W.A., Leadbeater, B.S.C., and Callow, M.E. (1996). The use of microelectrodes to study the precipitation of calcite upon algal biofilms. *J. Colloid Interface Sci.* 183: 498–505.

Hatch, L.K., Reuter, J.E., and Goldman, C.R. (1999). Daily phosphorus variation in a mountain stream. *Water Resour. Res.* 35: 3783–3791. https://doi.org/10.1029/1999WR900256.

Heathwaite, A.L. and Bieroza, M. (2021). Fingerprinting hydrological and biogeochemical drivers of freshwater quality. *Hydrol. Process.* 35: e13973. https://doi.org/10.1002/hyp.13973.

Heffernan, J.B. and Cohen, M.J. (2010). Direct and indirect coupling of primary production and diel nitrate dynamics in a subtropical spring-fed river. *Limnol. Oceanogr.* 55: 677.

Hensley, R.T. and Cohen, M.J. (2016). On the emergence of diel solute signals in flowing waters. *Water Resour. Res.* n/a–n/a. https://doi.org/10.1002/2015WR017895.

Hensley, R.T., Kirk, L., Spangler, M. et al. (2019). Flow extremes as spatiotemporal control points on river solute fluxes and metabolism. *J. Geophys. Res. Biogeosciences* 124: 537–555. https://doi.org/10.1029/2018JG004738.

Hill, W.R., Mulholland, P.J., and Marzolf, E.R. (2001). Stream ecosystem responses to forest leaf emergence in spring. *Ecology* 82: 2306–2319. https://doi.org/10.1890/0012-9658(2001)082[2306:SERTFL]2.0.CO;2.

Ho, D.T., Schlosser, P., and Orton, P.M. (2011). On factors controlling air–water gas exchange in a large tidal river. *Estuaries Coasts* 34: 1103–1116.

Hotchkiss, E.R. and Hall, R.O. (2014). High rates of daytime respiration in three streams: use of d 18 OO2 and O2 to model diel ecosystem metabolism. *Limnol. Ocean* 59: 798–810.

House, W.A. (1990). The prediction of phosphate coprecipitation with calcite in freshwaters. *Water Res.* 24: 1017–1023. https://doi.org/10.1016/0043-1354(90)90124-O.

Huang, J., Borchardt, D., and Rode, M. (2022). How do inorganic nitrogen processing pathways change quantitatively at daily, seasonal, and multiannual scales in a large agricultural stream? *Hydrol. Earth Syst. Sci.* 26: 5817–5833. https://doi.org/10.5194/hess-26-5817-2022.

Huisman, J.L., Weber, N., and Gujer, W. (2004). Reaeration in sewers. *Water Res.* 38: 1089–1100.

Jankowski, K.J., Mejia, F.H., Blaszczak, J.R., and Holtgrieve, G.W. (2021). Aquatic ecosystem metabolism as a tool in environmental management. *WIREs Water* 8: e1521. https://doi.org/10.1002/wat2.1521.

Jardim, W.F., Bisinoti, M.C., Fadini, P.S., and Silva, G.S.D. (2009). Mercury redox chemistry in the Negro River Basin, Amazon: the role of organic matter and solar light. *Aquat. Geochem.* 16: 267–278. https://doi.org/10.1007/s10498-009-9086-z.

Julian, J.P., Seegert, S.Z., Powers, S.M. et al. (2011). Light as a first-order control on ecosystem structure in a temperate stream. *Ecohydrology* 4: 422–432. https://doi.org/10.1002/eco.144.

Kaplan, L.A. and Bott, T.L. (1982). Diel fluctuations of DOC generated by algae in a piedmont stream. *Limnol. Oceanogr.* 27: 1091–1100.

Kendall, C., Elliott, E.M., and Wankel, S.D. (2007). Tracing anthropogenic inputs of nitrogen to ecosystems. *Stable Isot. Ecol. Environ. Sci.* 2: 375–449.

Kent, R., Belitz, K., and Burton, C.A. (2005). Algal productivity and nitrate assimilation in an effluent dominated concrete lined stream1. *JAWRA J. Am. Water Resour. Assoc.* 41: 1109–1128. https://doi.org/10.1111/j.1752-1688.2005.tb03788.x.

Kerkhoff, A.J., Enquist, B.J., Elser, J.J., and Fagan, W.F. (2005). Plant allometry, stoichiometry and the temperature-dependence of primary productivity. *Glob. Ecol. Biogeogr.* 14: 585–598. https://doi.org/10.1111/j.1466-822X.2005.00187.x.

Khamis, K., Blaen, P.J., Comer-Warner, S. et al. (2021). High-frequency monitoring reveals multiple frequencies of nitrogen and carbon mass balance dynamics in a headwater stream. *Front. Water* 3.

Khoshmanesh, A., Hart, B.T., Duncan, A., and Beckett, R. (2002). Luxury uptake of phosphorus by sediment bacteria. *Water Res.* 36: 774–778.

Kirchner, J.W., Feng, X., Neal, C., and Robson, A.J. (2004). The fine structure of water-quality dynamics: the(high-frequency) wave of the future. *Hydrol. Process.* 18: 1353–1359. https://doi.org/10.1002/hyp.5537.

Kirchner, J.W., Godsey, S.E., Solomon, M. et al. (2020). The pulse of a montane ecosystem: coupling between daily cycles in solar flux, snowmelt, transpiration, groundwater, and streamflow at Sagehen Creek and Independence Creek, Sierra Nevada, USA. *Hydrol. Earth Syst. Sci.* 24: 5095–5123. https://doi.org/10.5194/hess-24-5095-2020.

Kirchner, J.W. and Neal, C. (2013). Universal fractal scaling in stream chemistry and its implications for solute transport and water quality trend detection. *Proc. Natl. Acad. Sci.* 110: 12213–12218. https://doi.org/10.1073/pnas.1304328110.

Knapp, J.L., Osenbrück, K., Brennwald, M.S., and Cirpka, O.A. (2019). In-situ mass spectrometry improves the estimation of stream reaeration from gas-tracer tests. *Sci. Total Environ.* 655: 1062–1070.

Knapp, J.L., Osenbrück, K., and Cirpka, O.A. (2015). Impact of non-idealities in gas-tracer tests on the estimation of reaeration, respiration, and photosynthesis rates in streams. *Water Res.* 83: 205–216.

Knapp, J.L.A., 2017. *Advancing stream-tracer techniques and their mathematical analysis* (PhD Thesis). Eberhard Karls Universität Tübingen.

Knapp, J.L.A., González-Pinzón, R., and Haggerty, R. (2018). The resazurin-resorufin system: insights from a decade of "smart" tracer development for hydrologic applications. *Water Resour. Res.* 54: 6877–6889. https://doi.org/10.1029/2018WR023103.

Koenig, L.E., Helton, A.M., Savoy, P. et al. (2019). Emergent productivity regimes of river networks. *Limnol. Oceanogr. Lett.* 4: 173–181. https://doi.org/10.1002/lol2.10115.

Krems, P., Rajfur, M., Wacławek, M., and Kłos, A. (2013). The use of water plants in biomonitoring and phytoremediation of waters polluted with heavy metals. *Ecol. Chem. Eng. S* 20: 353–370. https://doi.org/10.2478/eces-2013-0026.

Kunz, J.V., Hensley, R., Brase, L. et al. (2017). High frequency measurements of reach scale nitrogen uptake in a fourth order river with contrasting hydromorphology and variable water chemistry (Weiße Elster, Germany). *Water Resour. Res.* 53: 328–343. https://doi.org/10.1002/2016WR019355.

Kurz, M.J., de Montety, V., Martin, J.B. et al. (2013). Controls on diel metal cycles in a biologically productive carbonate-dominated river. *Chem. Geol.* 358: 61–74. https://doi.org/10.1016/j.chemgeo.2013.08.042.

Kuserk, F.T., Kaplan, L.A., and Bott, T.L. (1984). In situ measures of dissolved organic carbon flux in a rural stream. *Can. J. Fish. Aquat. Sci.* 41: 964–973. https://doi.org/10.1139/f84-110.

Kuwabara, J.S. (1992). Associations between benthic flora and diel changes in dissolved arsenic, phosphorus, and related physico-chemical parameters. *J. North Am. Benthol. Soc.* 11: 218–228.

Laursen, A.E. and Seitzinger, S.P. (2004). Diurnal patterns of denitrification, oxygen consumption and nitrous oxide production in rivers measured at the whole-reach scale. *Freshw. Biol.* 49: 1448–1458. https://doi.org/10.1111/j.1365-2427.2004.01280.x.

Ledford, S.H., Diamond, J.S., and Toran, L. (2021a). Large spatiotemporal variability in metabolic regimes for an urban stream draining four wastewater treatment plants with implications for dissolved oxygen monitoring. *PLOS ONE* 16: e0256292. https://doi.org/10.1371/journal.pone.0256292.

Ledford, S.H., Kurz, M.J., and Toran, L. (2021b). Contrasting Raz–Rru stream metabolism and nutrient uptake downstream of urban wastewater effluent sites. *Freshw. Sci.* 40: 103–119. https://doi.org/10.1086/712932.

Liu, Z., Liu, X., and Liao, C. (2008). Daytime deposition and nighttime dissolution of calcium carbonate controlled by submerged plants in a karst spring-fed pool: insights from high time-resolution monitoring of physico-chemistry of water. *Environ. Geol.* 55: 1159–1168. https://doi.org/10.1007/s00254-007-1062-6.

López-Urrutia, Á., San Martin, E., Harris, R.P., and Irigoien, X. (2006). Scaling the metabolic balance of the oceans. *Proc. Natl. Acad. Sci.* 103: 8739–8744.

Lupon, A., Martí, E., Sabater, F., and Bernal, S. (2016). Green light: gross primary production influences seasonal stream N export by controlling fine-scale N dynamics. *Ecology* 97: 133–144. https://doi.org/10.1890/14-2296.1.

Martí, E., Feijoó, C., Vilches, C. et al. (2020). Diel variation of nutrient retention is associated with metabolism for ammonium but not phosphorus in a lowland stream. *Freshw. Sci.* 39: 268–280. https://doi.org/10.1086/708933.

Marzolf, E.R., Mulholland, P.J., and Steinman, A.D. (1994). Improvements to the diurnal upstream–downstream dissolved oxygen change technique for determining whole-stream metabolism in small streams. *Can. J. Fish. Aquat. Sci.* 51: 1591–1599. https://doi.org/10.1139/f94-158.

McKnight, D.M. and Duren, S.M. (2004). Biogeochemical processes controlling midday ferrous iron maxima in stream waters affected by acid rock drainage. *Appl. Geochem.* 19: 1075–1084. https://doi.org/10.1016/j.apgeochem.2004.01.007.

Mulholland, P.J., Roberts, B.J., Hill, W.R., and Smith, J.G. (2009). Stream ecosystem responses to the 2007 spring freeze in the southeastern United States: unexpected effects of climate change. *Glob. Change Biol.* 15: 1767–1776. https://doi.org/10.1111/j.1365-2486.2009.01864.x.

Nielsen, E.S. (1951). Measurement of the production of organic matter in the sea by means of carbon-14. *Nature* 167: 684–685.

Nimick, D.A., Gammons, C.H., and Parker, S.R. (2011). Diel biogeochemical processes and their effect on the aqueous chemistry of streams: a review. *Chem. Geol.* 283: 3–17. https://doi.org/10.1016/j.chemgeo.2010.08.017.

NOAA. (2010). Earth system research laboratory, global monitoring division. *Carbon Cycle Greenhouse Gases Group, U.S. Department of Commerce, NOAA/ ESRL*, Boulder, CO. NOAA.

O'Connor, D.J. and Dobbins, W.E. (1958). Mechanism of reaeration in natural streams. *Trans. Am. Soc. Civ. Eng.* 123: 641–666.

O'Donnell, B. and Hotchkiss, E.R. (2022). Resistance and resilience of stream metabolism to high flow disturbances. *Biogeosciences* 19: 1111–1134. https://doi.org/10.5194/bg-19-1111-2022.

Odum, H.T. (1956). Primary production in flowing waters. *Limnol. Oceanogr.* 1: 102–117.

Ohte, N., Sebestyen, S.D., Shanley, J.B. et al. (2004). Tracing sources of nitrate in snowmelt runoff using a high-resolution isotopic technique. *Geophys. Res. Lett.* 31: L21506. https://doi.org/10.1029/2004GL020908.

Oviedo-Vargas, D., Peipoch, M., and Dow, C. (2022). Metabolism and soil water viscosity control diel patterns of nitrate and DOC in a low order temperate stream. *J. Geophys. Res. Biogeosciences* 127: e2021JG006640. https://doi.org/10.1029/2021JG006640.

Parker, S.R., Gammons, C.H., Poulson, S.R., and DeGrandpre, M.D. (2007). Diel variations in stream chemistry and isotopic composition of dissolved inorganic carbon, upper Clark Fork River, Montana, USA. *Appl. Geochem.* 22: 1329–1343. https://doi.org/10.1016/j.apgeochem.2007.02.007.

Parker, S.R., Gammons, C.H., Poulson, S.R. et al. (2010a). Diel behavior of stable isotopes of dissolved oxygen and dissolved inorganic carbon in rivers over a range of trophic conditions, and in a mesocosm experiment. *Chem. Geol.* 269: 22–32. https://doi.org/10.1016/j.chemgeo.2009.06.016.

Parker, S.R., Poulson, S.R., Smith, M.G. et al. (2010b). Temporal variability in the concentration and stable carbon isotope composition of dissolved inorganic and organic carbon in Two Montana, USA Rivers. *Aquat. Geochem.* 16: 61–84. https://doi.org/10.1007/s10498-009-9068-1.

Pellerin, B.A., Downing, B.D., Kendall, C. et al. (2009). Assessing the sources and magnitude of diurnal nitrate variability in the San Joaquin River (California) with an in situ optical nitrate sensor and dual nitrate isotopes. *Freshw. Biol.* 54: 376–387. https://doi.org/10.1111/j.1365-2427.2008.02111.x.

Pellerin, B.A., Stauffer, B.A., Young, D.A. et al. (2016). Emerging tools for continuous nutrient monitoring networks: sensors advancing science and water resources protection. *JAWRA J. Am. Water Resour. Assoc.* 52: 993–1008. https://doi.org/10.1111/1752-1688.12386.

Pennington, R., Argerich, A., and Haggerty, R. (2018). Measurement of gas-exchange rate in streams by the oxygen–carbon method. *Freshw. Sci.* 37: 222–237. https://doi.org/10.1086/698018.

Pushnik, J.C., Miller, G.W., and Manwaring, J.H. (1984). The role of iron in higher plant chlorophyll biosynthesis, maintenance and chloroplast biogenesis. *J. Plant Nutr.* 7: 733–758. https://doi.org/10.1080/01904168409363238.

Rathbun, R.E., Stephens, D.W., Shultz, D.J., and Tai, D.Y. (1978). Laboratory studies of gas tracers for reaeration. *J. Environ. Eng. Div.* 104: 215–229.

Reddy, K.R., Kadlec, R.H., Flaig, E., and Gale, P.M. (1999). Phosphorus retention in streams and wetlands: a review. *Crit. Rev. Environ. Sci. Technol.* 29: 83–146. https://doi.org/10.1080/10643389991259182.

Reid, S.E., Mackinnon, P.A., and Elliot, T. (2007). Direct measurements of reaeration rates using noble gas tracers in the River Lagan, Northern Ireland. *Water Environ. J.* 21: 182–191.

Riley, A.J. and Dodds, W.K. (2013). Whole-stream metabolism: strategies for measuring and modeling diel trends of dissolved oxygen. *Freshw. Sci.* 32: 56–69.

Roberts, B.J. and Mulholland, P.J. (2007). In-stream biotic control on nutrient biogeochemistry in a forested stream, West Fork of Walker Branch. *J. Geophys. Res.* 112. https://doi.org/10.1029/2007JG000422.

Roberts, B.J., Mulholland, P.J., and Hill, W.R. (2007). Multiple scales of temporal variability in ecosystem metabolism rates: results from 2 years of continuous monitoring in a forested headwater stream. *Ecosystems* 10: 588–606. https://doi.org/10.1007/s10021-007-9059-2.

Rode, M., Halbedel Née Angelstein, S., Anis, M.R. et al. (2016a). Continuous in-stream assimilatory nitrate uptake from high-frequency sensor measurements. *Environ. Sci. Technol.* 50: 5685–5694. https://doi.org/10.1021/acs.est.6b00943.

Rode, M., Wade, A.J., Cohen, M.J. et al. (2016b). Sensors in the stream: the high-frequency wave of the present. *Environ. Sci. Technol.* 50: 10297–10307. https://doi.org/10.1021/acs.est.6b02155.

Ross, M.R., Nippgen, F., Hassett, B.A. et al. (2018). Pyrite oxidation drives exceptionally high weathering rates and geologic CO2 release in mountaintop-mined landscapes. *Glob. Biogeochem. Cycles* 32: 1182–1194.

Rusjan, S. and Mikoš, M. (2010). Seasonal variability of diurnal in-stream nitrate concentration oscillations under hydrologically stable conditions. *Biogeochemistry* 97: 123–140. https://doi.org/10.1007/s10533-009-9361-5.

Rutherford, J.E. and Hynes, H.B.N. (1987). Dissolved organic carbon in streams and groundwater. *Hydrobiologia* 154: 33–48. https://doi.org/10.1007/BF00026829.

Savoy, P., Appling, A.P., Heffernan, J.B. et al. (2019). Metabolic rhythms in flowing waters: an approach for classifying river productivity regimes. *Limnol. Oceanogr.* 64: 1835–1851. https://doi.org/10.1002/lno.11154.

Schade, J.D., Espeleta, J.F., Klausmeier, C.A. et al. (2005). A conceptual framework for ecosystem stoichiometry: balancing resource supply and demand. *Oikos* 109: 40–51.

Schmadel, N.M., Ward, A.S., Lowry, C.S., and Malzone, J.M. (2016). Hyporheic exchange controlled by dynamic hydrologic boundary conditions. *Geophys. Res. Lett.* 43: 4408–4417. https://doi.org/10.1002/2016GL068286.

Scott, D.T., Runkel, R.L., McKnight, D.M. et al. (2003). Transport and cycling of iron and hydrogen peroxide in a freshwater stream: influence of organic acids. *Water Resour. Res.* 39: 1308. https://doi.org/10.1029/2002WR001768.

Segatto, P.L., Battin, T.J., and Bertuzzo, E. (2021). The metabolic regimes at the scale of an entire stream network unveiled through sensor data and machine learning. *Ecosystems* 24: 1792–1809. https://doi.org/10.1007/s10021-021-00618-8.

Shangguan, Q., Lai, C.-Z., Beatty, C.M. et al. (2021). Autonomous in situ measurements of freshwater alkalinity. *Limnol. Oceanogr. Methods* 19: 51–66. https://doi.org/10.1002/lom3.10404.

Simonsen, J.F. and Harremoës, P. (1978). Oxygen and pH fluctuations in rivers. *Water Res.* 12: 477–489.

Soares, P.A., Faht, G., Pinheiro, A. et al. (2013). Determination of reaeration-rate coefficient by modified tracer gas technique. *Hydrol. Process.* 27: 2710–2720.

Spencer, R.G.M., Pellerin, B.A., Bergamaschi, B.A. et al. (2007). Diurnal variability in riverine dissolved organic matter composition determined byin situ optical measurement in the San Joaquin River (California, USA). *Hydrol. Process.* 21: 3181–3189. https://doi.org/10.1002/hyp.6887.

Spiro, B. and Pentecost, A. (1991). One day in the life of a stream—a diurnal inorganic carbon mass balance for a travertine-depositing stream (waterfall beck, Yorkshire). *Geomicrobiol. J.* 9: 1–11. https://doi.org/10.1080/01490459109385981.

Staehr, P.A., Bade, D., Van de Bogert, M.C. et al. (2010). Lake metabolism and the diel oxygen technique: state of the science. *Limnol. Oceanogr. Methods* 8: 628–644.

Staehr, P.A. and Sand-Jensen, K. (2007). Temporal dynamics and regulation of lake metabolism. *Limnol. Oceanogr.* 52: 108–120.

Sterner, R., Clasen, J.L., Lampert, W., and Weisse, T. (1998). Carbon: Phosphorus stoichiometry and food chain production. *Ecol. Lett.* 1: 146–150. https://doi.org/10.1046/j.1461-0248.1998.00030.x.

Tobias, C.R., Böhlke, J.K., and Harvey, J.W. (2007). The oxygen-18 isotope approach for measuring aquatic metabolism in high productivity waters. *Limnol. Oceanogr.* 52: 1439–1453. https://doi.org/10.4319/lo.2007.52.4.1439.

Triska, F.J., Kennedy, V.C., Avanzino, R.J. et al. (1989). Retention and transport of nutrients in a third-order stream in northwestern California: hyporheic processes. *Ecology* 70: 1893–1905. https://doi.org/10.2307/1938120.

Tromboni, F., Hotchkiss, E.R., Schechner, A.E. et al. (2022). High rates of daytime river metabolism are an underestimated component of carbon cycling. *Commun. Earth Environ.* 3: 1–10. https://doi.org/10.1038/s43247-022-00607-2.

Tsivoglou, E.C. and Wallace, J.R., 1972. *Characterization of stream reaeration capacity.* US Government Printing Office.

Ulseth, A.J., Hall, R.O., Jr, Boix Canadell, M. et al. (2019). Distinct air–water gas exchange regimes in low-and high-energy streams. *Nat. Geosci.* 12: 259–263.

Vachon, D., Sadro, S., Bogard, M.J. et al. (2020). Paired O2–CO2 measurements provide emergent insights into aquatic ecosystem function. *Limnol. Oceanogr. Lett.* 5: 287–294. https://doi.org/10.1002/lol2.10135.

Vannote, R.L., Minshall, G.W., Cummins, K.W. et al. (1980). The river continuum concept. *Can. J. Fish. Aquat. Sci.* 37: 130–137. https://doi.org/10.1139/f80-017.

Vaughan, M.C.H., Bowden, W.B., Shanley, J.B. et al. (2018). Using in situ UV-Visible spectrophotometer sensors to quantify riverine phosphorus partitioning and concentration at a high frequency. *Limnol. Oceanogr. Methods* 16: 840–855. https://doi.org/10.1002/lom3.10287.

Vesper, D.J. and Smilley, M.J. (2010). Attenuation and diel cycling of coal-mine drainage constituents in a passive treatment wetland: a case study from Lambert Run, West Virginia, USA. *Appl. Geochem.* 25: 795–808. https://doi.org/10.1016/j.apgeochem.2010.02.010.

Vincent, W.F. and Downes, M.T. (1980). Variation in nutrient removal from a stream by watercress (*Nasturtium officinale* R. Br.). *Aquat. Bot.* 9: 221–235. https://doi.org/10.1016/0304-3770(80)90024-8.

Volk, C.J., Volk, C.B., and Kaplan, L.A. (1997). Chemical composition of biodegradable dissolved organic matter in streamwater. *Limnol. Oceanogr.* 42: 39–44. https://doi.org/10.4319/lo.1997.42.1.0039.

Vrede, T., Dobberfuhl, D.R., Kooijman, S., and Elser, J.J. (2004). Fundamental connections among organism C: n: p stoichiometry, macromolecular composition, and growth. *Ecology* 85: 1217–1229.

Wanninkhof, R., Mulholland, P.J., and Elwood, J.W. (1990). Gas exchange rates for a first-order stream determined with deliberate and natural tracers. *Water Resour. Res.* 26: 1621–1630.

Ward, A.S., Packman, A., Bernal, S. et al. (2022). Advancing river corridor science beyond disciplinary boundaries with an inductive approach to catalyse hypothesis generation. *Hydrol. Process.* 36: e14540. https://doi.org/10.1002/hyp.14540.

Warwick, J.J. (1986). Diel variation of in-stream nitrification. *Water Res.* 20: 1325–1332.

Westhorpe, D.P., Mitrovic, S.M., and Woodward, K.B. (2012). Diel variation of dissolved organic carbon during large flow events in a lowland river. *Limnol. - Ecol. Manag. Inland Waters* 42: 220–226. https://doi.org/10.1016/j.limno.2011.12.003.

Wilson, H.F. and Xenopoulos, M.A. (2013). Diel changes of dissolved organic matter in streams of varying watershed land use. *River Res. Appl.* 29: 1330–1339. https://doi.org/10.1002/rra.2606.

Winkler, L.W. (1914). Über die Bestimmung des im Wasser gelösten Sauerstoffs. *Z. Für Anal. Chem.* 53: 665–672.

Withers, P.J.A. and Jarvie, H.P. (2008). Delivery and cycling of phosphorus in rivers: a review. *Sci. Total Environ.* 400: 379–395. https://doi.org/10.1016/j.scitotenv.2008.08.002.

Worrall, F., Howden, N.J.K., and Burt, T.P. (2015). Understanding the diurnal cycle in fluvial dissolved organic carbon – the interplay of in-stream residence time, day length and organic matter turnover. *J. Hydrol.* 523: 830–838. https://doi.org/10.1016/j.jhydrol.2015.01.075.

Wright, J.C. and Mills, I.K. (1967). Productivity studies on the Madison River, Yellowstone National Park 1. *Limnol. Oceanogr.* 12: 568–577.

Young, R.G. and Huryn, A.D. (1999). Effects of land use on stream metabolism and organic matter turnover. *Ecol. Appl.* 9: 1359–1376.

Yvon-Durocher, G., Jones, J.I., Trimmer, M. et al. (2010). Warming alters the metabolic balance of ecosystems. *Philos. Trans. R. Soc. B Biol. Sci.* 365: 2117–2126.

Zappa, C.J., McGillis, W.R., Raymond, P.A. et al. (2007). Environmental turbulent mixing controls on air-water gas exchange in marine and aquatic systems. *Geophys. Res. Lett.* 34.

Ziegler, S.E. and Fogel, M.L. (2003). Seasonal and diel relationships between the isotopic compositions of dissolved and particulate organic matter in freshwater ecosystems. *Biogeochemistry* 64: 25–52. https://doi.org/10.1023/A:1024989915550.

14

Evolving Molecular Methodologies for Monitoring Pathogenic Viruses in Ecohydrological Interfaces

Katarina Kovač[1], Mukundh N. Balasubramanian[1], Matjaž Hren[1], Ion Gutierrez Aguirre[2], and Valentina Turk[3]

[1] *BioSistemika LLC, Ljubljana, Slovenia*
[2] *Department of Biotechnology and Systems Biology, National Institute of Biology, Ljubljana, Slovenia*
[3] *National Institute of Biology, Marine Biology Station, Piran, Slovenia*

14.1 Introduction

Aquatic microbial ecology is a fast development field integrating "classical" microbial techniques with promising new technologies. The major questions in microbial ecology are which type of microorganisms are present and what is their impact on the environment, as well as how a wide range of natural and/or anthropogenic perturbations influence the microbial structure and function. Multiple studies in the last two decades have demonstrated the importance of viruses in marine waters (Fuhrman 1999; Suttle 2007; Wommack and Colwell 2000) and freshwater systems (see Peduzzi 2015 and references therein). Viruses are the most abundant and diverse biological entities, outnumbering any other living form. The abundance of viruses varies between 10^7 and 10^9/L and are frequently higher inland than in marine waters (Peduzzi and Luef 2008). Viruses are omnipresent in aquatic environments, including the most extreme environments, water, sediment, suspended particles (Luef et al. 2009), or different kinds of biofilms (Danovaro et al. 2008; Pinto et al. 2013).

Viruses exhibit various lifestyles, and are replicated by members of Bacteria, Eukarya and Archaea (Breitbart and Rohwer 2005; Edwards and Rohwer 2005). Recent studies in freshwater environments showed rich and diverse viral communities (Hewson et al. 2013; Mohiuddin and Schellhorn 2015; Roux et al. 2012). The majority of viruses identified are bacteriophages, and to a lesser extent animal, plant, and human viruses (Fancello et al. 2013; Tseng et al. 2013). Bacteriophages are viruses of heterotrophic bacteria, the most frequent host microorganisms of viruses in aquatic environments (Jacquet et al. 2010; Wommack and Colwell 2000). A significant relationship between viral numbers and phytoplankton biomass has been detected in lake ecosystems (Vrede et al. 2003), and involvement of viruses in the bloom collapse of marine phytoplankton species has been reported (Brussaard et al. 2005), or freshwater filamentous cyanobacterium and prochlorophyte (Gons et al. 2002). The role of viruses in regulation of the abundance of cyanobacterium *Synechococcus* sp. has also been examined, showing low mortality and stable virus-host coexistence in the

Ecohydrological Interfaces, First Edition. Edited by Stefan Krause, David M. Hannah, and Nancy B. Grimm.

environment (Waterbury and Valois 1993). Recently, genome sequences of viruses infecting eukaryotic algae (Jeanniard et al. 2013), or single-stranded DNA viruses involved in the seasonal dynamics of *Daphnia* spp. (Hewson et al. 2013) have also been analysed.

According to the high frequency of virus infection in aquatic environments, viruses are the major cause of bacteria, cyanobacteria, and phytoplankton mortality (Wilhelm and Suttle 1999). Empirical studies and models have estimated that 20–60% of the marine microbial community is infected each day (e.g. reviewed in Fuhrman 1999; Suttle 2007). Through their lytic activities, viruses influence the microbial community's composition (Fuhrman and Schwalbach 2003) and consequently the flow of carbon, nitrogen, and other nutrients in the aquatic environment (Brussaard et al. 2008). The latest results suggest that viral lysis of microbes change the quantity and quality of dissolved organic (Lønborg et al. 2013; Weitz and Wilhelm 2012) and inorganic pools (Shelford et al. 2012). The release of organic cellular contents and nutrients influence the growth and activity of other uninfected autotrophic and heterotrophic microbes (Lønborg et al. 2013; Odić et al. 2007). Ankrah et al. (2014) demonstrated that phage infections of *Sulfitobacter* sp. are physiologically distinct from their uninfected counterparts, since labile nutrients were being utilized by surviving cells. Recently, published results of multi-level models confirmed that aquatic ecosystems with high virus abundance have increased organic matter recycling, reduced transfer to higher trophic levels, and increased net primary production (Weitz et al. 2015). The released dissolved organic matter (DOM) is a major contributor to carbon pool and biogeochemical cycles (Weinbauer et al. 2010) within the microbial loop. Through bacterial respiration, dissolved organic carbon returns to the atmosphere as CO_2 (Pollard and Ducklow 2011). Viral lysis contributes significantly to a microbial loop and pool of rapidly cycling carbon, since on average 33.6% of globally respired carbon passes through a viral loop from fluvial systems (Peduzzi 2015).

Accompanying the rise of anthropogenic burden on water bodies is the concurrent increase in risk of infection by microbial water pathogens of several classes including bacteria, viruses, and protozoal parasites. These pathogens are primarily transmitted by the faecal-oral route impacting the safe use of water for drinking, aquaculture, irrigation, and recreational use. The most common pathogens and their risk profiles are listed in Table 14.1. Even though indigenous viruses outnumber human pathogenic viruses in aquatic environments, human viruses that are a major cause of water-related diseases are the primary focus of public health concern (Belkin and Colwell 2006). Pathogenic viruses are routinely introduced into the marine, freshwater, or groundwater matrices with the discharge of treated or untreated waste waters. Some pathogenic viruses that are normally detected in the human population might be introduced into the water and subsequently infect other humans via accidental contact or ingestion. Such viruses are also termed enteric viruses, a collective term describing viruses that are usually transmitted via the faecal-oral route and primarily infect and replicate in the gastrointestinal tract of the host (Carter 2005; Greening 2006). Humans are exposed to enteric viruses through various routes like contaminated drinking water, food crops (salad, fruits) grown in land irrigated with contaminated water or fertilized with sewage, or contaminated bivalves molluscs (oysters, clams, mussels). Major challenges to detect and characterize pathogenic viruses in water samples encompass viral diversity, occurrence of low virus particle numbers, and technical challenges of virus detection assays. Consequently, a critical first step in the application of detection approaches to virus ecology is obtaining a concentrate of viruses from an environmental sample.

Table 14.1 List of waterborne pathogens.

Pathogen[a]	Persistence[b]	Infectivity[c]	Associated disease[d]
Bacteria			
Campylobacter jejuni, *C. coli*	Moderate	Moderate	Diarrhoea, gastroenteritis
E. coli, particularly enterohemorrhagic *E. coli* (EHEC),	Moderate	High	Acute diarrhoea, bloody diarrhoea, and gastroenteritis
others like enteropathogenic (EPEC), enterotoxigenic			
(ETEC), and enteroinvasive (EIEC)			
Legionella spp.	May multiply	Moderate	Acute respiratory illness, pneumonia
Salmonella spp., *S. typhi*	Moderate	Low	Typhoid fever, paratyphoid fever, serious salmonellosis
Shigella spp.	Short	High	Dysentery
Vibrio cholerae	Short to Long	Low	Gastroenteritis, cholera
Yersinia enterocolitica	Long	Low	Diarrhoea
Viruses			
Rotavirus (RoV)	Long	High	Gastroenteritis
Norovirus (NoV)	Long	High	Gastroenteritis
Adenovirus (AdV)	Long	High	Gastroenteritis
Enterovirus (EV)	Long	High	Gastroenteritis
Astrovirus (AstV)	Long	High	Gastroenteritis
Sapovirus (SAV)	Long	High	Gastroenteritis
Hepatitis virus A (HAV) and E (HEV)	Long	High	Hepatitis
Protozoa			
Cryptosporidium parvum	Long	High	Cryptosporidiosis
Giardia intestinalis	Moderate	High	Diarrhoea
Toxoplasma gondii	Long	High	Toxoplasmosis
Acanthamoeba spp.	May multiply	High	Amoebic meningoencephalitis
Entamoeba histolytica	Moderate	High	Amoebic dysentery
Naegleria fowleri	May multiply	Moderate	Primary amoebic meningoencephalitis
Cyclospora cayetanensis	Long	High	Diarrhoea

a) Adapted from Table 7.1 in WHO Guidelines for drinking water quality (*WHO | Guidelines for drinking-water quality – Volume 1*, 2008).
b) Detection period for infective stage in water at 20 °C: short, up to 1 week; moderate, 1 week to 1 month; long, over 1 month.
c) From experiments with human volunteers, from epidemiological evidence, and from animal studies. High means infective doses can be $1–10^2$ organisms or particles, moderate $10^2–10^4$ and low $>10^4$.

This chapter presents an overview of the available methods that can be used to concentrate and detect viruses from water, including novel methods for fast detection of pathogenic viruses. The methods discussed are assessed according to their advantages, limitations, relevance, and applicability to pathogenic virus detection in water and ecosystem interfaces. In Section 14.2, we will highlight the isolation strategies for pathogenic viruses from various water environments. In Section 14.3, we summarize the evolution of currently used molecular detection approaches.

The main pathogenic enteric viruses belong to the families Adenoviridae (adenovirus), Astroviridae (astrovirus), Caliciviridae (norovirus, sapovirus), Hepeviridae (hepatitis E virus), Picornaviridae (hepatitis A virus, enteroviruses), and Reoviridae (rotavirus) (Greening 2006; Rodríguez-Lázaro et al. 2012). These viruses are generally non-enveloped and therefore very stable in the environment (Rzeżutka and Cook 2004). Their genome consists of either RNA (e.g. hepatitis A and E, norovirus, rotavirus) or DNA (e.g. adenovirus) and is protected by a protein capsid (Greening 2006). Most of these viruses are highly species-specific, infecting humans and other primates (Richards 2005). However, some of them, such as hepatitis E virus (HEV), can be zoonotic agents, meaning that they are naturally transmitted between vertebrate animals and humans (Greening 2006; Rodríguez-Lázaro et al. 2012).

Enteric viruses can enter the water environment by different ways; e.g. through leaking sewage or septic systems, urban or agricultural run-off and, in the case of estuarine and marine waters, sewage outfall and vessel waste water discharge (Fong and Lipp 2005). Human waste is processed in sewage treatment plants before discharged into surface water. However, the treatment procedures do not entirely remove enteric viruses from the water effluents leaving the treatment plants (Le Guyader et al. 2008; Victoria et al. 2010). There is also direct faecal contamination of the environment from humans and animals, for example, by bathers, or by defecation of free-range or wild animals onto soil or surface waters (Rodríguez-Lázaro et al. 2012). Drinking water is abstracted from surface water in many countries. This water is treated by sedimentation, filtration, and/or disinfection, which, if done effectively, can produce a virus-free end product, although this may be dependent on the quality of the source water (Steyer et al. 2011; Teunis et al. 2009). For the aforementioned reasons, enteric viruses were linked to several outbreaks originating from contaminated drinking water supplies (e.g. Breitenmoser et al. 2011; Dura et al. 2010) and recreational waters (e.g. water for swimming, canoeing, surfing) (e.g. Podewils et al. 2007; Sartorius et al. 2007).

Enteric viruses are inert particles and do not replicate in water because, like all other viruses, they are obligate pathogens which need live cells to replicate. Therefore, the number of viruses does not increase in water. However, enough viruses can survive to represent a high public health threat due to their low infectious dose (10–100 viruses or even less) and high level of resistance (e.g. too high or low temperatures, drying, acidity) (Greening 2006; Rzeżutka and Cook 2004). Three main types of diseases can be caused by enteric viruses: gastroenteritis, enterically transmitted hepatitis, and illnesses that can affect other parts of the body such as eye, respiratory system, and the central nervous system including conjunctivitis, poliomyelitis, meningitis, and encephalitis (Greening 2006). High concentrations of viruses are excreted in the faeces of infected individuals, i.e. up to 1013 virus particles per gram of stool (Bosch et al. 2008). As a consequence, raw sewage can contain high concentrations of these viruses (Lodder et al. 2010).

14.2 Concentration of Pathogenic Viruses from Environmental Water

As pathogenic viruses are usually present in water in very low numbers, the first step of virus detection in water is their recovery from the sample. Virus recovery from water matrices is based on concentration methods listed in Table 14.2 which include virus precipitation, ultracentrifugation, ultrafiltration, or one of the adsorption–elution methods (Fong and Lipp 2005). When environmental waters are concentrated, sometimes pre-filtration of the sample is needed to remove big organic particles which could later interrupt the concentration process (Gutiérrez-Aguirre et al. 2011; Lambertini et al. 2008). Another additional consideration is that different concentration techniques are sometimes combined synergistically considering the nature of the target pathogen and type of environmental water sample to be processed (Boxman 2010; Greening 2006).

Viruses are also present in filter feeding shellfish, especially when they are growing in water contaminated with sewage or faecal material (Boxman 2010; Greening 2006). Shellfish naturally concentrate viruses from polluted water, which can lead to concentrations of virus in shellfish that are 100–1000 times higher than that in the surrounding water (Carter 2005). Therefore, the recovery method applies different approaches in comparison to recovery directly from water. The most commonly used method to release viruses from the shellfish is proteinase K treatment (Table 14.2). The combination of proteinase K and heat treatment damages the virus capsid and releases virus nucleic acid (Nuanualsuwan and Cliver 2002). This approach using 65 °C heat treatment (Comelli et al. 2008; Jothikumar et al. 2005) has been used and is observed to be reliable for the detection of NoV in several food-borne outbreaks related to shellfish (reviewed in Stals et al. 2012) and has been selected by the CEN/TC275/WG6/TAG4 (European Committee for Standardization, Technical Committee 275 for Food analysis – Horizontal methods, Working Group 6 for Microbial contaminants, Technical Advisory Group 4 for Detection of viruses in food) working group for the extraction of the most common enteropathogenic viruses from shellfish digestive tissue (Lees and Tag 2010). Rarely are viruses from shellfish recovered by one of the virus concentration methods used for water matrices.

Another important consideration during virus concentration and subsequent nucleic acid extraction from heavily contaminated matrices such as raw sewage is minimizing the effect of inhibitory compounds (e.g. polysaccharides, proteins, fatty acids), which could interfere with molecular detection of the target pathogens. This is performed by either filtration through cheesecloth or 0.45 μm and 0.20 μm filters, or by treatment of virus suspension with Freon 13, Vertrel® XF, or chloroform:butanol (Stals et al. 2012).

14.2.1 Ultracentrifugation

Ultracentrifugation concentrates viruses in the sample using centrifugation forces from 100 000 × g to 235 000 × g during an adequate period of time (Casas et al. 2007; Hmaïed et al. 2015; Steyer et al. 2011; Sylvain et al. 2009). Due to difficulties in the processing of large volumes, ultracentrifugation is usually used as a secondary concentration method for water matrices (Fout et al. 2003; Hmaïed et al. 2015; Steyer et al. 2011; Sylvain et al. 2009), but was also efficiently applied as a direct method to concentrate viruses from waste

Table 14.2 Methods for concentration of viruses.

Concentration approach	Method/conditions	Target organism[1]	Sample	Selected references
Ultracentrifugation	Viruses are concentrated using centrifugation forces from 100 000 × g to 235 000 × g during an adequate period of time. Supernatant is removed and pellet resuspended in PBS	EVs, NoV, AdV, Bacteriophages X174 and MS2	Surface, ground, and drinking water	(Sylvain et al. 2009)
		NoV	Waste water	(Nordgren et al. 2009)
		RoV	Raw sewage	(Fumian et al. 2010)
		RoV, NoV, AstV	Surface and ground water	(Steyer et al. 2011)
		EVs	Raw and treated sewage	(Hmaïed et al. 2015)
		HAV	Shellfish[2]	(Casas et al. 2007)
Ultrafiltration	Ultrafiltration uses a size-exclusion-based mode of concentration where molecules smaller than the pore size of the filter (water and ions) pass through the membrane and out of the system while larger particles (microorganisms) are concentrated in the retentate	EVs, NoV, AdV, Bacteriophages X174 and MS2	Surface, ground, and drinking water	(Sylvain et al. 2009)
		EVs, bacteriophage MS2	Tap water	(Polaczyk et al. 2008)
		NoV, HAV	Estuarine water	(Hernandez-Morga et al. 2009)
		Rov	Raw and treated sewage, surface water and sea water	(Grassi et al. 2010)
		Bacteriophages MS2 and ΦX174	Reclaimed water samples	(Liu et al. 2012)
Adsorption–elution				
(a) Positively-charged filters	Viruses are attached to the filter and then eluted by elution buffer (beef extract)	AdV, EVs, bacteriophage MS2	Tap water	(Ikner et al. 2011)
		NoV, EVs	Tap and river water	(Karim et al. 2009)
		RoV, NoV, AstV	Surface and ground water	(Steyer et al. 2011)
		Nov, AdV, coliphage Qb	Sea water	(Gibbons et al. 2010)
		Bacteriophages	Surface, ground and drinking water	(Helmi et al. 2011)
		NoV	Tap water	(Haramoto et al. 2004)

(b) Negatively-charged filters	Viruses are attached to the filter and then eluted by the elution buffer (NaOH or beef extract). At ambient conditions enteric viruses adsorb to a negatively-charged membrane only under acidic conditions, when their net charge becomes positive, or in the presence of multivalent cations	NoV, EVs	River water and treated waste water	(Maunula et al. 2012)
		NoV, EVs	Sea water	(Katayama et al. 2002)
(c) Glass wool	Viruses are attached to the glass wool and then eluted from the glass wool by glycine/beef extract solutions	AdV	Treated drinking and river water	(van Heerden et al. 2005)
		NoV, AdV, EVs	Tap and ground water	(Lambertini et al. 2008)
		NoV, AdV	River water and treated waste water	(Maunula et al. 2012)
(d) Glass powder	Viruses are attached to borosilicate glass powder/beads and then eluted by glycine buffer containing beef extract.	EVs	Treated waste water	(Gantzer et al. 1997)
		EVs, AdV	Treated waste water	(Hugues et al. 1993)
		HAV	Tap, fresh, sea water and raw sewage	(Gajardo et al. 1991)
(e) Monolithic chromatographic supports	Viruses are attached to the ligands on the surface of the monolithic chromatographic supports allowing their use for anionic/cationic exchange, affinity, and hydrophobic interaction chromatography. The viruses are eluted by 1 M NaCl	RoV	Tap and river water	(Gutiérrez-Aguirre et al. 2009, 2011)
(f) Magnetic beads	Viral particles are attached to the magnetic beads, coated with antibodies or other molecules that allow viral binding and then bead–virus complexes are magnetically separated from the sample and virus particles eluted from the beads using an appropriate buffer	HAV	Tap water and sewage	(Jothikumar et al. 1998)
		NoV	Surface water and sewage	(Cannon and Vinjé 2008)
		NoV	Sewage	(Tian et al. 2012)
		EVs	Surface water	(Hwang et al. 2007)
		NoV	Oyster	(Tian et al. 2008)

(Continued)

Table 14.2 (Continued)

Concentration approach	Method/conditions	Target organism[1]	Sample	Selected references
Virus precipitation	Viruses are precipitated by polyethylene glycol (PEG), celite, or organic flocculation at acid or neutral pH. Then viruses are centrifuged or filtered and the virus pellet is resuspended in PBS or sodium phosphate	EVs, bacteriophage MS2	Tap water	(Rhodes et al. 2011)
		EVs, bacteriophage MS2	Tap water	(Polaczyk et al. 2008)
		AdV	Seawater	(Calgua et al. 2008)
		EV	Raw and treated sewage	(Hmaïed et al. 2015)
Proteinase K + heat treatment	The combination of proteinase K and heat treatment damages the virus capsid and releases virus nucleic acid	NoV	Shelfish	(Comelli et al. 2008; Jothikumar et al. 2005; Nuanualsuwan and Cliver 2002)

1) Abbreviations: PEG – polyethylene glycol; PBS – phosphate buffered saline; NoV – Norovirus; AdV – Adenovirus; HAV – Hepatitis A virus; RoV – Rotavirus; EV – Enterovirus.

2) Before ultracentrifugation, viruses are extracted from shellfish using alkaline elution with a glycine buffer. Then solids are removed by slow speed centrifugation and viruses are purified by chloroform extraction.

waters and recreational waters (Fumian et al. 2010; Nordgren et al. 2009; Prata et al. 2012). Ultracentrifugation has been used also to concentrate viruses from shellfish. However, prior to ultracentrifugation, viruses need to be extracted from the shellfish tissue using an alkaline elution with a glycine buffer, slow speed centrifugation to remove solids and purification by chloroform extraction (Casas et al. 2007).

14.2.2 Ultrafiltration

Ultrafiltration methods, such as vortex flow filtration or tangential flow filtration, use a size-exclusion-based mode of concentration where molecules smaller than the pore size of the filter (water and ions) pass through the membrane and out of the system while larger particles (microorganisms) are concentrated in the retentate (Fong and Lipp 2005; Grassi et al. 2010; Hernandez-Morga et al. 2009; Paul et al. 1991). Large volumes of water can be concentrated and a minimal manipulation of water is required as samples can be processed at natural pH and an elution step is not needed (Fong and Lipp 2005; Jiang et al. 2001). However, elution of the ultrafiltration membrane with a beef extract based buffer at high pH was shown to increase virus recovery (Sylvain et al. 2009). For water matrices, ultrafiltration may be used also as a secondary concentration step (Brassard et al. 2005; Gibson and Schwab 2011; Ikner et al. 2011; Villar et al. 2007). For this purpose, Centricon Plus-70 ultrafilters (30-kDa or 100-kDa cut-off; Millipore, Billerica, MA) are frequently used (Gibson and Schwab 2011; Hill et al. 2007; Ikner et al. 2011). Treatment of filters with bovine serum albumin (BSA) or sonication of the purified virus eluate can also increase virus recovery (Jones et al. 2009). The generic size-exclusion principle of ultrafiltration is particular amenable for simultaneous concentration of multiple classes of waterborne pathogens including viruses, bacteria, and protozoal parasites (Liu et al. 2012; Smith and Hill 2009).

14.2.3 Adsorption–Elution Methods

The virus adsorption–elution procedure has been the most commonly used method to concentrate viruses from water for decades. This approach effectively recovers viruses from water samples, but can be affected by water quality conditions such as pH, ionic strength, and organic content (Ikner et al. 2011; Lambertini et al. 2008; Lukasik et al. 2000). Viruses are concentrated from large volumes of water first passing the sample through microporous adsorptive media (e.g. filters, wool, and monolithic columns) to which the viruses adsorb on the principle of ionic charge. The second step is elution of viruses by buffer into a small volume (Bosch et al. 2008). Because of the fairly high final concentrate volumes (~15 to 40 ml) a secondary concentration step is normally needed. Recovered virus concentrates can still contain Polymerase Chain Reaction (PCR)-inhibiting substances (Ikner et al. 2011).

14.2.4 Positively-charged Filters

Positively-charged filters do not require any manipulation of pH as most enteric viruses are negatively charged at ambient pH (Lipp et al. 2001). However, electropositive filters are easily clogged and have low recovery rates for viruses in marine water because of the salt presence and alkalinity of seawater which cause low absorption of viruses to the filter

(Lukasik et al. 2000). The most commonly used positively-charged filter for the concentration of viruses from water is the Zeta Plus 1MDS filter (CUNO Inc., Meriden, CT) (Ikner et al. 2011). Good alternatives are cheaper NanoCeram filters (Argonide, Sanford, FL), which have been efficiently used for the concentration of various enteric viruses from tap and river water samples (Ikner et al. 2011; Karim et al. 2009).

Viruses that are more electronegative adsorb more strongly to positively-charged filter surfaces, which may subsequently impact the efficiency of their elution. Solutions consisting of various amino acids (e.g. glycine, which exhibits both acidic and basic properties due to the presence of carboxylic acid and amine functional groups and provides buffering) and complex proteinaceous solutions (e.g. beef extract) have been used to elute viruses from filters (Ikner et al. 2011; Steyer et al. 2011). The most commonly used elution solution is beef extract at pH 9.0–9.5 and concentration of 1.5–3.0% containing 0.05–0.1 M glycine (Gibbons et al. 2010; Ikner et al. 2011).

Viruses can be also captured to aluminium nanopore (Anodisc) filters (with pore size 0.02 μm) and stained with a nucleic acid binding fluorochrome stains like SYBR Green I or SYBR Gold (usually dsDNA). This enables subsequent virus visualization at high magnification by light emission of the fluorochrome stain (Noble and Fuhrman 2001). This method is mainly used in ecological studies but has been also applied, for example, for concentration of bacteriophages as bio-indicators to highlight faecal pollution (Helmi et al. 2011).

14.2.5 Negatively-charged Filters

Electronegative filters show higher virus recoveries from marine water and waters of high turbidity than electropositive filters (Katayama et al. 2002; Lukasik et al. 2000). As enteric viruses are negatively charged at ambient conditions they will adsorb to a negatively-charged membrane only under acidic conditions, when their net charge becomes positive, or in the presence of multivalent cations (e.g. Mg^{2+}) (Sobsey et al. 1973; Wallis and Melnick 1967). However, this electrical attractive interaction can affect the elution of viruses from the filter and results in poor recovery yield (Haramoto et al. 2004).

A modified virus concentration from seawater, where viruses are adsorbed to a negatively-charged HA membrane (Millipore, Billerica, MA) and then cations and other inhibitors are removed by acid rinse step, while viruses remain attached, was developed by Katayama and colleagues (Katayama et al. 2002). For elution of viruses from the membrane an organic elution medium was used (NaOH) which has less inhibitory effects in PCR assays than the commonly used organic elution medium (beef extract) (Haramoto et al. 2004; Katayama et al. 2002). Haramoto and colleagues improved the method for recovery of viruses from freshwater by precoating a type HA membrane with AlCl3 prior to filtering (Haramoto et al. 2004).

14.2.6 Glass Wool

Sodocalcic glass wool is a promising alternative as an adsorptive material for virus concentration. Glass wool, held together by a binding agent and coated with mineral oil, presents both hydrophobic and electropositive sites on its surface. While a virus suspension flows through the pore space of the column packed material, the fibre surface is able to attract and retain negatively-charged virus particles at near neutral pH without the addition of cations (Lambertini et al. 2008; Wyn-Jones and Sellwood 2001). Viruses are normally eluted from the glass wool by glycine/beef extract solutions (Gassilloud et al. 2003; Lambertini et al. 2008; van Heerden et al. 2005). Passing water samples through glass wool seems to diminish PCR inhibition (Lambertini et al. 2008; van Heerden et al. 2005). Glass wool recovery efficiency depends on the type of virus, water pH, and water matrix (Lambertini et al. 2008).

14.2.7 Glass Powder

Borosilicate glass beads of 100 ± 200 μm diameter form a good adsorbent support for viruses under conditions similar to those used for glassfibre filters. As they form a fluidized bed, the filter matrix cannot become clogged as with glassfibre systems (Wyn-Jones and Sellwood 2001). However, in common with the performance for glassfibre tube filters, the recovery varies widely with the type of sample (Joret et al. 1980).

14.2.8 Monolithic Chromatographic Supports

CIM® (Convective Interaction Media) is a monolithic chromatographic support made from a single block of porous material, with highly interconnected channels that enable very rapid transfer, based on convective flow, of the sample molecules between the mobile and the stationary phase (Strancar et al. 2002). The methacrylate monoliths can bear different ligands on the surface allowing their use for anionic/cationic exchange, affinity, and hydrophobic interaction chromatography (Barut et al. 2005). CIM QA (positively-charged quaternary amine) anion-exchange monolithic supports were highly efficient in binding RVs from stool samples, tap and river water samples (Gutiérrez-Aguirre et al. 2009). The method was also successfully applied for the concentration and detection of RoVs on-site (Gutiérrez-Aguirre et al. 2011) and RoV, NoV, SAV, AstV, HAV, and HEV from waste water samples (Steyer et al. 2015). CIM butyl (C4) hydrophobic interaction monolithic supports effectively concentrated NoV and RoV from water matrices with high salinity (seawater). The protocol was also successfully deployed in an on-site application (Balasubramanian et al. 2016).

14.2.9 Magnetic Beads

Viral particles can be concentrated and purified using magnetic beads, coated with antibodies or other molecules that allow viral binding, following which bead–virus complexes are magnetically separated from the sample and virus particles eluted from the beads using an appropriate buffer (Mattison and Bidawid 2009; Stals et al. 2012; Tian et al. 2008). Cationic separation is based on the binding of the virus capsid proteins (negatively charged) to the magnetic particles (positively charged) (Stals et al. 2012). The immunoconcentration method uses magnetic beads covered with antibodies, e.g. HBGAs, porcine gastric mucin, or rabbit anti-NoV, or anti-HAV polyclonal antibodies, to concentrate NoV or HAV, respectively (Cannon and Vinjé 2008; Jothikumar et al. 1998; Tian et al. 2008, 2012), or mouse anti-enterovirus monoclonal antibody to concentrate EV (Hwang et al. 2007). Immunomagnetic capture enables very specific extraction of viruses and efficient removal of PCR inhibitors. However, long-term use of antibodies due to immunogenetic drift could be problematic (Stals et al. 2012).

14.2.10 Virus Precipitation

Virus precipitation methods are frequently used as a secondary concentration step for water matrices (Polaczyk et al. 2008; van Heerden et al. 2005). Polyethylene glycol (PEG) is a well-validated technique which has been used with different viruses and a variety of environmental samples (Hmaïed et al. 2015; Mattison and Bidawid 2009) and shellfish (Baert et al. 2007). It easily allows the precipitation of viruses at neutral pH (7.0–7.4) and high ionic concentrations (~0.5 M NaCl) without precipitation of other organic material. After centrifugation of the sample, the virus pellet is resuspended in phosphate buffered saline (PBS) (Polaczyk et al. 2008). Organic flocculation using beef extract precipitates viruses by reducing the pH to 3.5. The virus pellet formed during centrifugation is dissolved in sodium phosphate or PBS (pH 9.0). The dissolved pellet is centrifuged again and the pH of the collected supernatant is adjusted to 7.0–7.5 (Rhodes et al. 2011). Celite precipitates viruses by reducing pH to 4. After stirring, the mixture is collected on the pre-filters by vacuum filtration. The adsorbed viruses are eluted by sodium phosphate or PBS (pH 9.0–9.5) resulting in a concentrate with a pH range of 7.0–7.5 (Karim et al. 2009; Rhodes et al. 2011). Viruses can be also concentrated by adsorption to pre-flocculated skimmed milk proteins. The flocs then sediment by gravity and the separated sediment is dissolved in PBS (Calgua et al. 2008).

14.3 Detection of Pathogenic Viruses from Environmental Water

Virus detection in water bodies can be broadly split into two categories on the basis of the intended application: enumeration and characterization of viral diversity for ecological information; or monitoring of waterborne viral pathogens to assess the potential risk of disease. For microbial detection to ascertain the risk posed by waterborne pathogens, various approaches detailed in Table 14.3 have been used, ranging from culture-dependent methods, to indirect diagnostic signals using immunological or molecular methods that are rapidly evolving in scope and sophistication (Rodríguez-Lázaro et al. 2012). As discussed in Section 14.2, detection of pathogens, especially viruses, is particularly challenging in water bodies due to their low concentrations, high infectivity, and presence of substances that interfere with detection techniques. As a result, every technique has a few limitations for effective detection of pathogens in water which will be briefly discussed.

14.3.1 Electron Microscopy

Transmission electron microscopy (TEM) is a widely used method to screen faecal specimens for enteric viruses in the public health laboratories of many countries. However, this method is subjective, laborious, painstaking, and time-consuming with a limited sensitivity (Atmar and Estes 2001). Therefore, it has been largely dismissed for food and water samples (Mattison and Bidawid 2009). Easier visualization of virus particles is enabled by immune electron microscopy (IEM), which uses antibodies that attach to the viruses which causes formation of aggregates (Atmar and Estes 2001).

Table 14.3 Methods for detection of pathogenic viruses.

Detection method	Target qrganism	Sample	Reference(s)
Media-based culture methods	Male specific somatic coliphages	Groundwater, recreational fresh or marine water	(USEPA 2001a, 2001b)
Enzyme linked immunosorbent assay (ELISA)	RoV	Waste water	(Guttman-Bass et al. 1987)
Polymerase chain reaction (PCR)	RoV, EV, HAV	Sewage and ocean water	(Tsai et al. 1994)
	NoV	Faecal contaminated surface water	(Schwab et al. 1996)
	AstV	Spiked tap water	(Abad et al. 1997)
	AdV	Sewage and polluted river water	(Puig et al. 1994)
Quantitative PCR (qPCR)	RoV	Stream and tap water	(Gutiérrez-Aguirre et al. 2008)
	NoV	Waste water effluent	(Haramoto et al. 2006; Hellmér et al. 2014; Kageyama et al. 2003)
	AstV	Spiked water samples	(Grimm et al. 2004)
	AdV	Spiked sterile water, creek water, brackish estuarine water, ocean water, and secondary sewage effluent	(Jiang et al. 2005)
Droplet digital PCR	RoV	Waste water treatment plant effluent, river and stream water	(Rački et al. 2013)
Loop mediated isothermal amplification (LAMP)	NoV	Municipal waste water	(Suzuki et al. 2011)
	AstV	Reclaimed water samples	(B.-Y. Yang et al. 2014a)
	RoV	Stool samples	(Malik et al. 2013)
Nucleic acid sequence based amplification (NASBA)	RoV	Spiked water and sewage effluent	(Jean et al. 2002)
	NoV	River water	(Rutjes et al. 2006)
	RoV, NoV, AdV, AstV	Stool samples	(Mo et al. 2015)
Microarray	NoV, RoV, AdV, AstV	Stool samples	
Next generation sequencing (NGS)	HAdVs (B, C and F), EV, PV (JC and BK), HPV	Untreated sewage samples	(Aw et al. 2014)

14.3.2 Cell Culture Methods

Traditionally, coliform bacteria and male-specific coliphages have been enumerated using culture-based methods as proxies for faecal and sewage contamination of water bodies. Most regulatory agencies continue to rely on these well characterized culture-based protocols for microbial risk assessment. Detection of viruses is also possible by adding the samples that may contain viruses to mammalian cell cultures. Specifically, viruses cause the formation of cytopathic effects (CPE), which can be visually observed as damaged cells, or rounding of cells, and sloughing of the monolayer. Using cell culture viruses can be quantified by plaque assays, the most probable number or tissue culture infectious dose 50 (TCID50) assay (Bosch et al. 2008; Fong and Lipp 2005). However, the detection of infectious viruses in food and water samples is laborious and time-consuming as well as difficult since it greatly depends on the assay conditions, i.e. duration of exposure to host cells, volume of inoculum, age of the cells, and the presence of inhibitory or toxic substances in the sample (Rodríguez et al. 2009). It requires days to weeks of incubation and several cell culture passages to confirm both positive and negative results. Also, a universal cell line that can be used for culturing all enteric viruses has not been established so far, and there are still some viruses, such as NoV and HEV, that cannot be detected as they either do not produce CPE, are extremely slow growing, or do not grow on established cell lines (Remick et al. 1990).

14.3.3 Immunological Tests

Several immunological tests like enzyme immunoassay (EIA), radioimmunoassay (RIA) or enzyme-linked immunosorbent assay (ELISA) exist and many are commercially available for the main enteric viruses (Dimitriadis et al. 2006; Wilhelmi de Cal et al. 2007). Even though immunological tests are widely used in the clinical setting, their analytical sensitivity is still too poor for effective testing of environmental water samples (Rodríguez-Lázaro et al. 2012).

14.3.4 Molecular Methods

Molecular methods have been the most commonly applied techniques for the detection of viruses in food and water samples as they have the highest degree of sensitivity and retain good specificity. For molecular detection of viruses, genomic nucleic acids (DNA or RNA) must be first extracted from the viral concentrates using in-house or commercial procedures (Mattison and Bidawid 2009). In the next step, the virus-specific DNA or RNA target is detected using various approaches described in the next section.

For extraction of viral genomic nucleic acids a wide variety of commercial kits have been applied, mostly based on a guanidinium thiocyanate lysis of viruses followed by silica-based capture of nucleic acids (Boom et al. 1990; Haramoto et al. 2005; Le Guyader et al. 2009; Rutjes et al. 2005). These kits offer reliability, reproducibility, and are easy to use. Other methods for viral nucleic acid extraction include proteinase K treatment followed by phenol-chloroform extraction and ethanol precipitation, sonication and heat treatment (Comelli et al. 2008; Guévremont et al. 2006; Jothikumar et al. 2005; Le Guyader et al. 2009). In the last few years, automated nucleic acid extraction platforms have been developed by commercial companies and have been efficiently applied for the analysis of viruses in food and water samples (Comelli et al. 2008; Perelle et al. 2009; Stals et al. 2011).

14.3.5 Conventional and Real-time PCR

Conventional PCR was the first molecular technique used for the detection of viral genomes. The main principle of detection with PCR is to first amplify a virus-specific section of viral genome (DNA or RNA). This process uses a pair of short oligonucleotides (primers) that direct the PCR polymerase to amplify the target sequence of the viral genome. For RNA viruses, reverse transcription (RT) of the viral RNA to a complementary DNA (cDNA) strand is necessary prior to the PCR, as PCR is able to amplify only DNA. PCR is normally used to detect highly conserved regions of the viral genome within multiple members of a particular viral family using specific sets of primers (Jothikumar et al. 1993; Kojima et al. 2002). After PCR is complete, PCR products (amplified virus-specific sections of viral genome) are typically visualized by agarose gel electrophoresis. Results can additionally be confirmed by hybridization of the PCR product using an internal oligonucleotide probe or by sequencing the DNA product (Ando et al. 1995; Deng et al. 1994; Jothikumar et al. 1993; Ma et al. 1994). The sensitivity and specificity can be improved using semi-nested or nested PCR, where a second, internal primer, or primer set are used. This is sometimes used as a conformation step (Allard et al. 1992; Van Heerden et al. 2003).

Real-time PCR or quantitative PCR (qPCR) can be considered as PCR's "next generation", as it shares the basic principle of amplification of DNA/RNA targets. It is one of the most widely used molecular biology techniques due to its robustness, speed, miniaturization of reactions, and excellent possibility of automation (high throughput). It has been around for almost 20 years and is therefore well established in scientific research, industrial development, and has become an indispensable tool for service companies, quality control applications, diagnostics, and other areas of life sciences.

qPCR combines primer amplification with detection of the amplified product in the single reaction mix in real time during the reaction (hence the name real-time PCR). A signal is detected using intercalating fluorescent dyes that bind to the amplified PCR products, by using fluorescently-labelled primers or by using an oligonucleotide probe in addition to the primer set. The latter approach allows for very specific detection of one or more targets in the same reaction. In addition to simply confirming the presence or absence of a target in a sample, qPCR also allows the quantification of the amount of the target template initially present in the sample, which is one of its greatest properties (Gibson et al. 1996; Heid et al. 1996; Mattison and Bidawid 2009).

As is the case for many enzymatic reactions, PCR is also prone to inhibition. Unfortunately, complex samples such as food or environmental samples frequently contain inhibitory compounds that can, in extreme situations, lead to false negative results. We can minimize this effect by using low sample volumes or dilution of samples (Rodríguez et al. 2009). The positive side of qPCR is that we can detect when we have inhibitors present and can be cautious about that.

Similarly to PCR, qPCR is able to amplify only DNA, therefore if our target is RNA (e.g. a virus that has the RNA genome), we first have to perform reverse transcription (RT) of the RNA to a complementary DNA (cDNA). In this case the process is caller RT-qPCR (or qRT-PCR). Viral genomes frequently lack universally conserved sequences (Rohwer and Edwards 2002), which presents a big challenge for their detection with PCR and qPCR. For these reasons, sequencing of viral genomes is becoming an increasingly important approach as described in the following section.

14.3.6 Digital PCR

Digital PCR (dPCR) is a variant of PCR where a single bulk PCR reaction (μl scale) is partitioned into thousands or even millions of reactions of nanolitre or picolitre scale (Huggett et al. 2013). The partitioning is enabled by either fabricated microfluidic chips housing the multiple partitions, which is known as chamber dPCR, or through generation of multitudes of water-in-oil droplets (droplet dPCR, ddPCR). The partitioning of the sample strives to assure that each individual partition contains only one or no target molecules. Each of the partitions (chambers or droplets) is then subjected to PCR reaction. The quantification in dPCR is obtained by counting partitions housing amplified PCR products (positive) and those where no amplification products are detected (negative) with the application of the Poisson distribution. Ideally, each positive partition would represent an individual PCR reaction. Since some partitions may house multiple amplified products, a Poisson correction coefficient is applied to the fraction of the positive partitions for compensation (Huggett et al. 2013). The principle of detection is also fluorescence.

There are many advantages of dPCR over traditional qPCR including precise quantification of nucleic acids (targets are literally counted), non-reliance on a standard curve for quantification, high reproducibility, and lower susceptibility to environmental PCR inhibitors (due to very small-volume reactions). The last characteristic of dPCR makes it particularly attractive for waterborne pathogen monitoring applications. Indeed, the superiority of dPCR over qPCR as a detection tool has been observed for enteric viruses (Rački et al. 2013), enterococcus bacteria (Bian et al. 2015), and protozoa (Yang et al. 2014a). The trends of decreased reagent costs as well as wider availability of equipment indicates higher utilization of dPCR as a detection tool for water quality monitoring (Cao et al. 2015).

14.3.7 Conventional and Next Generation Sequencing

In the section describing PCR, we mentioned that positive PCR results can be confirmed by conventional (Sanger) sequencing of the PCR products (Arraj et al. 2008). Slightly more than a decade ago the next generation of sequencing started to be established, conveniently referred to as Next Generation Sequencing (NGS) or Massive parallel sequencing. The information NGS provides is the same as in conventional (Sanger) sequencing – a sequence of base pairs of DNA. Why has NGS become so popular? NGS offers sequencing of a vast number of short DNA fragments in a single sequencing reaction producing base pair reads that can encompass the entire human genome in as little as a few days. NGS is an expression that combines several different technological variants/platforms, all of which share the same goal: to obtain large numbers of nucleic acid sequences in a short time (Dijk et al. 2014; Loman et al. 2012) .

To address the limitation of target-based detection methods like PCR, qPCR, and isothermal methods, which require that sequences of at least parts of viral genomes are known in advance, viral fractions can be analysed using NGS sequencing. Since these samples frequently represent environmental concentrates and contain an entire virome, this approach is also called viral metagenomics and provides an exceptional insight into viral diversity (Aw et al. 2014; Breitbart et al. 2002; Hurwitz and Sullivan 2013; Ogorzaly et al. 2015; Tan et al. 2015). Traditionally, for virome studies, the particles that pass through the 0.2 μm and m pore-size filters (Rohwer and Thurber 2009) are considered as viruses. Therefore, filtration is an important step as it excludes most cellular fractions from viruses.

It is also worth pointing out some of the downsides of NGS: most platforms are very costly but there are a lot of service providers available; processing and interpretation of NGS results relies heavily on bioinformatics and therefore requires specially trained personnel.

14.3.8 Isothermal Methods (NASBA and LAMP)

Isothermal techniques have a basic approach very similar to PCR: to amplify the target nucleic acid to the level that we are able to detect it. However, instead of using two or three different temperatures that drive the PCR reaction, isothermal reactions require a single constant temperature, hence the name (Yan et al. 2014; Zanoli and Spoto 2012). This does represent a significant difference to PCR and means that the instruments can be much less complicated (and cheaper) as they only need to hold one temperature. Some of them require different enzymes than DNA polymerase or even several enzymes. This is one of the reasons why isothermal techniques such as NASBA and LAMP arc increasingly popular.

Nucleic acid sequence based amplification (NASBA) and loop-mediated isothermal amplification (LAMP) are alternatives to PCR (Houde et al. 2006; Lamhoujeb et al. 2008; Rutjes et al. 2006). Two primers, an avian myeloblastosis virus reverse transcriptase and a T7 bacteriophage RNA polymerase, can be used to produce multiple copies of an RNA sequence in an isothermal reaction of the NASBA method. Molecular beacon probes are used to detect and confirm the amplification of the targeted viral gene segment during the real-time NASBA reactions (Lamhoujeb et al. 2008). At least four primers are used for the LAMP method, two of which are the loop primers that recognize two regions each in the target genetic sequence. The target sequence is amplified using a strand displacing DNA-dependent DNA polymerase and detection is accomplished by measuring an increase in turbidity or the binding of a fluorescent detection reagent; this can be monitored in real time (Fukuda et al. 2007; Iturriza-Gómara et al. 2008; Yoneyama et al. 2007). The specificity of the reaction in the LAMP system can be increased using additional primers (Mattison and Bidawid 2009).

14.3.9 On-site Approaches: Lateral Flow and Point-of-care Devices

Research that relies heavily on processing environmental samples has always been interested in on-site approaches that would give results fast. Not only in case of routine analysis, such as official environmental monitoring, but increasingly also in research. In some cases, samples deteriorate rapidly or the transport is very expensive, therefore the laboratory is not even an option, especially if it is located far away. As a rule of thumb, speed of getting results means a compromise on something else. When it comes to on-site approaches the compromise lies either in the number or volume of samples that can be processed in a unit of time, in reduced sensitivity, reduced scope of measurements or targets, and sometimes reduction of quantitative output to semi-quantitative or completely qualitative output. Why then, are on-site approaches useful? Sometimes analysing samples in the lab is very expensive. Therefore, any reduction in the number of samples is welcome. On-site approaches become useful in such cases as we are able to reduce the number of samples based on actual experimental data and not based on experience or best guesses. Instruments to measure physical and chemical properties of water samples are readily available. In this field also sensors for long-term measurements are available. In the last decade, the availability of such approaches to detect microorganisms have become available. Some are discussed here.

Since the advent of qPCR technology, the goal of developing detection devices for portable deployment has been widely pursued (Almassian et al. 2013). These portable solutions based on qPCR still rely on inherent complex thermal cycling which restricts the amount of miniaturization possible in practice and the requirement of highly skilled operators restricts their use in the field by non-specialists. In contrast, portable devices based on isothermal amplification of nucleic acids appear to be more amenable to in situ deployment and indeed a number of such devices have been commercially developed for use in clinical and environmental settings (Craw and Balachandran 2012).

An additional layer of portability is added by the use of lateral flow devices for end point detection of waterborne pathogens in simple formats, as have been widely adopted for testing biochemical parameters like blood glucose levels or protein makers for confirming pregnancy (St John and Price 2014). Such paper-based microfluidic devices made of simple, lightweight components are fabricated from materials like cotton, glass, nitrocellulose, polystyrene, polyesters, and plastics to produce a diagnostic device in disposable and portable format like dipsticks (Yetisen et al. 2013). One such device has been recently developed for detection of norovirus particles using phage nanoparticle reporters albeit with a lower sensitivity of detection than techniques like qPCR (Hagström et al. 2015). This remains true of most other such point-of-care devices as their intended application is for clinical or food samples, where a lower sensitivity of detection may still be acceptable due to higher pathogen load. In contrast, to achieve a lower level of detection required for environmental samples, some degree of target amplification prior to end point detection becomes essential (Craw and Balachandran 2012).

14.3.10 Optical Sensor Methods

Moving beyond end-point fluorescence-based detection of qPCR and ELISA are optical biosensors. Surface Plasmon Resonance (SPR) relies on detection based on refractive index changes in biomacromolecules immobilized in the vicinity of a thin-film metal surface and can be a sensitive pathogen detection technique, but its use is limited by high costs and complexity for adoption in portable applications (Lazcka et al. 2007). Pineda et al. (2009) developed a photonic crystal biosensor that quantifies the optical resonant reflection of virus particles as a peak wavelength value. The authors successfully combined immobilization using anti-rotavirus antibodies with an optical biosensor in a microplate format to detect porcine rotavirus from partially purified faecal suspensions with a sensitivity similar to commercially available ELISA kits. Though optical sensors like these have yet to be evaluated in environmental samples as extensively as the other molecular methods, their advantage of rapid detection and ease of operation does provide potentially useful characteristics for pathogen detection in water samples (Altintas et al. 2015).

14.3.11 Molecular-based Methods to Determine Viral Infectivity

Even though molecular methods provide rapid results with high specificity and sensitivity, they do not provide any information about virus infectivity. Therefore, several strategies have been developed, as highlighted in Table 14.4, to adapt molecular methods to quantify infective virus particles which would greatly enhance application for the monitoring of water and food quality and for treatment processes (Rodríguez et al. 2009).

Table 14.4 Molecular-based methods to determine viral infectivity.

Method	Principle of assessing loss of viral infectivity[1]	Reference(s)
Molecular		
5′ NTR RT-PCR	Targeting the 5′ NTR of the viral RNA which is more susceptible to degradation	(Bhattacharya et al. 2004; Li et al. 2002, 2004; Simonet and Gantzer 2006a)
LTR (q)RT-PCR	Analysing the LTR of the viral genome as a correlation between the length of the region amplified and sensitivity of the qRT-PCR to detect damage in genome has been found	(Li et al. 2002; Pecson et al. 2011; Simonet and Gantzer 2006a; Simonet and Gantzer 2006a)
Enzymatic + molecular		
Treatment with RNase + RT–(q) PCR	RNase degrades RNA of viruses with damaged viral capsid and causes loss of PCR signal	(Topping et al. 2009)
Treatment with proteinase and RNase + RT-(q) PCR/qNASBA	Proteinase degrades damaged capsid and then RNase degrades RNA of viruses with damaged capsid and causes loss of PCR signal	(Baert et al. 2008; Diez-Valcarce et al. 2011; Lamhoujeb et al. 2008; Nuanualsuwan and Cliver 2002, Nuanualsuwan and Cliver 2003)
Immunological + molecular		
Antibody capture of virus + RT–q(PCR)	Damage in the viral capsid may change the antigenic properties of the virus, and specific viral antigen-antibody complexes may not form resulting in a loss of PCR signal	(Abd El Galil et al. 2004; Gilpatrick et al. 2000; Myrmel et al. 2000; Nuanualsuwan and Cliver 2003; Schwab et al. 1996)
Cell culture + molecular		
Virus attachment to cell monolayer + RT–PCR	Damage in the viral capsid may not allow attachment to the cell monolayer resulting in a loss of PCR signal	(Nuanualsuwan and Cliver 2003)
Virus replication in cell culture + (RT)–(q)PCR	Damage in the viral capsid may not allow attachment and subsequent replication of the virus in cell culture resulting in a loss of PCR signal	(Blackmer et al. 2000; Chapron et al. 2000; Jiang et al. 2004; Ko et al. 2003, Ko et al. 2005; Lee and Jeong 2004; Lee and Kim 2002; Li et al. 2009; Nuanualsuwan and Cliver 2003; Reynolds 2004; Reynolds et al. 1996; Shieh et al. 2008)

1) NTR – non-translated region; LTR – long terminal region; RT-PCR – reverse transcriptase PCR; RT-qPCR – reverse transcriptase real-time PCR; qNASBA – real-time nucleic acid sequence based amplification.

Two different approaches have been used. In the first approach, the presence of an intact genome or amplifiable undamaged genome is determined by direct molecular detection. Using the second approach, the integrity of the viral capsid is determined by coupling either an enzymatic, immunological, or cell culture step with subsequent molecular detection (Table 14.4). However, it is unclear whether any of these molecular-based approaches can satisfactorily assess viral infectivity.

14.4 Conclusion

The prevalence of pathogenic viruses resulting from the release of faecal waste from humans and domesticated animals poses a severe burden on water bodies, including coastal water used for recreation and aquaculture, and ground or surface sources of potable water. Globally, as population density averages continue to rise, so does the potential of short-circuits between waste water and drinking water. Additionally, viral pathogens may be present in extremely high concentrations in waste waters (often >108/100 ml), while very low levels (sometimes <1/100 ml) are known to cause disease. Despite considerations of viral pathogens being subjected to inactivation measures and high dilutions in environmental water bodies, the analysis of the persistence, survival, and transport of these microbial pathogens in water and water interfaces is crucial to track the population of these pathogens and to limit their detrimental impact.

These concerns have been driving the constant evolution of techniques for microbial pathogen detection in water bodies ranging from detecting specific pathogens to broader studies of entire microbial communities. While culture-based techniques have the advantage of being the most well characterized reflecting their continued widespread adoption by environmental regulatory agencies across the world, newer molecular techniques, especially qPCR, are being accepted as well. Over the past decade, innovation in NGS technologies has enabled innovative insights into entire microbial communities in various ecological niches. In parallel, advances in nucleic acid based detection technologies such as dPCR, LAMP, NASBA, and other on-site detection techniques offer viable and alternatives to qPCR.

Many bottlenecks continue to impede the adoption of the innovative molecular detection techniques by regulatory agencies. Higher sensitivity of the molecular techniques is often negated by their susceptibility to the presence of inhibitor substances commonly encountered in water samples. Constant application-specific innovation in sample preparation and processing needs to be prioritized for successful deployment of the molecular detection methods in routine waterborne pathogen monitoring. New advances in molecular techniques such as dPCR and NGS, although promising, require complex instrumentation and well-equipped laboratories with highly-trained personnel, which is not practical, especially in developing countries and remote locations.

In the case of NGS, the fact that the cost of sequencing per sample is decreasing rapidly is a good sign that this technique will most probably grow in its use and application and potentially replace other molecular techniques. Another exciting approach is the development of portable solutions such as portable PCR/qPCR/LAMP instruments in combination with refrigeration-free reagents including sample processing reagents. These will empower ecologists by increasing the quantity and quality of results obtained on-site without relying on remote laboratories to obtain valuable information that may improve further sampling.

References

Abad, F.X., Pintó, R.M., Villena, C. et al. (1997). Astrovirus survival in drinking water. *Appl. Environ. Microbiol.* 63: 3119–3122.

Abd El Galil, K.H., El Sokkary, M.A., Kheira, S.M. et al. (2004). Combined immunomagnetic separation-molecular beacon-reverse transcription-PCR assay for detection of hepatitis a virus from environmental samples. *Appl. Environ. Microbiol.* 70: 4371–4374. https://doi.org/10.1128/AEM.70.7.4371-4374.2004.

Allard, A., Albinsson, B., and Wadell, G. (1992). Detection of adenoviruses in stools from healthy persons and patients with diarrhea by two-step polymerase chain reaction. *J. Med. Virol.* 37: 149–157.

Almassian, D.R., Cockrell, L.M., and Nelson, W.M. (2013). Portable nucleic acid thermocyclers. *Chem. Soc. Rev.* 42: 8769–8798. https://doi.org/10.1039/C3CS60144G.

Altintas, Z., Gittens, M., Pocock, J., and Tothill, I.E. (2015). Biosensors for waterborne viruses: detection and removal. *Biochimie* 115: 144–154. https://doi.org/10.1016/j.biochi.2015.05.010.

Ando, T., Monroe, S.S., Gentsch, J.R. et al. (1995). Detection and differentiation of antigenically distinct small round-structured viruses (Norwalk-like viruses) by reverse transcription-PCR and southern hybridization. *J. Clin. Microbiol.* 33: 64–71.

Ankrah, N.Y.D., May, A.L., Middleton, J.L. et al. (2014). Phage infection of an environmentally relevant marine bacterium alters host metabolism and lysate composition. *ISME J.* 8: 1089–1100. https://doi.org/10.1038/ismej.2013.216.

Arraj, A., Bohatier, J., Aumeran, C. et al. (2008). An epidemiological study of enteric viruses in sewage with molecular characterization by RT-PCR and sequence analysis. *J. Water Health* 6: 351–358. https://doi.org/10.2166/wh.2008.053.

Atmar, R.L. and Estes, M.K. (2001). Diagnosis of noncultivatable gastroenteritis viruses, the human caliciviruses. *Clin. Microbiol. Rev.* 14: 15–37. https://doi.org/10.1128/CMR.14.1.15-37.2001.

Aw, T.G., Howe, A., and Rose, J.B. (2014). Metagenomic approaches for direct and cell culture evaluation of the virological quality of wastewater. *J. Virol. Methods* 210: 15–21. https://doi.org/10.1016/j.jviromet.2014.09.017.

Baert, L., Uyttendaele, M., and Debevere, J. (2007). Evaluation of two viral extraction methods for the detection of human noroviruses in shellfish with conventional and real-time reverse transcriptase PCR. *Lett. Appl. Microbiol.* 44: 106–111. https://doi.org/10.1111/j.1472-765X.2006.02047.x.

Baert, L., Wobus, C.E., Van Coillie, E. et al. (2008). Detection of murine norovirus 1 by using plaque assay, transfection assay, and real-time reverse transcription-PCR before and after heat exposure. *Appl. Environ. Microbiol.* 74: 543–546. https://doi.org/10.1128/AEM.01039-07.

Balasubramanian, M.N., Rački, N., Gonçalves, J. et al. (2016). *Enhanced detection of pathogenic enteric viruses in coastal marine environment by concentration using monolithic chromatographic supports paired with quantitative PCR*. Submitted.

Barut, M., Podgornik, A., Brne, P., and Štrancar, A. (2005). Convective interaction media short monolithic columns: enabling chromatographic supports for the separation and purification of large biomolecules. *J. Sep. Sci.* 28: 1876–1892. https://doi.org/10.1002/jssc.200500246.

Belkin, S. and Colwell, R.R. (Eds.) (2006). *Oceans and Health: Pathogens in the Marine Environment*. New York, NY: Springer Science+Business Media.

Bhattacharya, S.S., Kulka, M., Lampel, K.A. et al. (2004). Use of reverse transcription and PCR to discriminate between infectious and non-infectious hepatitis A virus. *J. Virol. Methods* 116: 181–187.

Bian, X., Jing, F., Li, G. et al. (2015). A microfluidic droplet digital PCR for simultaneous detection of pathogenic Escherichia coli O157 and Listeria monocytogenes. *Biosens. Bioelectron.* 74: 770–777. https://doi.org/10.1016/j.bios.2015.07.016.

Blackmer, F., Reynolds, K.A., Gerba, C.P., and Pepper, I.L. (2000). Use of integrated cell culture-PCR to evaluate the effectiveness of poliovirus inactivation by chlorine. *Appl. Environ. Microbiol.* 66: 2267–2268.

Boom, R., Sol, C.J., Salimans, M.M. et al. (1990). Rapid and simple method for purification of nucleic acids. *J. Clin. Microbiol.* 28: 495–503.

Bosch, A., Guix, S., Sano, D., and Pintó, R.M. (2008). New tools for the study and direct surveillance of viral pathogens in water. *Curr. Opin. Biotechnol.* 19: 295–301. https://doi.org/10.1016/j.copbio.2008.04.006.

Boxman, I.L.A. (2010). Human enteric viruses occurrence in shellfish from European markets. *Food Environ. Virol.* 2: 156–166. https://doi.org/10.1007/s12560-010-9039-0.

Brassard, J., Seyer, K., Houde, A. et al. (2005). Concentration and detection of hepatitis A virus and rotavirus in spring water samples by reverse transcription-PCR. *J. Virol. Methods* 123: 163–169. https://doi.org/10.1016/j.jviromet.2004.09.018.

Breitbart, M. and Rohwer, F. (2005). Here a virus, there a virus, everywhere the same virus? *Trends Microbiol.* 13: 278–284. https://doi.org/10.1016/j.tim.2005.04.003.

Breitbart, M., Salamon, P., Andresen, B. et al. (2002). Genomic analysis of uncultured marine viral communities. *Proc. Natl. Acad. Sci.* 99: 14250–14255. https://doi.org/10.1073/pnas.202488399.

Breitenmoser, A., Fretz, R., Schmid, J. et al. (2011). Outbreak of acute gastroenteritis due to a washwater-contaminated water supply, Switzerland, 2008. *J. Water Health* 9: 569–576. https://doi.org/10.2166/wh.2011.158.

Brussaard, C.P.D., Mari, X., Bleijswijk, J.D.L.V., and Veldhuis, M.J.W. (2005). A mesocosm study of Phaeocystis globosa (Prymnesiophyceae) population dynamics. *Harmful Algae* 4: 875–893. https://doi.org/10.1016/j.hal.2004.12.012.

Brussaard, C.P.D., Wilhelm, S.W., Thingstad, F. et al. (2008). Global-scale processes with a nanoscale drive: the role of marine viruses. *ISME J.* 2: 575–578. https://doi.org/10.1038/ismej.2008.31.

Calgua, B., Mengewein, A., Grunert, A. et al. (2008). Development and application of a one-step low cost procedure to concentrate viruses from seawater samples. *J. Virol. Methods* 153: 79–83. https://doi.org/10.1016/j.jviromet.2008.08.003.

Cannon, J.L. and Vinjé, J. (2008). Histo-blood group antigen assay for detecting noroviruses in water. *Appl. Environ. Microbiol.* 74: 6818–6819. https://doi.org/10.1128/AEM.01302-08.

Cao, Y., Raith, M.R., and Griffith, J.F. (2015). Droplet digital PCR for simultaneous quantification of general and human-associated fecal indicators for water quality assessment. *Water Res.* 70: 337–349. https://doi.org/10.1016/j.watres.2014.12.008.

Carter, M.J. (2005). Enterically infecting viruses: pathogenicity, transmission and significance for food and waterborne infection. *J. Appl. Microbiol.* 98: 1354–1380. https://doi.org/10.1111/j.1365-2672.2005.02635.x.

Casas, N., Amarita, F., and de Marañón, I.M. (2007). Evaluation of an extracting method for the detection of hepatitis A virus in shellfish by SYBR-green real-time RT-PCR. *Int. J. Food Microbiol.* 20th International ICFMH Symposium on FOOD MICRO 2006 120: 179–185. https://doi.org/10.1016/j.ijfoodmicro.2007.01.017.

Chapron, C.D., Ballester, N.A., Fontaine, J.H. et al. (2000). Detection of astroviruses, enteroviruses, and adenovirus types 40 and 41 in surface waters collected and evaluated by the information collection rule and an integrated cell culture-nested PCR procedure. *Appl. Environ. Microbiol.* 66: 2520–2525.

Comelli, H., Rimstad, E., Larsen, S., and Myrmel, M. (2008). Detection of norovirus genotype I.3b and II.4 in bioaccumulated blue mussels using different virus recovery methods. *Int. J. Food Microbiol.* 127: 53–59. https://doi.org/10.1016/j.ijfoodmicro.2008.06.003.

Craw, P. and Balachandran, W. (2012). Isothermal nucleic acid amplification technologies for point-of-care diagnostics: a critical review. *Lab. Chip* 12: 2469–2486. https://doi.org/10.1039/C2LC40100B.

Danovaro, R., Corinaldesi, C., Filippini, M. et al. (2008). Viriobenthos in freshwater and marine sediments: a review. *Freshw. Biol.* 53: 1186–1213. https://doi.org/10.1111/j.1365-2427.2008.01961.x.

Deng, M.Y., Day, S.P., and Cliver, D.O. (1994). Detection of hepatitis A virus in environmental samples by antigen-capture PCR. *Appl. Environ. Microbiol.* 60: 1927–1933.

Diez-Valcarce, M., Kovač, K., Raspor, P. et al. (2011). Virus genome quantification does not predict norovirus infectivity after application of food inactivation processing technologies. *Food Environ.* 3: 141–146. https://doi.org/10.1007/s12560-011-9070-9.

Dijk, E.L.V., Auger, H., Jaszczyszyn, Y., and Thermes, C. (2014). Ten years of next-generation sequencing technology. *Trends Genet.* 30: 418–426. https://doi.org/10.1016/j.tig.2014.07.001.

Dimitriadis, A., Bruggink, L.D., and Marshall, J.A. (2006). Evaluation of the Dako IDEIA norovirus EIA assay for detection of norovirus using faecal specimens from Australian gastroenteritis outbreaks. *Pathology (Phila.)* 38: 157–165. https://doi.org/10.1080/00313020600559645.

Dura, G., Pándics, T., Kádár, M. et al. (2010). Environmental health aspects of drinking water-borne outbreak due to karst flooding: case study. *J. Water Health* 8: 513–520. https://doi.org/10.2166/wh.2010.099.

Edwards, R.A. and Rohwer, F. (2005). Opinion: viral metagenomics. *Nat. Rev. Microbiol.* 3: 504–510. https://doi.org/10.1038/nrmicro1163.

Fancello, L., Trape, S., Robert, C. et al. (2013). Viruses in the desert: a metagenomic survey of viral communities in four perennial ponds of the Mauritanian Sahara. *ISME J.* 7: 359–369. https://doi.org/10.1038/ismej.2012.101.

Fong, T.-T. and Lipp, E.K. (2005). Enteric viruses of humans and animals in aquatic environments: health risks, detection, and potential water quality assessment tools. *Microbiol. Mol. Biol. Rev.* 69: 357–371. https://doi.org/10.1128/MMBR.69.2.357-371.2005.

Fout, G.S., Martinson, B.C., Moyer, M.W.N., and Dahling, D.R. (2003). A multiplex reverse transcription-PCR method for detection of human enteric viruses in groundwater. *Appl. Environ. Microbiol.* 69: 3158–3164. https://doi.org/10.1128/AEM.69.6.3158-3164.2003.

Fuhrman, J.A. (1999). Marine viruses and their biogeochemical and ecological effects. *Nature* 399: 541–548.

Fuhrman, J.A. and Schwalbach, M. (2003). Viral influence on aquatic bacterial communities. *Biol. Bull.* 204: 192. https://doi.org/10.2307/1543557.

Fukuda, S., Sasaki, Y., Kuwayama, M., and Miyazaki, K. (2007). Simultaneous detection and genogroup-screening test for norovirus genogroups I and II from fecal specimens in single tube by reverse transcription-loop-mediated isothermal amplification assay. *Microbiol. Immunol.* 51: 547–550. https://doi.org/10.1111/j.1348-0421.2007.tb03932.x.

Fumian, T.M., Leite, J.P.G., Castello, A.A. et al. (2010). Detection of rotavirus A in sewage samples using multiplex qPCR and an evaluation of the ultracentrifugation and adsorption-elution methods for virus concentration. *J. Virol. Methods* 170: 42–46. https://doi.org/10.1016/j.jviromet.2010.08.017.

Gajardo, R., Díez, J.M., Jofre, J., and Bosch, A. (1991). Adsorption-elution with negatively and positively-charged glass powder for the concentration of hepatitis A virus from water. *J. Virol. Methods* 31: 345–351.

Gantzer, C., Senouci, S., Maul, A. et al. (1997). Enterovirus genomes in wastewater: concentration on glass wool and glass powder and detection by RT-PCR. *J. Virol. Methods* 65: 265–271. https://doi.org/10.1016/S0166-0934(97)02193-9.

Gassilloud, B., Duval, M., Schwartzbrod, L., and Gantzer, C. (2003). Recovery of feline calicivirus infectious particles and genome from water: comparison of two concentration techniques. *Water Sci. Technol. J. Int. Assoc. Water Pollut. Res.* 47: 97–101.

Gibbons, C.D., Rodríguez, R.A., Tallon, L., and Sobsey, M.D. (2010). Evaluation of positively charged alumina nanofibre cartridge filters for the primary concentration of noroviruses, adenoviruses and male-specific coliphages from seawater. *J. Appl. Microbiol.* 109: 635–641. https://doi.org/10.1111/j.1365-2672.2010.04691.x.

Gibson, K.E. and Schwab, K.J. (2011). Tangential-flow ultrafiltration with integrated inhibition detection for recovery of surrogates and human pathogens from large-volume source water and finished drinking water. *Appl. Environ. Microbiol.* 77: 385–391. https://doi.org/10.1128/AEM.01164-10.

Gibson, U.E., Heid, C.A., and Williams, P.M. (1996). A novel method for real time quantitative RT-PCR. *Genome Res.* 6: 995–1001. https://doi.org/10.1101/gr.6.10.995.

Gilpatrick, S.G., Schwab, K.J., Estes, M.K., and Atmar, R.L. (2000). Development of an immunomagnetic capture reverse transcription-PCR assay for the detection of Norwalk virus. *J. Virol. Methods* 90: 69–78.

Gons, H.J., Ebert, J., Hoogveld, H.L. et al. (2002). Observations on cyanobacterial population collapse in eutrophic lake water. *Antonie Van Leeuwenhoek* 81: 319–326. https://doi.org/10.1023/A:1020595408169.

Grassi, T., Bagordo, F., Idolo, A. et al. (2010). Rotavirus detection in environmental water samples by tangential flow ultrafiltration and RT-nested PCR. *Environ. Monit. Assess.* 164: 199–205. https://doi.org/10.1007/s10661-009-0885-x.

Greening, G.E. (2006). Human and animal viruses in food (including taxonomy of enteric viruses). In: *Viruses in Foods, Food Microbiology and Food Safety* (ed. S.M. Goyal), 5–42. Springer US.

Grimm, A.C., Cashdollar, J.L., Williams, F.P., and Fout, G.S. (2004). Development of an astrovirus RT–PCR detection assay for use with conventional, real-time, and integrated cell culture/RT–PCR. *Can. J. Microbiol.* 50: 269–278. https://doi.org/10.1139/w04-012.

Guévremont, E., Brassard, J., Houde, A. et al. (2006). Development of an extraction and concentration procedure and comparison of RT-PCR primer systems for the detection of hepatitis A virus and norovirus GII in green onions. *J. Virol. Methods* 134: 130–135. https://doi.org/10.1016/j.jviromet.2005.12.009.

Gutiérrez-Aguirre, I., Banjac, M., Steyer, A. et al. (2009). Concentrating rotaviruses from water samples using monolithic chromatographic supports. *J. Chromatogr. A* 1216: 2700–2704. https://doi.org/10.1016/j.chroma.2008.10.106.

Gutiérrez-Aguirre, I., Steyer, A., Banjac, M. et al. (2011). On-site reverse transcription-quantitative polymerase chain reaction detection of rotaviruses concentrated from environmental water samples using methacrylate monolithic supports. *J. Chromatogr. A* 1218: 2368–2373. https://doi.org/10.1016/j.chroma.2010.10.048.

Gutiérrez-Aguirre, I., Steyer, A., Boben, J. et al. (2008). Sensitive detection of multiple rotavirus genotypes with a single reverse transcription-real-time quantitative PCR assay. *J. Clin. Microbiol.* 46: 2547–2554. https://doi.org/10.1128/JCM.02428-07.

Guttman-Bass, N., Tchorsh, Y., and Marva, E. (1987). Comparison of methods for rotavirus detection in water and results of a survey of Jerusalem wastewater. *Appl. Environ. Microbiol.* 53: 761–767.

Hagström, A.E.V., Garvey, G., Paterson, A.S. et al. (2015). Sensitive detection of norovirus using phage nanoparticle reporters in lateral-flow assay. *PLOS ONE* 10: e0126571. https://doi.org/10.1371/journal.pone.0126571.

Haramoto, E., Katayama, H., Oguma, K., and Ohgaki, S. (2005). Application of cation-coated filter method to detection of noroviruses, enteroviruses, adenoviruses, and torque teno viruses in the Tamagawa River in Japan. *Appl. Environ. Microbiol.* 71: 2403–2411. https://doi.org/10.1128/AEM.71.5.2403-2411.2005.

Haramoto, E., Katayama, H., Oguma, K. et al. (2006). Seasonal profiles of human noroviruses and indicator bacteria in a wastewater treatment plant in Tokyo, Japan. *Water Sci. Technol.* 54: 301–308. https://doi.org/10.2166/wst.2006.888.

Haramoto, E., Katayama, H., and Ohgaki, S. (2004). Detection of noroviruses in tap water in Japan by means of a new method for concentrating enteric viruses in large volumes of freshwater. *Appl. Environ. Microbiol.* 70: 2154–2160. https://doi.org/10.1128/AEM.70.4.2154-2160.2004.

Heid, C.A., Stevens, J., Livak, K.J., and Williams, P.M. (1996). Real time quantitative PCR. *Genome Res.* 6: 986–994. https://doi.org/10.1101/gr.6.10.986.

Hellmér, M., Paxéus, N., Magnius, L. et al. (2014). Detection of pathogenic viruses in sewage provided early warnings of hepatitis a virus and norovirus outbreaks. *Appl. Environ. Microbiol.* 80: 6771–6781. https://doi.org/10.1128/AEM.01981-14.

Helmi, K., Jacob, P., Charni-Ben-Tabassi, N. et al. (2011). Comparison of two filtration-elution procedures to improve the standard methods ISO 10705-1 & 2 for bacteriophage detection in groundwater, surface water and finished water samples. *Lett. Appl. Microbiol.* 53: 329–335. https://doi.org/10.1111/j.1472-765X.2011.03112.x.

Hernandez-Morga, J., Leon-Felix, J., Peraza-Garay, F. et al. (2009). Detection and characterization of hepatitis A virus and norovirus in estuarine water samples using ultrafiltration–RT-PCR integrated methods. *J. Appl. Microbiol.* 106: 1579–1590. https://doi.org/10.1111/j.1365-2672.2008.04125.x.

Hewson, I., Ng, G., Li, W. et al. (2013). Metagenomic identification, seasonal dynamics, and potential transmission mechanisms of a *Daphnia*-associated single-stranded DNA virus in two temperate lakes. *Limnol. Oceanogr.* 58: 1605–1620. https://doi.org/10.4319/lo.2013.58.5.1605.

Hill, V.R., Kahler, A.M., Jothikumar, N. et al. (2007). Multistate evaluation of an ultrafiltration-based procedure for simultaneous recovery of enteric microbes in 100-liter tap water samples. *Appl. Environ. Microbiol.* 73: 4218–4225. https://doi.org/10.1128/AEM.02713-06.

Hmaïed, F., Jebri, S., Saavedra, M.E.R. et al. (2015). Comparison of two concentration methods for the molecular detection of enteroviruses in raw and treated sewage. *Curr. Microbiol.* 72: 12–18. https://doi.org/10.1007/s00284-015-0909-4.

Houde, A., Leblanc, D., Poitras, E. et al. (2006). Comparative evaluation of RT-PCR, nucleic acid sequence-based amplification (NASBA) and real-time RT-PCR for detection of noroviruses in faecal material. *J. Virol. Methods* 135: 163–172. https://doi.org/10.1016/j.jviromet.2006.03.001.

Huggett, J.F., Foy, C.A., Benes, V. et al. (2013). The digital MIQE guidelines: minimum information for publication of quantitative digital PCR experiments. *Clin. Chem.* 59: 892–902. https://doi.org/10.1373/clinchem.2013.206375.

Hugues, B., Andre, M., Plantat, J.L., and Champsaur, H. (1993). Comparison of glass wool and glass powder methods for concentration of viruses from treated waste waters. *Zentralblatt Für Hyg. Umweltmed. Int. J. Hyg. Environ. Med.* 193: 440–449.

Hurwitz, B.L. and Sullivan, M.B. (2013). The pacific ocean virome (POV): a marine viral metagenomic dataset and associated protein clusters for quantitative viral ecology. *PLoS ONE* 8. https://doi.org/10.1371/journal.pone.0057355.

Hwang, Y.-C., Leong, O.M., Chen, W., and Yates, M.V. (2007). Comparison of a reporter assay and immunomagnetic separation real-time reverse transcription-PCR for the detection of enteroviruses in seeded environmental water samples. *Appl. Environ. Microbiol.* 73: 2338–2340. https://doi.org/10.1128/AEM.01758-06.

Ikner, L.A., Soto-Beltran, M., and Bright, K.R. (2011). New method using a positively charged microporous filter and ultrafiltration for concentration of viruses from tap water. *Appl. Environ. Microbiol.* 77: 3500–3506. https://doi.org/10.1128/AEM.02705-10.

Iturriza-Gómara, M., Xerry, J., Gallimore, C.I. et al. (2008). Evaluation of the loopamp (loop-mediated isothermal amplification) kit for detecting norovirus RNA in faecal samples. *J. Clin. Virol. Off. Publ. Pan Am. Soc. Clin. Virol.* 42: 389–393. https://doi.org/10.1016/j.jcv.2008.02.012.

Jacquet, S., Miki, T., Noble, R. et al. (2010). Viruses in aquatic ecosystems: important advancements of the last 20 years and prospects for the future in the field of microbial oceanography and limnology. *Adv. Oceanogr. Limnol.* 1: 97–141. https://doi.org/10.1080/19475721003743843.

Jean, J., Blais, B., Darveau, A., and Fliss, I. (2002). Rapid detection of human rotavirus using colorimetric nucleic acid sequence-based amplification (NASBA)–enzyme-linked immunosorbent assay in sewage treatment effluent. *FEMS Microbiol. Lett.* 210: 143–147. https://doi.org/10.1111/j.1574-6968.2002.tb11173.x.

Jeanniard, A., Dunigan, D.D., Gurnon, J.R. et al. (2013). Towards defining the chloroviruses: a genomic journey through a genus of large DNA viruses. *BMC Genomics* 14: 158. https://doi.org/10.1186/1471-2164-14-158.

Jiang, S., Dezfulian, H., and Chu, W. (2005). Real-time quantitative PCR for enteric adenovirus serotype 40 in environmental waters. *Can. J. Microbiol.* 51: 393–398. https://doi.org/10.1139/w05-016.

Jiang, S., Noble, R., and Chu, W. (2001). Human adenoviruses and coliphages in urban runoff-impacted coastal waters of Southern California. *Appl. Environ. Microbiol.* 67: 179–184. https://doi.org/10.1128/AEM.67.1.179-184.2001.

Jiang, Y.-J., Liao, G.-Y., Zhao, W. et al. (2004). Detection of infectious hepatitis A virus by integrated cell culture/strand-specific reverse transcriptase-polymerase chain reaction. *J. Appl. Microbiol.* 97: 1105–1112. https://doi.org/10.1111/j.1365-2672.2004.02413.x.

Jones, T.H., Brassard, J., Johns, M.W., and Gagné, M.-J. (2009). The effect of pre-treatment and sonication of centrifugal ultrafiltration devices on virus recovery. *J. Virol. Methods* 161: 199–204. https://doi.org/10.1016/j.jviromet.2009.06.013.

Joret, J.C., Block, J.C., Lucena-Gutierrez, F. et al. (1980). Virus concentration from secondary wastewater: comparative study between epoxy fiberglass and glass powder adsorbents. *Eur. J. Appl. Microbiol. Biotechnol.* 10: 245–252. https://doi.org/10.1007/BF00508611.

Jothikumar, N., Aparna, K., Kamatchiammal, S. et al. (1993). Detection of hepatitis E virus in raw and treated wastewater with the polymerase chain reaction. *Appl. Environ. Microbiol.* 59: 2558–2562.

Jothikumar, N., Cliver, D.O., and Mariam, T.W. (1998). Immunomagnetic capture PCR for rapid concentration and detection of hepatitis A virus from environmental samples. *Appl. Environ. Microbiol.* 64: 504–508.

Jothikumar, N., Lowther, J.A., Henshilwood, K. et al. (2005). Rapid and sensitive detection of noroviruses by using TaqMan-based one-step reverse transcription-PCR assays and application to naturally contaminated shellfish samples. *Appl. Environ. Microbiol.* 71: 1870–1875. https://doi.org/10.1128/AEM.71.4.1870-1875.2005.

Kageyama, T., Kojima, S., Shinohara, M. et al. (2003). Broadly reactive and highly sensitive assay for norwalk-like viruses based on real-time quantitative reverse transcription-PCR. *J. Clin. Microbiol.* 41: 1548–1557. https://doi.org/10.1128/JCM.41.4.1548-1557.2003.

Karim, M.R., Rhodes, E.R., Brinkman, N. et al. (2009). New electropositive filter for concentrating enteroviruses and noroviruses from large volumes of water. *Appl. Environ. Microbiol.* 75: 2393–2399. https://doi.org/10.1128/AEM.00922-08.

Katayama, H., Shimasaki, A., and Ohgaki, S. (2002). Development of a virus concentration method and its application to detection of enterovirus and norwalk virus from coastal seawater. *Appl. Environ. Microbiol.* 68: 1033–1039. https://doi.org/10.1128/AEM.68.3.1033-1039.2002.

Ko, G., Cromeans, T.L., and Sobsey, M.D. (2003). Detection of infectious adenovirus in cell culture by mRNA reverse transcription-PCR. *Appl. Environ. Microbiol.* 69: 7377–7384. https://doi.org/10.1128/AEM.69.12.7377-7384.2003.

Ko, G., Cromeans, T.L., and Sobsey, M.D. (2005). UV inactivation of adenovirus type 41 measured by cell culture mRNA RT-PCR. *Water Res.* 39: 3643–3649. https://doi.org/10.1016/j.watres.2005.06.013.

Kojima, S., Kageyama, T., Fukushi, S. et al. (2002). Genogroup-specific PCR primers for detection of Norwalk-like viruses. *J. Virol. Methods* 100: 107–114.

Lambertini, E., Spencer, S.K., Bertz, P.D. et al. (2008). Concentration of enteroviruses, adenoviruses, and noroviruses from drinking water by use of glass wool filters. *Appl. Environ. Microbiol.* 74: 2990–2996. https://doi.org/10.1128/AEM.02246-07.

Lamhoujeb, S., Fliss, I., Ngazoa, S.E., and Jean, J. (2008). Evaluation of the persistence of infectious human noroviruses on food surfaces by using real-time nucleic acid sequence-based amplification. *Appl. Environ. Microbiol.* 74: 3349–3355. https://doi.org/10.1128/AEM.02878-07.

Lazcka, O., Campo, F.J.D., and Muñoz, F.X. (2007). Pathogen detection: a perspective of traditional methods and biosensors. *Biosens. Bioelectron.* 22: 1205–1217. https://doi.org/10.1016/j.bios.2006.06.036.

Le Guyader, F.S., Le Saux, J.-C., Ambert-Balay, K. et al. (2008). Aichi virus, norovirus, astrovirus, enterovirus, and rotavirus involved in clinical cases from a French oyster-related gastroenteritis outbreak. *J. Clin. Microbiol.* 46: 4011–4017. https://doi.org/10.1128/JCM.01044-08.

Le Guyader, F.S., Parnaudeau, S., Schaeffer, J. et al. (2009). Detection and quantification of noroviruses in shellfish. *Appl. Environ. Microbiol.* 75: 618–624. https://doi.org/10.1128/AEM.01507-08.

Lee, H.K. and Jeong, Y.S. (2004). Comparison of total culturable virus assay and multiplex integrated cell culture-PCR for reliability of waterborne virus detection. *Appl. Environ. Microbiol.* 70: 3632–3636. https://doi.org/10.1128/AEM.70.6.3632-3636.2004.

Lee, S.-H. and Kim, S.-J. (2002). Detection of infectious enteroviruses and adenoviruses in tap water in urban areas in Korea. *Water Res.* 36: 248–256.

Lees, D. and Tag, C.W. (2010). International standardisation of a method for detection of human pathogenic viruses in molluscan shellfish. *Food Environ. Virol.* 2: 146–155. https://doi.org/10.1007/s12560-010-9042-5.

Li, D., Gu, A.Z., He, M. et al. (2009). UV inactivation and resistance of rotavirus evaluated by integrated cell culture and real-time RT-PCR assay. *Water Res.* 43: 3261–3269. https://doi.org/10.1016/j.watres.2009.03.044.

Li, J.W., Xin, Z.T., Wang, X.W. et al. (2002). Mechanisms of inactivation of hepatitis A virus by chlorine. *Appl. Environ. Microbiol.* 68: 4951–4955. https://doi.org/10.1128/AEM.68.10.4951-4955.2002.

Li, J.W., Xin, Z.T., Wang, X.W. et al. (2004). Mechanisms of inactivation of hepatitis A virus in water by chlorine dioxide. *Water Res.* 38: 1514–1519. https://doi.org/10.1016/j.watres.2003.12.021.

Lipp, E.K., Lukasik, J., and Rose, J.B. (2001). Human enteric viruses and parasites in the marine environment. In: *Marine Microbiology* (ed. B.-M. Microbiology), 559–588. Academic Press.

Liu, P., Hill, V.R., Hahn, D. et al. (2012). Hollow-fiber ultrafiltration for simultaneous recovery of viruses, bacteria and parasites from reclaimed water. *J. Microbiol. Methods* 88: 155–161. https://doi.org/10.1016/j.mimet.2011.11.007.

Lodder, W.J., van den Berg, H.H.J.L., Rutjes, S.A., and de Roda Husman, A.M. (2010). Presence of enteric viruses in source waters for drinking water production in the Netherlands. *Appl. Environ. Microbiol.* 76: 5965–5971. https://doi.org/10.1128/AEM.00245-10.

Loman, N.J., Misra, R.V., Dallman, T.J. et al. (2012). Performance comparison of benchtop high-throughput sequencing platforms. *Nat. Biotech.* 30: 434–439.

Lønborg, C., Middelboe, M., and Brussaard, C.P.D. (2013). Viral lysis of Micromonas pusilla: impacts on dissolved organic matter production and composition. *Biogeochemistry* 116: 231–240. https://doi.org/10.1007/s10533-013-9853-1.

Luef, B., Neu, T.R., Zweimuller, I., and Peduzzi, P. (2009). Structure and composition of aggregates in two large European rivers, based on confocal laser scanning microscopy and image and statistical analyses. *Appl. Environ. Microbiol.* 75: 5952–5962. https://doi.org/10.1128/AEM.00186-09.

Lukasik, J., Scott, T.M., Andryshak, D., and Farrah, S.R. (2000). Influence of salts on virus adsorption to microporous filters. *Appl. Environ. Microbiol.* 66: 2914–2920.

Ma, J.F., Straub, T.M., Pepper, I.L., and Gerba, C.P. (1994). Cell culture and PCR determination of poliovirus inactivation by disinfectants. *Appl. Environ. Microbiol.* 60: 4203–4206.

Malik, Y.S., Sharma, K., Kumar, N. et al. (2013). Rapid detection of human rotavirus using NSP4 gene specific reverse transcription loop-mediated isothermal amplification assay. *Indian J. Virol.* 24: 265–271. https://doi.org/10.1007/s13337-013-0147-y.

Mattison, K. and Bidawid, S. (2009). Analytical methods for food and environmental viruses. *Food Environ. Virol.* 1: 107–122. https://doi.org/10.1007/s12560-009-9017-6.

Maunula, L., Söderberg, K., Vahtera, H. et al. (2012). Presence of human noro- and adenoviruses in river and treated wastewater, a longitudinal study and method comparison. *J. Water Health* 10: 87–99. https://doi.org/10.2166/wh.2011.095.

Mo, Q.-H., Wang, H.-B., Dai, H.-R. et al. (2015). Rapid and simultaneous detection of three major diarrhea-causing viruses by multiplex real-time nucleic acid sequence-based amplification. *Arch. Virol.* 160: 719–725. https://doi.org/10.1007/s00705-014-2328-4.

Mohiuddin, M. and Schellhorn, H.E. (2015). Spatial and temporal dynamics of virus occurrence in two freshwater lakes captured through metagenomic analysis. *Front. Microbiol.* 6. https://doi.org/10.3389/fmicb.2015.00960.

Myrmel, M., Rimstad, E., and Wasteson, Y. (2000). Immunomagnetic separation of a Norwalk-like virus (genogroup I) in artificially contaminated environmental water samples. *Int. J. Food Microbiol.* 62: 17–26.

Noble, R.T. and Fuhrman, J.A. (2001). Enteroviruses detected by reverse transcriptase polymerase chain reaction from the coastal waters of Santa Monica Bay, California: low correlation to bacterial indicator levels. *Hydrobiologia* 460: 175–184. https://doi.org/10.1023/A:1013121416891.

Nordgren, J., Matussek, A., Mattsson, A. et al. (2009). Prevalence of norovirus and factors influencing virus concentrations during one year in a full-scale wastewater treatment plant. *Water Res.* 43: 1117–1125. https://doi.org/10.1016/j.watres.2008.11.053.

Nuanualsuwan, S. and Cliver, D.O. (2002). Pretreatment to avoid positive RT-PCR results with inactivated viruses. *J. Virol. Methods* 104: 217–225.

Nuanualsuwan, S. and Cliver, D.O. (2003). Capsid functions of inactivated human picornaviruses and feline calicivirus. *Appl. Environ. Microbiol.* 69: 350–357. https://doi.org/10.1128/AEM.69.1.350-357.2003.

Odić, D., Turk, V., and Stopar, D. (2007). Environmental stress determines the quality of bacterial lysate and its utilization efficiency in a simple microbial loop. *Microb. Ecol.* 53: 639–649. https://doi.org/10.1007/s00248-006-9143-8.

Ogorzaly, L., Walczak, C., Galloux, M. et al. (2015). Human adenovirus diversity in water samples using a next-generation amplicon sequencing approach. *Food Environ. Virol.* 7: 112–121. https://doi.org/10.1007/s12560-015-9194-4.

Paul, J.H., Jiang, S.C., and Rose, J.B. (1991). Concentration of viruses and dissolved DNA from aquatic environments by vortex flow filtration. *Appl. Environ. Microbiol.* 57: 2197–2204.

Pecson, B.M., Ackermann, M., and Kohn, T. (2011). Framework for using quantitative PCR as a nonculture based method to estimate virus infectivity. *Environ. Sci. Technol.* 45: 2257–2263. https://doi.org/10.1021/es103488e.

Peduzzi, P. (2015). Virus ecology of fluvial systems: a blank spot on the map?: virus ecology of fluvial systems. *Biol. Rev.* n/a–n/a. https://doi.org/10.1111/brv.12202.

Peduzzi, P. and Luef, B. (2008). Viruses, bacteria and suspended particles in a backwater and main channel site of the Danube (Austria). *Aquat. Sci.* 70: 186–194. https://doi.org/10.1007/s00027-008-8068-3.

Perelle, S., Cavellini, L., Burger, C. et al. (2009). Use of a robotic RNA purification protocol based on the NucliSens easyMAG for real-time RT- PCR detection of hepatitis A virus in bottled water. *J. Virol. Methods* 157: 80–83. https://doi.org/10.1016/j.jviromet.2008.11.022.

Pineda, M.F., Chan, L.L.-Y., Kuhlenschmidt, T. et al. (2009). Rapid specific and label-free detection of porcine rotavirus using photonic crystal biosensors. *IEEE Sens. J.* 9: 470–477. https://doi.org/10.1109/JSEN.2009.2014427.

Pinto, F., Larsen, S., and Casper, P. (2013). Viriobenthos in aquatic sediments: variability in abundance and production and impact on the C-cycle. *Aquat. Sci.* 75: 571–579. https://doi.org/10.1007/s00027-013-0301-z.

Podewils, L.J., Zanardi Blevins, L., Hagenbuch, M. et al. (2007). Outbreak of norovirus illness associated with a swimming pool. *Epidemiol. Infect.* 135: 827–833. https://doi.org/10.1017/S0950268806007370.

Polaczyk, A.L., Narayanan, J., Cromeans, T.L. et al. (2008). Ultrafiltration-based techniques for rapid and simultaneous concentration of multiple microbe classes from 100-L tap water samples. *J. Microbiol. Methods* 73: 92–99. https://doi.org/10.1016/j.mimet.2008.02.014.

Pollard, P.C. and Ducklow, H. (2011). Ultrahigh bacterial production in a eutrophic subtropical Australian river: does viral lysis short-circuit the microbial loop? *Limnol. Oceanogr.* 56: 1115–1129.

Prata, C., Ribeiro, A., Cunha, Â. et al. (2012). Ultracentrifugation as a direct method to concentrate viruses in environmental waters: virus-like particle enumeration as a new approach to determine the efficiency of recovery. *J. Environ. Monit. JEM* 14: 64–70. https://doi.org/10.1039/c1em10603a.

Puig, M., Jofre, J., Lucena, F. et al. (1994). Detection of adenoviruses and enteroviruses in polluted waters by nested PCR amplification. *Appl. Environ. Microbiol.* 60: 2963–2970.

Rački, N., Morisset, D., Gutierrez-Aguirre, I., and Ravnikar, M. (2013). One-step RT-droplet digital PCR: a breakthrough in the quantification of waterborne RNA viruses. *Anal. Bioanal. Chem.* 406: 661–667. https://doi.org/10.1007/s00216-013-7476-y.

Remick, D.G., Kunkel, S.L., Holbrook, E.A., and Hanson, C.A. (1990). Theory and applications of the polymerase chain reaction. *Am. J. Clin. Pathol.* 93: S49–54.

Reynolds, K.A. (2004). Integrated cell culture/PCR for detection of enteric viruses in environmental samples. *Methods Mol. Biol. Clifton NJ* 268: 69–78. https://doi.org/10.1385/1-59259-766-1:069.

Reynolds, K.A., Gerba, C.P., and Pepper, I.L. (1996). Detection of infectious enteroviruses by an integrated cell culture-PCR procedure. *Appl. Environ. Microbiol.* 62: 1424–1427.

Rhodes, E.R., Hamilton, D.W., See, M.J., and Wymer, L. (2011). Evaluation of hollow-fiber ultrafiltration primary concentration of pathogens and secondary concentration of viruses from water. *J. Virol. Methods* 176: 38–45. https://doi.org/10.1016/j.jviromet.2011.05.031.

Richards, G.P. (2005). Food- and waterborne enteric viruses. In: *Foodborne Pathogens: Microbiology and Molecular Biology* (ed. P.M. Fratamico, A.K. Bhunia, and J.L. Smith), 121–143. Caister Academic Press.

Rodríguez, R.A., Pepper, I.L., and Gerba, C.P. (2009). Application of PCR-based methods to assess the infectivity of enteric viruses in environmental samples. *Appl. Environ. Microbiol.* 75: 297–307. https://doi.org/10.1128/AEM.01150-08.

Rodríguez-Lázaro, D., Cook, N., Ruggeri, F.M. et al. (2012). Virus hazards from food, water and other contaminated environments. *FEMS Microbiol. Rev.* 36: 786–814. https://doi.org/10.1111/j.1574-6976.2011.00306.x.

Rohwer, F. and Edwards, R. (2002). The phage proteomic tree: a genome-based taxonomy for phage. *J. Bacteriol.* 184: 4529–4535. https://doi.org/10.1128/JB.184.16.4529-4535.2002.

Rohwer, F. and Thurber, R.V. (2009). Viruses manipulate the marine environment. *Nature* 459: 207–212. https://doi.org/10.1038/nature08060.

Roux, S., Enault, F., Robin, A. et al. (2012). Assessing the diversity and specificity of two freshwater viral communities through metagenomics. *PLoS ONE* 7: e33641. https://doi.org/10.1371/journal.pone.0033641.

Rutjes, S.A., Berg, H.H.J.L.V.D., Lodder, W.J., and Husman, A.M.D.R. (2006). Real-time detection of noroviruses in surface water by use of a broadly reactive nucleic acid sequence-based amplification assay. *Appl. Environ. Microbiol.* 72: 5349–5358. https://doi.org/10.1128/AEM.00751-06.

Rutjes, S.A., Italiaander, R., van den Berg, H.H.J.L. et al. (2005). Isolation and detection of enterovirus RNA from large-volume water samples by using the NucliSens miniMAG system and real -time nucleic acid sequence-based amplification. *Appl. Environ. Microbiol.* 71: 3734–3740. https://doi.org/10.1128/AEM.71.7.3734-3740.2005.

Rzeżutka, A. and Cook, N. (2004). Survival of human enteric viruses in the environment and food. *FEMS Microbiol. Rev.* 28: 441–453. https://doi.org/10.1016/j.femsre.2004.02.001.

Sartorius, B., Andersson, Y., Velicko, I. et al. (2007). Outbreak of norovirus in Västra Götaland associated with recreational activities at two lakes during August 2004. *Scand. J. Infect. Dis.* 39: 323–331. https://doi.org/10.1080/00365540601053006.

Schwab, K.J., Leon, R.D., and Sobsey, M.D. (1996). Immunoaffinity concentration and purification of waterborne enteric viruses for detection by reverse transcriptase PCR. *Appl. Environ. Microbiol.* 62: 2086–2094.

Shelford, E., Middelboe, M., Møller, E., and Suttle, C. (2012). Virus-driven nitrogen cycling enhances phytoplankton growth. *Aquat. Microb. Ecol.* 66: 41–46. https://doi.org/10.3354/ame01553.

Shieh, Y.C., Wong, C.I., Krantz, J.A., and Hsu, F.C. (2008). Detection of naturally occurring enteroviruses in waters using direct RT-PCR and integrated cell culture-RT-PCR. *J. Virol. Methods* 149: 184–189. https://doi.org/10.1016/j.jviromet.2007.12.013.

Simonet, J. and Gantzer, C. (2006a). Degradation of the Poliovirus 1 genome by chlorine dioxide. *J. Appl. Microbiol.* 100: 862–870. https://doi.org/10.1111/j.1365-2672.2005.02850.x.

Smith, C.M. and Hill, V.R. (2009). Dead-end hollow-fiber ultrafiltration for recovery of diverse microbes from water. *Appl. Environ. Microbiol.* 75: 5284–5289. https://doi.org/10.1128/AEM.00456-09.

Sobsey, M.D., Wallis, C., Henderson, M., and Melnick, J.L. (1973). Concentration of enteroviruses from large volumes of water. *Appl. Microbiol.* 26: 529–534.

St John, A. and Price, C.P. (2014). Existing and emerging technologies for point-of-care testing. *Clin. Biochem. Rev.* 35: 155–167.

Stals, A., Baert, L., De Keuckelaere, A. et al. (2011). Evaluation of a norovirus detection methodology for ready-to-eat foods. *Int. J. Food Microbiol.* 145: 420–425. https://doi.org/10.1016/j.ijfoodmicro.2011.01.013.

Stals, A., Baert, L., Van Coillie, E., and Uyttendaele, M. (2012). Extraction of food-borne viruses from food samples: a review. *Int. J. Food Microbiol.* 153: 1–9. https://doi.org/10.1016/j.ijfoodmicro.2011.10.014.

Steyer, A., Gutiérrez-Aguirre, I., Rački, N. et al. (2015). The detection rate of enteric viruses and clostridium difficile in a waste water treatment plant effluent. *Food Environ. Virol.* 7: 164–172. https://doi.org/10.1007/s12560-015-9183-7.

Steyer, A., Torkar, K.G., Gutiérrez-Aguirre, I., and Poljšak-Prijatelj, M. (2011). High prevalence of enteric viruses in untreated individual drinking water sources and surface water in Slovenia. *Int. J. Hyg. Environ. Health* 214: 392–398. https://doi.org/10.1016/j.ijheh.2011.05.006.

Strancar, A., Podgornik, A., Barut, M., and Necina, R. (2002). Short monolithic columns as stationary phases for biochromatography. In: *Modern Advances in Chromatography, Advances in Biochemical Engineering/Biotechnology* (ed. P.D.R. Freitag), 49–85. Berlin Heidelberg: Springer.

Suttle, C.A. (2007). Marine viruses — major players in the global ecosystem. *Nat. Rev. Microbiol.* 5: 801–812. https://doi.org/10.1038/nrmicro1750.

Suzuki, Y., Narimatsu, S., Furukawa, T. et al. (2011). Comparison of real-time reverse-transcription loop-mediated isothermal amplification and real-time reverse -transcription polymerase chain reaction for detection of noroviruses in municipal wastewater. *J. Biosci. Bioeng.* 112: 369–372. https://doi.org/10.1016/j.jbiosc.2011.06.012.

Sylvain, S., Christophe, G., Karim, H. et al. (2009). Simultaneous concentration of enteric viruses and protozoan parasites: a protocol based on tangential flow filtration and adapted to large volumes of surface and drinking waters. *Food Environ. Virol.* 1: 66–76. https://doi.org/10.1007/s12560-009-9011-z.

Tan, B., Ng, C., Nshimyimana, J.P. et al. (2015). Next-generation sequencing (NGS) for assessment of microbial water quality: current progress, challenges, and future opportunities. *Front. Microbiol.* 6. https://doi.org/10.3389/fmicb.2015.01027.

Teunis, P.F.M., Rutjes, S.A., Westrell, T., and de Roda Husman, A.M. (2009). Characterization of drinking water treatment for virus risk assessment. *Water Res.* 43: 395–404. https://doi.org/10.1016/j.watres.2008.10.049.

Tian, P., Engelbrektson, A., and Mandrell, R. (2008). Two-log increase in sensitivity for detection of norovirus in complex samples by concentration with porcine gastric mucin conjugated to magnetic beads. *Appl. Environ. Microbiol.* 74: 4271–4276. https://doi.org/10.1128/AEM.00539-08.

Tian, P., Yang, D., Pan, L., and Mandrell, R. (2012). Application of a receptor-binding capture quantitative reverse transcription-PCR assay to concentrate human norovirus from sewage and to study the distribution and stability of the virus. *Appl. Environ. Microbiol.* 78: 429–436. https://doi.org/10.1128/AEM.06875-11.

Topping, J.R., Schnerr, H., Haines, J. et al. (2009). Temperature inactivation of feline calicivirus vaccine strain FCV F-9 in comparison with human noroviruses using an RNA exposure assay and reverse transcribed quantitative real-time polymerase chain reaction-A novel method for predicting virus infectivity. *J. Virol. Methods* 156: 89–95. https://doi.org/10.1016/j.jviromet.2008.10.024.

Tsai, Y.L., Tran, B., Sangermano, L.R., and Palmer, C.J. (1994). Detection of poliovirus, hepatitis A virus, and rotavirus from sewage and ocean water by triplex reverse transcriptase PCR. *Appl. Environ. Microbiol.* 60: 2400–2407.

Tseng, C.-H., Chiang, P.-W., Shiah, F.-K. et al. (2013). Microbial and viral metagenomes of a subtropical freshwater reservoir subject to climatic disturbances. *ISME J.* 7: 2374–2386. https://doi.org/10.1038/ismej.2013.118.

USEPA. (2001a). *Method 1601: Detection of Male-specific (F+) and Somatic Coliphage in Water by Two-step Enrichment Procedure (No. EPA 821-R-01-030).* Washington, DC: Office of Water, Engineering and Analysis Division.

USEPA. (2001b). *Method 1602: Detection of Male-specific (F+) and Somatic Coliphage in Water by Single Agar Layer (SAL) Procedure (No. EPA 821-R-01-029)*. Washington, DC: Office of Water, Engineering and Analysis Division.

van Heerden, J., Ehlers, M.M., Heim, A., and Grabow, W.O.K. (2005). Prevalence, quantification and typing of adenoviruses detected in river and treated drinking water in South Africa. *J. Appl. Microbiol.* 99: 234–242. https://doi.org/10.1111/j.1365-2672.2005.02617.x.

Van Heerden, J., Ehlers, M.M., Van Zyl, W.B., and Grabow, W.O.K. (2003). Incidence of adenoviruses in raw and treated water. *Water Res.* 37: 3704–3708. https://doi.org/10.1016/S0043-1354(03)00245-8.

Victoria, M., Guimarães, F.R., Fumian, T.M. et al. (2010). One year monitoring of norovirus in a sewage treatment plant in Rio de Janeiro. *Brazil. J. Water Health* 8: 158–165. https://doi.org/10.2166/wh.2009.012.

Villar, L.M., de Paula, V.S., Diniz-Mendes, L. et al. (2007). Molecular detection of hepatitis A virus in urban sewage in Rio de Janeiro. *Brazil. Lett. Appl. Microbiol.* 45: 168–173. https://doi.org/10.1111/j.1472-765X.2007.02164.x.

Vrede, K., Stensdotter, U., and Lindström, E.S. (2003). Viral and bacterioplankton dynamics in two lakes with different humic contents. *Microb. Ecol.* 46: 406–415. https://doi.org/10.1007/s00248-003-2009-4.

Wallis, C. and Melnick, J.L. (1967). Concentration of enteroviruses on membrane filters. *J. Virol.* 1: 472–477.

Waterbury, J.B. and Valois, F. (1993). Resistance to co-occurring phages enables marine synechococcus communities to coexist with cyanophages abundant in seawater. *Appl. Enviromental Microbiol.* 59: 3393–3399.

Weinbauer, M., Kerros, M., Motegi, C. et al. (2010). Bacterial community composition and potential controlling mechanisms along a trophic gradient in a barrier reef system. *Aquat. Microb. Ecol.* 60: 15–28. https://doi.org/10.3354/ame01411.

Weitz, J. and Wilhelm, S. (2012). Ocean viruses and their effects on microbial communities and biogeochemical cycles. *F1000 Biol. Rep.* https://doi.org/10.3410/B4-17.

Weitz, J.S., Stock, C.A., Wilhelm, S.W. et al. (2015). A multitrophic model to quantify the effects of marine viruses on microbial food webs and ecosystem processes. *ISME J.* 9: 1352–1364. https://doi.org/10.1038/ismej.2014.220.

WHO. (2008). *Guidelines for Drinking-water Quality - Volume 1: Recommendations*, 3e. Geneva: WHO.

Wilhelm, S.W. and Suttle, C.A. (1999). Viruses and nutrient cycles in the sea. *BioScience* 49: 781. https://doi.org/10.2307/1313569.

Wilhelmi de Cal, I., Revilla, A., Del Alamo, J.M. et al. (2007). Evaluation of two commercial enzyme immunoassays for the detection of norovirus in faecal samples from hospitalised children with sporadic acute gastroenteritis. *Clin. Microbiol. Infect. Off. Publ. Eur. Soc. Clin. Microbiol. Infect. Dis.* 13: 341–343. https://doi.org/10.1111/j.1469-0691.2006.01594.x.

Wommack, K.E. and Colwell, R.R. (2000). Virioplankton: viruses in aquatic ecosystems. *Microbiol. Mol. Biol. Rev.* 64: 69–114. https://doi.org/10.1128/MMBR.64.1.69-114.2000.

Wyn-Jones, A.P. and Sellwood, J. (2001). Enteric viruses in the aquatic environment. *J. Appl. Microbiol.* 91: 945–962.

Yan, L., Zhou, J., Zheng, Y. et al. (2014). Isothermal amplified detection of DNA and RNA. *Mol. Biosyst.* 10: 970–1003. https://doi.org/10.1039/C3MB70304E.

Yang, B.-Y., Liu, X.-L., Wei, Y.-M. et al. (2014a). Rapid and sensitive detection of human astrovirus in water samples by loop-mediated isothermal amplification with hydroxynaphthol blue dye. *BMC Microbiol.* 14: 38. https://doi.org/10.1186/1471-2180-14-38.

Yetisen, A.K., Akram, M.S., and Lowe, C.R. (2013). Paper-based microfluidic point-of-care diagnostic devices. *Lab. Chip* 13: 2210–2251. https://doi.org/10.1039/C3LC50169H.

Yoneyama, T., Kiyohara, T., Shimasaki, N. et al. (2007). Rapid and real-time detection of hepatitis A virus by reverse transcription loop-mediated isothermal amplification assay. *J. Virol. Methods* 145: 162–168. https://doi.org/10.1016/j.jviromet.2007.05.023.

Zanoli, L.M. and Spoto, G. (2012). Isothermal amplification methods for the detection of nucleic acids in microfluidic devices. *Biosensors* 3: 18–43. https://doi.org/10.3390/bios3010018.

Section 4

15

Global Environmental Pressures

Glenn Watts

Environment Agency, Bristol, UK

15.1 Introduction: What Is a Global Environmental Pressure?

There can be no doubt that the environment throughout the world is affected by human activity. It has been suggested that human influence is now so great that the world has entered a new geological era: the Anthropocene (Crutzen 2002). The human activities that cause environmental change are often called pressures on the environment. In physics, pressure is defined as a force applied over an area. In more general use, a pressure is thought of as a stress, usually unwelcome, that produces a response. For example, people can be pressured to change their minds or act in a different way, often against their own opinions or judgement. Similarly, an environmental pressure may produce a response that moves the environmental system from one state to another. This change in state may be undesirable: environmental pressures are generally thought of as negative forces that cause unnecessary environmental stress and unwanted change.

This chapter explores how global environmental pressures affect catchments, and how improved understanding of catchment processes can help with the management of pressures on the environment. The chapter starts with a discussion of the concepts of planetary boundaries, a safe operating space for humanity, and virtual water. This helps to frame the interventions, both deliberate and accidental, that humans make in in hydroecological processes. The chapter concludes with a discussion of how a better understanding of hydrological processes and interfaces can inform and improve catchment management. This chapter makes no claim to be a complete survey of environmental pressures, but concentrates on how a conceptual understanding of interfaces and boundaries can help with the practical management of pressures on the water environment.

Ecohydrological Interfaces, First Edition. Edited by Stefan Krause, David M. Hannah, and Nancy B. Grimm.

15.2 Planetary Boundaries

Rockström et al. (2009) introduced the concept of a "safe operating space for humanity", identifying nine global biophysical thresholds that, if crossed, would shift the earth's system to a different state with negative effects on humans (Table 15.1). Underlying the concept of planetary boundaries is the idea that the relatively stable environmental conditions of the last 10 000 years, the Holocene, have provided the right environment for human society to develop. Moving away from these conditions, it is argued, may mean that the earth can no longer support our modern society.

The idea of planetary boundaries has been highly influential, though not without controversy. It has been argued that the boundaries are effectively arbitrary. For example, the atmospheric carbon dioxide (CO_2) concentration boundary is identified by Rockström et al. (2009) at 350 parts per million (ppm). This level was set to reduce the risk of irreversible climate change on a timescale of 50–100 years or more. Given the uncertainties in modelling the future global climate, it is not currently possible to know whether returning atmospheric CO_2 concentration to this level would manage the probability of irreversible change to an acceptable level or whether a lower threshold would be more appropriate. Part of the problem here is in knowing how to define a globally acceptable probability of irreversible change, because even a small change in climate can have a serious impact on people living in vulnerable locations.

Some critics of planetary boundaries suggest that many systems do not exhibit tipping points beyond which their state changes completely. In many systems, it is argued, change is rather gradual and even if the end point can be recognized as a different state, it is not possible to identify the moment when the change in state took place, or when progress towards this new state became irreversible. Phillips (2006) suggests that this view is one of the "widely-held (mis)perceptions... in the earth science community" (p. 733) as "geomorphic systems are overwhelmingly non-linear" (p. 733) with thresholds being particularly

Table 15.1 Planetary boundaries identified by Rockström et al. (2009).

Process	Parameters
Climate change	Atmospheric carbon dioxide concentration
	Change in radiative forcing
Rate of biodiversity loss	Extinction rate
Nitrogen cycle	Nitrogen removed from the atmosphere for human use
Phosphorus cycle	Phosphorus flowing into the oceans
Stratospheric ozone depletion	Ozone concentration
Ocean acidification	Level of ocean acidity
Global freshwater use	Volume used
Change in land use	Land area used for crops
Atmospheric aerosol loading	Particulate concentration
Chemical pollution	Chemical concentrations in the environment

important. Thresholds or tipping points are, then, common in natural systems; the question is whether the global planetary boundaries can be related to tipping points in natural systems.

A third criticism is that it is hard to apply the concepts of planetary boundaries at a scale that is relevant to policy or processes. This is evidently true: even where pressures are widely accepted to be truly global, it remains difficult to find agreed global action. The 2015 Paris agreement was the 21st attempt to agree a global approach to limiting climate change (Rockström 2015). However, this is not really a problem with the concept of planetary boundaries but rather a recognition that managing global environmental pressures requires international co-operation, and reaching a shared, agreed outcome is extremely difficult.

Despite these misgivings about planetary boundaries, it is still worth focusing on the fundamental concept: that local activities can accumulate to cause global problems. In recent decades this has become obvious for atmospheric pollution. The discovery of the Antarctic ozone hole, caused by anthropogenic release of chlorine- and bromine-containing compounds, prompted global action to reduce emissions of these gases, in the form of the Montreal protocol signed in 1989 (Chipperfield 2015). Climate change, the result of the release of greenhouse gases into the atmosphere, presents, by the end of the twenty-first century, a risk of "severe, pervasive and irreversible impacts globally" (IPCC 2014, p. 17). The atmosphere is well-mixed: cumulative emissions of persistent gases like CO_2 affect the whole world. This means that the concept of a planetary boundary for greenhouse gas concentrations is rational and logical, even if it is hard to negotiate the precise value that should be set as the boundary. There is global agreement on the need to avoid ozone depletion, a boundary also identified by Rockström et al. (2009). Notably, the release of ozone-depleting and greenhouse gases generally has no immediate adverse local environmental impact: it is the accumulation of these releases that affects the whole world. This means that for these two boundaries there is no conflict between global and local environmental limits because the reason to set limits for these gases is to manage global atmospheric concentrations.

What of the other planetary boundaries? Rockström et al. (2009) present a strong case for the global relevance of the other planetary boundaries, often through linkages between the coupled ocean–land–atmosphere system. For example, functioning ecosystems, rich in biodiversity, are thought to be important for resilience to change in the face of pressures from climate change. Land-use change and the associated loss of biodiversity affect the global climate: this link is well-known for large areas like the Amazon (e.g. Werth and Avissar 2002), but earth-surface–atmosphere interactions have an impact on local weather. The very hot August in central Europe in 2003 is thought to have been partly a response to reduced surface latent heat fluxes because of very dry soils and wilted vegetation, perhaps with feedback between reduced soil moisture and reduced convection and cloudiness (Fink et al. 2004). Nitrates and phosphates, used to fertilize land, are released through rivers, where they can damage ecosystems, to the oceans. Phosphorus inflow to oceans is thought to have caused large-scale anoxic events that may have led to mass extinctions of marine life in the mid-Cretaceous (Handoh and Lenton 2003). Reducing phosphorus inflow to oceans is intended to reduce the risk of a future mass extinction.

These other global boundaries are different, though, because they are also manifested directly as local pressures on the environment, human health, well-being, or all of these. Phosphates and nitrates cause eutrophication in rivers, with algal blooms threatening the quality of water supplies and damaging ecosystems (for example, Whitehead et al. 2009). Locally, there may be a need for limits that are much more stringent than any global restriction. The freshwater withdrawals limit has been a particular focus of debate (e.g. Heistermann 2017). Taking too much water from rivers, lakes, or groundwater causes environmental damage in all regions of the world. However, leaving additional water in the environment in some regions will not compensate for the damage resulting from over-abstraction in other regions. This applies at all geographical scales, from river reach to continent. Reducing abstraction in the lower reaches of a river will not negate environmental damage in the upper reaches. Returning European rivers to a more natural state will not make excessive water use in African or North American rivers any more acceptable. Freshwater abstraction limits are set with reference to local environmental needs and values: this means that it can be hard to see how the existence of global limits or boundaries could have any direct influence on setting local abstraction rules. Gerten et al. (2013) propose that the freshwater withdrawals boundary should be calculated from a "bottom up" quantification of local water availability that considers environmental flow requirements. This addresses the problem of local relevance, but may reduce the value of the global boundary for decision-making. Gleeson et al. (2020b) argue that a global boundary remains relevant because it can consider flows beyond traditional river basin boundaries, including the flow of virtual water (Section 15.3). Additionally, Gleeson et al. (2020a) suggest that local modifications to the water cycle can accumulate to cause regional or global regime changes. This idea provides the theoretical link between local and global boundaries. Gleeson et al. (2020b) propose that further research in this area will lead to effective sub-boundaries for six different components of the hydrological cycle: evapotranspiration, precipitation, soil moisture, surface water, groundwater, and frozen water.

15.3 Blue Water, Green Water, Small Water, Big Water, Virtual Water

Planetary boundaries try to establish global limits for human activities: beyond these limits the earth may move permanently to a different state. Planetary boundaries are a useful way to think about the accumulated pressures that humans are imposing on the environment, and have been extremely influential in making both scientists and policy makers think about the cumulative impacts of local and regional changes. However, planetary boundaries are perhaps best seen as a large-scale framing of the problem of environmental pressures. They do not help directly with understanding the local impact of human activities or the interconnected nature of different actions. For example, the freshwater withdrawals limit as currently formulated draws attention to the problem of water use but does not give much assistance in illuminating local or regional problems, or in identifying where more water use may be acceptable.

The idea of "virtual water" (Allan 1996, 1998, 2011) was introduced as a way to help people to understand and examine their collective impact on global water resources. Virtual water is the water embedded in food: most of it is not physically present in the food we buy, but has been used in its production. Allan (2011) calls the water we see when we drink it, wash in, or cook with, "small water" because this is a minor part of human water use: the "big water" is the much greater volume of water associated with food. For example, in the

UK people use on average 150 litres of water every day at home (Watts and Anderson 2016), which adds up to about 1 cubic metre, or 1 tonne, of water each week. Allan (2011) estimates that it takes 140 litres of water to make a single cup of coffee, and 1000 litres (1 cubic metre) of water for a single litre of milk. He also points out that virtually every nation is self-sufficient in "small water", but many, especially in the developed world, do not have nearly enough "big water" to meet their food requirements.

Allan (2011) also distinguishes between blue and green water. "Blue water" is water in rivers, lakes, and groundwater; "green water" is the water in vegetation and crops that we do not really see. Water is essential for plant growth: green water flows back into the atmosphere as transpiration and evaporation (Falkenmark and Röckstrom 2006). Partitioning water between blue and green reminds us that water management, especially for food production, is not just about the relatively small volume of water in rivers, lakes, and groundwater, but also about land management for rain-fed crops and vegetation. Falkenmark and Röckstrom (2006) argue that integrated water management must consider all of the water in the environment rather than just the water directly under human control.

Allan (2011) explores the way that virtual water is moved around the world. Virtual water is traded in the form of food, and tends to flow from poor countries to rich countries. The way that people in rich countries consume food has a direct impact on water use, and therefore the environment, in poor countries. On a positive note, the trade in virtual water feeds the world and provides income for people in poorer countries. However, the virtual nature of this trade in water hides the global context of the problem of over-exploitation of water from the people whose decisions determine how this trade works. This means that decisions about food are made without full knowledge of the consequences, at a time when feeding the growing world population has never been more difficult.

Virtual water provides a valuable way of understanding the links between people, food, and water, and makes the global nature of water exploitation much clearer. Allan (2011) also identifies a solution to the problem of water exploitation at a transboundary scale that would allow more food to be produced and support global water security: waste less food and change diets to consume less water. While simple, Allan also recognizes that this is not easy, requiring political will and actions through the human food production chain.

Planetary boundaries and the ideas of embedded water and the trade in virtual water help us to see that local actions have global consequences through links that are often invisible. However, within this global context it is still necessary to manage water at the local, catchment scale, where people and the physical environment interact, noting that this interaction is a response not only to the environment but also relations of social power (Lave et al. 2014). To understand, manage, and remediate the impact of environmental pressures requires a focused, process-based understanding of the effect of human action on catchments, grounded in a socio-political context.

15.4 Human Interventions in Catchment Hydrological Processes

Why do humans intervene in catchment processes? Measures to reduce flood risk are directly designed to change the way catchments respond to rainfall, but most interventions in hydrological processes are a by-product of other activities intended to offer benefits for some part of human society. Other common human interventions include agriculture and forestry, land-use change, provision of water supply and waste water treatment, and water use in industry and commerce. Understanding the hydrological future requires an understanding of the interaction between these interventions and the physical catchment (Sivapalan et al. 2012; Thompson et al. 2013).

15.4.1 Flood Risk Management

Flood risk management activities aim to reduce the damage caused to humans and property by flooding. Risk is conventionally defined as the product of probability and consequence: in the context of floods, this means that risk can be reduced either by reducing the probability of flooding or by lowering the consequences of flood events. Globally, flooding is the most frequent natural hazard (World Bank and United Nations 2010). Climate change is expected to increase the frequency and intensity of extreme rainfall in many parts of the world (Wilby and Keenan 2012), though this does not necessarily translate into an increased flood risk, because the resulting flood risk depends on the success of the interventions that have been made.

For much of the twentieth century, flood risk management relied on what are often called "hard engineering" solutions: flood defence structures that contain water within channels or divert it away from vulnerable areas. This type of structure can be very successful in reducing flood risk, though this is often at the cost of increasing flood risk at a downstream location. It is also rarely affordable to build flood structures that are sufficient to protect against very extreme floods. This means that areas defended by structures can still be subject to flooding; this is a real-world example of a threshold or tipping point, where there is a sudden switch from no impact to what can be quite a considerable impact with only a slight increase in flood size (Figure 15.1).

Flood control structures constrain flow and shape channel processes, preventing meandering and changing erosion and deposition patterns. They tend to hasten flow, moving water away from the protected areas to reach other downstream areas more quickly. This can increase downstream flood magnitude, potentially increasing flood risk in the lower reaches of a river. Alternatively, risk may be transferred upstream, as flows are held back to avoid downstream flooding. Flood risk management can produce conflict between those protected by defences and people upstream or downstream whose flood risk is increased by the measures taken (see Thorne (2014) for a description of tensions between urban and rural in the English floods of winter 2013–2014). For transboundary rivers this increased flood risk can be transferred from one country to another, with the potential to cause disagreement and conflict between different locations unless a negotiated, shared solution can be found.

Flood management interventions are driven by the need to protect people and property from floods. Mitigation of the environmental impact of flood management is often an afterthought, considered after the main design of the structure. An alternative approach known as natural flood management (Lane 2017) describes catchment interventions that are intended to reduce the risk of flooding without major engineering works. Natural flood management measures include tree planting to dry soils and increase infiltration, increasing bankside roughness to slow flows, and allowing the build-up of woody debris in channels to hold flows back, reducing peak flows. Reconnecting rivers with their flood plains

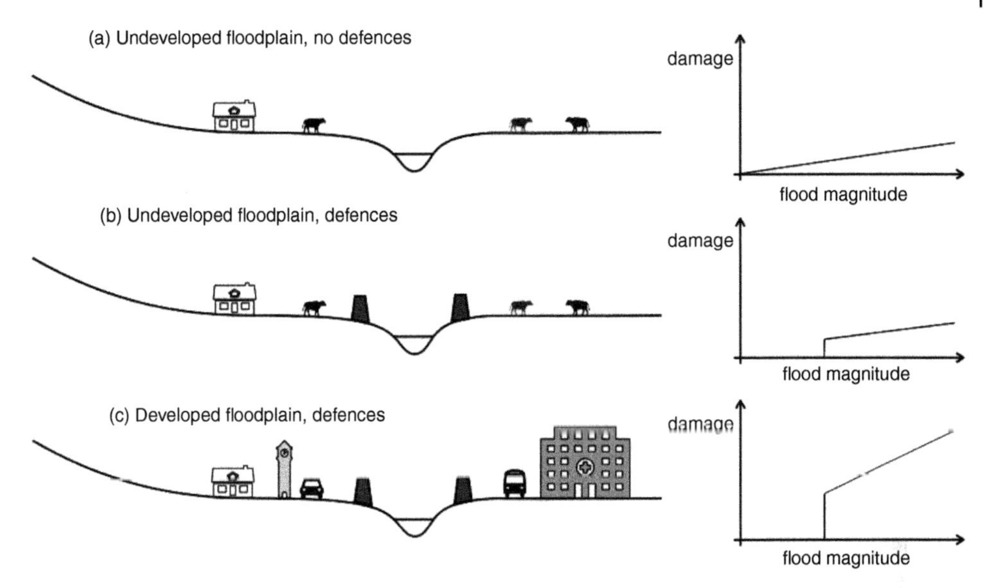

Figure 15.1 Thresholds in damage caused by floods. (a) In an undeveloped flood plain with no defences, damage is proportional to the size of flood. (b) Defences protect from all damage until the flood is large enough to overtop the defences. (c) As people take advantage of the protection offered by defences, the flood plain is developed. Now the overtopping flood causes much more damage.

allows water to spread out of channels, slowing flow and reducing the height of flood peaks. However, flood plains are often developed or used for high-value agricultural crops: allowing rivers to spread into their flood plains involves a trade-off between cost, environmental protection, and reduction in flood risk (Guida et al. 2016).

15.4.2 Agriculture and Forestry

Cropping land for food and plant-based materials affects catchment hydrological processes in many different ways, creating new boundaries and thresholds.

Different plants extract different quantities of water from the soil and have different rooting structures and depths, changing soil structure and levels of soil organic matter. The way the soil is worked can encourage or inhibit infiltration to groundwater. Land drainage, aimed at reducing water-logging, increases run-off to rivers. In arable cultivation, bare soil is susceptible to erosion, increasing sediment loads and ultimately reducing soil productivity. New crops, for example biofuels, can lead to increased water use (Vanloocke et al. 2010). Fertilizer use alters the nutrient status of receiving waters, changing aquatic ecosystems. In pasture systems, grazing animals can also cause soil erosion, with particular problems where large animals like cattle have direct access to water courses, where trampling mobilizes sediments. Intensive grazing can increase run-off through soil poaching and degradation (Weatherhead and Howden 2009).

Irrigated agriculture is a significant use of water in many countries. Irrigation improves the quality, yield, and profitability of crops. Irrigation is most needed in the summer, when river flows are lowest, so irrigation can have the effect of reducing flows dramatically. Irrigation water is taken up by crops or evaporated, with very little return to rivers or groundwater.

It is widely accepted that forestry activities have a particular impact on catchment hydrological response. In the UK, the Plynlimon study was established in the 1960s specifically to address the question of the water use of coniferous forests (Kirby et al. 1991). Binns (1979) examined the hydrological impact of commercial forests in the UK, finding reduced water yield downstream, but also setting out in detail the impact of different stages of forest operations. Binns saw four phases of forest operations: establishment (clearing old vegetation and planting saplings); thicket (as the canopy closes but before thinning); pole (periodic removal of part of the crop); and finally, clear-felling, when the final crop of trees is removed. There is a broad cycle with increased run-off and sediment yield in the establishment phase, with additional nutrients if fertilizers are used. During the thicket and pole phase, the forest uses more water. In the clear-felling phase there can be sudden changes in run-off patterns and greatly increased sediment yield. More recent work confirms the impact of forestry in reducing water yield in many parts of the world (Trabucco et al. 2008, Van Dijk and Keenan 2008) and reducing annual recharge to groundwater (Zhang and Hiscock 2010). The overall reduction in flow is not necessarily negative, because forests can regulate flow with the effect of increasing low flows even with a reduction of total flow (Trabucco et al. 2008). Newly-planted forests also have the effect of removing carbon from the atmosphere, helping to reduce greenhouse gas concentrations (IPCC 2000).

15.4.3 Land-use Change

Land-use change is a broad term that encompasses all changes in land use at all scales. Permanent or semi-permanent changes in land use include changes from pasture to arable agriculture, clearing natural or old forests for agriculture, moving agricultural land into commercial forestry, and urbanization. All of these activities change catchment hydrology, often inducing a shock into the system at the time of change, after which the hydrological response may recover towards a previous state or adjust to something new.

Urban areas have a high proportion of impermeable surfaces, for example in roads, pavements, and roofs. They also often have efficient drainage systems, aimed at moving water quickly into rivers and water courses. This can lead to increases in run-off (Haase 2009) and reductions in infiltration, and recharge to groundwater (O'Driscoll et al. 2010) though poorly-maintained leaking pipes and sewers can, in some cities, lead to increased recharge (O'Driscoll et al. 2010). Rapid run-off of water from roads and sewers has long been recognized as a factor in reduced water quality in urban water courses (e.g. Ellis 1979). In recent years, Sustainable Urban Drainage Systems ("SUDS") have received much attention. The aim of SUDS is to encourage infiltration and detention of water where it falls, rather than moving it away to water courses (Charlesworth 2010). Measures include porous pavements, soakaways, detention ponds, and vegetated areas. SUDS can not only capture and slow water, but can also reduce the run-off of pollutants into rivers.

15.4.4 Provision of Water Supply and Waste Water Treatment

Water supply is perhaps the most obvious way that humans interfere in the water cycle. Provision of reliable, clean water for drinking, washing, and cooking is an essential part of modern society, securing public health and underpinning economic activity. The water supply problem is essentially one of smoothing spatial and temporal variability in the availability of water to make sure that adequate volumes of water are available at all times. Highly populated areas need to draw on sources that are often distant from the centre of demand. By drawing from groundwater or from water stored in reservoirs, water suppliers can maintain the availability of water even through extended droughts. However, solving this apparently simple problem requires major infrastructure. Reservoirs dam river valleys, retaining water but necessarily reducing downstream flow. Treated water flows through pipes to centres of demand, where it is distributed to neighbourhoods and households. These may be in the same river catchment in which the rain fell, but this is not always the case: water transfers often move water away from the original catchment.

Mains water is used in many ways. At home, people use water for drinking, cooking, washing themselves and their clothes, toilet flushing, and for many other purposes including garden watering and car washing. In most of these activities, little water is actually consumed: water used for washing, showering, bathing, or toilet flushing is then released back into the environment. In the UK, it is estimated that about 90% of the water entering households is returned to sewers: most of the remaining 10% is used outdoors, for example, for garden watering. Offices, schools, and businesses also need water. People need water for washing, hygiene, and toilet flushing, whether at work or at leisure. Many businesses use water as part of their activities, for example, in manufacturing.

The water left after people have finished with it is usually released to sewers, from where it is gathered, treated, and returned to rivers or the sea. The returned water will normally be downstream of the water source, though this is not always the case. During the summer, returned water is usually water that has been stored since the winter in reservoirs or groundwater, so summer discharges can increase river flows above natural levels. In coastal catchments, water may be returned to the sea, sometimes because a lower treatment standard is required.

The overall effect of water supply and treatment is to reduce river flows downstream of reservoirs throughout the year. Returned flows from waste water treatment tend to increase summer river flows but inject this extra water at a single point, sometimes causing a step change in flow volume and temperature. In heavily used catchments, natural variability is smoothed between winter and summer and from year to year (Figure 15.2). Proportionally, the greatest hydrological effect of water supply is when flows are lowest (usually summer, except in river flow regimes dominated by snow-melt), where upper river reaches may be depleted and lower reaches have increased flow with reduced variability compared to the natural flow regime. When flows are high, the volume of water abstraction and discharge is often only a very small part of the total flow, though in some places discharges can add to flood risk. This is a particular problem where sewers carry not only waste water but also storm flows from surface drainage. This presents an additional risk of overwhelming the capacity of waste water treatment, leading to discharges of untreated sewage into flood waters. This is an additional hazard to human health, and the flooding of property with contaminated water adds significantly to the problems of recovering after floods.

15.4.5 Other Uses of Water

Large volumes of water are taken from rivers and groundwater for other uses that include power generation and manufacturing. In contrast to public water supply, where water is often transferred over large distances, other major uses of water tend to use water from sources near to the production site, but the impact on catchment hydrology can be significant.

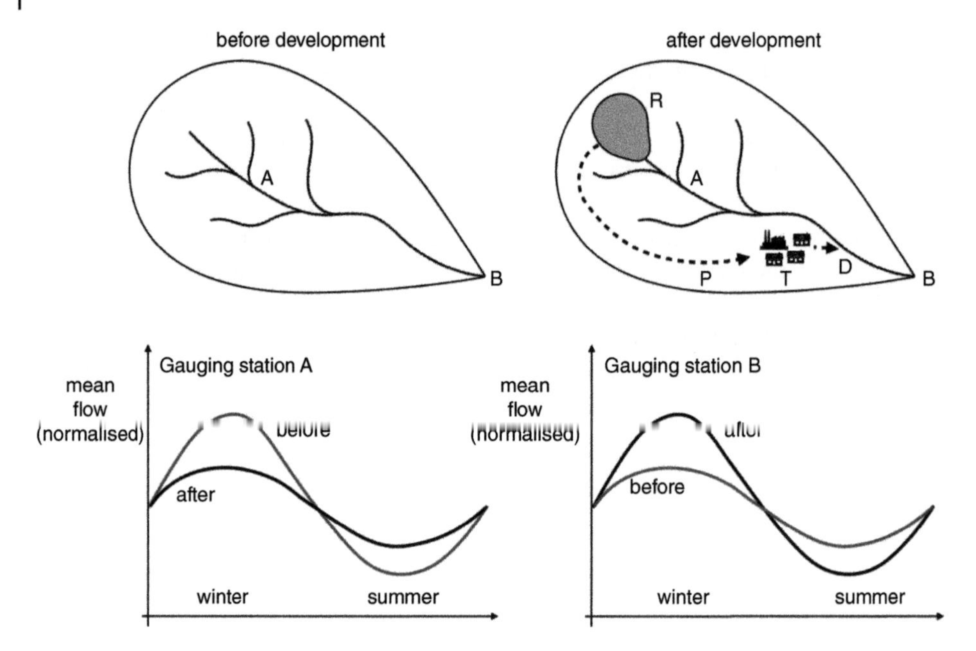

Figure 15.2 The impact of catchment development on river flows. Reservoir R feeds town T through pipe P. Treated effluent is discharged at point D. At point A in the upper part of the catchment, the reservoir reduces high flows, especially in winter, and increases low flows (reservoir operating rules usually ask for releases to maintain summer flows). At point B, below town T, flows are smoothed, with notably higher summer flows as a result of discharges of treated effluent.

Power generation is a major use of water. Hydropower schemes use water directly, with big hydropower schemes usually supported by a reservoir from which releases can be made. Water passes through the turbines and then flows into the river. As an easily controlled source of power, hydropower schemes can be used to manage peak demand for electricity by releasing more water at times of high demand, such as in the early evening. This can lead to daily cycles of high and low releases and hence downstream flows: this is a very unnatural pattern for many rivers and can change the characteristics of downstream reaches.

Most large thermal power plants need water for cooling. Cooling may be evaporative (plumes of water vapour are often seen above the huge chimneys of coal-fired power stations) but is often non-evaporative, with warmer water returned to rivers. Thermal pollution from cooling water from power stations can alter river ecology for some distance downstream. Nuclear power stations also need adequate supplies of cooling water; while many nuclear plants are on the coast, some are inland. In France in the very hot summer of 2003, nuclear power plants had to be switched off as river flows fell and river water temperatures rose, threatening the ability to cool power stations sufficiently (Poumadère et al. 2005).

Water is an important part of the manufacturing process for many items: it is not only food that contains embedded water. Large industrial users of water often abstract water from rivers or groundwater and treat it to the standard that they need for their process, which is not necessarily drinking water standard. These large abstractions can disrupt catchment hydrology, and the water returned is of a different quality from the river water.

15.5 Predicting Catchment Response: Approaches to Hydrological Modelling

In much of the world the consequences of human activities disturb catchment hydrological processes. All of these activities are intended to achieve a benefit for people, with the impact on catchments generally incidental to the activity. Actions tend to be at a point, but the effects are often spread much more widely. The results of some activities accumulate through the catchment, but for others the impact may be local. For example, water abstraction and discharges accumulate down the catchment, so that the total reduction in flow at the catchment outflow is the sum of all the upstream net abstraction. On the other hand, point sources of pollution may be important locally but have little widespread effect. An example of this is thermal pollution, for example, from a power station. In the immediate vicinity of the outfall the water temperature will increase, but as the warmer water moves down the catchment it will cool as it mixes with other water and through heat exchange with the river bed and the atmosphere.

Predicting and understanding the combined impact of human interventions, as well as the effectiveness of mitigation measures, usually requires a numerical catchment model. Broadly, models can be lumped, distributed, or semi-distributed (Watts 1997) (Figure 15.3).

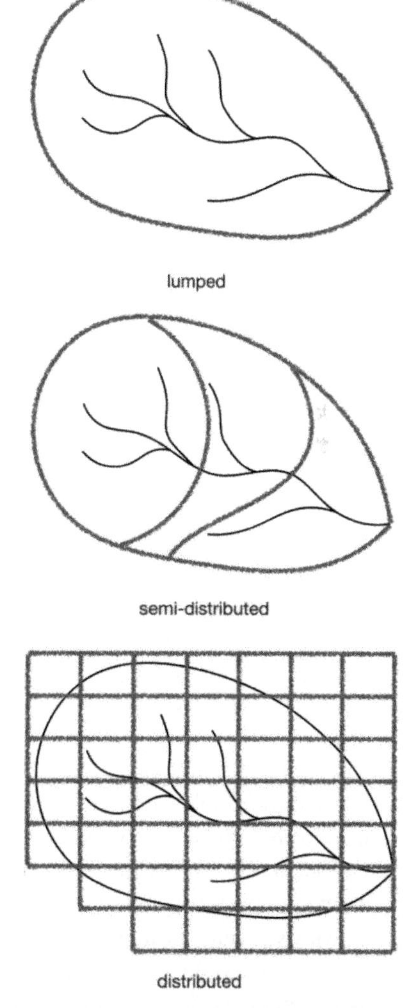

Figure 15.3 Conventional approaches to catchment hydrological modelling.

The simplest catchment models are known as "lumped" models. The catchment is represented as a single entity. Typically, hydrological lumped models are conceptualized as a series of stores that route rainfall to the end of the catchment. Model parameters determine the rate of flow between the stores: parameter values are adjusted to make the model output match measured river flows. While it is easy to think about routing rainfall to river flow, lumped models can also be used to model water quality, as long as a suitable representation of catchment processes can be identified and incorporated in the model. Lumped models are often surprisingly effective at modelling river flows, despite the simple way that they represent catchments. This is partly because lumped models often have a large number of parameters: with enough parameters the models can be tuned to match the real world closely. Lumped models are also successful precisely because they are lumped: representing the whole catchment simultaneously means that the model only needs to simulate the important, dominant factors that really affect the way the catchment behaves. However, lumped models are very broad conceptualizations of catchments that do not represent catchment processes. This means not only that lumped models may perform poorly outside their range of calibration, but also that it is very hard to diagnose poor model performance. Perhaps the biggest difficulty in using lumped models to understand the effect of human interventions is that actions are at points, but lumped models can only represent the total impact across the catchment. For impacts that accumulate down the catchment, like net abstraction, this can be an appropriate and acceptable way to understand total catchment impact. However, where impacts do not accumulate, lumped models will fail to identify the scale or nature of the change that human interventions have caused.

Lumped models may be based on a conceptual understanding of catchment behaviour, but distributed models try to represent the physics of catchment processes. A typical distributed model represents the catchment on a rectangular grid, routing water across and under the surface using equations derived from a theoretical appraisal of catchment processes: for example, the Richards equation (Richards 1931) is often used to route water in the unsaturated and saturated zone below the surface. In theory these physically-based models offer a good representation of catchment processes: their realistic physical basis means that they can be used to examine conditions outside historical experience, because the same physical processes will still occur. In practice, though, physically-based models still need to be parameterized to represent processes that occur within the scale of the grid-cell or within the model timestep. A very high resolution physically-based hydrological model may discretize the catchment into grid cells of 1 km by 1 km, or even 500 m by 500 m. In a relatively small catchment of around 10 000 km^2 like the Thames (England) or the Júcar (Spain), a 1 km resolution model would have 10 000 grid cells and at a 500 m resolution there would be 40 000 cells. Yet it is clear that in the real world, even a 500 m by 500 m area will not be homogenous: topography, soils, and vegetation will vary and at any time some parts of the area will be wetter than others. Beven and Germann (2013) argue that these so-called physically-based models may not even be representing the right processes: for example, the Richards equation represents subsurface flow as a diffuse, slow process, but we have known since the 1980s that macropores are a significant vector of flow in many soils, bypassing the soil matrix and moving water quickly in preferential pathways (for example, Beven and Germann 1981).

Semi-distributed models are conceptually between lumped and distributed models, but in practice they are more like a series of lumped models, with individual models representing discrete areas of the bigger catchment. Semi-distributed models can offer a better sense of differential impacts across the catchment, but they are still conceptual models, with little explanatory power.

Lumped and semi-distributed models are easy and quick to set up, tend to perform well, at least at the whole-catchment scale, and run quickly, meaning that they are particularly suited to probabilistic approaches, where many thousands of model runs may be necessary. Distributed physically-based models are hard to set up, requiring much information about catchment properties, like soil type and soil hydraulic characteristics. They are slow to run and with so many parameters for each grid cell it is difficult to perform comprehensive sensitivity analyses or develop probabilistic projections.

No model is perfect: the modeller's challenge is to identify useful models that help to illuminate the problem under investigation. In principle, physically-based models offer the best approach to understanding the complex interactions of different interventions without needing to impose assumptions about the way that they work together. Coding the fundamental physics and letting the model route water according to these physical rules can lead to "emergent" behaviour: processes can combine to produce non-linear outcomes that are physically realistic and plausible but were not anticipated by the modeller. This can lead to the identification of thresholds, boundaries, or "hot spots": times, places, or points that act as an important control on the way the catchment behaves. However, conventional hydrological models are essentially static: they route water through an unchanging physical environment. In the real world, catchments are always changing, with human intervention sometimes imposing very rapid changes. In a rapidly changing world, hydrological modelling must also consider feedbacks between human activities and catchment response (Thompson et al. 2013) as well as the non-stationary climate (Milly et al. 2008). The next section considers some geomorphological concepts of change and looks at how a consideration of these may improve the way that we understand environmental pressures in our changing world. These geomorphological concepts could inform approaches to catchment modelling that can consider boundaries and interfaces more explicitly than many current hydrological models.

15.6 Geomorphological Understanding of Catchment Change

Hydrology is the study of the way that water circulates through the hydrological cycle, from oceanic evaporation to rainfall, to rivers and back to the ocean. Geomorphology is the study of how the land changes. The two disciplines are closely linked: water is one of the main agents of geomorphological change in many parts of the world, and on a more practical level, many students and academics study both hydrology and geomorphology. However, much hydrological analysis assumes that the catchment presents an unvarying response between rainfall and evapotranspiration and river flow. This is a useful and sensible simplifying assumption in many circumstances. For example, in real-time flood forecasting ignoring catchment change during a flood event will definitely add to uncertainty in the projection of flood magnitude or duration, but this will be a very small part of the total uncertainty in flood projections: uncertainty in forecasts and spatial distribution of rainfall is certain to be several orders of magnitude greater than the uncertainty from ignoring catchment change over this short time period. However, ignoring catchment change is unlikely to be sensible in long-term planning. Water supply planning tends to consider a minimum time horizon of 25–30 years, but major water supply infrastructure is often in place for a century or more. Many water supply reservoirs in Europe and north America are already over 100 years old, and are expected to continue to provide water through the rest of this century and beyond. Most new water supply schemes will need to function through the rest of the twenty-first century, a time in which we expect accelerated climate change as well as increasing population. Similarly, any flood defence scheme designed now must consider changing rainfall volumes and intensities through the rest of the century, as well as the inevitable sea level rise that climate change will bring.

What does geomorphology offer to people making plans and decisions that need to consider long time periods? The origins of geomorphology are in theoretical approaches to long-term evolution of landforms. For example, Davis (1899), a Harvard professor and one of the fathers of geomorphology, considers the way that landforms change over time and how "all forms, however high and however resistant, must be laid low, and thus destructive process gains rank equivalent to that of structure in determining the shape of a land-mass" (p. 482). Davis's "geographical cycles" operate on what he acknowledges to be geological timescales: he observes that "all historic time is hardly more than a negligible fraction of so vast a duration" (p. 483). While hydrology's assumption of an effectively unchanging catchment response is increasingly untenable, Davis's timescales are far beyond anything that today's decision-maker needs to consider.

Brunsden and Thornes (1979) introduced the concept of landscape sensitivity, where "landscape stability is a function of the temporal and spatial distribution of the resisting and disturbing forces" (p. 476). In this conceptualization, measurable change depends on the scale of the disturbing force and the capacity of the system to absorb this force. Brundsen (2001), in a critical appraisal of this concept, lists disturbing forces as tectonic, climatic, biotic, and anthropogenic. In the terms used in this chapter, these disturbing forces might also be called "pressures". Resisting forces are a system's "ability to resist displacement from its initial state following a disturbance" (p. 101). Brunsden identifies resisting forces as strength resistance, morphological resistance, structural resistance, filter resistance, and state resistance. **Strength resistance** is a measure of the physical and chemical characteristics of the landform, often derived from the characteristics of the underlying geology. Catchments where underlying rocks are soft (for example, sedimentary rocks such as chalk, limestone, and sandstone) change more quickly than hard rocks (for example, igneous or metamorphosed rocks like granite, gneiss, or marble). **Morphological resistance** reflects the existing shape of the system: the energy available for change depends on elevation and slope. **Structural resistance** is about how well linked the system is: a well-organized system of river channels will route water and sediment quickly, transmitting energy through the catchment. **Filter resistance** is how the system absorbs and removes energy that would otherwise cause change: Brunsden's example is the way that sandy beaches absorb wave energy. **System state resistance** is perhaps the most interesting type of resistance. It is a function of the history of the system, because how the catchment has reached its current state affects how it will respond to future change. This implies that the current state of the system cannot be determined simply by measuring its present characteristics, but also needs an understanding of whether it is still recovering from previous change. This may help to explain why two apparently similar catchments can respond to pressures in different ways.

Thinking about the ratio of disturbing and restraining forces helps to explain how catchments can respond to different pressures (Figure 15.4). Some pressures may be absorbed by the catchment, leading to no change in state. Other pressures may lead to gradual change, while others can lead to a rapid change to what is effectively a new state. The system may recover from the new state back to something resembling the old state, or alternatively the new state may become a new normal for the catchment. Importantly, the response to a disturbing pressure will depend on the scale, location, magnitude, and timing of the disturbance.

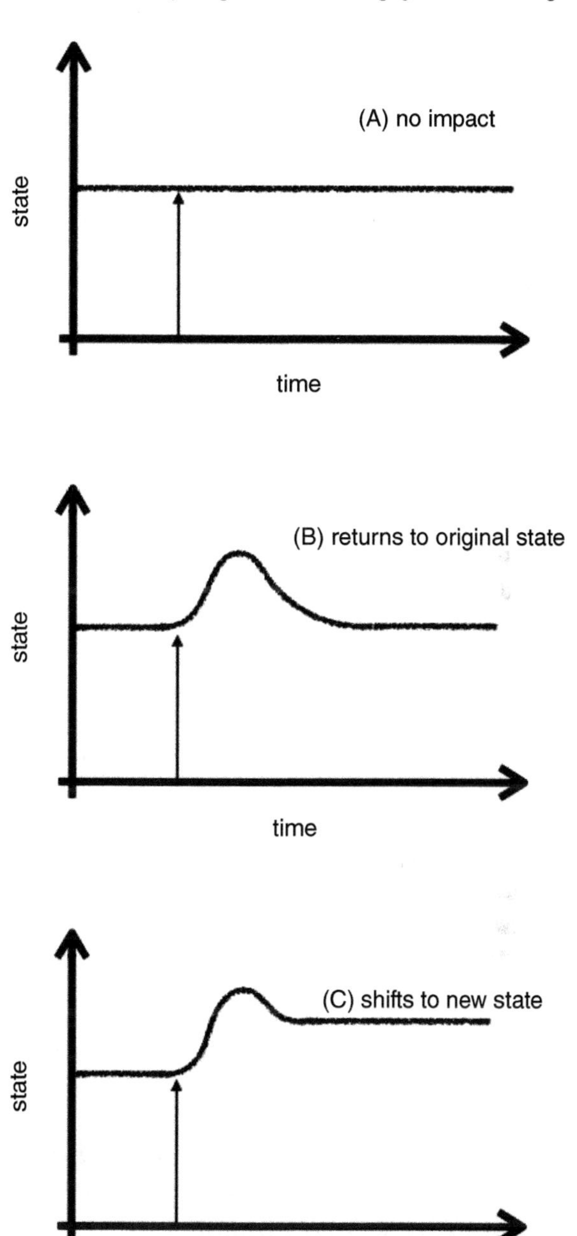

Figure 15.4 The impact of an instantaneous pressure on catchment state. (a) The catchment resists the pressure and there is no change. (b) The catchment responds to the pressure but over time returns to its original state. (c) The catchment responds to the pressure and moves to a new, different state.

How might these concepts help the catchment manager? Mapping the ecological state of the water environment is now carried out across Europe in response to the European Water Framework Directive. For example, Raven et al. (1998) show how river habitat surveys can be used to map the quality of the water environment. However, current state, usually measured as a deviation from a theoretical natural state, is, in the main, a response of the catchment to historical pressures. Might it be possible to evaluate the current resistance of different parts of the catchment to disturbing pressures? Brundsen (2001) suggests that this is feasible, though still in need of further research, particularly in establishing and synthesizing the data. Sánchez-Pinillos et al. (2016) propose that the "persistence capacity" of ecological communities can be mapped based on their natural traits. Their Persistence Index (PI) helps to indicate the way that a community can continue to function as an ecosystem after disturbance, although when applied to forests in the Iberian Peninsula, non-native species were found to demonstrate the most resistance.

The idea of geomorphological resistance has remained broadly theoretical, but the allied concept of resilience has been increasingly applied to environmental systems. Holling (1973) argued that the behaviour of ecological systems can be defined by two distinct properties: resilience and stability. Holling (1973) defined resilience as the ability of a system to absorb changes and still persist. Holling saw stability as the ability of a system to return to an equilibrium state after a temporary disturbance. Holling took the view that an ecological system could be highly unstable and still resilient. Furthermore, instability, for example, driven by climate shocks, could actually drive increased resilience in an ecological system, because the system adjusts to cope with disturbance.

In recent years there has been increased interest in catchment resilience, defined variously as an ability to resist or absorb change or the capacity to return to a previous state after a disturbance. Discussions of resilience are confused by the multiple definitions in circulation, and further by the use of different terms for these concepts in different disciplines. Resilience is sometimes considered to be a measurable attribute of a system (e.g. Yi and Jackson 2021) but is increasingly used as a system target or goal (Adger et al. 2021), with increased resilience a desired outcome of flood management strategies (Fekete et al. 2019), water supply planning (POST 2021), or climate change adaptation action (Murgatroyd et al. 2022). Thinking about resilience, however defined, does encourage a consideration of the catchment as a dynamic social and ecological system (Adger et al. 2021).

Clearly there is more work to do on understanding how resistance to change and resilience varies across catchments, but understanding the vulnerability of different locations would seem to offer new approaches to understanding where either protection from new pressures or measures to reduce existing pressures might be most effective.

15.7 Boundaries, Thresholds, and Hot Spots: Towards Better Prediction and Management of the Impact of Pressures on the Water Environment

The central theme of this book is that an improved understanding of boundaries, interfaces, thresholds, and hotspots will provide new insights into the way that catchments function and change. How do these concepts help with catchment management?

The consideration of ecohydrological interfaces provides an alternative approach to understanding catchment processes, putting particular emphasis on identifying the locations and times that are important in determining ecohydrological system behaviour. Interfaces are dynamic boundaries that can appear and disappear, and can occur in different locations at different times. The implication of the existence of hotspots and thresholds is that catchment interventions can be particularly effective at some times and in some places. This presents an exciting prospect for catchment managers, because it suggests a new way of managing catchments. Important interfaces could be protected, while it may be possible to relax catchment controls in relatively insensitive locations. If pressures could introduce undesirable interfaces that magnify the impact of change, perhaps the pressures can be managed to avoid this consequence.

New approaches based on the concepts of interfaces, hotspots, and boundaries are exciting, but they also present challenges for decision-makers. Decision-makers work with boundaries all the time, but these are usually fixed. Administrative units are defined politically, often with little physical meaning, although some administrative boundaries follow topographic divides, and borders are often formed by rivers. The fluidity of boundaries implied by the interface concept does not fit with the static boundaries that define the areas in which decisions are made. Within political boundaries, though, could concepts of interfaces and hotspots translate into better catchment management? Identifying locations, times, or conditions where different management measures are needed could enhance environmental protection while at the same time allowing people to benefit more from using the environment.

Decision-makers are rarely water specialists, though all have some degree of experience with water simply because water is part of everyday life and it is impossible not to develop some understanding, however flawed. Lane (2014) refers to people who have experienced flooding as "non-certified hydrological experts": people whose experience makes them expert in some aspect of hydrology, even though they are without formal training. Decision-makers are also experts in their own domain, and may not like scientists stepping too far into the decision-maker's own responsibilities (Gluckman 2014). Decision-makers often have to work to short deadlines and deal with conflicting interests. In such circumstances, any complex scientific advice may feel unwelcome: decision-makers are used to dealing with uncertainty but also like clarity. When scientific advice is unclear or ambivalent, decision-makers may fall back on their own water expertise, however lacking that may be (Watts 2016).

Much environmental regulation is rule-based. Activities that could cause environmental damage are controlled by permits. For example, abstraction of water may have conditions that not only set out the maximum volume of water that can be taken but also reduce or stop abstraction if river flows or groundwater levels are unusually low. Permits that allow the discharge of treated water into rivers or the sea will usually set chemical limits that must not be exceeded. Rules need to be relatively simple, partly because they need to be understood by the water user but perhaps more importantly because they must be enforceable. This means that it must be clear when the rules are being broken. In general, environmental regulations must also be accepted to be fair: for example, scarce water resources need to be shared equitably between different sectors and different users within sectors. This is important for credibility with water users and the people who are affected by their actions.

Could new approaches to water management based on the concepts of interfaces help to manage increasing environmental pressures? The science behind the interfaces concept is complicated and not altogether intuitive: this may make it difficult to persuade decision-makers and the people they serve that new, more dynamic approaches are an improvement on existing methods, and also that they are fair and equitable. But the benefits could be massive. In the Anthropocene, we face unprecedented pressures on the water environment. New understanding of catchment processes could help us to use the environment more wisely, bringing benefits in food productivity, flood protection, and water availability. The challenge for the scientist is to bring these complex but exciting ideas to life with evidence of the improvements that can be achieved.

Acknowledgements

Much of this chapter was informed by discussions in the Interfaces ITN project meetings, and I am especially grateful to David Hannah and Stefan Krause. I consider myself extremely lucky to have known Tony Allan (1937–2021), whose seminal work on virtual water is central to the understanding of the global impacts of water use. I thank my colleagues Harriet Orr and Stuart Allen for many useful conversations about the practical aspects of catchment resilience. I also thank Robert Bradburne Chief Scientist at the Environment Agency, for permission to publish this chapter. The views are my own and do not represent the position of the Environment Agency.

References

Adger, W.N., Brown, K., Butler, C., and Quinn, T. (2021). Social ecological dynamics of catchment resilience. *Water* 13 (3): 349.

Allan, J.A. (1996). Policy response to the closure of water resources. In: *Water Policy: Allocation and Management in Practice* (ed. P. Howsam and R. Carter). Chapman and Hall.

Allan, J.A. (1998). Virtual water: a strategic resource. Global solutions to regional deficits. *Groundwater* 36 (4): 545–546.

Allan, J.A. (2011). *Virtual Water: Tackling the Threat to Our Planet's Most Precious Resource*. London: IB Tauris.

Beven, K. and Germann, P. (1981). Water flow in soil macropores II. A combined flow model. *Journal of Soil Science* 32: 15–29.

Beven, K. and Germann, P. (2013). Macropores and water flow in soils revisited. *Water Resources Research* 49: 3017–3092.

Binns, W.O. (1979). The hydrological impact of afforestation in Great Britain. In: *Man's Impact on the Hydrological Cycle in the United Kingdom* (ed. G.E. Hollis), 55–70. Norwich: Geo Abstracts.

Brundsen, D. (2001). A critical assessment of the sensitivity concept in geomorphology. *Catena* 42: 99–123.

Brunsden, D. and Thornes, J.B. (1979). Landscape sensitivity and change. *Transactions of the Institute of British Geographers* 4: 463–484.

Charlesworth, S. (2010). A review of the adaptation and mitigation of global climate change using sustainable drainage in cities. *Journal of Water and Climate Change* 1 (3): 165–180.

Chipperfield, M.P. (2015). Global atmosphere - the Antarctic ozone hole. In: *Still Only One Earth: Progress in the 40 Years since the First UN Conference on the Environment* (ed. R.M. Harrison and R.E. Hester). https://doi.org/10.1039/9781782622178.

Crutzen, P.J. (2002). Geology of mankind. *Nature* 415: 23.

Davis, W.M. (1899). The geographical cycle. *The Geographical Journal* 14 (5): 481–504.

Ellis, J.B. (1979). The nature and sources of urban sediments and their relation to water quality: a case study from north-west London. In: *Man's Impact on the Hydrological Cycle in the United Kingdom* (ed. G.E. Hollis), 199–216. Norwich: Geo Abstracts.

Falkenmark, M. and Rockström, J. (2006). The new blue and green water paradigm: breaking new ground for water resources planning and management. *Journal of Water Resources Planning and Management* 132 (2): 129–132.

Fekete, A., Hartmann, T., and Jüpner, R. (2019). Resilience: on-going wave or subsiding trend in flood risk research and practice? *WIREs Water* 7 (1): e1397.

Fink, A.H., Brücher, T., Krüger, A. et al. (2004). The 2003 European summer heatwaves and drought - synoptic diagnosis and impacts. *Weather* 59: 209–216.

Gerten, D., Hoff, H., Rockström, J. et al. (2013). Towards a revised planetary boundary for consumptive freshwater use: role of environmental flow requirements. *Current Opinion in Environmental Sustainability* 5: 551–558.

Gleeson, T., Wang-Erlandsson, L., Porkka, M. et al. (2020a). Illuminating water cycle modifications and Earth system resilience in the Anthropocene. *Water Resources Research* 56 (4): e2019WR024957.

Gleeson, T., Wang-Erlandsson, L., Zipper, S.C. et al. (2020b). The water planetary boundary: interrogation and revision. *One Earth* 2 (3): 223–234.

Gluckman, P. (2014). The art of science advice to government. *Nature* 507: 163–165.

Guida, R.J., Remo, J.W.F., and Secchi, S. (2016). Tradeoffs of strategically reconnecting rivers to their floodplains: the case of the Lower Illinois River (USA). *Science of the Total Environment* 572: 43–55.

Haase, D. (2009). Effects of urbanisation on the water balance - a long-term trajectory. *Environmental Impact Assessment Review* 29: 2111–2219.

Handoh, I.C. and Lenton, T.M. (2003). Periodic mid-Cretaceous oceanic anoxic events linked by oscillations of the phosphorus and oxygen biogeochemical cycles. *Global Biogeochemical Cycles* 17 (4). https://doi.org/10.1029/2003GB002039.

Heistermann, M. (2017). HESS opinions: a planetary boundary on freshwater use is misleading. *Hydrology and Earth System Sciences* 21: 3455–3461.

Holling, C.S. (1973). Resilience and stability of ecological systems. *Annual Review of Ecology and Systematics* 4: 1–23.

IPCC. (2000). *Special report: land use, land-use change, and forestry*. Summary for policy makers. Intergovernmental Panel on Climate Change.

IPCC. (2014). *Climate change 2014. Synthesis report: summary for policy makers.* Intergovernmental Panel on Climate Change.

Kirby, C., Newson, M.D., and Gilman, K. (1991). *Plynlimon research: the first two decades.* Institute of Hydrology Report 109.

Lane, S.N. (2014). Acting, predicting and intervening in a socio-hydrological world. *Hydrology and Earth System Science* 18: 927–952.

Lane, S.N. (2017). Natural flood management. *WIREs Water* 4: 3 e1211.

Lave, R., Wilson, M.W., Barron, E.S. et al. (2014). Intervention: critical physical geography. *The Canadian Geographer / Le Géographie canadien* 58 (1): 1–10.

Milly, P.C.D., Betancourt, J., Falkenmark, M. et al. (2008). Climate change – stationarity is dead: whither water management? *Science* 319 (5863).

Murgatroyd, A. et al. (2022). Strategic analysis of the drought resilience of water supply systems. *Phil. Trans. R. Soc. A* 380: 20210292. https://doi.org/10.1098/rsta.2021.0292

O'Driscoll, M., Clinton, S., Jefferson, A. et al. (2010). Urbanization effects on watershed hydrology and in-stream processes in the southern United States. *Water* 2: 605–648.

Phillips, J.D. (2006). Evolutionary geomorphology: thresholds and nonlinearity in landform response to environmental change. *Hydrology and Earth System Science* 10: 731–742.

POST (Parliamentary Office of Science and Techology). (2021). *POSTbrief 40, Water Supply Resilience and Climate Change*. UK Parliament.

Poumadère, M., Mays, C., Le Mer, S., and Blong, R. (2005). The 2003 heat wave in France: dangerous climate change here and now. *Risk Analysis* 25 (6): 1483–1494.

Raven, P.J., Holmes, N.T.H., Dawson, F.H., and Everard, M. (1998). Quality assessment using river habitat survey data. *Aquatic Conservation: Marine and Freshwater Ecosystems* 8: 477–499.

Richards, L.A. (1931). Capillary conduction of liquids through porous mediums. *Physics* 1 (5): 318–333.

Rockström, J. (2015). A "perfect" agreement in Paris is not essential. *Nature* 527: 411.

Rockström, J., Steffe, W., Noone, K. et al. (2009). A safe operating space for humanity. *Nature* 461: 472–475.

Sánchez-Pinillos, M., Lluís, C., Cáceres M, D., and Ameztegui, A. (2016). Assessing the persistence capacity of communities facing natural disturbances on the basis of species response traits. *Ecological Indicators* 66: 76–85.

Sivapalan, M., Savenije, H.H.G., and Blöschl, G. (2012). Socio-hydrology: a new science of people and water. *Hydrological Processes* 26: 1270–1276.

Thompson, S.E., Sivapalan, M., Harman, C.J. et al. (2013). Developing predictive insight into changing water systems: use-inspired hydrologic science for the Anthropocene. *Hydrology and Earth System Science* 17: 5013–5039.

Thorne, C. (2014). Geographies of UK flooding in 2013/4. *The Geographical Journal* 180: 297–309.

Trabucco, A., Zomer, R.J., Bossio, D.A. et al. (2008). Climate change mitigation through afforestation/reforestation: a global analysis of hydrologic impacts with four case studies. *Agriculture, Ecosystems and Environment* 126: 81–97.

Van Dijk, A.I.J.M. and Keenan, R.J. (2008). Overview: planted forests and water in perspective. *Forest Ecology and Management* 251: 1–9.

Vanloocke, A., Berneschi, C.J., and Twine, T.E. (2010). The impact of *Miscanthus giganteus* production on the Midwest US hydrologic cycle. *Global Change Biology Bioenergy* 2: 180–191.

Watts, G. (1997). Hydrological modelling in practice. In: *Contemporary Hydrology: Towards Holistic Environmental Science* (ed. R.L. Wilby), 151–193. John Wiley.

Watts, G. (2016). Hydrology with impact: how does hydrological science inform decision-makers? *Hydrology Research* 47 (3): 545–551.

Watts, G. and Anderson, M. (2016). *Water climate change impacts report card 2016 edition*. Living With Environmental Change. ISBN 978-0-9934074-1-3.

Weatherhead, E.K. and Howden, N.J.K. (2009). The relationship between land use and surface water resources in the UK. *Land Use Policy* 265: 5243–5250.

Werth, D. and Avissar, R. (2002). The local and global effects of Amazon deforestation. *Journal of Geophysical Research* 107. https://doi.org/10.1029/2001JD000717.

Whitehead, P.G., Wilby, R.J., Battarbee, R.W. et al. (2009). A review of the potential impacts of climate change on surface water quality. *Hydrological Sciences Journal / Journal Des Sciences Hydrologiques* 54 (1): 101–124.

Wilby, R.L. and Keenan, R. (2012). Adapting to flood risk under climate change. *Progress in Physical Geography* 36 (3): 348–378.

World Bank and United Nations. (2010). *Natural hazards, unnatural disasters: the economics of effective prevention.* https://doi.org/10.1596/978-0-8213-8050-5.

Yi, C. and Jackson, N. (2021). A review of measuring ecosystem resilience to disturbance. *Environmental Research Letters* 16: 053008.

Zhang, H. and Hiscock, K.M. (2010). Modelling the impact of forest cover on groundwater resources: a case study of the Sherwood Sandstone aquifer in the East Midlands, UK. *Journal of Hydrology* 392: 136–149.

16

Restoring the Liver of the River

Actionable Research Insights to Guide the Restoration of the Hyporheic Zone for the Improvement of Water Quality

Ben Christopher Howard[1,2], Ian Baker[3], Mike Blackmore[4], Nicholas Kettridge[1,2], Sami Ullah[1,2], and Stefan Krause[1,2]

[1] *School of Geography, Earth & Environmental Sciences, University of Birmingham, Birmingham, B15 2TT, UK*
[2] *Birmingham Institute of Forest Research, University of Birmingham, Birmingham, B15 2TT, UK*
[3] *Small Woods Association, Station Road, Coalbrookdale, Telford, TF8 7DR, UK*
[4] *Wessex Rivers Trust, Phillips Lane, Salisbury, SP1 3YR, UK*

16.1 Introduction

Society faces significant challenges in sustainably managing water resources to meet increasing demand amidst disrupted supply, leading to increasing water scarcity (He et al. 2021). The potential of rivers to provide nature-based solutions to some of these challenges is increasingly recognized and, as such, the interest and investment in the restoration of degraded river systems has become substantial (Logar et al. 2019; UNESCO 2018). Poor water quality is a particularly prevalent and persistent driver of water scarcity, reducing ecosystem health, threatening food production, and thwarting human potential (Damania et al. 2019). However, poor water quality in river networks may be particularly amenable to nature-based solutions because, when in good condition, streams and rivers have a significant capacity to retain and process many of the pollutants they receive, for example nutrients, heavy metals, pathogens, and emerging contaminants (Herzog et al. 2016; Tesoriero et al. 2013; UNESCO 2018). Therefore, the improvement of water quality is one of the most commonly cited objectives of river restoration (Wohl et al. 2015). Despite this focus, there is little evidence to support the efficacy of restoration to deliver improvements to water quality, and it remains a pervasive problem (Hannah et al. 2022; Palmer 2008).

The efficacy of functional river restoration is limited by the failure to adopt a process-based approach, i.e. to apply process understanding to design and implementation (Palmer 2008; Palmer et al. 2014). Whilst process-based restoration has been adopted to deliver some river restoration objectives, like natural flood management (NFM) and habitat improvement, it is not yet thoroughly applied to realize improvements to water quality (Beechie et al. 2010; Wohl et al. 2015). For example, research is unequivocal about the

Ecohydrological Interfaces, First Edition. Edited by Stefan Krause, David M. Hannah, and Nancy B. Grimm.

importance of the hyporheic zone for ecosystem function, but it is seldom considered in river restoration design (Hester and Gooseff 2010; Lewandowski et al. 2019; Moren et al. 2017). The hyporheic zone is particularly important for regulating water quality because it hosts hotspots of the biogeochemical reactions that remove many pollutants (Boano et al. 2014). It can remediate a range of pollutants from various sources, including point, diffuse, and groundwater sources (Krause et al. 2010).

The poor uptake of hyporheic research in river restoration practice may be in part due to limited efforts by researchers to synthesize research outcomes and frame them saliently to contemporary agendas and within the frameworks and constraints in which practitioners must work. There have been some attempts to summarize hyporheic research (e.g. Krause et al. 2022; Lewandowski et al. 2019; UK Environment Agency 2009) but, in general, research outcomes have not been communicated in a manner that is actionable, which is to say offers clear insight that may be used to directly inform decisions (Mach et al. 2020). To communicate insight in a way which is accessible and actionable, it is necessary to summarize, generalize, and contextualize evidence, and present it in practical guidelines (Grabowski et al. 2019). For example, insight should be presented in the context of the environmental settings in which it is likely to be applied and the nature of degradation which it may be adopted to restore.

Whilst the techniques and designs used in restoration are important, the influence of river restoration on the hyporheic zone may be determined more by the environmental setting in which it occurs (Hester et al. 2018; Krause et al. 2014). In this work, the most essential controls of the environmental setting on the function of the hyporheic zone have been outlined, which should allow the reader to formulate hypotheses about how restoration may influence specific properties of the hyporheic zone (outlined in Section 2.2), and the likely implications for water quality objectives. In doing so, the reader may identify sites with high restoration potential and improve the design of restoration schemes to enhance the function of the hyporheic zone. It is not our aim to provide a comprehensive review of hyporheic zone controls – readers seeking this should refer to the many existing reviews, such as Boano et al. (2014), Harvey and Gooseff (2015) and Krause et al. (2022). These works review decades of work on hyporheic zones, but some uncertainties still exist about exactly how coupled hydrological and biogeochemical processes in the hyporheic zone determine its effect on water quality (Boano et al. 2014; Ward 2015). This certainly limits our ability to exactly predict the response of the hyporheic zone to restoration in different settings. However, most river restoration is conducted with similar uncertainty, and therefore practitioners are experts at working with this uncertainty to maximize the chance of success. It does not eliminate our ability to offer guidelines of practical benefit that, if applied, will lead to better designs than those that are currently implemented.

The aim of this work is to synthesize and communicate insight from research on restoring hyporheic zone functions in an actionable format, with a focus on river corridor restoration strategies that are familiar to river restoration professionals. This represents a challenge not least because the specific foci of research and practice tend not to align. In general, practitioners adopt a holistic approach that aims to maximize multiple benefits and mitigate risks. On the contrary, researchers are typically concerned only with the performance of restoration for a single objective, usually failing to consider trade-offs, conflicts, or potential co-benefits, as well as practical constraints. Furthermore, whilst practice evolves quickly, research is slow to recognize and adopt contemporary practices, often focusing on techniques that have not been applied in practice for several years. As a result, hyporheic zone research often delivers highly engineered solutions that are easily characterized and replicable but not directly relevant to the frameworks in which practitioners work. For example, research tends to measure the performance of log weirs or engineered deflectors, but in practice these solutions are no longer considered restoration – arguably, creating such features to harness the power of the river for pollutant processing is no different than harnessing a river for hydropower, or modifying it for land drainage or navigation. River restoration is the undoing of such modifications to reinstate natural hydrology, geomorphology, and ecology. This is delivered using process-based concepts like biomic restoration, stage zero restoration, naturalization, nature-based solutions, or nature-culture hybrids (Cluer and Thorne 2013; Eden et al. 2000; Johnson et al. 2019). Closer collaboration between researchers and practitioners is essential to close this mismatch between research and practice but in the meantime, this chapter attempts to apply insights generated by studying highly engineered features to contemporary practice-relevant concepts. This work is aimed at river restoration professionals (practitioners, engineers, managers, and planners) and is designed to complement tools that are already used in decision making, along with the values, local knowledge, and often significant expertise of the individual (Dicks et al. 2014).

In Section 16.2, we briefly explain the function of the hyporheic zone and how it is driven by processes in the stream and in the streambed, as well as how its properties effect the retention and transformation of pollutants. In Section 16.3, we outline the characteristics of the environmental setting that should be considered when planning hyporheic restoration. Section 16.4 sets out in more detail how the architecture and arrangement of restoration strategies and designs effect their influence on the hyporheic zone. Section 16.5 provides a series of scenarios, including environmental setting, pressure on the river system, and a solution to restore the hyporheic zone. In Section 16.6, potential synergies and conflicts between restoration of the hyporheic zone for improvements to water quality and for other objectives are briefly discussed. Finally, the discussion (Section 16.7) and conclusions (Section 16.8) position hyporheic restoration within a larger spatial context (i.e. catchment) and within the context of the current frameworks for river restoration, making recommendations on how to improve river restoration in the future.

16.2 Drivers and Controls of Pollution Remediation in the Hyporheic Zone

16.2.1 What is the Hyporheic Zone?

Many definitions of the hyporheic zone are adopted by different disciplines and communities (Ward 2015), but simply put it may be defined as the area surrounding the river channel (in the bed and banks) where surface water and groundwater interact. Hyporheic exchange can be thought of as a "parcel" of water moving from the channel into the streambed and back into the channel (Zarnetske et al. 2012). This exchange is driven by energy gradients along the riverbed that arise from hydrostatic and hydrodynamic forces (Sawyer et al. 2011). Hydrostatic forces are usually caused by differences in elevation between two points (like from upstream to downstream of a logjam) and the level of the overlying surface water – typically represented by water flowing over something like a logjam or a step, or through something like a meander (Boano et al. 2014). Hydrodynamic forces are caused by flow velocity and turbulent energy transferring momentum into the riverbed – typically represented by water flow around something (or over something submerged) like a boulder or instream wood. Both hydrostatic and hydrodynamic forces usually contribute to driving hyporheic exchange but one or the other may dominate depending on the environmental setting (Krause et al. 2022).

16.2.2 Properties of Hyporheic Exchange

The properties of hyporheic exchange can vary and this has important implications for the way the hyporheic zone functions and its effects on water quality. The most important properties of hyporheic exchange with regard to its influence on water quality are its volume and residence time in the hyporheic zone (Ward 2015). The volume of hyporheic exchange is the volume (or proportion) of the surface water in the channel that enters the hyporheic zone and re-enters the channel in a given reach length. It can also be thought of as the length of reach required for 100% of the water in the channel to enter the hyporheic zone and re-enter the channel (Wondzell 2011). The residence time is the length of time the surface water remains in the hyporheic zone before it returns to the channel (Boano et al. 2014). Every parcel of water might have a different residence time in the hyporheic zone, but here residence time refers to the distribution of residence times of all the parcels of water (i.e. the average residence time) (see Figure 16.1). Other useful variables include the length of the hyporheic flow paths and the vertical and lateral extent of the hyporheic zone (i.e. distance from the main channel), which can give an indication of the exposure to different environments (i.e. biotic assimilation by plant roots in riparian zones) (Krause et al. 2010). In general, hydrostatic forces induce hyporheic exchange of higher residence times and lower volume, and hydrodynamic forces induce hyporheic exchange of lower residence times and higher volume (Boano et al. 2014).

16.2.3 Pollution Retention and Attenuation in the Hyporheic Zone

Pollutants can be retained in the river corridor by sedimentation or by transport into the hyporheic zone where they are deposited (Boano et al. 2014). Some pollutants (e.g. synthetic particles like microplastics) will remain until they are remobilized, often by flood events (Drummond et al. 2022; Lewandowski et al. 2019). Some pollutants may be transformed in the river corridor, and especially in the hyporheic zone which hosts hotspots of reactions because it contains a matrix of different conditions (Krause et al. 2010). It also hosts the diversity and density of microbes, and high content of organic matter, which are often necessary to facilitate biogeochemical reactions (Zarnetske et al. 2012). Different pollutants are transformed under different conditions and the complete removal of some pollutants requires a series of conditions (Quick et al. 2019). The most important distinction is whether reactions occur in the presence of high dissolved oxygen concentration (aerobic) or low dissolved oxygen concentration (anaerobic). Both of these conditions are found in the hyporheic zone (Zarnetske et al. 2012). Aerobic conditions are generally observed in

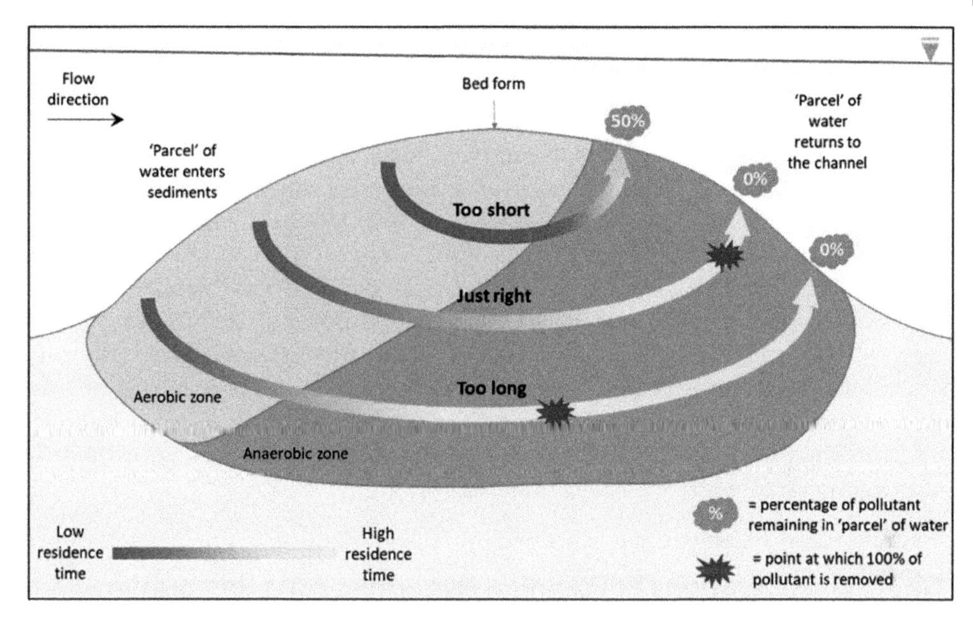

Figure 16.1 Hyporheic exchange through a bedform (e.g. sand dune). Thick arrows represent the general movement of water from upstream of the bedform (left) to downstream (right), and do not necessarily represent the pathway of water through the bedform. The colour of the arrow represents the residence time distribution of water in the hyporheic zone. The "Too short" arrow represents water that returns to the channel before all the pollutants are transformed. The "Just right" arrow represents water that returns to the channel just as but not before all the pollutants are transformed. The "Too long" arrow represents water that returns to the channel a long time after all the pollutants are transformed. The purple star represents the point at which all the pollutant has been removed, which is well before the water returns to the channel for the "too long" flow path. The percentage in the orange cloud represents the remaining percentage of pollutant when the water is returned to the channel, which is still 50% for the "too short" flow path.

shallow hyporheic exchange (close to the channel) with short residence times. When water is in the hyporheic zone oxygen is consumed by aerobic respiration and the concentration of dissolved oxygen decreases leading to anaerobic conditions, which are usually observed in deeper hyporheic exchange with longer residence times (Boano et al. 2014). Sometimes pockets of anaerobic conditions can be found in predominantly aerobic zones, for example within biofilms or lenses of low permeability organic matter in sediments, which can host anaerobic reactions (Krause et al. 2010).

Reactions take time. As well as determining reaction conditions, the residence time also largely determines how much pollutant is removed from a parcel of water in one visit to the hyporheic zone. This creates a "Goldilocks dilemma": residence times that are too short remove only a small proportion of the total pollutant in the water, but residence times that are too long retain water in the hyporheic zone long after all the pollutant is removed (see Figure 16.1) (Herzog et al. 2016). Using this Goldilocks model, the most efficient residence time is one where water re-enters the channel from the hyporheic zone as soon as, but not before, all the pollutant is removed. However, the reality is probably more complex. For example, reaction rate is sometimes proportional to the concentration of the pollutant, in which case flow paths of a medium residence time might be optimal (Le Traon et al. 2021).

16.2.4 Designing Restoration to Target Pollution

Restoration may target specific properties of the hyporheic zone in order to induce the necessary reactions that remove a specific pollutant (Ward et al. 2011). For example, where nitrate (NO_3) is identified as a problematic pollutant, restoration may be designed to induce hyporheic exchange of long residence times because the removal of nitrate (usually by denitrification) requires anaerobic conditions (Zarnetske et al. 2012). In systems high in mine-derived pollutants, especially metals such as iron, manganese, copper, and zinc, restoration may be designed to induce a high volume of shallow hyporheic exchange with low residence times, because attenuation is achieved in aerobic conditions (Gandy et al. 2007). The removal of many pollutants requires a series of reactions for which a variety of conditions are necessary, for example for chlorinated ethenes (Weatherill et al. 2018). In this case, and where water quality objectives are less specific (e.g. where detailed water quality data are not available) restoration may be designed to induce a variety of hyporheic properties, for example by adopting a range of restoration strategies and designs.

16.3 Considerations in Planning Hyporheic Zone Restoration

The environmental setting determines the properties of hyporheic exchange more than the restoration strategy (Hester et al. 2018). Therefore, identifying the most suitable locations for restoration within the stream network is probably the most important factor in determining outcomes, especially given limited resources which may restrict the scale of restoration. Other works have focused specifically on identifying suitable sites for hyporheic restoration on a catchment scale, for example Magliozzi et al. (2019) and Kasahara et al. (2009), which is useful for those coordinating restoration in a top-down approach (e.g. regional planners). However, many river restoration schemes are organized in a bottom-up approach, whereby sites are identified based on convenience or opportunity (Koontz and Newig 2014). In this section, we outline the characteristics of environmental setting and the river corridor that are important controls of hyporheic exchange and that should be considered in restoration planning. This should allow professionals to identify the most suitable locations for restoration within a reach, to identify the causes of degradation that limit the function of the hyporheic zone, and to identify suitable solutions (i.e. restoration strategies).

16.3.1 Sediment Hydraulic Conductivity

The hydraulic conductivity of the sediment is one of the most important controls on hyporheic exchange (Herzog et al. 2016; Hester et al. 2016). The hydraulic conductivity describes the ease at which water moves through the sediment. It is primarily determined by the ratio of sediment grains to pore space; sand has a lower hydraulic conductivity than gravel, for example, because there is less space between sediment grains. High hydraulic conductivity sediments (coarse sediments like cobbles and boulders) are likely to result in a high volume of hyporheic exchange but a low residence time. Low hydraulic conductivity sediments (fine sediments like sands and silts) are likely to result in a low volume of hyporheic exchange but high residence times. Hester et al. (2016) found that a sediment hydraulic conductivity below 10^{-4} m/s (equivalent to clean sand) led to residence times that were too long (i.e. water remained after all the nitrate was removed) and above 10^{-3} m/s (equivalent to gravel) led to residence times that were too short (i.e. not all the nitrate was removed before water returned to the channel), though this depends on background conditions (especially gradient and discharge) and could be different for different pollutants. Sediment thickness is also an important control of hyporheic change, and shallow sediments (i.e. short distance to bedrock or inundation layer) will likely limit hyporheic exchange volume and residence time (Toran et al. 2013).

16.3.2 Gaining or Losing Conditions

Water can flow from the groundwater into the channel (gaining conditions) or from the channel into the groundwater (losing conditions). This is different from hyporheic exchange because it is unidirectional, i.e. the water does not return to its source (Ward 2015). Strongly gaining conditions can limit or eliminate hyporheic exchange (Cardenas 2009; Fox et al. 2014). Losing conditions can either limit or enhance the properties of hyporheic exchange (Hester et al. 2018; Krause et al. 2022). Locations where neither strong losing or gaining conditions exist (neutral conditions) are optimal for hyporheic restoration (Hester et al. 2018). Strong gaining conditions (often observed in chalk streams) may make hyporheic restoration less effective but shallow hyporheic exchange of short residence times may still be generated, which could have a positive effect on water quality. Most commonly, losing conditions are observed in streams with a lower stream gradient and gaining conditions are observed in streams with high gradient – streams of medium gradients are most likely to afford neutral conditions (see Section 3.3) (Hester et al. 2018; Kasahara and Wondzell 2003). Losing or gaining conditions can be observed for part of the year only, for example when the groundwater table is high after winter (Azinheira et al. 2014). This could result in hyporheic exchange being active for only part of the year, but this can still lead to water quality improvements (Hester et al. 2016).

It is difficult to estimate gaining/losing conditions, which presents a major barrier for professionals in designing hyporheic restoration strategies (Lewandowski et al. 2019). Some fairly simple and affordable tools are available for use in the reach scale, such as mini-piezometers (Rivett et al. 2008) and biological indicators (Graillot et al. 2014), and on a landscape scale, such as statistical methods (Magliozzi et al. 2019). However, more accessible techniques need to be developed to allow professionals to estimate this vital ecosystem control, which would be useful well beyond hyporheic restoration (Krause et al. 2011).

16.3.3 Stream Gradient

The stream gradient is the slope or change in elevation in a given distance. Upland streams typically have higher gradients and lowland streams, and rivers have lower gradients. The stream gradient controls the properties of the hyporheic zone in three main ways. Firstly, it largely determines the stream energy, including reach-scale velocity (Hester et al. 2018). The river restoration centre in the UK uses reach-scale stream gradient as a surrogate for stream energy, where >1% is high energy, 0.125–1% is medium energy, and <0.125% is low energy, but a larger range of gradients are observed in more mountainous regions. Stream energy is one of the key controls of hyporheic exchange, especially where the primary driver is hydrodynamic forces, in which case a higher stream energy leads to a higher volume of hyporheic exchange and longer residence times (Rana et al. 2017). Secondly, the stream gradient determines the capacity to induce hydrostatic forces because, in a high gradient system, it is easier to create a difference in elevation over short distances, for example from upstream to downstream of a logjam. Finally, stream gradient also largely determines the relationship between the water level in the channel and that in the adjacent groundwater, which controls the losing or gaining conditions (Hester et al. 2018). In general, hyporheic exchange increases but residence time decreases with gradient (Herzog et al. 2016). In most circumstances, restoration should aim to induce hydrostatic forces in high gradient systems and hydrodynamic forces in low gradient systems.

16.3.4 Flow Conditions

Hyporheic exchange is controlled by discharge, and its associated hydromorphological conditions like water level and velocity. For example, Hester et al. (2019) found that increasing the surface water level from 0.1–1 m (e.g. summer to winter conditions) increased aerobic reactions by 270% and anaerobic reactions by 78%. In general, higher discharge leads to a higher volume and residence time of hyporheic exchange, but the relationship between discharge, restoration features, and hyporheic exchange is not fully understood (Ward et al. 2018). Variations in discharge can change the function of a restoration technique in the channel (Krause et al. 2014). For example, a feature that emerges from the surface water during low flows, primarily driving hydrodynamic hyporheic exchange, may be submerged during high flows, creating a difference in elevation from upstream to downstream and driving hydrostatic hyporheic exchange. It is vital, therefore, to consider the impact of restoration in the full range of flow conditions that the site is subject to (Wohl et al. 2015).

It may be possible to design restoration strategies to target specific conditions. For example, if groundwater conditions only allow hyporheic exchange during certain times of the year (e.g. during summer when the groundwater table is low), then restoration may be designed to induce hyporheic exchange during summer baseflow conditions. For example, Hester et al. (2016) used channel-spanning features in an upland setting to achieve 30% nitrogen removal during summer baseflow in a 900 m reach, but the average annual removal was only 3.7%. It seems that whatever the impact of restoration on hyporheic, it is exaggerated by an increase in discharge. The range of flow conditions could also have important practical implications for restoration design, for example in terms of feature stability and longevity.

16.3.5 Biogeochemical Opportunity

Biogeochemical opportunity refers to the potential for biogeochemical reactions to occur if transport (i.e. hyporheic exchange) is not limiting, though in most cases both transport and biogeochemical opportunity control reaction rates in the hyporheic zone (Findlay 1995). Most reactions occurring in the hyporheic zone that can lead to the improvement of water quality are facilitated by microbes. These microorganisms get their energy by metabolizing chemical compounds and in doing so transforming these into different compounds, sometimes from polluting forms to environmentally benign forms (Krause et al. 2010). For example, aqueous nitrate (NO_3), which can cause eutrophication in high concentrations, can be transformed by a series of reactions to gaseous dinitrogen (N_2), which is environmentally benign. Most of the microbes that perform these functions are heterotrophs, meaning that they require a source of available organic carbon, primarily dissolved organic carbon, for metabolism (Quick et al. 2019). Therefore, the availability and quality (bioavailability) of dissolved organic carbon is a primary control of reaction rates in the hyporheic zone (Zarnetske et al. 2011).

Dissolved organic carbon can be produced in the stream (autochthonously) by primary production, which is easily transported to the hyporheic zone by hyporheic exchange and diffusion. This production is largely dependent on light and nutrient availability, but it can contribute a large part of the total carbon stock and produce carbon which is especially high quality (Berggren and Giorgio 2015). Dissolved organic carbon may also be introduced to the hyporheic zone by external (allochthonous) sources, which are transported in the stream as particulate organic matter and may have terrestrial (e.g. soils) or aquatic (e.g. dead microbes) origins (Wohl et al. 2017). Furthermore, leaves, wood, and fine sediments deposited in the hyporheic zone can decompose and provide a persistent and bioavailable (autochthonous) source of dissolved organic carbon (Howard et al. 2023; Krause et al. 2014). Overall, a higher availability of dissolved organic carbon can be expected in lowland settings compared to upland because the lower stream energy allows its capture and retention, and its sources (e.g. instream wood) are afforded longer times for decomposition (Krause et al. 2014). Furthermore, there is a larger upstream stream length and catchment area contributing to the introduction of dissolved organic carbon to the stream. The availability of dissolved organic carbon may be increased by introducing wood to the stream, directly or by the enhancement of riparian vegetation, and by the capture of fine sediments (Wohl et al. 2017).

Biogeochemical opportunity is ultimately controlled by the availability of dissolved organic carbon together with the availability of reactive chemical compounds for use in metabolism (such as metals, nutrients, and chlorinated solvents), and the ratio between the two. Therefore, the biogeochemical opportunity may be especially high near to sources of reactive compounds, such as near mine drainage or downstream of waste water treatment plants (Duvert et al. 2019).

16.3.6 Sediment Load

The sediment load is the volume of sediment transported or carried by the surface water. Upstream bank erosion or scouring, periodic damn releases, and overland flow (especially over erosion of vulnerable land surface, like heavily cultivated fields) can result in substantial sediment load. The deposition of fine sediments like silts and clays can lead to clogging of the spaces between larger sediments (called the interstitial spaces); this is often called siltation or colmation (Kasahara and Hill 2008). Siltation reduces the sediment hydraulic conductivity which is likely to limit the volume of hyporheic exchange that may be induced. This is especially common upstream of channel-spanning features where the water velocity is decreased leading to the deposition of sediments and consequently a reduction in water depth (Fanelli and Lautz 2008; Krause et al. 2014). Sedimentation may be the primary reason why hyporheic restoration schemes can become quickly ineffective, even if they are initially effective (Pander et al. 2014). Where the sediment load is high, restoration schemes should probably avoid channel-spanning features and sediment reintroduction, unless they are designed specifically to limit sedimentation, as discussed in sections 4.1 and 4.4.

16.3.7 Sites with the Highest Restoration Potential

The sites with the greatest potential to improve hyporheic zone function for water quality objectives are those where the groundwater conditions are neutral (i.e. not strongly gaining or losing), usually observed in mid-order streams, and where existing hyporheic exchange is limited by stream degradation (Hester et al. 2018). These are likely to include streams which have received high fine sediment loads, those which have been incised, straightened, or channelized, and those with low geomorphic complexity, for example as result of the removal of wood (Beechie et al. 2010). This is not dissimilar to identifying sites for river restoration with other objectives and therefore practitioners will already possess significant expertise in this area, which is easily transferred to hyporheic restoration.

16.4 Designing Hyporheic Restoration Strategies

Although the environmental setting primarily controls the properties of the hyporheic zone, these properties are amenable to modification by river restoration, particularly at the feature scale. As with all river restoration, it is vital to adopt a technique that is suitable for the environmental setting and that is informed by the probable historic condition of the river and the variability in its conditions in time and space (Wohl et al. 2015). It is the interaction between restoration and the environmental setting that determines the properties of the hyporheic zone, so the form of restoration should complement environmental drivers (i.e. hydrostatic and hydrodynamic forces discussed in Section 2). A technique that enhances the functioning of the hyporheic zone in one setting could have a negative effect in another. For example, the occurrence of channel-spanning wood may decrease hyporheic exchange in low-gradient streams (Rana et al. 2017; Wondzell et al. 2009) but increases it in high-gradient streams (Kasahara and Wondzell 2003).

The restoration techniques discussed here will be familiar to river restoration practitioners as they have been commonly employed to deliver many different objectives. As discussed in the introduction, research tends to focus on highly engineered solutions which do not necessarily align with contemporary river restoration practices. To capture the full extent of research on hyporheic zone restoration, and to represent its insights authentically, it has been necessary to continue to represent restoration techniques in engineering formats to some extent, for example by categorizing research into single techniques. Additionally, techniques are necessarily simplified in schematics (e.g. Figure 16.2) in order to represent important considerations and processes clearly, resulting in engineering-style blueprints. However, we aim to present research findings in a way that allows their integration into contemporary restoration concepts, for example highlighting to which concepts research insights might apply. With their significant experience adapting to quickly evolving practices in river restoration, practitioners are likely to be adept at recognizing where applications of research insights to novel concepts are appropriate.

The design of the selected technique is important and can determine the influence it exerts on the hyporheic zone. The design may refer to the architectural characteristics of the restoration like size, form, and material, or its arrangement in the channel, like spacing and orientation. The effect of restoration is likely to differ in different environmental conditions (e.g. discharge) and its effect should be considered in the range of conditions it is likely to be subject to (Wohl et al. 2015). In this section, we briefly outline the mechanisms by which

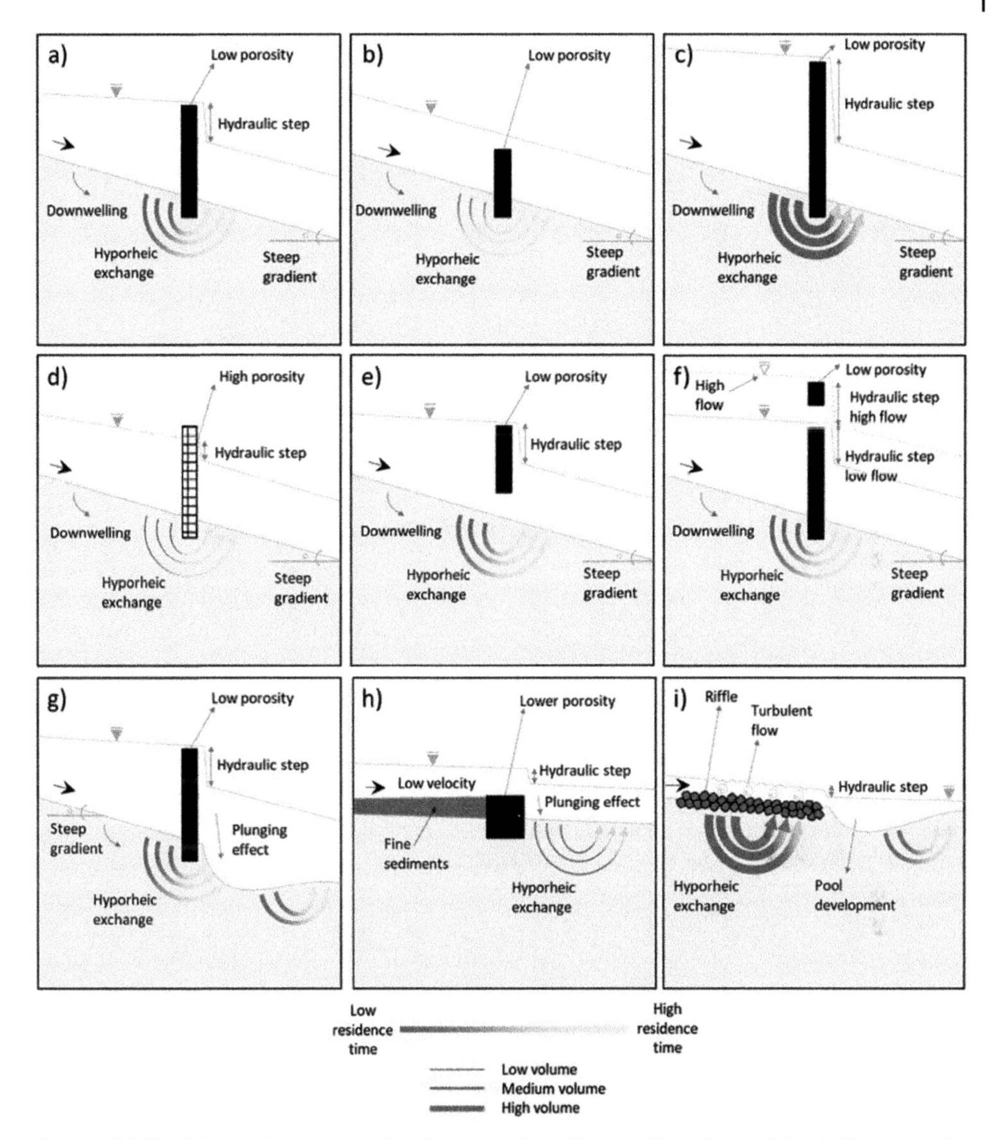

Figure 16.2(i) Schematics representing the properties of hyporheic exchange (blue/yellow arrows) generated by different restoration techniques and designs in rivers with permeable sediments. Diagrams are not necessarily scalable. (a) Impermeable channel-spanning feature of perfect height in an upland setting. (b) Impermeable channel-spanning feature of short height in an upland setting. (c) Impermeable channel-spanning feature of tall height in an upland setting. (d) Permeable channel-spanning feature of perfect height in an upland setting. (e) Impermeable channel-spanning feature of perfect height with a bottom gap in an upland setting. (f) Impermeable channel-spanning feature of perfect height with two outflow heights in an upland setting. (g) Impermeable channel-spanning feature of perfect height with pool development in an upland setting. (h) Impermeable step with fine sediment upstream in a lowland setting. (i) Constructed riffle and pool in a lowland setting.

common river restoration techniques are likely to influence the hyporheic zone and how their characteristics are likely to affect their influence. Whilst the efficacy of some techniques to restore the hyporheic zone has received considerable attention, others have hardly been researched at all. Therefore, it has sometimes been necessary to speculate based on related research; these instances have been clearly highlighted.

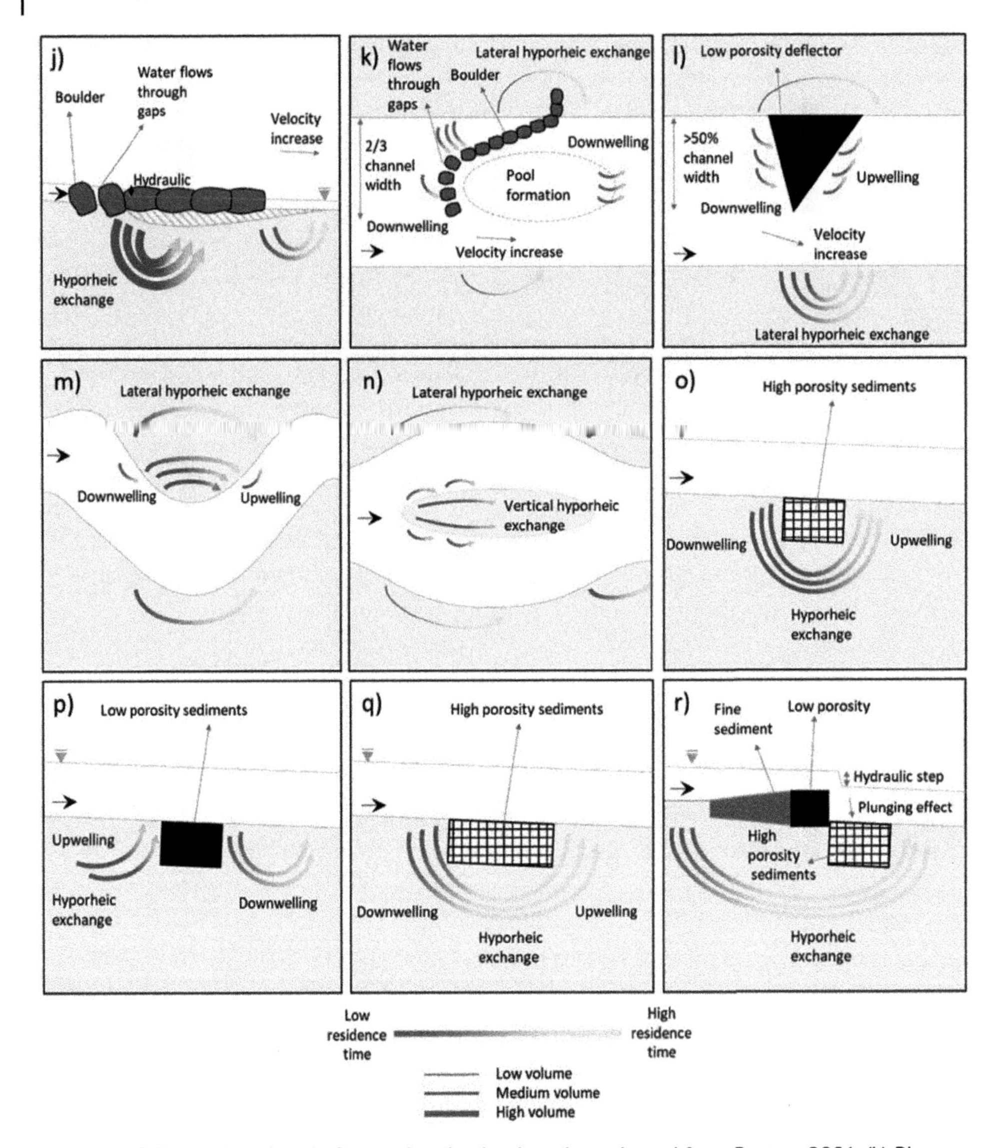

Figure 16.2(ii) (j) J-hook style feature in a lowland setting, adapted from Rosgen 2001. (k) Plan view of a J-hook style feature and pool in a lowland setting, adapted from Rosgen 2001. (l) Plan view of a deflector-like feature in a lowland setting. (m) Plan view of channel reconfiguration (re-meandering) in a lowland setting. (n) Plan view of channel reconfiguration (island construction) in a lowland setting. (o) High porosity sediment introduction in a lowland setting, adapted from Ward et al. 2011. (p) Low porosity sediment introduction in a lowland setting, adapted from Ward et al. 2011. (q) Long low porosity sediment introduction in a lowland setting, adapted from Ward et al. 2011. (q) Long low porosity sediment introduction in a lowland setting, adapted from Ward et al. 2011. (r) Low porosity sediment introduction coupled with upstream step in a lowland setting, adapted from Bakke et al. 2020.

16.4.1 Channel-spanning Features

Channel-spanning features may include weirs, steps, dams, and in some cases large wood. Whilst they can be associated with negative consequences for river health (discussed in

Section 6), they are probably the most effective means of inducing hyporheic exchange in upland settings, especially where simplification has occurred such as the removal of instream wood (Hester et al. 2018). Dams and weirs may be impractical in most settings, but the effects of channel-spanning features may be produced in line with contemporary concepts, for example by felling trees into and across channels, by using boulders to create steps over longer longitudinal distances, and by beaver dam analogues. Furthermore, deflector-like features may produce similar effects (discussed in Section 4.2).

Usually, the aim of a channel-spanning feature is to impound water on the upstream side, creating a difference in elevation in the surface water from upstream to downstream of the feature, often called a hydraulic step (see Figure 16.2a). This generates hydrostatic forces which induce areas of downwelling upstream and upwelling downstream, resulting in hyporheic exchange of high residence times (Hester and Doyle 2008). This is particularly important for the anaerobic removal of pollutants (e.g. nitrate) and is not realized to the same extent by non-channel spanning features, which can fail to induce sufficient hydrostatic force (Hester and Doyle 2008). The impounding of water upstream also increases the area of the wetted channel, which could increase the volume of hyporheic exchange, but potentially at the cost of reducing the proportion of the water in contact with the riverbed, which could decrease the ratio of the volume of water in the hyporheic zone compared to that in the channel. Additionally, the water flowing over the feature may "plunge" into the downstream streambed, forcing hydrodynamic downwelling and creating pools (Endreny et al. 2011).

The characteristics of channel-spanning features that chiefly determine their effect on the hyporheic zone are porosity and height. Porosity refers to the ratio of material to open space, where zero porosity is impermeable. A high porosity (e.g. in some large wood dams) may limit the volume of water that is impounded and the difference in elevation in the surface water profile, leading to a lower volume and residence time of hyporheic exchange (see Figure 16.2d) (Manners et al. 2007). For low porosity features, feature height is the most important characteristic. Taller features induce a higher volume of hyporheic exchange but a lower residence time (see Figure 16.2c) (Feng et al. 2022; Hester and Doyle 2008). The optimal feature height maximizes both hyporheic exchange volume and residence time. This optimal height is determined by the environmental setting but for upland sites a height of around 0.5 m has been suggested, which may also lead to the highest anaerobic reaction rates (see Figure 16.2a) (Hester and Doyle 2008; Liu and Chui 2020). Similarly, Wade et al. (2020) observed that beaver dam analogues function with size-dependent behaviours, where small features had little effect but larger features – which created a larger hydraulic step – were able to induce high volumes of hyporheic exchange with sufficient residence time to host anaerobic reactions. Designs that allow water to exit at different heights (for example, Figure 16.2f) may maintain a high hyporheic exchange volume and residence time during different discharges.

The most common failure of channel-spanning features is siltation of sediments upstream (Hester et al. 2018). This could be avoided by introducing a gap at the base of the feature (see Figure 16.2e), which would capture less sediment but does not reduce effectiveness (Sawyer et al. 2011), or by using features which concentrate stream power at the centre of the channel (e.g. cross-vanes or W-weirs) thereby flushing sediments through (see Figure 16.2j, k) (Rosgen 2001). However, this design feature could introduce additional challenges for fish passage, creating a flume of fast flowing water that will only be passable to larger, stronger swimming species (such as *salmonids*), whilst simultaneously complicating attraction flow and potentially compromising other fish passage easement options such as technical passes or rock ramps. Furthermore, the interruption to sediment transport as a result of impounding is likely to disrupt natural geomorphological processes which themselves can generate features that induce hyporheic exchange.

Features in close proximity can interact, sometimes leading to increased residence times and sometimes to a reduction in the volume of hyporheic exchange (Herzog et al. 2019). The interactions of features are complex and so it is unlikely, given current understanding and monitoring techniques, that restoration can be reasonably designed to target positive interactions that increase residence times. Therefore, the reader should aim to avoid reducing hyporheic exchange by feature interaction. This reduction in hyporheic exchange is caused by a reduction in the difference in elevation from upstream to downstream of the feature due to the increased water level upstream of the feature immediately downstream (Li et al. 2022). Therefore, features should be spaced a sufficient distance as to not be within the impoundment of the feature immediately downstream. This distance will be determined by discharge, channel width and gradient, and the feature height, but is typically in the region of 2–5 channel widths (Han et al. 2015). In practice, the spacing of features will be primarily determined by the associated geomorphological changes that are appropriate to the planform, gradient, and cross-sectional profile of channel, and features are increasingly installed in a more randomized and naturalistic manner.

In some settings, channel-spanning features can lead to a reduction in hyporheic exchange, at least in the short-term (Rana et al. 2017). This usually occurs in low-gradient streams with sediment of a low hydraulic conductivity. In this case, the reduction in water velocity decreases existing hydrodynamic hyporheic exchange, which is not replaced by hydrostatic hyporheic exchange because of the low stream gradient. This suggests that removing channel-spanning features in lowland streams could increase hyporheic exchange, though to our knowledge this has not been explicitly investigated.

16.4.2 Deflectors

In research literature, deflectors may include vanes, J-hooks, groynes, croys, and baffles, but these features are now rarely implemented in practice. Instead, contemporary concepts may include deflector-type elements, such as large woody material introduced opportunistically to create "fallen tree analogues", installed or induced gravel bars, and some beaver dam analogues. They are structures that cover part of the channel laterally, usually between one and two thirds, and have typically been adopted to reduce the scour and erosion of banks by redirecting the stream power to the centre of the channel, often creating a deep pool downstream (Rosgen 2001; Toran et al. 2013). A series of these features can create pool-riffle sequences in certain settings, which can prove powerful generators of hyporheic exchange and be effective pollutant remediators (see Figure 16.2i) (Crispell and Endreny 2009). Deflectors can generate hydrostatic hyporheic exchange but usually to a lesser extent than channel-spanning features (Hester and Doyle 2008). Similarly to channel-spanning features, downwelling occurs upstream of the feature and upwelling occurs downstream (see Figure 16.2l) (Crispell and Endreny 2009; Han et al. 2015). On top of this though, deflectors can increase the water velocity, stream energy, and turbulence, all of which are likely to induce hydrodynamic hyporheic exchange and could serve to flush out gravels downstream, helping to maintain high hydraulic conductivity and effective hyporheic exchange (Crispell and Endreny 2009; Han et al. 2015; Hester and Doyle 2008). This combination of drivers may provide a good balance between hyporheic exchange volume and residence time, which could prove an effective means of delivering a variety of water quality objectives. Furthermore, deflector-type features can largely avoid some of the concerns associated with channel-spanning features, such as a reduction in longitudinal connectivity for fish passage or reduced efficacy by colmation.

The effect of porosity is, in general, similar as for channel-spanning features, where a lower porosity is likely to lead to a larger impact, but in some cases gaps in the feature near the centre of the channel can increase velocity and flush out sediments, for example in J-hooks (see Figure 16.2j, k) (Rosgen 2001). To have the greatest impact, the height of deflectors should exceed the water level during baseflow and be sufficient to create an elevation difference in the surface water profile from upstream to downstream during flood (Han et al. 2015). Laterally, the feature should cover at least 50% of the channel but exceeding this is unlikely to have a greater effect on hyporheic exchange (Hester and Doyle 2008). The angle of the deflector to the flow is likely to be one of the most important design characteristics. Little research has focused explicitly on the effect of different angles on hyporheic exchange, but reasonable speculation can be made. A larger difference in elevation from upstream to downstream is probably created by features which are orientated perpendicular to the direction of flow (i.e. 90° from the bank) – as indicated in the work by Biron et al. (2004) – which will likely induce a higher hyporheic exchange volume and residence time. The pools that can develop as a result of deflector installation may contribute significantly to the total hyporheic exchange in a reach (Crispell and Endreny 2009). Therefore, features should be spaced sufficiently as to not limit pool development, usually 2–5 times deflector length, but this is determined by stream gradient, where closer spacing is suitable in steeper gradients, and by the geomorphological setting (Mohammed et al. 2016; Rosgen 2001). There are practical considerations in the design of deflectors, such as managing bank erosion, which are outlined in Rosgen (2001).

16.4.3 Channel Reconfiguration

Channel reconfiguration is a common restoration technique for rivers that have been channelized, straightened, lowered below or cut off from the floodplain, or simplified (e.g. consolidation of many channels into one), all of which can lead to a reduction in hyporheic exchange (Sparacino et al. 2019). Channel reconfiguration may include restoring meanders and increasing sinuosity, creating islands and braided channels, reconnecting the channel to the floodplain, biomic restoration, stage zero restoration, and re-naturalization of the channel. This broad technique may include many different designs which could affect hyporheic exchange in a variety of ways. However, in general, channel reconfiguration results in lower stream gradients, reduced reach-scale velocity (though local velocities could be increased), reduced stream energy, and deposition of sediments (Becker et al. 2013; Bukaveckas 2007). This will tend to result in a shift in the dominant driver of hyporheic exchange from hydrodynamic to hydrostatic, and subsequently in its properties from high volume and low residence time to lower volume and higher residence time (Mason et al. 2012). Channel reconfiguration is also likely to result in the induction of lateral hyporheic exchange (i.e. from one side of a meander to another, called intrameander flow) (see Figure 16.2m), which can have high residence times and lead to high potential for anaerobic removal of pollutants (Boano et al. 2014; Kasahara and Hill 2007; Krause et al. 2022). More research is required to quantify the effects of promising contemporary concepts, in particular, stage zero restoration, on hyporheic exchange and water quality (Cluer and Thorne 2013).

Channel reconfiguration schemes aiming to enhance the function of the hyporheic zone have delivered mixed results; many detected no significant changes to hyporheic exchange or the capacity for pollutant removal, suggesting channel configuration may not be an especially effective means to restore the hyporheic zone (Bukaveckas 2007; Martín et al. 2018; Wohl et al. 2015). Due to the scale of these programmes, they may be especially vulnerable to controls of environmental setting, such as sediment hydraulic conductivity and stream gradient (Kasahara and Hill 2007; Gabriele et al. 2013). Furthermore, the full benefits may only be realized during or after an *evolutionary cycle* of the river system, following the stream evolution model which underpins stage zero restoration (Cluer and Thorne 2013). What is clear is that schemes must be well considered and acutely targeted to suit the environmental setting; simply widening incised or restricted channels (sometimes referred to as de-channelization) is probably not sufficient (Martín et al. 2018). Some studies have observed a reduction in the volume of hyporheic exchange after channel reconfiguration, which has mostly been attributed to the loss of hydrodynamic drivers due to velocity decreases and vertical bed simplification (e.g. loss of bedforms) (e.g. Mason et al. 2012). To avoid this, lateral modifications, like meanders and islands, should be coupled with vertical modifications, like steps, deflectors, or gravel bars (see Figure 16.2h) (Kasahara and Hill 2007; Kurth et al. 2015).

As a generalization, where the stream gradient is relatively high (>1%), schemes should use large meanders and braiding which will create large differences in elevation from upstream to downstream and could generate substantial hydrostatic hyporheic exchange with long residence times (Boano et al. 2014; Krause et al. 2022). Where the stream gradient is low, schemes should instead adopt designs that maintain high velocity (locally and on the reach scale) and generate stream power, such as gravel bars, small islands, and steps (see Figure 16.2n) (Kurth et al. 2015). Coupling channel reconfiguration with floodplain reconnection activities and stage zero restoration could prove effective when employed collectively (Cluer and Thorne 2013; Wohl et al. 2015, 2021).

16.4.4 Sediment Restoration

Sediments in many river systems have been degraded from their natural composition, for example by removal of gravels for construction, dredging of rivers for navigation, or siltation as a result of changes to sediment transport dynamics (Ward et al. 2018). The removal of sediments can lead to a reduction in the sediment depth, and siltation reduces the hydraulic conductivity of the sediment, both of which can limit the function of the hyporheic zone (Fanelli and Lautz 2008; Toran et al. 2013). Given that the depth and hydraulic conductivity of the river sediments are critically important controls on the function of the hyporheic zone, sediment restoration offers a promising hyporheic restoration technique. Research has focused on modifications such as fine sediment removal or gravel cleaning (e.g. Ward et al. 2018), reintroduction of gravels (e.g. Pander et al. 2014), and more complex engineered streambeds (e.g. Herzog et al. 2019; Ward et al. 2011). Some modifications aim to simply increase the sediment hydraulic conductivity to increase the volume of hyporheic exchange. More complex designs aim to disrupt or divert hyporheic flow paths (the route a parcel of water takes through the hyporheic zone), similarly to natural heterogeneities in the hydraulic conductivity of the sediments, such as peat and clay lenses (Krause et al. 2014). This can generate hydrostatic forcing, manifesting in patches of upwelling and downwelling, and increase residence times, and consequently the potential for anaerobic reactions (Ward et al. 2011).

Sediment restoration may aim to alter or restore sediments across the entire streambed (e.g. Ward et al. 2011; Pander et al. 2014) or in smaller patches (e.g. Bakke et al. 2020; Ward et al. 2018). At the streambed scale, Pander et al. (2014) modified 50 m^2 areas of streambed and found that introducing sediments with a higher grain size (i.e. coarser gravel) led to a higher volume of hyporheic exchange, though probably of short residence times. Sediment depth is probably important, with deeper sediments allowing higher residence times, but this has not been explicitly tested. It may be necessary to dredge fine sediments and remove macrophytes before sediment introduction to avoid siltation (Pander et al. 2014). Where coarser sediments remain but siltation has occurred, gravel cleaning may be a viable restoration technique, though this has yielded mixed results. Mathers et al. (2021) observed an increase in the volume of hyporheic exchange after fine sediment removal, but Ward et al. (2018) found that fine sediment removal had little effect on transient storage (which includes hyporheic exchange), which was attributed to the associated loss of macrophytes.

On the patch scale, Ward et al. (2011) investigated the effect of installing 3 m × 1.5 m areas of sediment with high or low hydraulic conductivity in the streambed. High hydraulic conductivity structures created areas of downwelling upstream and upwelling downstream (similar to channel-spanning features) (see Figure 16.2o), whilst low hydraulic conductivity structures created areas of upwelling upstream and downwelling downstream (see Figure 16.2p). High hydraulic conductivity structures positioned adjacent to the stream bed maximized hyporheic exchange, and residence time was increased by increasing feature length (see Figure 16.2q). These kinds of engineered hyporheic zones can lead to significant improvements to many aspects of water quality, but they conflict with contemporary restoration agendas (Peter et al. 2019). Similar effects could be generated within contemporary restoration concepts, for example trees could be buried during sediment reintroductions, or sediments of different grain sizes could be introduced in patches.

Some field studies found that the effect of sediment introduction on the hyporheic zone did not persist for more than 1 year, primarily due to siltation of sediments (Mathers et al. 2021; Pander et al. 2014). To avoid this, sediment introductions could be coupled with scouring features or vertical steps (e.g. created by boulders or fallen tree analogues), which can vastly increase hyporheic exchange (by as much as 89 times) and prevent siltation (see Figure 16.2r) (Bakke et al. 2020). In general, sediment introductions are probably not suitable in systems with a high sediment load. Caution should be exercised to limit the negative impact on downstream environments caused by the significant mobilization of fine sediments following sediment introductions (Pander et al. 2014).

16.4.5 Development

Most instream restoration is likely to induce an evolution of the local geomorphology, and the ecosystem as a whole, to a particular equilibrium state or an evolutionary cycle (Cluer and Thorne 2013; Wohl et al. 2015). The geomorphological state that develops following restoration could have a more substantial effect on the hyporheic zone than the restoration itself. For example, Fanelli and Lautz (2008) observed the development of a small pool followed by a glide upstream of a channel-spanning feature, and a plunge pool followed by a riffle downstream, generating patchy spatial patterns of hyporheic exchange (see Figure 16.2g). Furthermore, the introduction of deflectors could lead to meandering, which can result in significant lateral hyporheic exchange (Krause et al. 2022). Restoration could be designed to induce a geomorphic condition that generates the desired properties of hyporheic exchange. For example, Kobayashi et al. (2022) designed wood "training features" to facilitate pond formation during flood events, which probably led to increased hyporheic exchange as indicated by decreased temperatures in the streambed relative to the rest of the channel. Some restoration designs may take more than a decade to reach their maximum effectiveness, and this should be considered in monitoring strategies (Wondzell et al. 2009). Conversely, the evolution of the river system post-intervention could decrease the efficacy of restoration and degrade the function of the hyporheic zone, such as siltation, which is discussed in Section 4.1.

Much uncertainty exists about how most restoration techniques develop, and in general about how changes to river systems affect the hyporheic zone. For example, the establishment of macrophytes has been shown to effect hyporheic exchange and instream storage both negatively (Romeijn et al. 2021) and positively (Vereecken et al. 2006). It may be necessary in some cases to design a maintenance plan, for example the removal of silt behind channel spanning features, removal or cutting of macrophytes, or the regular introduction of large wood material. However, this kind of maintenance does not meet accepted standards for successful river restoration and is, in general, not in line with contemporary restoration concepts (Cluer and Thorne 2013; Palmer et al. 2005). A network of hyporheic restoration demonstration sites, with long-term monitoring programmes, is necessary to improve our understanding of exactly how hyporheic restoration schemes evolve and how their efficacy changes over time, as discussed further in Section 16.7.

16.5 Scenarios of Hyporheic Restoration

To make research actionable its insights must be directly applicable to the environmental setting in which it will be applied, which as we have seen is the most important control of the function of the hyporheic zone. However, there are few demonstrations of hyporheic restoration that river restoration professionals are able to draw upon to directly inform the design of hyporheic restoration in a range of environmental settings and to abate a range of pressures. On the contrary, there has been substantial dedicated research into the function of the hyporheic zone in different settings and its interaction with various river restoration techniques, as has been outlined in sections 16.3 and 16.4. In this section, this research has been used to inform the design of hyporheic restoration in a series of theoretical scenarios. These scenarios are designed to represent typical situations where hyporheic restoration may be appropriate and possible. They have been constructed to include the considerations that have been made when selecting and designing the restoration technique, and they are structured in the way that the design of hyporheic restoration should be considered: setting, pressure, solution. Pressure refers specifically to the origins of hyporheic zone degradation and not of the stream function in general. It is important to identify and treat the root source or cause of the degradation as well as to mitigate the manifestation of degradation (Kasahara et al., 2009). In some cases, restoration designs are speculative, but they still allow the reader to formulate sensible hypotheses about the influence of restoration on the hyporheic zone and to select the most suitable designs for restoration.

16.5.1 Scenario 1

Setting

An upland 2nd order stream with a steep gradient of 5.5%, a width of 2 m, and gravel and cobble sediment (Figure 16.3). Riparian trees have been removed over a century ago and are prevented from regenerating by intensive sheep grazing. The discharge responds

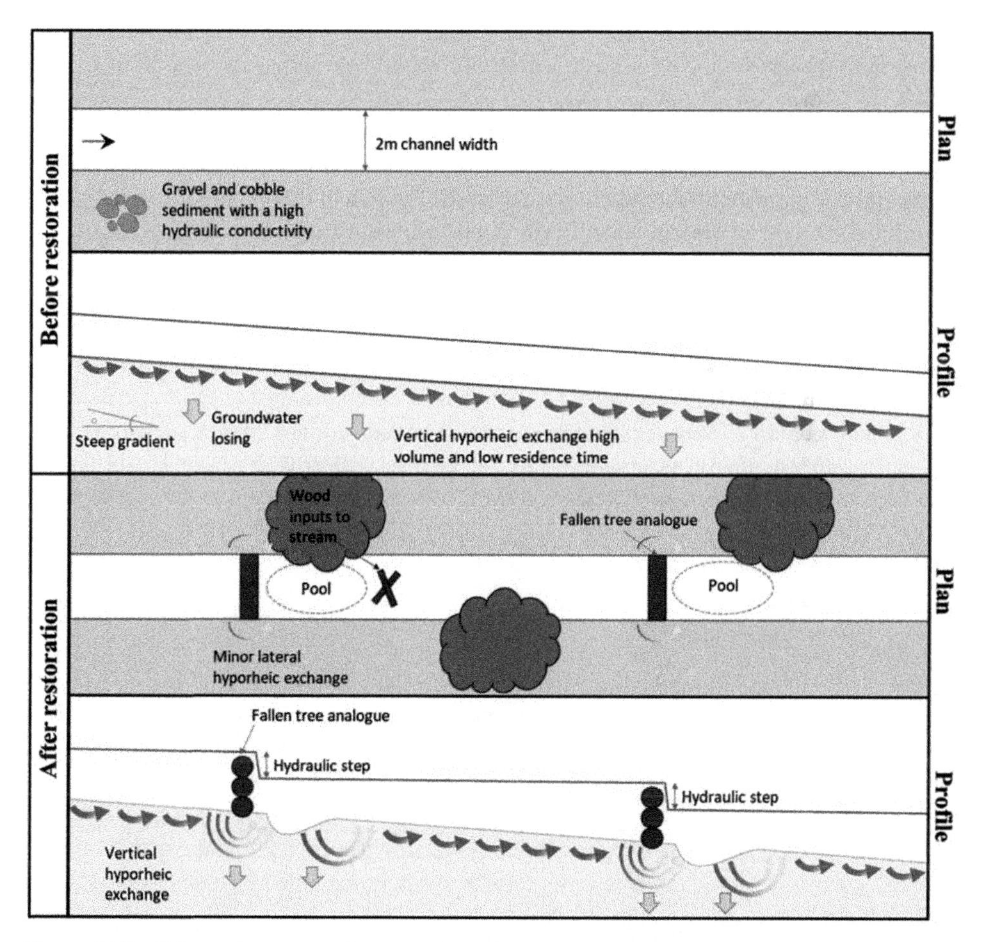

Figure 16.3 Schematic representing scenario 1. Curved arrows represent hyporheic exchange. Arrow thickness represents relative volume (i.e. change from before to after restoration) – thicker arrows represent relatively higher volume. The arrow colour represents residence time distribution, where blue represents shorter times and yellow represents longer times.

quickly to rainfall and has a summer baseflow of 20 l/s and a winter baseflow of 50 l/s. There are slightly losing conditions throughout the reach, especially in summer. There is little input of sediment into the stream but high input of sheep excrement.

Pressure
The lack of a riparian zone has resulted in a low loading of instream wood and ultimately a simplification of the channel. Water travels downstream almost unimpeded and has a high velocity and energy. The flow around and over cobbles creates hydrodynamic forces at the streambed interface and, along with the high hydraulic conductivity of the sediment, allows a high volume of hyporheic exchange but of only short residence times. The mildly losing conditions leads to some loss of water to the groundwater because there are few areas of low pressure that return water back into the channel. Sheep excrement enters the channel leading to potential pathogen input and moderate nutrient concentrations, but the biogeochemical opportunity is limited by the low carbon availability.

Solution
Sheep should be excluded from the stream and its banks by fencing, allowing riparian vegetation to recover, which may be accelerated by tree planting. This will eventually facilitate the recovery and maintenance of the instream wood loading, but it could take several centuries to achieve (Beechie et al. 2000; Stout et al. 2018). To realize hyporheic restoration more quickly, more direct intervention is required. The aim of this restoration should be to increase the residence time of water in the hyporheic zone and reduce the loss to groundwater. By generating hydrostatic forces, water will be driven deeper into the hyporheic zone by areas of lower pressure and returned to the channel by areas of higher pressure (Sawyer et al. 2011). This is best achieved by introducing channel-spanning features, or deflector-like features if fish passage is of concern. The features should have a low porosity and a height of 0.5 m. They should be spaced 4–10 m apart and be installed to capitalize on natural stream topography. Some existing boulders can be repositioned, and large woody material can be pinned into the bed and the bank and secured with wire or rope to create a fallen tree analogue. Using large woody material could also introduce carbon to the system.

16.5.2 Scenario 2

Setting
A 2nd order stream with a medium gradient of 0.9%, a width of 3m, and fine gravel and sand sediment (Figure 16.4). The groundwater conditions are neutral, and the stream discharge responds quickly to rainfall due to extensive overland flow in the catchment. The mean discharge is 10 l/s, but maximum flows reach 150 l/s. There is a small storage reservoir upstream which keeps the stream flowing all year round and releases high discharges periodically which transport a high sediment load. The upper catchment consists of deep organic soils and the land use is intensive arable agriculture, which includes high fertilizer addition and heavy cultivation practices.

Pressure
The stream has been straightened, deeply incised (channelized), and cleared of obstacles like wood and boulders. Agricultural activities encroach within 2 m of the stream and the landowner is not willing to sacrifice land for floodplain reconnection or stage zero

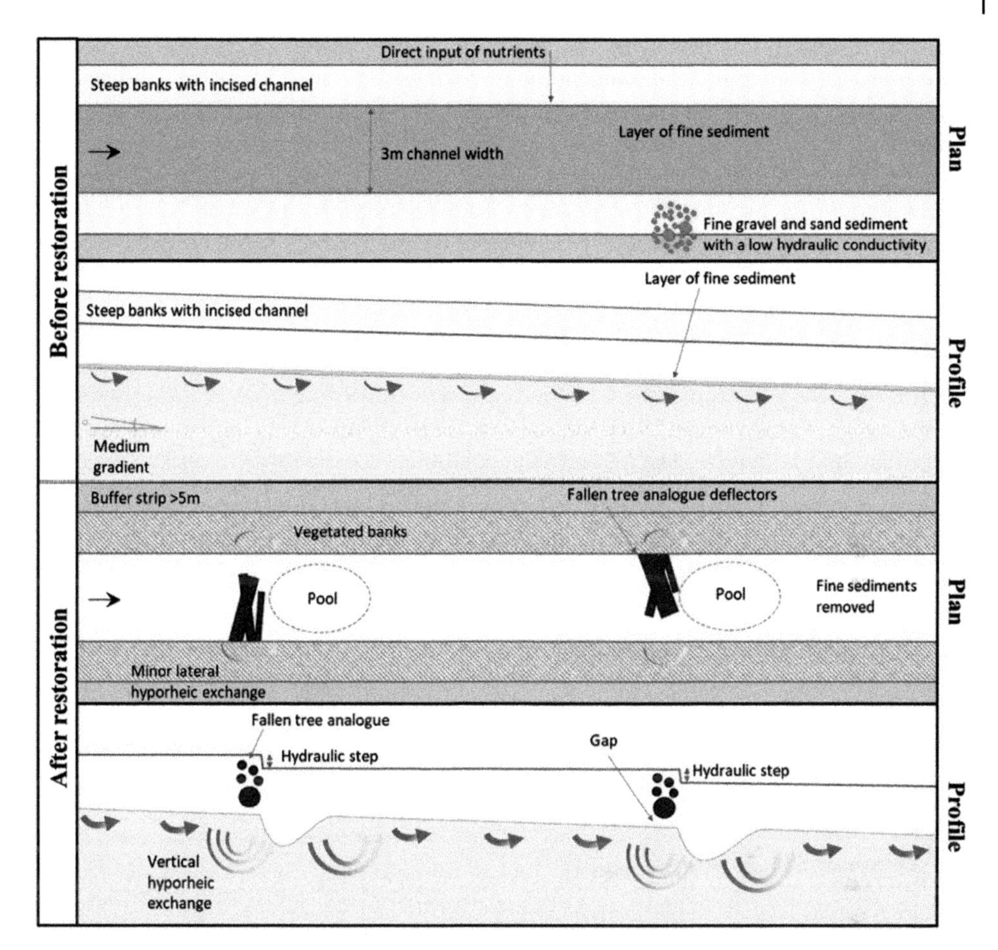

Figure 16.4 Schematic representing scenario 2. Curved arrows represent hyporheic exchange. Arrow thickness represents relative volume (i.e. change from before to after restoration) – thicker arrows represent relatively higher volume. The arrow colour represents residence time distribution, where blue represents shorter times and yellow represents longer times.

restoration. Due to the reservoir releases and sediment entering the stream from overland flow, some gravels have become clogged, and the streambed profile is buried in a fine layer of silt, resulting in a smooth surface. As a result, almost no hyporheic exchange occurs. Overland flow directly into the stream, as well as interaction with polluted groundwaters, result in high nutrient concentrations. Organic matter content is also high as a result of the organic soils transported to the stream in overland flow.

Solution

Buffer strips of at least 5 m should be established either side of the stream, containing trees and long grasses, which will intercept run-off and help reduce the direct input of sediment and nutrients to the stream. The challenges of restoration are to avoid inducing bank erosion of soft organic soils and to manage the high sediment load. The medium gradient and neutral groundwater conditions allow for the induction of hydrostatic forces. Raised deflector-like features with natural gaps at the streambed could generate hyporheic exchange of high residence times but prevent the deposition of sediment. The medium hydraulic conductivity of the sediment could further facilitate high residence times and the organic matter from organic soils could lead to a high biogeochemical opportunity. Because of the high variability in mean and maximum discharge, a complex feature with gaps at different heights could generate hyporheic exchange during baseflow and high flow conditions. Trees could be used to construct the fallen tree analogues.

16.5.3 Scenario 3

Setting
A 3rd order urban stream with a medium gradient of 0.9% and coarse gravel sediment (Figure 16.5). The channel width is around 2 m and is enclosed by roads on both sides in an area of around 10 m wide. The groundwater conditions are slightly gaining, and the mean

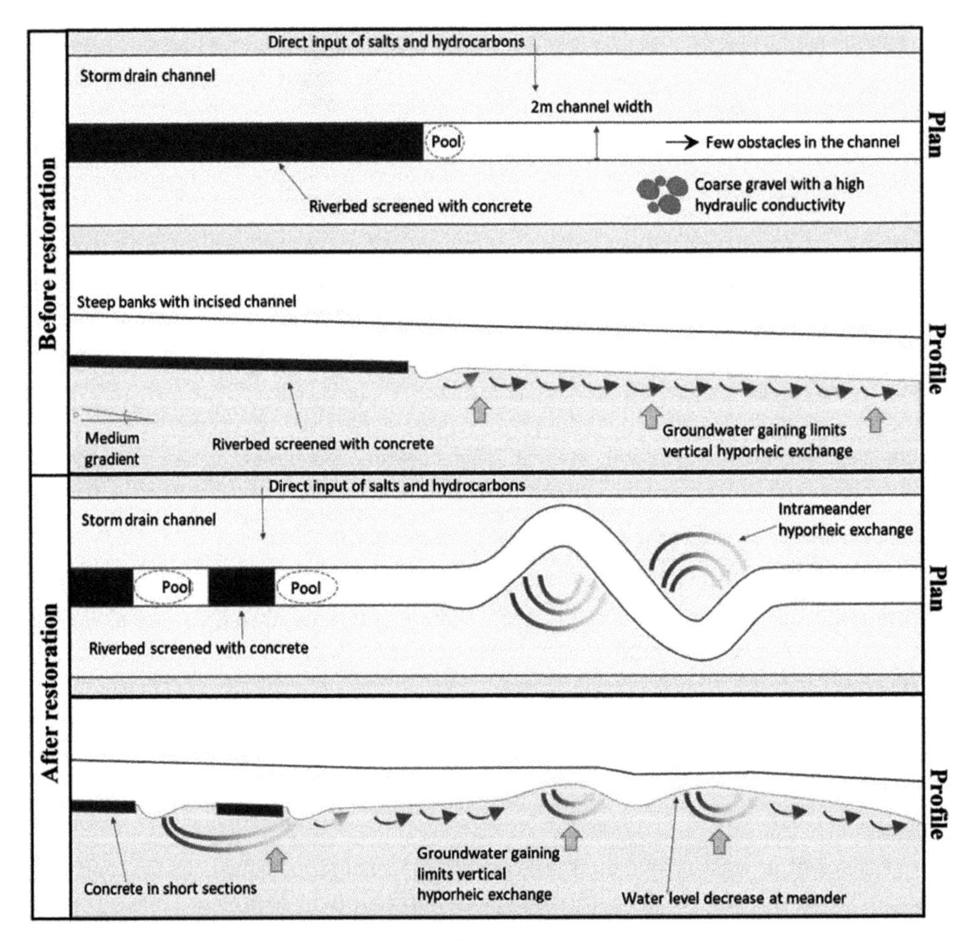

Figure 16.5 Schematic representing scenario 3. Curved arrows represent hyporheic exchange. Arrow thickness represents relative volume (i.e. change from before to after restoration) – thicker arrows represent relatively higher volume. The arrow colour represents residence time distribution, where blue represents shorter times and yellow represents longer times.

annual discharge is 40 l/s. The stream discharge responds quickly to rainfall due to fast overland flow over impermeable surfaces. There is a waste water treatment plant 300 m upstream which releases waste water during peak flows.

Pressure
The channel has been straightened and simplified, and in parts the streambed has been sealed with concrete. Where the streambed is sealed, no hyporheic exchange occurs, and elsewhere in the stream hyporheic exchange is limited in volume and residence time by the lack of features and the gaining conditions. The waste water treatment discharges high concentrations of nutrients and carbon into the stream, and other pollutants are introduced to the stream by overland flow.

Solution
Most of the concrete that seals the streambed should be removed, but retaining some sections could induce hyporheic exchange of long residence times and help maintain high water velocity. This should be carefully considered because the remaining sections could be vulnerable to undercutting and mobilizing, potentially becoming a risk downstream. The channel could be reconfigured to meander throughout the enclosed 10 m – the medium gradient could allow a hydraulic head across meanders, generating hydrostatically-driven hyporheic exchange with long residence times, facilitating the removal of nutrients by anaerobic reactions. As hydrostatic hyporheic exchange may be limited, it is important to maintain stream energy to enhance hydrodynamically-driven hyporheic exchange. This may be achieved by features that create a gradient in the surface water profile, like steps, or that concentrate the stream energy like large woody material.

16.5.4 Scenario 4

Setting
A 4th order lowland river with a low gradient of 0.1% and thin silty sediment overlying chalk bedrock (Figure 16.6). The channel is 8 m wide and gently meanders through a lowland meadow, with a thin riparian zone. The groundwater conditions are gaining, and the mean discharge is 40 l/s, which responds slowly and mildly to rainfall events. Regenerative farming practices, the high porosity of the underlying chalk aquifer (which limits overland flow), and the maintenance of riparian buffer strips limit sediment inputs to the stream. Groundwaters are polluted with nutrients due to historic organic fertilizer addition which quickly percolated to the porous chalk aquifer. Wood is contributed to the stream by the riparian zone.

Pressure
The river sediments, which were originally fine and coarse gravels, have been removed by the land owner for road construction and have been replaced by a fine layer of silts which directly overlays chalk bedrock, from which groundwater is discharged. The removal of sediments is associated with the loss of instream macrophytes. Together, this has eliminated any opportunity for vertical hyporheic exchange, and only limited hydrostatic hyporheic exchange exists across meander bends because of the shallow stream gradient.

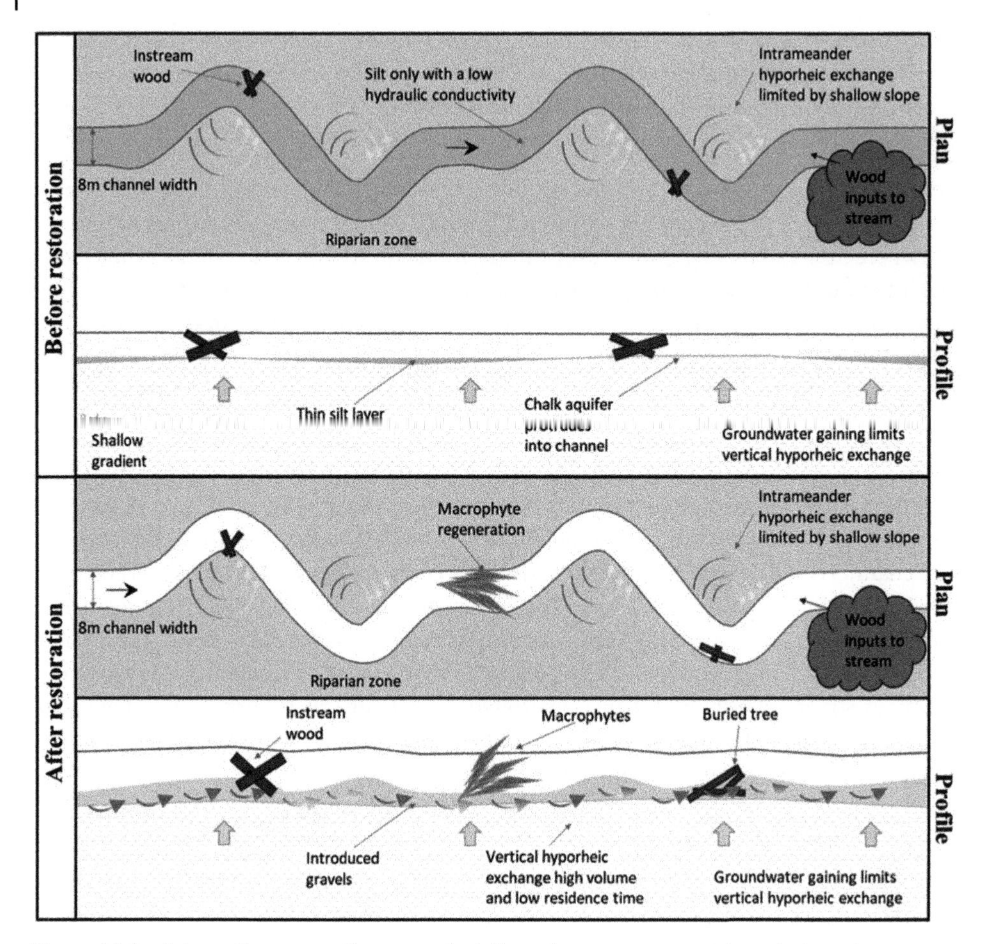

Figure 16.6 Schematic representing scenario 4. Curved arrows represent hyporheic exchange. Arrow thickness represents relative volume (i.e. change from before to after restoration) – thicker arrows represent relatively higher volume. The arrow colour represents residence time distribution, where blue represents shorter times and yellow represents longer times.

Solution

Sediments should be restored by the reintroduction of flint or chalk gravels of high hydraulic conductivity. A thicker sediment layer is preferable, which could raise the riverbed and reconnect the channel to the floodplain (as in the stage zero concept), but this will likely be limited by cost and practicality. Burying fallen trees in the sediment could introduce heterogeneity in the subsurface hydraulic conductivity, inducing a range of hyporheic flow paths. Where silt is especially thick it may be necessary to remove it first to prevent siltation of the new gravels. However, the capture of silt using large woody material and bankside trees that interact with the flow (such as willow) is preferable. Silt and fine sediment can protect large woody material features from decay, could maintain high carbon availability, and create heterogeneity in the hydraulic conductivity of the sediment, thus increasing hyporheic residence times. A programme of gravel cleaning following restoration may be necessary to maintain a high hydraulic conductivity of sediments, until some equilibrium is reached. Macrophyte regeneration should be encouraged (which may require planting), which can generate instream storage and hyporheic exchange, and trap fine sediments and introduce carbon, ultimately increasing biogeochemical reaction rates. Furthermore, macrophytes can remove nutrients by direct assimilation (Levi et al., 2015). As a result of this restoration, a higher volume of hyporheic exchange can be expected, but with short residence times.

16.5.5 Scenario 5

Setting

A 4th order lowland river with a low gradient of 0.04% and a sandy streambed (Figure 16.7). The channel is around 10 m wide and meanders through a wooded floodplain.

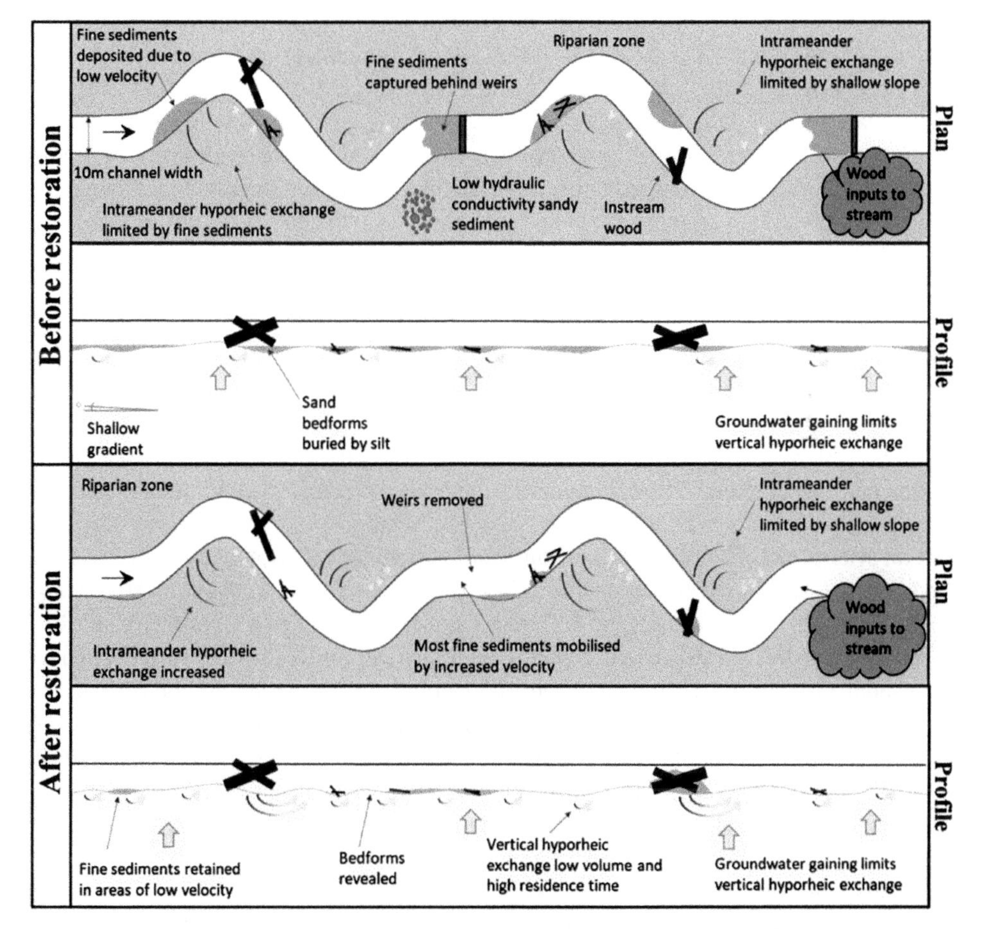

Figure 16.7 Schematic representing scenario 5. Curved arrows represent hyporheic exchange. Arrow thickness represents relative volume (i.e. change from before to after restoration) – thicker arrows represent relatively higher volume. The arrow colour represents residence time distribution, where blue represents shorter times and yellow represents longer times.

The groundwater conditions are gaining, and the river has a mean discharge of 50 l/s, which responds slowly and mildly to rainfall. Wood has fallen from the surrounding riparian woodland littering the bed with small features which are not long enough to span the channel-width.

Pressure
Several weirs block the channel and impound water upstream. This reduces the reach-scale water velocity and the stream energy, dampening hydrodynamic hyporheic exchange. Furthermore, the low velocity encourages deposition of sediments which bury bedforms that readily form in sand streambeds, further limiting hydrodynamic forces. Hyporheic exchange is not induced by hydrostatic forces at the weirs because of the low stream gradient, which prevents the creation of a sizeable hydraulic head from upstream to downstream of the weir, and because of the gaining groundwater conditions. The result is a limited volume of hyporheic exchange.

Solution
The removal of weirs will increase the water velocity, generating hydrodynamic hyporheic exchange around existing wood features and over bedforms. The number and design of wood features may be enhanced by felling bank side trees into the stream and pinning them perpendicular to flow, in a J-hook style design. This could create patches of high velocities, driving hyporheic exchange, and mobilizing fine sediments. Finer sediments will be demobilized in regions of lower velocity, which are likely to be colonized by macrophytes. The gaining conditions are likely to limit hyporheic residence times in any case, but water quality objectives could still be realized.

16.6 Synergies and Conflicts

River ecosystem processes are intrinsically intertwined and interconnected. Modifications to one process are likely to lead to changes in many aspects of ecosystem function and subsequently the delivery of river ecosystem services and dis-services. This presents the potential for synergies (i.e. co-benefits) and conflicts (i.e. trade-offs) between different objectives and different stakeholders (Hester and Gooseff, 2010; Alves et al., 2019). In other words, hyporheic restoration for water quality could have side-effects that are considered positive or negative by different stakeholder groups. All river restoration presents similar opportunities and risks, which river restoration professionals become adept at considering and managing (Dicks, Walsh & Sutherland, 2014). In this section, some of the more likely and most important side-effects of the restoration of hyporheic zone function are discussed. The discussion is by no means comprehensive, and the reader should use this section alongside the tools and methods they usually employ to aid in decision making and balancing trade-offs.

16.6.1 Natural Flood Management

One of the most common motivations of river restoration in the twenty-first century is NFM, which seeks to restore or enhance catchment processes in order to reduce the frequency and severity of floods (Dadson et al., 2017; Grabowski et al., 2019). Interventions for NFM usually aim to "slow the flow" (i.e. increase the time it takes for water to travel downstream), which is sometimes achieved by creating instream storage, for example behind leaky dams and large woody material features. Similar techniques have been suggested here for the restoration of hyporheic zone function (e.g. scenario 1), especially in upland settings where most restoration for NFM is conducted. Therefore, in upland settings the techniques employed for either NFM or the improvement of water quality are likely to have co-benefits for both. In lowland settings, restoration for NFM that reduces water velocity during baseflow conditions could have negative consequences for the function of the hyporheic zone (Rana et al. 2017). However, most NFM measures are designed to only affect flow during flood, as in Figure 16.2e.

16.6.2 Ecology

The hyporheic zone was originally identified by biologists who recognized it as a distinct ecotone, hosting organisms that spend some or all of their time there (named hyporheos) (Ward 2015). Much of the research assessing the effect of river restoration on hyporheic exchange was conducted with the objective of improving this habitat, especially for fish spawning and invertebrates (Hester and Gooseff 2010). Hyporheic restoration for ecological objectives particularly focuses on vertical hyporheic exchange, often targeting high volume and low residence time, with features like riffles and steps, which deliver high concentrations of dissolved oxygen to the hyporheic zone (Hester and Gooseff 2010). Most hyporheic restoration measures for the improvement of water quality are likely to also improve hyporheic habitats, as a result of the increased delivery of dissolved oxygen to the subsurface (Noack et al. 2016). Sediment introductions, and other measures that reduce colmation or increase sediment hydraulic conductivity (e.g. Figure 16.2I and K, and scenarios 4 and 5) are likely to be especially beneficial for fish habitats. However, hyporheic restoration strategies that target hyporheic exchange of lower volume and higher residence time in replacement of hyporheic exchange of higher volume and lower residence time (e.g. Figure 16.2m or scenario 1) may be detrimental to some ecological objectives. Furthermore, increasing hyporheic exchange volume and extent can increase the buffering of temperature and hydrological extremes, providing refuge for a range of organisms (Klaar et al. 2020). These areas of refuge may become increasingly important in enabling organisms to adapt to climate change, for example as surface water temperatures in rivers increase (Thomas et al. 2015). Using wood in river restoration is likely to offer benefits for both water quality and ecology (Krause et al. 2014). On the contrary, some hyporheic restoration techniques, for example channel-spanning features, may reduce longitudinal connectivity which could disrupt ecological pathways and fluxes, for example preventing upstream fish passage.

16.6.3 Biogeochemical

The river corridor can play an important role in biogeochemical cycles, facilitating the exchange of carbon and nutrients between freshwater, terrestrial, atmospheric, and marine environments. Carbon can be captured and stored in the river corridor in instream wood, vegetation, aquatic biota, sediments, and in particle and dissolved forms (Wohl et al. 2017). The short-term storage of carbon could be increased by restoration measures that induce the deposition and storage of fine sediments, which may include channel-spanning features installed for the improvement of water quality (Fanelli and Lautz 2008). Furthermore, increasing vertical and lateral hyporheic exchange may transport particulate and dissolved organic carbon deeper into river sediments and riparian soils, respectively, increasing the amount and duration of carbon storage (Wohl et al. 2017). Overall, it is likely that hyporheic restoration of the hyporheic zone for improvements to water quality has co-benefits for carbon capture, though in some cases the opposite could be true (e.g. by inducing significant fine sediment mobilization) (Pander et al. 2014).

On the contrary, hyporheic restoration for water quality improvements may have associated risks. For example, the increase in biogeochemical reaction rates that hyporheic restoration for water quality improvement seeks to achieve could lead to an increase in the production of greenhouse gases (Romeijn et al. 2019), which presents a conflict given their contribution to climate change (Wohl et al. 2015). Furthermore, biogeochemical reactions in the hyporheic zone can sometimes lead to an increase in the concentration of pollutants, for example by the transformation of less polluting forms to more polluting forms, as is the case for the remineralization of nutrients (Krause et al. 2011).

16.7 Discussion

To communicate insight in an actionable format it has been necessary to simplify and generalize. For example, in Section 16.4, restoration techniques have been grouped into broad classifications which have been reviewed in isolation. However, in practice, restoration programmes should usually adopt a suite of restoration strategies that are applied within contemporary process-based frameworks. This creates heterogeneity in the properties of hyporheic exchange, facilitating different reactions and consequently allowing the removal of different pollutants, and of pollutants which are removed in a series of reactions (Weatherill et al. 2018). Furthermore, this could allow the activation of different zones of hyporheic exchange during different flow conditions, maintaining the reach-scale function of the hyporheic zone (Boano et al. 2014). This strategy could be especially useful in instances where the nature and dynamics of water quality are not well understood, for example where sampling/monitoring frequency is low, because it could maximize the potential of generating the properties of hyporheic exchange that are required to remove unknown pollutants. It is also likely to maximize the delivery of co-benefits (Alves et al. 2019).

Furthermore, restoration strategies have been further reduced to the smallest scale, by typically considering the design of individual features. However, the impact of an individual feature is likely to be negligible and restoration programmes will usually include a series of features employed in a catchment-based approach whose cumulative effect can realize significant improvements to water quality (Rana et al. 2017). For example, Hester et al. (2018) observed only a 0.28% increase in the rate of denitrification per feature but achieved a 25% increase using a series of features in a 200 m reach. The scale of restoration is likely to be a critical determinant of project success. Reach-scale restoration is not usually effective at delivering water quality objectives; instead, basin or catchment scale restoration is necessary (Wohl et al. 2015). However, restoration on this scale requires significant coordination and is usually prevented by practicalities like complicated land access and ownership (Kelly and Kusel 2015). River restoration professionals can apply this insight within practical constraints by linking up several smaller schemes to maximize the scale of restoration. This may require a level of communication between professionals and organizations which is not currently realized. Furthermore, top-down coordination of the typical bottom-up approach may be necessary to deliver larger scale restoration programmes (Koontz and Newig 2014), but it remains unclear to whom this responsibility might reasonably fall. Boundary organizations like the River Restoration Centre in the UK (https://www.therrc.co.uk) or the European Centre for River Restoration (https://www.ecrr.org) in the European Union could play an important role.

The focus of this work has been on river corridor restoration strategies, i.e. those which are conducted directly in, or close to, the channel. However, restoration of the floodplain, where it is practical, has the potential to deliver water quality benefits to an even larger extent (Azinheira et al. 2014). Reconnecting channels to their floodplains (e.g. in stage zero restoration) is a popular technique that has delivered water quality benefits (amongst many others) and may sometimes result in the restoration of the hyporheic zone, for example by increasing lateral hyporheic exchange (Azinheira et al. 2014; Wohl et al. 2021). The rewooding of riparian zones could offer benefits for water quality, hosting hot spots and hot moments of reactions, and increasing hyporheic exchange (Goeller et al. 2019; Vidon et al. 2010). It could also eventually lead to an increased instream wood loading which is likely to increase hyporheic exchange (Kasahara and Wondzell 2003). Passive management methods, such as rewilding, are likely to lead to the long-term restoration of the river corridor and the hyporheic zone, although these could take a long time to be realized. Allowing natural dynamics, for example in wood and sediment transport, to shape the evolution of the river corridor is probably the best strategy in delivering long-term, persistent improvements to water quality with minimum input (Krause et al. 2014). For example, the reintroduction of beavers (*Castor fiber*) in the UK could lead to the development of many of the restoration techniques described in this work (Larsen et al. 2021; Wade et al. 2020). Ultimately, employing a variety of techniques and approaches to river restoration is the best strategy in maximizing the chance of success, especially given the uncertainty surrounding the relative efficacy of different restoration techniques (Palmer et al. 2008).

Nature-based solutions, like restoring the function of the hyporheic zone, are a critical strategy in delivering widespread and long-term improvements to water quality. However, they are only one component of what must be a larger programme of efforts (Haas et al. 2017). Water quality objectives will not be achieved without significant reductions in the inputs of pollutants to freshwater bodies, including surface waters and groundwaters (Wohl et al. 2015). This may be achieved by tightening regulations to reduce the use and application of potential pollutants, for example inputs of nutrient fertilizer and pesticides to agricultural fields. It may also be achieved by reducing the transport of pollutants to freshwater bodies, for example by increasing retention of fertilizers on fields by using cover crops (Couëdel et al. 2018).

Despite decades of research and practice, and increasing commitment of resources, there is little evidence to suggest that river restoration is effective at delivering on water quality objectives or worse still, that its effectiveness is improving (Palmer 2008). The success of restoration is not limited by scientific understanding of ecosystem processes and function (Wohl et al. 2015). Instead, it is limited by the failure to apply this understanding in practice (Palmer et al. 2014). It is important that researchers continue to improve our understanding of the processes and mechanisms that underpin ecosystem function and the delivery of ecosystem (dis)services. However, if this work is to have real-world impact, it is equally important that researchers commit the same energy, if not more, to facilitating its uptake in practice. This requires researchers to communicate insight from their work in a format that is actionable, or to work closely with those who are able to do so, such as boundary organizations. By doing so, researchers can empower decision-makers with tools and knowledge to make more effective decisions. Similarly, researchers must capitalize on the opportunity to learn from the many restoration programmes that are designed and implemented annually. Fatally few of these programmes are currently monitored or evaluated – a missed opportunity which surely contributes to the lack of improvement of river restoration schemes.

16.8 Conclusions

This work aimed to synthesize and distil the substantial body of research which investigates the restoration of the hyporheic zone for improvements to water quality, communicating its complex insights in an actionable format for the first time. These insights can guide river restoration professionals in delivering more effective restoration programmes that are based on evidence. It is communicated in design considerations and scenarios that are centred around process-based restoration, seeking to re-establishing and enhance the function of the hyporheic zone to harness its capacity to improve water quality. These guidelines allow the application of process-based principles to the improvement of water quality, which until now has been difficult due to the paucity of actionable insight available to professionals.

Poor water quality remains a persistent and pervasive problem, degrading ecosystems, worsening water scarcity, and threatening the success of our global society. Whilst recognition of the importance of water quality is improving, actions and governance to manage water quality are still lacking in comparison to those to manage water quantity, which manifests in more visible events such as floods and droughts (Hannah et al. 2022). The importance of water quality for society must be elevated through education and by increasing public engagement with water quality challenges and solutions. Water quality and water quantity are not separate challenges – they are intrinsically intertwined in the processes that control them. Therefore, the successful management and governance of water resources requires an integrated approach that acknowledges this co-dependency.

References

Alves, A., Gersonius, B., Kapelan, Z. et al. (2019). Assessing the co-benefits of green-blue-grey infrastructure for sustainable urban flood risk management. *Journal of Environmental Management* 239: 244–254.

Azinheira, D., Scott, D., Hession, W., and Hester, E. (2014). Comparison of effects of inset floodplains and hyporheic exchange induced by in-stream structures on solute retention. *Water Resources Research* 50 (7): 6168–6190.

Bakke, P., Hrachovec, M., and Lynch, K. (2020). Hyporheic process restoration: design and performance of an engineered streambed. *Water* 12 (2): 425.

Becker, J., Endreny, T., and Robinson, J. (2013). Natural channel design impacts on reach-scale transient storage. *Ecological Engineering* 57: 380–392.

Beechie, T., Pess, G., Kennard, P. et al. (2000). Modeling recovery rates and pathways for woody debris recruitment in Northwestern Washington streams. *North American Journal of Fisheries Management* 20 (2): 436–452.

Beechie, T., Sear, D., Olden, J. et al. (2010). Process-based principles for restoring river ecosystems. *BioScience* 60 (3): 209–222.

Berggren, M. and Giorgio, P. (2015). Distinct patterns of microbial metabolism associated to riverine dissolved organic carbon of different source and quality. *Journal of Geophysical. Research: Biogeosciences* 120 (6): 989–999.

Biron, P.M., Robson, C., Lapointe, M.F., and Gaskin, S.J. (2004). Deflector designs for fish habitat restoration. *Environmental Management* 33 (1): 25–35.

Boano, F., Harvey, J., Marion, A. et al. (2014). Hyporheic flow and transport processes: mechanisms, models, and biogeochemical implications. *Reviews of Geophysics* 52 (4): 603–679.

Bukaveckas, P. (2007). Effects of channel restoration on water velocity, transient storage, and nutrient uptake in a channelized stream. *Environmental Science and Technology* 41 (5): 1570–1576.

Cardenas, M.B. (2009). Stream-aquifer interactions and hyporheic exchange in gaining and losing sinuous streams. *Water Resources Research* 45 (6).

Couëdel, A., Alletto, L., Tribouillois, H., and Justes, É. (2018). Cover crop crucifer-legume mixtures provide effective nitrate catch crop and nitrogen green manure ecosystem services. *Agriculture, Ecosystems & Environment* 254: 50–59.

Crispell, J. and Endreny, T. (2009). Hyporheic exchange flow around constructed in-channel structures and implications for restoration design. *Hydrological Processes* 23 (8): 1158–1168.

Cluer, B. and Thorne, C. (2013). A stream evolution model integrating habitat and ecosystem benefits. *River Research and Applications* 30: 135–154.

Dadson, S., Hall, J., Murgatroyd, A. et al. (2017). A restatement of the natural science evidence concerning catchment-based 'natural' flood management in the UK. *Proceedings of the Royal Society A: Mathematical, Physical and Engineering Sciences* 473 (2199): 20160706.

Damania, R., Desbureaux, S., Rodella, A.-S. et al. (2019). *Quality Unknown: The Invisible Water Crisis*. World Bank.

Dicks, L., Walsh, J., and Sutherland, W. (2014). Organising evidence for environmental management decisions: a '4S' hierarchy. *Trends in Ecology & Evolution* 29 (11): 607–613.

Drummond, J.D., Aquino, T., Davies-colley, R.J. et al. (2022). Modeling contaminant microbes in rivers during both baseflow and stormflow. *Geophysical Research Letters* 49 (8).

Duvert, C., Priadi, C., Rose, A. et al. (2019). Sources and drivers of contamination along an urban tropical river (Ciliwung, Indonesia): insights from microbial DNA, isotopes and water chemistry. *Science of the Total Environment* 682: 382–393.

Eden, S., Tunstall, S.M., and Tapsell, S.M. (2000). Translating nature: river restoration as nature - culture. *Environment and Planning D: Society and Space* 18: 257–273.

Endreny, T., Lautz, L., and Siegel, D. (2011). Hyporheic flow path response to hydraulic jumps at river steps: flume and hydrodynamic models. *Water Resources Research* 47 (2).

Environment Agency (2009). *The Hyporheic Handbook: Groundwater-surface Water Interface and Hyporheic Zone for Environment Managers* https://www.gov.uk/government/publications/

the-hyporheic-handbook-groundwater-surface-water-interface-and-hyporheic-zone-for-environment-managers.

Fanelli, R.M. and Lautz, L.K. (2008). Patterns of water, heat, and solute flux through streambeds around small dams. *Ground Water* 46 (5): 671–687.

Feng, J., Liu, D., Liu, Y. et al. (2022). Hyporheic exchange due to in-stream geomorphic structures. *Journal of Freshwater Ecology* 37 (1): 221–241.

Findlay, S. (1995). Importance of surface-subsurface exchange in stream ecosystems: the hyporheic zone. *Limnology and Oceanography* 40 (1): 159–164.

Fox, A., Boano, F., and Arnon, S. (2014). Impact of losing and gaining stream flow conditions on hyporheic exchange fluxes induced by dune-shaped bedforms. *Water Resources Research* 50 (3): 1895–190.

Gabriele, W., Welti, N., and Hein, T. (2013). Limitations of stream restoration for nitrogen retention in agricultural headwater streams. *Ecological Engineering* 60: 224–234.

Gandy, C.J., Smith, J.W.N., and Jarvis, A.P. (2007). Attenuation of mining-derived pollutants in the hyporheic zone: a review. *Science of the Total Environment* 373 (2–3): 435–446.

Goeller, B., Burbery, L., Febria, C. et al. (2019). Capacity for bioreactors and riparian rehabilitation to enhance nitrate attenuation in agricultural streams. *Ecological Engineering* 134: 65–77.

Grabowski, R., Gurnell, A., Burgess-Gamble, L. et al. (2019). The current state of the use of large wood in river restoration and management. *Water and Environment Journal* 33(3): 366–377.

Graillot, D., Paran, F., Bornette, G. et al. (2014). Coupling groundwater modeling and biological indicators for identifying river/aquifer exchanges. *SpringerPlus* 3 (1).

Haas, M., Guse, B., and Fohrer, N. (2017). Assessing the impacts of best management practices on nitrate pollution in an agricultural dominated lowland catchment considering environmental protection versus economic development. *Journal of Environmental Management* 196: 347–364.

Han, B., Chu, H., and Endreny, T. (2015). Streambed and water profile response to in-channel restoration structures in a laboratory meandering stream. *Water Resources Research* 51 (11): 9312–9324.

Hannah, D., Abbott, B., Khamis, K. et al. (2022). Illuminating the 'invisible water crisis' to address global water pollution challenges. *Hydrological Processes* 36 (3).

Harvey, J. and Gooseff, M. (2015). River corridor science: hydrologic exchange and ecological consequences from bedforms to basins. *Water Resources Research* 51 (9): 6893–6922.

He, C., Liu, Z., Wu, J. et al. (2021). Future global urban water scarcity and potential solutions. *Nature Communications* 12.

Herzog, S.P., Higgins, C.P., and McCray, J.E. (2016). Engineered streambeds for induced hyporheic flow: enhanced removal of nutrients, pathogens, and metals from urban streams. *Journal of Environmental Engineering* 142 (1).

Herzog, S., Ward, A., and Wondzell, S. (2019). Multiscale feature-feature interactions control patterns of hyporheic exchange in a simulated headwater mountain stream. *Water Resources Research* 55 (12): 10976–10992.

Hester, E. and Doyle, M. (2008). In-stream geomorphic structures as drivers of hyporheic exchange. *Water Resources Research* 44 (3).

Hester, E., Brooks, K.E., and Scott, D.T. (2018). Comparing reach scale hyporheic exchange and denitrification induced by instream restoration structures and natural streambed morphology. *Ecological Engineering* 115: 105–121.

Hester, E. and Gooseff, M. (2010). Moving beyond the banks: hyporheic restoration is fundamental to restoring ecological services and functions of streams. *Environmental Science Andamp; Technology* 44 (5): 1521–1525.

Hester, E., Eastes, L., and Widdowson, M. (2019). Effect of surface water stage fluctuation on mixing-dependent hyporheic denitrification in riverbed dunes. *Water Resources Research* 55 (6): 4668–4687.

Hester, E., Hammond, B., and Scott, D.T. (2016). Effects of inset floodplains and hyporheic exchange induced by in-stream structures on nitrate removal in a headwater stream. *Ecological Engineering* 97: 452–464.

Howard, B.C., Baker, I., Kettridge, N. et al. (2023). Wood increases greenhouse gas emissions and nitrate removal rates in river sediments: Risks and opportunities for instream wood restoration. *ACS EST Water*. https://doi.org/10.1021/acsestwater.3c00014

Johnson, M.F., Thorne, C.R., Castro, J.M. et al. (2019). Biomic river restoration: a new focus for river management. *River Research and Applications* 36: 3–12.

Kasahara, T. and Hill, A. (2007). Lateral hyporheic zone chemistry in an artificially constructed gravel bar and a re-meandered stream channel, Southern Ontario, Canada. *Journal of the American Water Resources Association* 43 (5): 1257–1269.

Kasahara, T. and Hill, A.R. (2008). Modelling the effects of lowland stream restoration projects on stream: subsurface water exchange. *Ecological Engineering* 32 (4): 310–319.

Kasahara, T. and Wondzell, S. (2003). Geomorphic controls on hyporheic exchange flow in mountain streams. *Water Resources Research* 39 (1): SBH 3-1-SBH 3-14.

Kasahara, T., Datry, T., Mutz, M., and Boulton, A. (2009). Treating causes not symptoms: restoration of surface - groundwater interactions in rivers. *Marine and Freshwater Research* 60 (9): 976.

Kelly, E. and Kusel, J. (2015). Cooperative, cross-boundary management facilitates large-scale ecosystem restoration efforts. *California Agriculture* 69 (1): 50–56.

Klaar, M., Shelley, F., Hannah, D., and Krause, S. (2020). Instream wood increases riverbed temperature variability in a lowland sandy stream. *River Research and Applications* 36 (8): 1529–1542.

Kobayashi, S., Kantoush, S., Al-mamari, M. et al. (2022). Local flow convergence, bed scour, and aquatic habitat formation during floods around wooden training structures placed on sand-gravel bars. *Science of the Total Environment* 817: 152992.

Koontz, T. and Newig, J. (2014). From planning to implementation: top-down and bottom-up approaches for collaborative watershed management. *Policy Studies Journal* 42 (3): 416–442.

Krause, S., Abbott, B., Baranov, V. et al. (2022). Organizational principles of hyporheic exchange flow and biogeochemical cycling in river networks across scales. *Water Resources Research* 58 (3).

Krause, S., Hannah, D., Fleckenstein, J. et al. (2010). Inter-disciplinary perspectives on processes in the hyporheic zone. *Ecohydrology* 4 (4): 481–499.

Krause, S., Klaar, M., Hannah, D. et al. (2014). The potential of large woody debris to alter biogeochemical processes and ecosystem services in lowland rivers. *Wiley Interdisciplinary Reviews: Water* 1 (3): 263–275.

Kurth, A., Weber, C., and Schirmer, M. (2015). How effective is river restoration in re-establishing groundwater–surface water interactions? – a case study. *Hydrology and Earth System Sciences* 19 (6): 2663–2672.

Larsen, A., Larsen, J., and Lane, S. (2021). Dam builders and their works: beaver influences on the structure and function of river corridor hydrology, geomorphology, biogeochemistry and ecosystems. *Earth-Science Reviews* 218: 103623.

Le Traon, C., Aquino, T., Bouchez, C. et al. (2021). Effective kinetics driven by dynamic concentration gradients under coupled transport and reaction. *Geochimica et Cosmochimica Acta* 306: 189–209.

Levi, P., Riis, T., Alnøe, A. et al. (2015). Macrophyte complexity controls nutrient uptake in lowland streams. *Ecosystems* 18 (5): 914–931.

Lewandowski, J., Arnon, S., Banks, E. et al. (2019). Is the hyporheic zone relevant beyond the scientific community? *Water* 11 (11): 2230–233.

Li, H., Liu, Y., Feng, J. et al. (2022). Influence of the in-stream structure on solute transport in the hyporheic zone. *International Journal of Environmental Research and Public Health* 19 (10): 5856.

Liu, S. and Chui, T. (2020). Optimal in-stream structure design through considering nitrogen removal in hyporheic zone. *Water* 12 (5): 1399.

Logar, I., Brouwer, R., and Paillex, A. (2019). Do the societal benefits of river restoration outweigh their costs? A cost-benefit analysis. *Journal of Environmental Management* 232: 1075–1085.

Mach, K.J., Lemos, M.C., Meadow, A.M. et al. (2020). Actionable knowledge and the art of engagement. *Current Opinion in Environmental Sustainability* 42: 30–37.

Magliozzi, C., Coro, G., Grabowski, R. et al. (2019). A multiscale statistical method to identify potential areas of hyporheic exchange for river restoration planning. *Environmental Modelling Andamp; Software* 111: 311–323.

Manners, R., Doyle, M., and Small, M. (2007). Structure and hydraulics of natural woody debris jams. *Water Resources Research* 43 (6).

Martín, E., Ryo, M., Doering, M., and Robinson, C. (2018). Evaluation of restoration and flow interactions on river structure and function: channel widening of the Thur River, Switzerland. *Water* 10 (4): 439.

Mason, S., McGlynn, B., and Poole, G. (2012). Hydrologic response to channel reconfiguration on Silver Bow Creek, Montana. *Journal of Hydrology* 438–439: 125–136.

Mathers, K.L., Robinson, C.T. and Weber, C. (2021). Artificial flood reduces fine sediment clogging enhancing hyporheic zone physicochemistry and accessibility for macroinvertebrates. *Ecological Solutions and Evidence* 2 (4). https://doi.org/10.1002/2688-8319.12103.

Mohammed, V., Yaser, S., and Shaker, H.S. (2016). Effects of distance between the T-Shaped spur dikes on flow and scour patterns in 90 °C bend using the SSIIM model. *Ain Shams Engineering Journal* 7 (1): 31–45.

Morén, I., Wörman, A., and Riml, J. (2017). Design of remediation actions for nutrient mitigation in the hyporheic zone. *Water Resources Research* 53 (11): 8872–8899.

Noack, M., Ortlepp, J., and Wieprecht, S. (2016). An approach to simulate interstitial habitat conditions during the incubation phase of gravel-spawning fish. *River Research and Applications* 33 (2): 192–201.

Palmer, M., Bernhardt, E., Allan, J. et al. (2005). Standards for ecologically successful river restoration. *Journal of Applied Ecology* 42 (2): 208–217.

Palmer, M.A. (2008). Reforming watershed restoration: science in need of application and applications in need of science. *Estuaries and Coasts* 32 (1): 1–17.

Palmer, M.A., Filoso, S., and Fanelli, R.M. (2014). From ecosystems to ecosystem services: stream restoration as ecological engineering. *Ecological Engineering* 65: 62–70.

Pander, J., Mueller, M., and Geist, J. (2014). A comparison of four stream substratum restoration techniques concerning interstitial conditions and downstream effects. *River Research and Applications* 31 (2): 239–255.

Peter, K., Herzog, S., Tian, Z. et al. (2019). Evaluating emerging organic contaminant removal in an engineered hyporheic zone using high resolution mass spectrometry. *Water Research* 150: 140–152.

Quick, A., Reeder, W., Farrell, T. et al. (2019). Nitrous oxide from streams and rivers: a review of primary biogeochemical pathways and environmental variables. *Earth-Science Reviews* 191: 224–262.

Rana, S., Scott, D., and Hester, E. (2017). Effects of in-stream structures and channel flow rate variation on transient storage. *Journal of Hydrology* 548: 157–169.

Rivett, M.O., Ellis, P.A., Greswell, R. et al. (2008). Cost-effective mini drive-point piezometers and multilevel samplers for monitoring the hyporheic zone. *Quarterly Journal of Engineering Geology and Hydrogeology* 41 (1): 49–60.

Romeijn, P., Comer-Warner, S., Ullah, S. et al. (2019). Streambed organic matter controls on carbon dioxide and methane emissions from streams. *Environmental Science & Technology* 53 (5): 2364–2374.

Romeijn, P., Hannah, D., and Krause, S. (2021). Macrophyte controls on urban stream microbial metabolic activity. *Environmental Science Andamp; Technology* 55 (8): 4585–4596.

Rosgen, D.L. (2001). The cross-vane, W-weir and J-hook vane structures...their description, design and application for stream stabilisation and river restoration. In: *Wetlands Engineering & River Restoration 2001*, 1–22. (ASCE). https://doi.org/10.1061/40581(2001)72.

Sawyer, A., Bayani Cardenas, M., and Buttles, J. (2011). Hyporheic exchange due to channel-spanning logs. *Water Resources Research* 47 (8).

Sparacino, M., Rathburn, S., Covino, T. et al. (2019). Form-based river restoration decreases wetland hyporheic exchange: lessons learned from the Upper Colorado River. *Earth Surface Processes and Landforms* 44 (1): 191–203.

Stout, J., Rutherfurd, I., Grove, J. et al. (2018). Passive recovery of wood loads in rivers. *Water Resources Research* 54 (11): 8828–8846.

Tesoriero, A.J., Duff, J.H., Saad, D.A. et al. (2013). Vulnerability of streams to legacy nitrate sources. *Environmental Science and Technology* 47 (8): 3623–3629.

Thomas, S., Griffiths, S., and Ormerod, S. (2015). Beyond cool: adapting upland streams for climate change using riparian woodlands. *Global Change Biology* 22 (1): 310–324.

Toran, L., Nyquist, J., Fang, A. et al. (2013). Observing lingering hyporheic storage using electrical resistivity: variations around stream restoration structures, Crabby Creek, PA. *Hydrological Processes* 27 (10): 1411–1425.

UNESCO (2018). *World water assessment programme. The United Nations world water development report 2018: nature-based solutions for water.* ISBN: 978-92-3-100264-9.

Vereecken, H., Baetens, J., Viaene, P. et al. (2006). Ecological management of aquatic plants: effects in lowland streams. *Hydrobiologia* 570 (1): 205–210.

Vidon, P., Allan, C., Burns, D. et al. (2010). Hot spots and hot moments in riparian zones: potential for improved water quality management. *JAWRA Journal of the American Water Resources Association* 46 (2): 278–298.

Wade, J., Lautz, L., Kelleher, C. et al. (2020). Beaver dam analogues drive heterogeneous groundwater–surface water interactions. *Hydrological Processes* 34 (26): 5340–5353.

Ward, A. (2015). The evolution and state of interdisciplinary hyporheic research. *WIREs Water* 3 (1): 83–103.

Ward, A., Gooseff, M., and Johnson, P. (2011). How can subsurface modifications to hydraulic conductivity be designed as stream restoration structures? Analysis of Vaux's conceptual models to enhance hyporheic exchange. *Water Resources Research* 47 (8).

Ward, A., Morgan, J., White, J., and Royer, T. (2018). Streambed restoration to remove fine sediment alters reach-scale transient storage in a low-gradient fifth-order river, Indiana, USA. *Hydrological Processes* 32 (12): 1786–1800.

Weatherill, J., Atashgahi, S., Schneidewind, U. et al. (2018). Natural attenuation of chlorinated ethenes in hyporheic zones: a review of key biogeochemical processes and in-situ transformation potential. *Water Research* 128: 362-382.

Wohl, E., Castro, J., Cluer, B. et al. (2021). Rediscovering, reevaluating, and restoring lost river-wetland corridors. *Frontiers in Earth Science* 9.

Wohl, E., Hall, R., Lininger, K. et al. (2017). Carbon dynamics of river corridors and the effects of human alterations. *Ecological Monographs* 87 (3): 379–409.

Wohl, E., Lane, S.N., and Wilcox, A.C. (2015). The science and practice of river restoration. *Water Resources Research* 51 (8): 5974–5997.

Wondzell, S. (2011). The role of the hyporheic zone across stream networks. *Hydrological Processes* 25 (22): 3525–3532.

Wondzell, S.M., LaNier, J., Haggerty, R. et al. (2009). Changes in hyporheic exchange flow following experimental wood removal in a small, low-gradient stream. *Water Resources Research* 45 (5).

Zarnetske, J., Haggerty, R., Wondzell, S., and Baker, M. (2011). Dynamics of nitrate production and removal as a function of residence time in the hyporheic zone. *Journal of Geophysical Research* 116 (G1).

Zarnetske, J., Haggerty, R., Wondzell, S. et al. (2012). Coupled transport and reaction kinetics control the nitrate source-sink function of hyporheic zones. *Water Resources Research* 48 (11).

Index

Ecohydrological Interfaces, First Edition. Edited by Stefan Krause, David M. Hannah, and Nancy B. Grimm.

b

c

d

e

g

h

i

n

o

u

v